Christopher J. Willis

University of Western Ontario, Canada

Problem-Solving in General Chemistry

Houghton Mifflin Company   *Boston*

Atlanta   Dallas   Hopewell, New Jersey   Geneva, Illinois   Palo Alto   London

To my parents

The excerpts from the stories of Sherlock Holmes are reprinted
by permission of International Creative Management as agents
for the estate of Sir Arthur Conan Doyle. All rights reserved.

Printed in the U.S.A.

Library of Congress Catalog Card Number: 76–14004

ISBN: 0–395–24532-X

Contents

Preface

This is a problem-solving guide intended to supplement introductory chemistry textbooks. It covers most of the topics included in a full-year course, and is suitable for use either in tutorial periods or by students working alone or with a minimum of supervision.

Each chapter is arranged with the most elementary concepts first and gradually works up to the more sophisticated problems. The overall coverage is broad, ranging from the fundamental topics of the mole, the equation, and stoichiometry, up to more advanced chapters on chemical kinetics, solution equilibrium, and free energy. Flexible organization among the chapters allows topics to be omitted or the order of study to be changed at will.

Although the emphasis throughout is on numerical problems, there are two chapters of a more descriptive nature on atomic and molecular structure demonstrating, without going into any detailed theory of bonding, how the shape of a covalent molecule may be worked out.

All new terms are defined as they are introduced, and a review listing is given at the end of each chapter, together with some general questions on concepts covered in the chapter. Answers to these general points will be found in the body of the chapter (with a little additional reasoning now and then!) and no explicit list of solutions to them is given. The reader should try and understand these pages before tackling the problems that follow at the end of the chapter.

There are in the text a total of about 240 fully explained examples, arranged in increasing order of difficulty, and the end-of-chapter problems give about 440 more. Although the latter are also provided with detailed solutions, the maximum benefit to the student will come from working them through first *without* reading the answers.

In keeping with the common practice among chemists, modified System International units of measure are used rather than strictly metric units. Thus kilojoules are used as energy units, but pressures are measured in atmospheres rather than in kilopascals.

The quotation at the head of each chapter is taken from literature's most famous amateur chemist, Sherlock Holmes.

It is a pleasure to acknowledge the assistance given by my colleagues, J. W. Lorimer and E. O. Sherman, toward the production of this book through revision, corrections, and criticism, and the numerous students who have added their comments during class use of three preliminary editions. I would also like to thank readers who have provided many helpful suggestions strengthening the book: Ernest I. Becker of the University of Massachusetts, Darrell Ebbing of Wayne State University, Mildred Johnson of City College of San Francisco, Lee G. Pederson of the University of North Carolina, William M. Ritchey of Case Western Reserve, Alan and Sharon Sherman of Middlesex County College, and Wayne E. Wentworth of the University of Houston.

Every care has been taken to keep the number of errors in this book to a minimum, but I should be glad to have my attention drawn to the inevitable mistakes which may remain.

C. J. Willis
Chemistry Department
University of Western Ontario
London, Canada.

Introduction
The General Approach to Solving Problems

"You see, my dear Watson"—he propped his test-tube in the rack and began to lecture with the air of a professor addressing his class—"it is not really difficult to construct a series of inferences, each dependent on its predecessor and each simple in itself ... every problem becomes very childish when once it is explained to you." THE ADVENTURE OF THE DANCING MEN

The problems that have to be solved in an introductory chemistry course do not involve any advanced knowledge of mathematics. Addition and subtraction, multiplication and division, square roots, logarithms, and exponentials; these are the few basic operations needed in common calculations. Any student who has mastered these few fundamentals can tackle a general chemistry problem, but your chances of success will be much better if you can make a habit of approaching any problem in a systematic, step-by-step manner. Although problems are of many different types, there are certain methods of tackling them which are generally useful and applicable, and these are outlined in the following steps.

1. *Think positively*! Tell yourself that the problem *can* be solved by a student at your level of achievement. It's very easy to take one look at a problem and think: "I can't solve this—I don't know the molecular weight—the density isn't given—I don't know what the pressure is—etc." Obviously, if you tell yourself the problem is not capable of solution, you are making life difficult for yourself.

Try instead to work with what you *do* know and the data which *are* given in the problem. Very likely the fact you first worried about will not enter into the calculation, or you will find you can work it out from some other information given. It will hardly ever happen that a given problem is impossible (certainly not in this book).

2. *Read the question carefully.* You'd be amazed at the number of marks lost on every test or problem assignment by students who have simply not read the question. It's no use going to a lot of trouble to calculate something that was not asked for in the question; you will receive no marks for that.

Even more common is the loss of marks through incorrectly copying down the given data. Was the formula N_2O or NO_2? Was the temperature 250° or 205°? Was the pressure 150 mm Hg or 150 cm Hg? Was the concentration 1.50 molar or 1.50 grams per liter? Only careful attention to the *exact* wording of the problem will remove these possible errors, with their catastrophic consequences for the success of the calculation.

3. *Remember your definitions.* There are a lot of different terms and concepts used in chemistry, and each has a definition that must be remembered. The purpose of memorizing definitions is *not*, of course, to be able to parrot them back at examination time. Instead, we remember definitions because these terms and concepts are the basic vocabulary of chemistry. They are the means by which chemists communicate.

Thus if a chemist writes: "The solution had a concentration of 1.85 molar" he or she is telling us that each liter of the solution contained 1.85 moles of solute. In turn, our understanding of this sentence depends on our comprehension of the terms "solution," "mole," and "solute." Without an immediate understanding of the basic terms, we cannot hope to solve any problems.

So your first reaction when you read the data from a given problem should be to define each term as you come to it (and if you don't know it, go and look it up, unless you've left it too late and find yourself in the exam room). Write it down, not only to help your thinking, but because it will very likely be worth a mark or two, even if you get no further with the calculation.

4. *Write a balanced equation for any reaction occurring.* A great many of the problems in general chemistry involve a chemical reaction or equilibrium. Any calculation will require a correctly balanced equation as a starting point, because, as we see later, an equation carries a great deal of useful information with it. But be sure that it *is* balanced; an unbalanced equation is worse than useless, since it will very often lead to the wrong answer.

5. *Identify the required quantity.* Look at the question again and decide exactly what it was that you were asked to find. If it isn't immediately obvious how you go from the starting data to the answer, decide on some symbol to represent the quantity you're going to calculate, so that you can use it in an equation. Sometimes the symbol will be self-evident, e.g., you can use V for volume or T for temperature, but it's just as well to write down an explanatory line, including *units*:
    "Let the final volume be V liter . . . ."
    "Suppose the initial temperature was T Kelvin . . . ."
At other times, it will be more convenient to use a general symbol such as x or y, especially when it is going to represent several related unknowns in the calculation. But you must *always* start out with a clear statement of what this symbol represents:
    "Suppose that x mole liter^{-1} have dissociated. Then the concentration of acid is $(0.10 - x)$ mole liter^{-1} and of hydrogen ion x mole liter^{-1} . . . ."

Never write down an algebraic equation in x, y, etc., without a preliminary sentence of this type. If you do, you run the risk of solving the equation, finding the value of x, then forgetting what quantity it was you were supposed to be calculating. You also make life very difficult for people trying to mark your work if you don't tell them what you are doing.

6. *Construct an algebraic equation that will enable you to find the unknown.* Of course, this is the key step in solving a problem, and the only way you can hope to succeed

in this is by practice and familiarity. However, you can get off to a good start by the steps we have already gone through above. If you have read the problem, written out the definitions, constructed a balanced chemical equation, and assigned symbols to any unknown quantities, you should have all the pieces necessary to construct an algebraic equation that you can solve.

Don't hesitate to assign symbols $(x, y, z \ldots)$ to any quantities you don't know. Eventually you will hope to end up with one equation and one unknown, but in the early stages of the calculation there may be other unknowns that will disappear along the way.

One simple technique often useful in making up an equation is to reverse the problem. There are many problems that students can easily solve in one direction, but have difficulty in solving when they are presented in reverse order. Here's a simple example:

Given a concentration in gram liter^{-1} and the solute's molecular weight, how do you find the molarity of a solution? *Answer:* Divide the gram liter^{-1} concentration by the molecular weight. Correct!

Given a concentration for a solution in molarity and in gram liter^{-1}, how do you find the molecular weight? *Answer:* You multiply the molarity by the volume . . . no, you divide the grams by . . . uh er . . . you work out the grams in . . .?

In the author's experience, many students can do the first of the above problems, but are baffled by the second. A moment's thought, however, shows that they are one and the same problem. They both depend on the relationship

$$\frac{\text{gram liter}^{-1}}{\text{molecular weight}} = \text{molarity}$$

If any *two* of these three interrelated quantities are known, the third can be found. In other words, if you have molecular weight as the unknown, you take the above equation and rearrange it in the opposite direction to read

$$\text{molecular weight} = \frac{\text{gram liter}^{-1}}{\text{molarity}}$$

Try to get into the habit of looking at any algebraic equation as a means of relating several quantities, any one of which may be the unknown in a particular problem, rather than thinking of an equation as a means of calculating a particular x from a particular y.

7. *Reduce the number of unknowns in your equation.* Sometimes, there will be only one unknown in an equation, so you can find its value by doing the necessary arithmetic. On other occasions, you will find that you have one equation with two or more unknowns in it, so some additional information has to be found before you can solve it. This will often be found in the original data or equation, or it may be deduced from the nature of the chemical system you are dealing with.

Somehow or other, you have to come down to one equation containing one unknown, either by finding the values of all other quantities involved, or by finding a second, independent, equation relating two unknowns. As with so many other aspects of problem solving, only practice and experience will show you how this is to be done.

8. *Insert numerical values for given constants, etc., and do the necessary arithmetic to solve the equation.* We started out by stating that the arithmetic operations of general chemistry are relatively simple, but they do have to be done—and done correctly! Many

an otherwise successful solution to a problem comes to grief in the final couple of lines over simple arithmetic mistakes. Although you will usually receive some marks for correct method in a problem, there is no reason why you should not complete the job with a correct numerical result.

Experience shows that numerical errors result mainly from carelessness. However, many students make life unnecessarily difficult for themselves by continuing to carry out long multiplication and long division, which are extremely time-consuming and error-prone. It should be considered absolutely essential for every student studying chemistry (or any other science) at university level to be familiar with the use of logarithmic tables. The principles are mastered in a few minutes, and will save hours of work later. Some account of the theory of logarithms is given as Appendix A.3 in this book.

For routine multiplication and division, the slide rule can be warmly recommended. The accuracy (1% or better) is quite good enough for the majority of chemical calculations, and the speed and convenience of the slide rule will well repay the hour or so spent learning how to use it. In recent years, a third aid to calculation has appeared, in the form of the miniature electronic calculator. These instruments are now fairly inexpensive. Although very fast and accurate (if used correctly), they should be used with care, particularly with reference to significant figures (Appendix A.1). The long line of digits appearing in colored lights may look impressive, but try to think how many of them are meaningful before copying them down. Remember always that *no computing aid can give an answer any more accurate than the information fed into it.*

9. *Look at the answer.* Is it reasonable? Sometimes you will have no means of knowing whether your answer is correct or not, but very often you can check it for yourself. Obviously, some answers will be chemically ridiculous. If you find a molecular weight of 0.33 or a carbon content of 170%, you can see at once that something is wrong, and you must check your calculation. For other calculations, you can feed back your answer into the original data or equation, and see if it fits. Many times, this simple additional step will show you that you have made a simple slip (perhaps dropped a factor of 10) in deriving your answer.

10. *Read the question again!* Have you answered it completely? Did you calculate the quantity that was actually asked for? Was there more than one part to the problem? Did you use all the data that were given? (If you didn't use them all, be very careful to check that you didn't need to. It may be that the problem was given with extra, unneeded data, but it's much more likely that you have neglected to include something as you went along). Is the answer in the right units? Have you written down the units beside your numerical answer?

It is not possible to give specific instructions on how to solve any possible problem, but the ten steps outlined above should aid you in tackling the majority of them. The most effective three steps towards success in solving problems will always be practice, practice, and more practice.

Problem-Solving in General Chemistry

Chapter 1
Units and Dimensions

A man should keep his little brain-attic stocked with all the furniture that he is likely to use. The rest he can put in the lumber-room of his library. THE FIVE ORANGE PIPS

1.1 The SI System

For many years, all scientific calculations have been carried out in the metric system of units. More recently, a very specific set of definitions within the metric system has been agreed upon, known as SI units (an abbreviation for Système International). Undoubtedly SI units will be used almost exclusively in the future, but at the moment we are in a transitional stage, with many existing textbooks and research papers using units outside the SI system. At the level of elementary chemistry, there are only a few points at which these differences are noticeable, and we will discuss these after first noting the standard units of the SI system.

The three fundamental quantities of length, mass, and time are arbitrarily chosen as the meter (m), the kilogram (kg), and the second (s). These are multiplied or subdivided by the use of prefixes, of which the more common are: mega (1,000,000 or 10^6), kilo (1,000 or 10^3), deci (1/10 or 10^{-1}), centi (1/100 or 10^{-2}), milli (1/1,000 or 10^{-3}), and micro (1/1,000,000 or 10^{-6}). Frequently encountered combinations of these prefixes with the fundamental units are:

length: kilometer, km (1,000 m); meter, m; centimeter, cm (0.01 m); millimeter, mm (0.001 m); micrometer, μm (10^{-6} m).

mass: kilogram, kg; gram, g (10^{-3} kg); milligram, mg (10^{-6} kg or 10^{-3} g).

(It is an oddity of SI nomenclature that the fundamental unit of mass, the kg, has itself the multiplicative prefix "kilo." Although the kg and its multiples are units of mass, we commonly express the weight of an object in grams, milligrams, etc.)

time: The second is subdivided into milliseconds (10^{-3} s) etc., but longer time intervals are still expressed in minutes and hours. The kilosecond has not yet arrived.

1

Units for describing any physical quantity may be constructed in terms of these fundamentals. We will discuss these briefly, with a note on their relation to the cgs (centimeter-gram-second) system and some of the deviations from strict SI nomenclature that are widely used by chemists and are therefore included in this book.

1.2 Length, Area, and Volume

It so happens that the sizes of atoms and molecules, and the lengths of chemical bonds, are of the order of 10^{-10} m (10^{-8} cm). Strictly speaking, these should be expressed in nanometers, nm (1 nm = 10^{-9} m), or in picometers, pm (1 pm = 10^{-12} m). However, for many years chemists have used the angstrom (Å) when referring to such quantities, and we will use the angstrom in this book.

$$1 \text{ angstrom unit} = 10^{-10} \text{ m} = 10^{-8} \text{ cm} = 0.1 \text{ nm} = 100 \text{ pm}$$

It will be seen that the angstrom is related to the standard units of length by exact powers of 10, so conversions are easy.

Area is a quantity that presents no difficulty, being expressed in m^2, cm^2, mm^2, etc., as appropriate.

Volume is another quantity where chemists deviate slightly from SI units. The unit of volume in the SI system is the cubic meter, but this is uncomfortably large for general use. The cubic decimeter (dm^3) is a more convenient size, but this is usually denoted by its familiar name of liter

$$1 \text{ liter} = 1 \text{ dm}^3 = 10^{-3} \text{ m}^3$$

(*Note*: The liter is now, by definition, *exactly* equal to the cubic decimeter. A previous definition of the liter, in terms of the volume of 1 kg of water under specified conditions, has been abandoned.)

The liter may be abbreviated to the letter "l," but the word will be spelled out in full in this book to avoid confusion with the use of *l* to denote liquid phase.

Subdivision of the liter gives the milliliter, ml (10^{-3} liter). This is, of course, exactly equal to the cubic centimeter (cm^3). Although the latter term is closer to the SI system, we will use the ml in this book because of its obvious relation to the liter.

1.3 Force and Energy

More complex quantities can be made by combining the fundamental SI units, e.g.,

$$\text{velocity} = \frac{\text{length}}{\text{time}} = \text{m s}^{-1} \quad \text{(meters per second)}$$

$$\text{acceleration} = \frac{\text{velocity}}{\text{time}} = \text{m s}^{-2} \quad \text{(meters per second per second)}$$

$$\text{force} = \text{mass} \times \text{acceleration} = \text{kg m s}^{-2} \quad \text{(kilogram meters per second per second)}$$

Note: We indicate division by a unit by raising that unit to a *negative* power.

$$\frac{1}{\text{second}} = (\text{second})^{-1} \qquad \frac{1}{(\text{second})^2} = (\text{second})^{-2}$$

This nomenclature is very useful when complicated units have to be given, and we shall use it throughout this book. Although a velocity in meters per second could be abbreviated m/s, the "oblique stroke" system becomes complex and ambiguous when several units are combined, whereas the above method can be expanded to as many units as are needed without confusion.

The SI unit of force is called the newton (N), which is defined as that force which will produce an acceleration of 1 m s^{-2} in a mass of 1 kg. A smaller unit of force is the dyne (g cm s^{-2}).

$$1 \text{ newton} = 10^5 \text{ dynes}$$

Pressure is defined as force per unit area, so the SI pressure unit will be in N m^{-2}. Since the dimensions of the newton are kg m s^{-2}, the dimensions of this pressure unit will be

$$(\text{kg m s}^{-2}) \times (\text{m}^{-2}) = \text{kg m}^{-1}\text{s}^{-2}$$

A pressure of 1 newton per square meter is known as a pascal (Pa), the official SI unit of pressure. However, a great many operations in the laboratory are conducted near atmospheric pressure, using the mercury barometer. It is not convenient to convert all laboratory pressure measurements into pascals, so we shall use atmospheres and the mercury barometer for pressure measurements in this book. For the record, we should note the relationship (under standard conditions):

$$1 \text{ standard atmosphere} = 76 \text{ cm Hg} = 760 \text{ mm Hg} = 1.01325 \times 10^5 \text{ Pa}$$

Work or energy has the units force × distance, so in SI units we have

$$\text{energy} = \text{force} \times \text{distance} = 1 \text{ newton} \times 1 \text{ meter} = \text{kg m}^2\text{s}^{-2}$$

This unit of energy is the joule (J). Relating to the erg, the cgs unit of energy, gives

$$1 \text{ joule} = 10^7 \text{ erg}$$

For many years, chemists used the calorie as a convenient measure of energy in expressing energies. Since the calorie is related to the joule by a factor which is *not* an exact power of 10, i.e.,

$$1 \text{ calorie} = 4.184 \text{ joule} \quad \text{(by definition)}$$

the calorie is not compatible with SI units, and is now passing out of use, so energies in this book are expressed in joules (see Chapter 4 for further discussion of this point). However, some calculations including calories are included in order to give familiarity with this unit.

At the moment, the major part of the English-speaking world does not use metric units for commercial transactions, although this situation is likely to change soon. The conversion factors for the familiar "English" units are based on the following:

1 pound = 453.59 g	1 kilogram = 2.205 pound
1 inch = 2.5400 cm	1 meter = 39.37 inch
1 gallon (Imperial) = 4.5460 liter	1 liter = 0.8799 quart (British)
1 gallon (U.S.) = 3.7853 liter	1 liter = 1.0567 quart (U.S.)

These figures are shown here to give some idea of the magnitude of metric quantities, without any suggestion that they should be memorized. It is better for the student to think throughout in metric units, rather than to attempt frequent conversions.

Temperature in SI units is measured in kelvins (K), which are numerically the same as degrees kelvin or degrees absolute. The zero on this scale is the absolute zero of temperature (discussed further in Chapter 8). Note that the degree symbol is not included in the SI system. Temperatures around the laboratory are still measured in degrees centigrade or Celsius (C), which are converted into kelvins by adding 273.15.

$$0°C = 273.15 \text{ K}$$

Fahrenheit temperatures are never used in scientific work. The conversion is

$$\text{temp (C)} = \frac{5}{9}[\text{temp (F)} - 32]$$

1.4   Dimensional Analysis

As we go through a chemical calculation, we are continually introducing quantities as factors in the calculation. These quantities will be of various types, such as experimental data, physical constants, or conversion factors to change the units of the answer. If we are careful to write down the units of each quantity as we introduce it, then we can be sure that we are doing the calculation correctly, because we may deduce the units of each intermediate, and the final answer, by combining the units of all the factors as we multiply or divide with them. This process is called *dimensional analysis.*

Note that a factor may have units without having dimensions, that is, it may relate two quantities that have the same dimensions. For example, to convert from m to mm, multiply by 1000 mm m^{-1} (both m and mm have the dimension of length); or to convert from calories to joules, multiply by 4.184 J cal^{-1} (both calories and joules have the dimension of energy). Nevertheless, even though the conversion factors used may be dimensionless ratios, their units must be included.

It is advisable to express *all* units by using numerical exponents, either positive or negative, to facilitate combining and canceling out units. Thus, density in grams per cubic centimeter should be written g cm^{-3} rather than g/cm^3. The oblique stroke format can become very confusing when complicated units arise.

Some examples will explain the process of dimensional analysis:

Example 1.1   A piece of gold weighing 9.86 g has a volume of 0.510 cm^3. What is the density of gold? What will be the mass of 1000 cm^3 of gold? What will be the volume of 100 mg of gold? Express your answers in reasonable units in each case.

SOLUTION   Density is, by definition, mass per unit volume; so, for this element,

$$\text{density} = \frac{9.86 \text{ g}}{0.510 \text{ cm}^3} = 19.3 \text{ g cm}^{-3}$$

These are the usual units for density. Thus, if

$$\text{density} = \frac{\text{mass}}{\text{volume}},$$

then

$$\text{mass} = \text{volume} \times \text{density}$$

$$\text{mass of 1000 cm}^3 = 1000 \text{ cm}^3 \times 19.3 \text{ g cm}^{-3} = 19,300 \text{ g}$$

(on multiplication, cm^3 and cm^{-3} cancel out). This large mass would be better expressed in kg, for which the conversion factor is 1000 g kg^{-1}:

$$\text{mass} = \frac{19{,}300 \text{ g}}{1000 \text{ g kg}^{-1}} = 19.3 \text{ kg}$$

(on division, g cancels out and kg^{-1} becomes kg).

To work out the volume of 100 mg of gold, we convert 100 mg to g:

$$100 \text{ mg} = \frac{100 \text{ mg}}{1000 \text{ mg g}^{-1}} = 0.100 \text{ g}$$

$$\text{volume} = \frac{\text{mass}}{\text{density}}$$

$$\text{volume} = \frac{0.100 \text{ g}}{19.3 \text{ g cm}^{-3}} = 5.17 \times 10^{-3} \text{ cm}^3$$

This small volume may be expressed in mm^3 rather than cm^3:

$$1 \text{ cm} = 10 \text{ mm} \quad \text{so} \quad 1 \text{ cm}^3 = 1000 \text{ mm}^3$$

$$\text{volume of 100 mg gold} = 5.17 \times 10^{-3} \text{ cm}^3 \times 1000 \text{ mm}^3 \text{ cm}^{-3} = 5.17 \text{ mm}^3$$

This simple calculation illustrates the general approach of dimensional analysis. As a further illustration of the value of this technique, suppose that we make a mistake in setting up this problem. Suppose we incorrectly write volume = mass × density! Then, putting in the numbers for 100 mg gold, gives us

$$\text{volume} = (0.100 \text{ g}) \times (19.3 \text{ g cm}^{-3}) = 1.93 \text{ g}^2 \text{ cm}^{-3}$$

At once, we notice that this answer does *not* have the dimensions of volume, so we are alerted to our error in the formulation.

Example 1.2 A car travels 15 miles to the gallon. Express this in km to the liter, using the conversion factors 1 in. = 2.54 cm and 1 gallon (U.S.) = 3.79 liter.

SOLUTION There are several conversion factors here. Let's first convert 15 miles to km:

$$15 \text{ mi} = \frac{15 \text{ mi} \times (1760 \text{ yd mi}^{-1}) \times (36 \text{ in. yd}^{-1}) \times (2.54 \text{ cm in.}^{-1})}{(100 \text{ cm m}^{-1}) \times (1000 \text{ m km}^{-1})} = 24.14 \text{ km}$$

(*Note*: All the intermediate units of length—yard, inch, centimeter, and meter—cancel each other out in the evaluation.) The volume conversion factor is $3.79 \text{ liter gal}^{-1}$:

$$15 \text{ mi gal}^{-1} = 24.14 \text{ km gal}^{-1} = \frac{24.14 \text{ km gal}^{-1}}{3.79 \text{ liter gal}^{-1}} = 6.37 \text{ km liter}^{-1}$$

Clearly, we *divide* by 3.79 in the last step so that gal^{-1} cancels out and leaves the answer in $km \text{ liter}^{-1}$. We notice, incidentally, that "mileage" figures are going to look a whole lot worse after metrication!

Example 1.3 The standard atmosphere is 76.0 cm of mercury. Express this in pascals ($N \text{ m}^{-2}$). Take the density of mercury as 13.60 g cm^{-3} and $g = 9.80 \text{ m s}^{-2}$ (g is the acceleration due to gravity, and is used as a conversion factor to find the downward force exerted by an object of given mass. Force = mass × acceleration.)

SOLUTION   Take a barometric column of mercury 1 cm² in cross-section. Its volume will be

$$76.0 \text{ cm} \times 1 \text{ cm}^2 = 76.0 \text{ cm}^3$$

Its total mass will be

$$\frac{76.0 \text{ cm}^3 \times 13.60 \text{ g cm}^{-3}}{1000 \text{ g kg}^{-1}} = 1.034 \text{ kg}$$

At the bottom of the column, this mass rests on 1 cm² cross-section area. It exerts a force of

$$1.034 \text{ kg} \times 9.80 \text{ m s}^{-2} = 10.13 \text{ kg m s}^{-2}$$

The unit of force in the SI system is defined as the newton, 1 kg m s⁻², so we have a force of 10.13 N. This is exerted over a 1 cm² area, so the pressure is

$$(10.13 \text{ N cm}^{-2}) \times (10^4 \text{ cm}^2 \text{ m}^{-2}) = 1.013 \times 10^5 \text{ N m}^{-2}$$

(Remember there are 10^4 cm² in 1 m²).

Remember always to get these conversion factors the right way round!

If there are 1000 mm in 1 m, then the factor will be 1000 mm m⁻¹ or 10^{-3} m mm⁻¹. Always say these units to yourself as you write them down: "1000 mm m⁻¹ means 1000 mm per m, or 1000 mm in each meter." Does that make sense? If each factor going into your calculation has its units examined from a commonsense standpoint, you stand a better chance of getting the right answer every time.

PROBLEMS

Using dimensional analysis, carry out the following conversions. Necessary data are on p. 3

1. Express 1 mile in (a) m and (b) km.

2. Express 2240 lb in kg.

3. Express in liters an engine capacity of 350 cubic inches.

4. Convert a tire pressure of 30 lb in.⁻² to kg cm⁻².

5. Express a density of 60 lb ft⁻³ in units of g cm⁻³.

6. Convert a velocity of 60 miles per hour to (a) km per hour and (b) m s⁻¹.

7. A field has an area of 40 acres (640 acres make 1 square mile). Express this in (a) km² and (b) m².

8. 1.0 pound of salt is dissolved in water and made up to 10 gallons (U.S.). What is the concentration of salt in g liter^{-1}?

SOLUTIONS TO PROBLEMS

1. (a) $1 \text{ mile} = \dfrac{1 \text{ mi} \times 1760 \text{ yd mi}^{-1} \times 36 \text{ in. yd}^{-1} \times 2.54 \text{ cm in.}^{-1}}{100 \text{ cm m}^{-1}} = 1609 \text{ m}$

    (b) $1609 \text{ m} = \dfrac{1609 \text{ m}}{1000 \text{ m km}^{-1}} = 1.609 \text{ km}$

2. $2240 \text{ lb} = \dfrac{2240 \text{ lb} \times 453.6 \text{ g lb}^{-1}}{1000 \text{ g kg}^{-1}} = 1016 \text{ kg}$

3. $350 \text{ in.}^3 = \dfrac{350 \text{ in.}^3 \times (2.54)^3 (\text{cm in.}^{-1})^3}{1000 \text{ cm}^3 \text{ liter}^{-1}} = 5.74 \text{ liter}$

4. $30 \text{ lb in.}^{-2} = \dfrac{30 \text{ lb in.}^{-2} \times 454 \text{ g lb}^{-1}}{1000 \text{ g kg}^{-1} \times (2.54)^2 (\text{cm in.}^{-1})^2} = 2.1 \text{ kg cm}^{-2}$

5. $60 \text{ lb ft}^{-3} = \dfrac{60 \text{ lb ft}^{-3} \times 454 \text{ g lb}^{-1}}{(12)^3 (\text{in. ft}^{-1})^3 \times (2.54)^3 (\text{cm in.}^{-1})^3} = 0.96 \text{ g cm}^{-3}$

6.   Using 1 mi = 1.609 km from solution 1,
    (a) $60 \text{ mi h}^{-1} = 60 \text{ mi h}^{-1} \times 1.609 \text{ km mi}^{-1} = 97 \text{ km h}^{-1}$

    (b) $97 \text{ km h}^{-1} = \dfrac{97 \text{ km h}^{-1} \times 1000 \text{ m km}^{-1}}{3600 \text{ s h}^{-1}} = 27 \text{ m s}^{-1}$

7.   Using 1 mi = 1.609 km,

    (a) $40 \text{ acre} = \dfrac{40 \text{ acre} \times (1.609)^2 (\text{km mi}^{-1})^2}{640 \text{ acre mi}^{-2}} = 0.16 \text{ km}^2$

    (b) $0.16 \text{ km}^2 = 0.16 \text{ km}^2 \times (1000)^2 (\text{m km}^{-1})^2 = 1.6 \times 10^5 \text{ m}^2$

8.   1.0 lb in 10 gal is 0.10 lb gal^{-1}.

    $0.10 \text{ lb gal}^{-1} = \dfrac{0.10 \text{ lb gal}^{-1} \times 454 \text{ g lb}^{-1}}{3.79 \text{ liter gal}^{-1}} = 12 \text{ g liter}^{-1}$

Chapter 2
The Mole Concept

In solving a problem of this sort, the grand thing is to be able to reason backward. This is a very useful accomplishment, and a very easy one, but people do not practice it much. A STUDY IN SCARLET

Chemistry is a quantitative science. We don't just ask *what* the product of a reaction is; we ask *how much* is produced from *how much* starting material. The most convenient way to describe quantities is by their masses, so we must have some standard of mass to which every other mass is ultimately related. In all scientific work, the standard is the kilogram and its subdivisions the gram, the milligram, etc. In chemistry, we need, in addition, a quantity that is related to the *number* of atoms, molecules, ions, or electrons in a sample of a substance.

2.1 The Mole

In the international system of units (SI), this fundamental base unit of quantity is called the *mole*. It is defined as follows:

The mole is the amount of substance of a system that contains as many elementary entities (atoms, molecules, etc.) as there are atoms in exactly 12 grams of the isotope carbon-12 (^{12}C).

The abbreviation for mole is mol, but we will spell it out in full in this book. Note that the actual number of elementary entities in the mole is not defined. It is a quantity that has to be determined by experiment, and it is known as *Avogadro's constant* or, more usually, *Avogadro's number*, symbol N. It has units $mole^{-1}$ (i.e., entities per mole) and the accepted value is

$$N = 6.0220 \times 10^{23} \text{ mole}^{-1}$$

We have to be careful to specify what we mean by "elementary entity" when talking about a mole of something. It is the smallest sample of a substance that still retains the identity of that substance. If we have one mole of iron, we have $N = 6.02 \times 10^{23}$ *atoms* of iron, since a piece of iron is a collection of many atoms. One mole of water, however, contains $N = 6.02 \times 10^{23}$ *molecules* of water, each of which contains three atoms, since the molecule is the smallest entity that is still recognizable as water. We could speak of a mole of sodium ions, Na^+, or of a mole of sulfate ions, SO_4^{2-}, meaning a collection of $N = 6.02 \times 10^{23}$ ions in each case. There is nothing to stop us talking about one mole of grains of sand, or one mole of stars, if we think of grains of sand or stars as "elementary entities" *and say what we mean* by elementary entities.

In this connection, some care is required when speaking of a substance that can exist in more than one form. Hydrogen, for example, is usually encountered as the molecule H_2, but can exist under other conditions as atomic hydrogen, H. To be exact, therefore, we should speak of a mole of molecular hydrogen (N H_2 molecules) or a mole of atomic hydrogen (N H atoms). A more subtle example may be found in the element phosphorus, which exists in one form as the molecule P_4. What would one commonly understand as a "mole" of phosphorus?

In practice, we usually understand a mole to refer to the substance as generally encountered under standard conditions. Thus, the gases hydrogen, nitrogen, oxygen, fluorine, chlorine, etc., occur as the diatomic molecules H_2, N_2, O_2, F_2, Cl_2, etc., so a mole would normally refer to N such *molecules*. If we want to refer specifically to the monatomic forms H, O, etc., then we must talk of "a mole of atomic hydrogen." In the case of phosphorus, we should speak of "a mole of P_4" if we want to refer to such a species, otherwise "a mole of phosphorus" would generally mean N atoms of the element.

The above principle enables us to calculate the actual masses of atoms and molecules. In the case of ^{12}C, we know that one mole has a mass of exactly 12 g and contains N atoms, so the mass of one atom is

$$\frac{12 \text{ g mole}^{-1}}{N \text{ mole}^{-1}} = \frac{12}{6.0220 \times 10^{23}} = 1.9927 \times 10^{-23} \text{ g}$$

according to the accepted value of N.

2.2   Atomic Weight

For elements other than ^{12}C, we are principally interested in their *relative* masses, rather than the absolute values of the masses in grams. The relative masses of all particles (i.e., fundamental particles, atoms, molecules, ions, etc.) are measured relative to the mass of one ^{12}C atom, usually by means of an instrument called a mass spectrometer. For instance, an atom of ^{69}Ga weighs 68.926/12 times as much as an atom of ^{12}C. It is convenient to express these relative masses in terms of 1/12 the mass of one ^{12}C atom (i.e., ^{12}C has a relative mass of 12), on which scale ^{69}Ga has a relative mass of 68.926. This is called the relative atomic mass of this species, or, more generally, its *atomic weight*.

$$\text{atomic weight of } ^{69}Ga = \frac{\text{mass of one atom of } ^{69}Ga}{1/12 \text{ the mass of a } ^{12}C \text{ atom}}$$

(*Note*: Despite the name "atomic weight," this quantity is only a *ratio*; it has *no units*.)

Sometimes the quantity of mass equal to 1/12 the mass of a ^{12}C atom is called one atomic mass unit, abbreviated to "amu." The conversion from amu to grams may be readily found by recalling that one ^{12}C atom, by definition, has a mass $12/N$ g, so $1/12$ of this is equal to

$$\frac{1}{12} \times \frac{12}{N} = \frac{1}{N} = \frac{1}{6.0220 \times 10^{23}} = 1.6606 \times 10^{-24} \text{ g}$$

Note that this is simply the reciprocal of Avogadro's number. Obviously the masses of fundamental particles, atoms, etc., could be expressed in amu (which have the dimension of mass), but it is generally more convenient to regard them simply as dimensionless ratios. The numerical values are, of course, the same in either convention.

2.3   Isotopes

Before further developing this, we should define and explain an important concept concerning atomic structure, that of *isotopes*. There are two important properties of an atomic nucleus that we have to note, its charge and its mass. These in turn reflect the composition of the nucleus in terms of the fundamental particles it contains.

The positive charge on the nucleus is always given as an exact multiple of the charge on the proton, which is the basic unit of positive charge on the atomic scale. It is this nuclear charge which establishes the nature of the element that is present; all atoms of that element have the same charge on the nucleus. We call this charge the *atomic number* of the element, and it is equal to the number of protons contained in the nucleus.

Thus, carbon has atomic number 6, which means each nucleus contains 6 protons and has a charge of $+6$. Since the specifying of an atomic number absolutely defines the identity of the element present, the name is actually redundant, but is retained for ease in remembering and referring to the element. It is easier to say "carbon" than to say "element number six."

The other important number associated with an atom is the number of neutrons it contains. With the exception of H, every nucleus contains one or more neutrons. We do not directly record the number of neutrons present, but we add together the total of neutrons and protons present and call this the *mass number* of the nucleus. Since we are only counting particles, the mass number must be an integer (whole number). Thus, fluorine has atomic number 9 and mass number 19—each nucleus contains 9 protons and 10 neutrons, for a total of 19 particles. We indicate this in the full symbol for the element by writing $^{19}_9F$. Atomic number goes at the lower left of the symbol, mass number at the upper left.

Unlike atomic number, mass number can vary without changing the identity of the element. All we are doing is changing the number of neutrons within the nucleus, without changing its charge. Atoms with the same atomic number (number of protons) but a differing mass number (number of neutrons) are known as *isotopes*. Thus, for chlorine, we have two common isotopes:

$^{35}_{17}Cl$ contains 17 protons in the nucleus and 18 neutrons.

$^{37}_{17}Cl$ contains 17 protons in the nucleus and 20 neutrons.

The chemical properties of different isotopes of an element are almost identical, the only difference is in mass. So for any isotope, we can express an atomic weight in terms of the relative mass of one atom of the isotope

$$\text{atomic weight of an isotope} = \frac{\text{mass of one atom of the isotope}}{1/12 \text{ the mass of one } ^{12}C \text{ atom}}$$

The familiar periodic table contains atomic weights for all the stable (nonradioactive) elements; and, in the cases of those containing only one naturally occurring isotope, the above definition is sufficient. For example, natural fluorine is 100% ^{19}F, atomic weight 18.9984; and natural sodium is 100% ^{23}Na, atomic weight 22.9898. Such elements, of which there are about 22, are said to be "isotopically pure."

2.4   Average Atomic Weight

The majority of elements, however, occur in nature as mixtures of several isotopes, the proportions of each isotope in the mixture being, in general, remarkably constant. How is atomic weight defined for these elements? The most useful method is to determine the atomic weight of each isotope separately, then combine the atomic weights of all the isotopes of a particular element to give an *average atomic weight*. In computing this average, we must take into account the relative abundance of each isotope.

average atomic weight = (atomic weight of isotope 1) × (fraction of isotope 1)

+ (atomic weight of isotope 2) × (fraction of isotope 2)

+ ...etc., for all isotopes present

(*Note*: By "fraction" of each isotope, we mean "number fraction," that is, the number of atoms of that isotope present as a fraction of the total number of atoms present. We do not mean "mass fraction," which is the mass of the isotope present as a fraction of the total mass present.)

These, then, are the quantities that are included in the familiar table of atomic weights of the elements—they are absolutely vital to quantitative chemical calculations! An example will show the method used to compute average atomic weights.

Example 2.1   Copper contains two isotopes:

^{63}Cu, atomic weight 62.9298, amounting to 69.09%
^{65}Cu, atomic weight 64.9278, amounting to 30.91%

What is the average atomic weight of copper?

SOLUTION   The relative amounts of the two isotopes are given as percentages. To convert them to fractions, we simply divide each by 100. The total of the fractions of different isotopes making up the whole must come to 1.00, just as the total of percentages must come to 100%.

fractions present: ^{63}Cu, 0.6909; ^{65}Cu, 0.3091

average atomic weight = (62.9298 × 0.6909) + (64.9278 × 0.3091)

= 43.478 + 20.069

= 63.547

Obviously, the average atomic weight comes between those of the two isotopes, but it is not exactly midway between the two values, because the lighter isotope preponderates.

If you could not follow the above example at first reading, try this one: "I have 100 marbles, 69 weighing 10 g and 31 weighing 12 g. What does an average marble weigh?" (*Answer*: 10.62 g.) The principle of this calculation is precisely the same.

The same type of calculation may be done for three or more isotopes, and it may also be done in reverse.

Example 2.2   Calculate the average atomic weight of magnesium, given the composition

^{24}Mg, atomic weight 23.9850, occurrence 78.70%
^{25}Mg, atomic weight 24.9858, occurrence 10.13%
^{26}Mg, atomic weight 25.9826, occurrence 11.17%

SOLUTION   The calculation is the same as before, except that we must add together three terms instead of two.

$$\text{average atomic weight} = (23.9850 \times 0.7870) + (24.9858 \times 0.1013) + (25.9826 \times 0.1117)$$

$$= 18.876 + 2.5311 + 2.9024$$

$$= 24.310$$

Example 2.3   Natural carbon is mainly ^{12}C, with a small amount of ^{13}C. If natural carbon has an atomic weight of 12.0111 and ^{13}C has an atomic weight of 13.00335, what percentage of natural carbon is ^{13}C?

SOLUTION   We need to know the atomic weight of ^{12}C, which is, of course, 12 exactly (by definition).

Our unknowns are the fractions of natural carbon which correspond to the two isotopes. Let's say that fraction of ^{13}C is x; fraction of ^{12}C is y. Then the average atomic weight (which we know to be 12.0111) will be derived from the atomic weights of the two isotopes as before:

$$13.00335x + 12y = 12.0111$$

We cannot solve this equation as it stands, because it contains two unknowns, but we can make use of the additional fact that the sum of the fractions of all isotopes present must be exactly 1. This gives us the second equation needed:

$$x + y = 1 \quad y = 1 - x$$

Now we can put $(1 - x)$ in the first equation in place of y, and solve for x:

$$13.00335x + 12(1 - x) = 12.0111$$

Rearrange to put all terms in x on the left:

$$13.00335x - 12x = 12.0111 - 12$$

$$1.00335x = 0.0111$$

$$x = 0.0111$$

Expressed as a percentage, the ^{13}C content is, therefore,

$$0.0111 \times 100 = 1.11\%$$

Note carefully the logic used in solving this problem. The method is similar to that in Examples 2.1 and 2.2, except that we used algebraic symbols for the fractions that were our unknowns.

2.5 Molecular Weight

If the fundamental entities with which we are dealing are *molecules*, then a collection of molecules will contain samples of all possible isotopes. Water molecules, as found naturally, contain $^1H_2^{16}O$, $^1H_2^{17}O$, $^1H_2^{18}O$, $^2H_2^{16}O$, $^2H_2^{17}O$, $^2H_2^{18}O$ molecules, and $^1H^2H^{16}O$, $^1H^2H^{17}O$, $^1H^2H^{18}O$ molecules as well! The mass of an *average* water molecule depends on the isotopic compositions of its constituent elements. Fortunately, these are very nearly constant, so we can use the average atomic weights of the elements, given in tables, and know that these will take into account the isotopic distribution found in our sample.

For any compound whose composition is known, we express the combining ratio of the elements by its *formula*. Subscript numbers following the symbols for each element indicate the number of atoms of that element that have entered into combination, except that, if only one atom of that element is present, the subscript 1 is omitted. For example,

hydrogen chloride: HCl, one H atom, one Cl atom

water: H_2O, two H atoms, one O atom

sulfuric acid: H_2SO_4, two H atoms, one S atom, four O atoms.

It oftens happens that a number of atoms are associated together within the compound to form a stable group, and we indicate this by enclosing them in parentheses in the formula. A subscript number following the parentheses multiplies all the atoms inside them. For example,

ammonium sulfate: $(NH_4)_2SO_4$, contains two N atoms, $(4 \times 2) =$ eight H atoms, one S and four O atoms.

calcium nitrate: $Ca(NO_3)_2$, contains one Ca, two N, and $(3 \times 2) =$ six O atoms.

For any compound, an important characteristic is its *molecular weight*, defined as follows:

The molecular weight is the sum of the average atomic weights of all the atoms in a molecule.

Since the average atomic weights are ratios having no dimension, the molecular weight (which we will abbreviate *M.W.*) is likewise dimensionless. Another way of looking at it would be as the ratio of the mass of an "average" molecule to the weight of one atom of ^{12}C.

Example 2.4 Calculate the molecular weights of (a) hydrogen bromide, HBr; and (b) acetic acid, CH_3COOH

SOLUTION

(a) Use the table of atomic weights:

1 hydrogen at	1.008
1 bromine at	79.904
molecular weight	80.912

(b) The formula of acetic acid is written as CH_3COOH to emphasize the structural arrangement of the atoms; but for the *M.W.* evaluation, we just need the total number of each kind of atom:

$$
\begin{array}{lr}
\text{2 carbon at } 12.011 & 24.022 \\
\text{4 hydrogen at } 1.008 & 4.032 \\
\text{2 oxygen at } 15.999 & \underline{31.998} \\
\text{molecular weight} & \overline{60.052}
\end{array}
$$

The above *M.W.* values have been calculated to three decimal places (5 significant figures; see Appendix A.1), using the accurate atomic weights. For most ordinary chemical work, a lower degree of accuracy (3 or 4 significant figures) is sufficient, and the table of atomic weights in this book (Appendix B.2) gives values to 4 figures, rather than to the limit of known accuracy.

2.6   Formula Weight

A concept closely related to molecular weight is that of *formula weight*. As the name implies, it is the sum of the atomic weights of all the atoms in the formula of the compounds, so it would be identical with *M.W.* in the above examples. Formula weight may be used, however, in cases where it would be inappropriate to talk about "molecules" of a substance, and this is particularly the case for ionic compounds. In sodium chloride, for example, the formula is NaCl but the structure is known to contain separate Na^+ and Cl^- ions, rather than NaCl molecules. We could therefore calculate the formula weight of NaCl as $23.0 + 35.5 = 58.5$.

As you will see, the same numerical result would be obtained if we had called this a molecular weight. Since it is often difficult to decide whether or not "molecules" are present in a compound, we will, in this book, use the term "molecular weight" as synonymous with "formula weight," with the reservation that this usage does not imply anything about the actual molecular structure of the compound.

2.7   Molar Mass

In dealing with quantities of substances in the laboratory, it is most convenient to know how many moles of each substance are present. Bearing in mind our definition of the mole, we define the *molar mass* as the mass of one mole of the substance in question. The number of moles of substance, mass of substance, and molar mass, are connected by the following relationship:

$$
\text{number of moles of substance} = \frac{\text{mass of substance present}}{\text{molar mass}}
$$

We usually measure molar mass in grams, when it will be numerically equal to the molecular weight, atomic weight, or formula weight, as appropriate. For example,

methane: CH_4, molecular weight 16.0; molar mass 16.0 g $mole^{-1}$

gallium: average atomic weight 69.72; molar mass 69.72 g $mole^{-1}$

sodium chloride: NaCl, formula weight 58.44; molar mass 58.44 g $mole^{-1}$

Note that molar mass has the dimension of mass and the units of grams per mole. Knowing the molar mass, we can easily work out the mass of any number of moles of a substance.

Example 2.5 What is the mass of the following?

(a) 1.00 mole of hydrogen peroxide, H_2O_2
(b) 15.00 mole of sulfuric acid, H_2SO_4
(c) 0.375 mole of sodium sulfate decahydrate, $Na_2SO_4 \cdot 10H_2O$

SOLUTION   We can arrange the above relationship to give

mass of substance = (number of moles present) × (molar mass)

(a) H_2O_2 has *M.W.* 34.0, molar mass 34.0 g mole^{-1}
   1.00 mole is 34.0 g of H_2O_2
(b) H_2SO_4 has *M.W.* 98.08, molar mass 98.08 g mole^{-1}
   15.00 mole is 15.00 mole × 98.08 g mole^{-1} = 1471 g
(c) $Na_2SO_4 \cdot 10H_2O$ has *M.W.* 322.2, molar mass 322.2 g mole^{-1}

(*Note*: The $10H_2O$ joined by a dot to the remainder of the formula means that 10 water molecules are loosely linked to each Na_2SO_4 unit in the crystal.)

   0.375 mole is 0.375 mole × 322.2 g mole^{-1} = 120.8 g

Example 2.6 How many moles are represented by the following quantities?

(a) 100 g of boron trichloride, BCl_3
(b) 1.28 g of benzene, C_6H_6
(c) 5.40×10^{-2} g of chromium trioxide, CrO_3

SOLUTION

$$\text{number of moles} = \frac{\text{mass of sample}}{\text{molar mass}}$$

(a) For BCl_3, the molar mass is 117.2 g mole^{-1}

$$\text{number of moles} = \frac{100 \text{ g}}{117.2 \text{ g mole}^{-1}} = 0.853 \text{ mole}$$

(b) For benzene, molar mass = 78.1 g mole^{-1}

$$\text{number of moles} = \frac{1.28 \text{ g}}{78.1 \text{ g mole}^{-1}} = 1.64 \times 10^{-2} \text{ mole}$$

(c) For CrO_3, molar mass = 100 g mole^{-1}

$$\text{number of moles} = \frac{5.40 \times 10^{-2} \text{ g}}{100 \text{ g mole}^{-1}} = 5.40 \times 10^{-4} \text{ mole}$$

Example 2.7 Calculate molecular weights from the following data.

(a) 12.6 mole weigh 1380 g
(b) 0.580 mole weighs 211 g
(c) 2.81×10^{-3} mole weighs 0.489 g

SOLUTION   Here we rearrange the relationship again to give

$$\text{molar mass} = \frac{\text{mass of sample}}{\text{number of moles}}$$

(a) 12.6 mole weigh 1380 g

$$\text{molar mass} = \frac{1380 \text{ g}}{12.6 \text{ mole}} = 110 \text{ g mole}^{-1}$$

$$M.W. = 110$$

(b) 0.580 mole weighs 211 g

$$\text{molar mass} = \frac{211 \text{ g}}{0.580 \text{ mole}} = 364 \text{ g mole}^{-1}$$

$$M.W. = 364$$

(c) 2.81×10^{-3} mole weighs 0.489 g

$$\text{molar mass} = \frac{0.489 \text{ g}}{2.81 \times 10^{-3} \text{ mole}} = 174 \text{ g mole}^{-1}$$

$$M.W. = 174$$

The above examples show that, when any three quantities are related, if any two of them are known, the third may be found. All we have to do is to rewrite the equation connecting them in the appropriate way. Note the use of dimensional analysis to make sure the above fractions are written the right way up. As far as simple calculation is concerned, we may substitute "molecular weight" for "molar mass" in the above expression (remember, the two terms are numerically equal) and remember it as:

To convert from grams to moles, *divide* by molecular weight.
To convert from moles to grams, *multiply* by molecular weight.

The interrelationship of these quantities is of the greatest importance in chemical calculations and in the laboratory, since we usually measure chemicals by weighing them in grams, but our subsequent calculations will generally involve mole quantities.

Problems involving the calculation of molecular weight will be found throughout this book.

2.8   Stoichiometry: The Composition of Compounds

The word "stoichiometry" has to do with the various components present and the quantitative relationships between them. We may use it in two senses: (1) to talk of *stoichiometry of composition,* by which we mean the proportions in which different elements are present in a compound, or (2) to talk of *stoichiometry of reaction,* by which we mean the proportions in which different elements or compounds react together, and the amount of products they yield. We can work out stoichiometry of composition from the formula of a compound, while to find the stoichiometry of reaction we need a balanced chemical equation. The first of these problems will be discussed in this chapter, the second in Chapter 3.

It is a general characteristic of a pure chemical compound that it has different elements present in a fixed, definite ratio which we indicate by its formula. (There are a few compounds in which a definite formula cannot be assigned; these are known as "non-stoichiometric.") Given the formula and a table of atomic weights, we can work out the weight ratios in which different elements are present. It is most convenient to do this through the mole concept, since this enables us to use reasonable, easily comprehended weights of material.

Example 2.8 Calculate the percentage composition, by weight, of the following compounds: (a) KBr, (b) $C_{10}H_{22}$, and (c) HNO_3.

SOLUTION By "percentage composition," we mean the ratio of the mass of a particular element present to the total mass of compound present, expressed as a percentage. If we work this out on a basis of one mole of the compound, we would obtain

$$\text{percent of element X} = \frac{\text{mass of X present in 1 mole} \times 100}{\text{molar mass}}$$

(a) One mole of KBr contains

1 mole of K, which weighs	39.1 g
1 mole of Br, which weighs	79.9 g
total mass	119.0 g

$$\text{percent of potassium} = \frac{39.1}{119.0} \times 100 = 32.9\%$$

$$\text{percent of bromine} = \frac{79.9}{119.0} \times 100 = 67.1\%$$

(b) One mole of $C_{10}H_{22}$ contains

10 mole of C, which weigh $10 \times 12.01 =$	120.1 g
22 mole of H, which weigh $22 \times 1.008 =$	22.2 g
total mass	142.3 g

$$\text{percent of carbon} = \frac{120.1 \times 100}{142.3} = 84.4\%$$

$$\text{percent of hydrogen} = \frac{22.2}{142.3} \times 100 = 15.6\%$$

(c) One mole of HNO_3 contains

1 mole of H, which weighs	1.01 g
1 mole of N, which weighs	14.0 g
3 moles of O, which weigh	48.0 g
total mass	63.0 g

$$\text{percent of hydrogen} = \frac{1.01}{63.0} \times 100 = 1.6\%$$

$$\text{percent of nitrogen} = \frac{14.0}{63.0} \times 100 = 22.2\%$$

$$\text{percent of oxygen} = \frac{48.0}{63.0} \times 100 = 76.2\%$$

Several points should be noted in these calculations. When speaking of the content of elements such as hydrogen, oxygen, or nitrogen, which normally exist as the diatomic molecules H_2, O_2, N_2, etc., we are careful to specify that the moles present in the compound are moles of the *atomic* form of the element. Thus, we write "1 mole of N," or "22 moles of H." In calculations of this type, it would be confusing to speak of the form in which the pure element exists, when we are only concerned with the number of atoms of it which are present in combination.

Obviously, the total of the percentages should always add up to 100% (or within a few tenths of 100, allowing for "rounding off" errors in the individual figures). Always work out each percentage separately and total them, as this gives a check on the accuracy of your work.

Sometimes the composition will be given in a manner which is less precise than an actual formula.

Example 2.9 A compound contains only carbon and hydrogen, the combining ratio being two hydrogen atoms to one carbon atom. What is the percentage composition?

SOLUTION We may take the formula as CH_2, which contains in each mole

$$
\begin{array}{ll}
\text{1 mole of C, which weighs} & 12.01 \text{ g} \\
\text{2 mole of H, which weigh} & \underline{2.02 \text{ g}} \\
\text{total mass} & 14.03 \text{ g}
\end{array}
$$

$$\text{percent of carbon} = \frac{12.01}{14.03} \times 100 = 85.6\%$$

$$\text{percent of hydrogen} = \frac{2.02}{14.03} \times 100 = 14.4\%$$

Clearly, the same result would have been found if we had taken C_2H_4, C_3H_6, or any other compound in which the C:H atom ratio was 1:2.

2.9 Simplest Formula

The last example gives us a clue to answering an important question; can we calculate a formula for a substance if we are given its percentage composition? The answer is a qualified "yes"; we can calculate the atomic combining ratio, but this will correspond to any number of possible formulas in which the atoms are present in this ratio.

Example 2.10 A compound is found to contain 87.5% nitrogen and 12.5% hydrogen. What formulas are possible for the compound?

SOLUTION We start by noting that the content of nitrogen and hydrogen add up to 100%, so these are the only two elements present.

In 100 g of compound, there are 87.5 g of N and 12.5 g of H present. To convert these to mole quantities, we must *divide* each by the appropriate atomic weight:

$$87.5 \text{ g of N} = \frac{87.5}{14.0} = 6.25 \text{ mole of N}$$

$$12.5 \text{ g of H} = \frac{12.5}{1.0} = 12.5 \text{ mole of H}$$

Therefore, the mole ratio $\dfrac{H}{N}$ present is $\dfrac{12.5}{6.25}$ or $\dfrac{2}{1}$.

This is, of course, the atomic combining ratio of the elements, so the simplest formula for this compound is NH_2; but any multiple of this will have the same composition, e.g., N_2H_4, N_3H_6, etc., or, in general, $(NH_2)_n$ where n is any integer.

Since there is no limit to the number of possible formulas that will fit a given percentage composition, it is only useful to give the *simplest formula* (sometimes called the empirical formula)—which was NH_2 in the above example.

One additional item of information would enable us to assign the complete, correct formula of the compound, and that is its molecular weight. By many different methods (see Chapters 8 and 9), molecular weights may be found experimentally, and a combination of molecular weight and percentage composition enables the complete, or *molecular*, formula to be found.

Example 2.11 The compound mentioned in Example 2.10 is found to have a *M.W.* of 32. What is its molecular formula?

SOLUTION We know the formula is a multiple of the simplest formula, NH_2, which we can denote $(NH_2)_n$. Thus, the *M.W.* of $(NH_2)_n$ will be

$$n(14 + 2) = 16n$$

But we know the *M.W.* is 32, and since

$$16n = 32 \qquad n = 2$$

the molecular formula is therefore $(NH_2)_2$ or N_2H_4.

Very often, the experimentally determined molecular weight will contain some degree of error, but even a rough value will enable us to calculate a molecular formula, from which an accurate *M.W.* can then be found.

Example 2.12 A compound is known to contain only carbon, hydrogen, and oxygen. Analysis gives the composition: carbon, 38.7%; hydrogen, 9.7%. The molecular weight is found to be 65 ± 5. Calculate the (a) simplest formula, (b) the molecular formula, and (c) the accurate *M.W.*

SOLUTION Knowing that oxygen is the only other element present, we can find the oxygen content by difference, since the three percentages total 100.

$$\text{percent of oxygen} = 100 - (38.7 + 9.7) = 51.6\%$$

(a) Knowing the relative weights of the three elements present, we convert these to mole ratios by dividing by the appropriate atomic weights:

Carbon	Hydrogen	Oxygen
38.7 g	9.7 g	51.6 g
$\dfrac{38.7}{12.0} = 3.23$ mole	$\dfrac{9.7}{1.0} = 9.7$ mole	$\dfrac{51.6}{16.0} = 3.23$ mole

To put these in a neater form, we divide each one by the *smallest* of the three numbers:

$\dfrac{3.23}{3.23} = 1.00$	$\dfrac{9.7}{3.23} = 3.0$	$\dfrac{3.23}{3.23} = 1.00$

So our ratio of elements is one carbon to three hydrogen to one oxygen, or CH_3O (simplest formula).

(b) The molecular formula will be $(CH_3O)_n$, giving a molecular weight of

$$n(12 + 3 + 16) = 31n.$$

The molecular weight is known to be about 65, so we can write

$$31n = 65 \pm 5 \quad \text{or} \quad n = \frac{65 \pm 5}{31} \approx 2$$

Knowing n must be a whole number, we see that $n = 2$ and the molecular formula is $(CH_3O)_2$ or $C_2H_6O_2$. So the accurate molecular weight, calculated for $C_2H_6O_2$, is 62.0.

Example 2.13   A compound analyzes as carbon, 18.0%; hydrogen, 2.3%; chlorine, 80.0%. The molecular weight is 130 ± 5. Calculate the simplest formula and molecular formula.

SOLUTION   We note that the given percentages add up to near 100%, so all elements are accounted for. Conversion to atomic ratios gives

Carbon	Hydrogen	Chlorine
18.0 g	2.3 g	80.0 g
$\dfrac{18.0}{12.0} = 1.50$ mole	$\dfrac{2.3}{1.0} = 2.3$ mole	$\dfrac{80.0}{35.5} = 2.25$ mole

Divide by the smallest:

$\dfrac{1.50}{1.50} = 1.00$	$\dfrac{2.3}{1.5} = 1.5$	$\dfrac{2.25}{1.50} = 1.50$

It would be wrong to take these numbers as the simplest formula. No compound could be $CH_{1.5}Cl_{1.5}$. Instead, we should take a multiple of these ratios which enables us to give whole-number ratios of the elements. Obviously, doubling the numbers will give $C_2H_3Cl_3$. This is the simplest formula of the compound. $C_2H_3Cl_3$ has $M.W.$ $(24 + 3 + 106.5) = 133.5$, which is very close to the value given for the $M.W.$ of the compound; so we conclude that this is also the molecular formula.

There are many compounds where the term "molecule" cannot be applied, as no discrete molecules exist (e.g., ionic salts). In this case, we always take the simplest formula as all that is needed to represent the composition.

Example 2.14   An oxide of chromium analyzes to be 68.4% of the metal. What is its formula?

SOLUTION   The percent of oxygen is $100 - 68.4 = 31.6\%$.

Chromium	Oxygen
68.4 g	31.6 g
$\dfrac{68.4}{52.0} = 1.32$ mole	$\dfrac{31.6}{16.0} = 1.98$ mole

Divide by the smaller:

$$\frac{1.32}{1.32} = 1.00 \qquad\qquad \frac{1.98}{1.32} = 1.50$$

As in Example 2.13, we double these numbers to give the whole-number relationship 2:3, i.e., the formula is Cr_2O_3.

Example 2.15 A nitrate of tin contains 11.5% nitrogen. What is its formula?

SOLUTION This problem cannot be solved by the method previously used, as we only know the content of one element out of the three present (Sn, N, and O). However, we do know that the compound will contain only two ingredients, tin atoms and nitrate ions, NO_3^-. (If you are not sure what a nitrate ion is, look ahead to Table 3.3 at the end of Chapter 3.) Our unknown is the ratio of nitrate to tin, so we can call this x and the formula will be $Sn(NO_3)_x$.

Now we can work out the *M.W.* and percentage composition in terms of x. We have altogether one Sn atom, x N atoms and $3x$ O atoms, so the *M.W.* is

$$119 + 14.0x + (16.0 \times 3)x = 119 + 62.0x$$

Out of this total, the nitrogen amounts to $14.0x$ so we can express the percent nitrogen in the usual way and put it equal to the given nitrogen content (11.5%).

$$\text{percent nitrogen} = \frac{14.0x}{119 + 62.0x} \times 100 = 11.5$$

This construction of an equation in which x is the only unknown is the key step in solving this problem. Multiplying it out, we have

$$1400x = 11.5(119 + 62.0x)$$

$$1400x = 1369 + 713x$$

$$1400x - 713x = 1369$$

$$687x = 1369$$

$$x = \frac{1369}{687} = 1.99$$

Knowing x must be a whole number, we see that $x = 2$ and the formula of the nitrate is $Sn(NO_3)_2$.

This problem is a good example of the "reversible calculation" approach to problem solving. Many students who could easily calculate the percent of nitrogen in $Sn(NO_3)_2$ would have trouble with the above problem, but a moment's thought shows that we are simply using exactly the same concepts in reverse order. What was the solution to one problem becomes the starting data for the other problem. The following example is very similar.

Example 2.16 Lead perchlorate crystallizes from water as a hydrate, $Pb(ClO_4)_2 \cdot xH_2O$. If the crystals contain 45.0% of lead, what is the value of x in the formula?

SOLUTION   Here we have our unknown, x, already in the formula, so we can work out the $M.W.$ and lead content in terms of x.

The anhydrous (that is, no water) material, $Pb(ClO_4)_2$, has $M.W.$ $207 + 71.0 + 128 = 406$. The additional x molecules of water ($M.W.$ 18.0) will add $18.0x$, to make the total $M.W.$ of the hydrated salt $406 + 18.0x$. Out of this, the lead made up 207, so the percent lead can be expressed as usual

$$\text{percent lead} = \frac{207}{406 + 18.0x} \times 100$$

We can put this equal to 45.0, the given lead content, to obtain our equation in x

$$\frac{207 \times 100}{406 + 18.0x} = 45.0$$

$$2.07 \times 10^4 = 45.0(406 + 18.0x)$$

$$2.07 \times 10^4 = (1.83 \times 10^4) + 810x$$

$$810x = (2.07 \times 10^4) - (1.83 \times 10^4) = 2.4 \times 10^3$$

$$x = \frac{2.4 \times 10^3}{810} = 3.0$$

The formula of the hydrate is $Pb(ClO_4)_2 \cdot 3H_2O$.

2.10   Solution Concentrations

We have so far dealt with the mole concept in connection with weighed quantities of elements or compounds. However, we frequently meet compounds in the form of solutions, since this is a very convenient method for handling them and allowing them to react. In solutions, there are various ways of measuring concentrations; we will consider only two in this chapter.

We must first define the terms *solute* and *solvent*. The solvent is the substance, generally liquid, present in the larger proportion, while the solute is the substance present in the smaller proportion. The solute may be solid, liquid, or gas. There is usually far more of the solvent than the solute, but we may occasionally mix two things in equal amounts, e.g., alcohol and water, so it is difficult to decide which is solvent and which solute. Obviously the terms are useful only for dilute solutions. Other ways of describing concentrations in such solutions are discussed in Chapter 9.

The commonest type of solution encountered in the laboratory consists of a solid or a liquid dissolved in water, e.g., NaCl (solute) in water (solvent). The two common methods for describing the concentration of such a solution are

1. Grams per liter, or similar units, in which a mass of *solute* is dissolved in a certain volume of *solution.*
2. Molarity, which is the number of moles of *solute* per liter of *solution,* given the symbol M.

Note that, in both cases, we measure the volume of solution, *not* the volume of the solvent. Clearly, there will be a relationship between grams per liter and molarity similar to that between the number of moles and the number of grams of a compound:

$$\text{concentration in g liter}^{-1} = \frac{\text{mass of solute in g}}{\text{volume of solution in liter}}$$

$$M = \text{molarity} = \frac{\text{number of moles of solute}}{\text{volume of solution in liter}}$$

$$M = \frac{\text{solution concentration in g liter}^{-1}}{\text{molar mass of solute}}$$

$$M = \frac{\text{mass of solute in g}}{(\text{molar mass of solute}) \times (\text{volume of solution in liter})}$$

The above relationships should be studied carefully, as their correct use is the key to calculations on solution concentrations. Remember that they can all be rearranged so that any one quantity can be found if the others are known, as the following examples demonstrate.

Example 2.17   Express the following solution concentrations in g liter^{-1}:

(a) 218 g dissolved in 6.83 liter of solution
(b) 5.66 g dissolved in 0.500 liter
(c) 0.128 g dissolved in 25.0 ml

SOLUTION

(a) 218 g in 6.83 liter $= \dfrac{218 \text{ g}}{6.83 \text{ liter}} = 31.9$ g liter^{-1}

(b) 5.66 g in 0.500 liter $= \dfrac{5.66 \text{ g}}{0.500 \text{ liter}} = 11.3$ g liter^{-1}

(c) 1000 ml $=$ 1 liter   so   25.0 ml $= \dfrac{25.0}{1000} = 0.0250$ liter

0.128 g in 0.0250 liter $= \dfrac{0.128 \text{ g}}{0.0250 \text{ liter}} = 5.12$ g liter^{-1}

You should realize that the concentration of a solution is *independent* of the amount of solution present. We might have one drop, one liter, or an oceanful, but in each case its concentration would be the same. There's no need to have an actual liter of solution in your hand, we simply use the liter as a convenient measure with which to compare solutions.

Example 2.18   What mass of solute will be present in the following?

(a) 2.23 liter of a solution which contains 12.1 g liter^{-1}
(b) 115 ml of a solution which contains 9.83 g liter^{-1}

SOLUTION   The relationship we need here is

mass of solute $=$ (volume of solution in liter) $\times$ (concentration in g liter^{-1})

(a) mass of solute $=$ 2.23 liter $\times$ 12.1 g liter^{-1} $=$ 27.0 g
(b) mass of solute $=$ 0.115 liter $\times$ 9.83 g liter^{-1} $=$ 1.13 g

(*Note*: Do not forget to convert 115 ml to liter by dividing by 1000.)

Example 2.19   A solution contains 15.7 g liter^{-1}. What volume will contain 2.00 g of solute?

SOLUTION   Once again we rearrange:

$$\text{volume of solution in liter} = \frac{\text{mass of solute}}{\text{concentration in g liter}^{-1}}$$

In this case,

$$\text{volume} = \frac{2.00 \text{ g}}{15.7 \text{ g liter}^{-1}} = 0.127 \text{ liter}$$

Multiply by 1000 to convert to ml:

$$0.127 \text{ liter} = (0.127 \text{ liter}) \times (1000 \text{ ml liter}^{-1}) = 127 \text{ ml}$$

Example 2.20   What is the molarity of the following solutions?

(a) 42.0 ml contain 1.83×10^{-2} mole
(b) 850 ml contain 0.662 mole

SOLUTION   All we have to do is take the ratio of the number of moles to the volume of the solution in liter.

(a) $42.0 \text{ ml} = \dfrac{42.0}{1000} = 0.0420 \text{ liter}$

$$\text{molarity} = \frac{1.83 \times 10^{-2} \text{ mole}}{0.0420 \text{ liter}} = 0.436 M$$

(b) $850 \text{ ml} = \dfrac{850}{1000} = 0.850 \text{ liter}$

$$\text{molarity} = \frac{0.662 \text{ mole}}{0.850 \text{ liter}} = 0.779 M$$

Example 2.21   How many moles of solute are present in the following?

(a) 16.3 liter of a solution which is 0.113M
(b) 15.6 ml of a solution which is 0.0250M

SOLUTION   The relationship here is

$$\text{mole of solute} = (\text{volume of solution in liter}) \times \text{molarity}$$

(a) mole of solute = 16.3 liter $\times$ 0.113 mole liter^{-1} = 1.84 mole
(b) 15.6 ml = 0.0156 liter

mole of solute = 0.0156 liter $\times$ 0.0250 mole liter^{-1} = 3.90×10^{-4} mole

Example 2.22   What volume of an 0.130M solution contains 0.275 mole of solute?

SOLUTION   Rearranging the relationship gives

$$\text{volume in liter} = \frac{\text{mole of solute}}{\text{molarity}}$$

$$\text{volume} = \frac{0.275 \text{ mole}}{0.130 \text{ mole liter}^{-1}} = 2.12 \text{ liter}$$

Example 2.23 Calculate the molarity of the following solutions:

(a) H_2SO_4, 196 g liter^{-1}
(b) acetic acid, CH_3COOH, 2.00 g in 75.0 ml solution

SOLUTION In every case where we convert from g to mole, we must calculate the molar mass.

(a) H_2SO_4 has *M.W.* 98.1; molar mass 98.1 g.

$$\text{molarity} = \frac{196 \text{ g liter}^{-1}}{98.1 \text{ g mole}^{-1}} = 2.00M \qquad (\text{mole liter}^{-1})$$

(b) Acetic acid has *M.W.* 60.1; molar mass 60.1 g.

$$\text{molarity} = \frac{2.00 \text{ g}}{0.0750 \text{ liter} \times 60.1 \text{ g mole}^{-1}} = 0.444M$$

Example 2.24 Express the following concentrations in g liter^{-1}.

(a) 0.104M hydrochloric acid, HCl
(b) 2.23 mole of potassium bromide, KBr, dissolved in 4.55 liter of solution

SOLUTION Again, we calculate molar masses:

(a) HCl has *M.W.* 36.5; molar mass 36.5 g.

concentration in g liter^{-1} = (0.104 mole liter^{-1}) × (36.5 g mole^{-1}) = 3.80 g liter^{-1}

(b) KBr has *M.W.* 119; molar mass 119 g.

$$\text{concentration in g liter}^{-1} = \frac{2.23 \text{ mole} \times 119 \text{ g mole}^{-1}}{4.55 \text{ liter}} = 58.3 \text{ g liter}^{-1}$$

Example 2.25 What volume of 0.225M sodium nitrate ($NaNO_3$) solution contains 5.00 g of solute?

SOLUTION $NaNO_3$ has *M.W.* 85.0, molar mass 85.0 g.

$$\text{volume of solution} = \frac{\text{mass of solute}}{\text{molar mass} \times \text{molarity}}$$

$$= \frac{5.00 \text{ g}}{(85.0 \text{ g mole}^{-1}) \times (0.225 \text{ mole liter}^{-1})}$$

$$= 0.261 \text{ liter} = 261 \text{ ml}$$

The above calculations may look different at first sight, but they are really very similar, being based on the definitions of molarity and the molar mass. Memorize the definitions and the basic relationships and get into the habit of turning a relationship around so that it fits the particular problem you are facing. Note the use of dimensional analysis to make sure you have things the right way up.

KEY WORDS

atomic mass unit	atomic weight
Avogadro's number	molar mass
average atomic mass	stoichiometry of composition
isotopes	atomic combining ratio
molecular weight	simplest formula
mole	molecular formula
molarity	

STUDY QUESTIONS

1. What are the units of each of the above?

2. Why are atomic and molecular weight dimensionless?

3. What is the unique definition of a mole of ^{12}C?

4. Why is the atomic weight of carbon 12.01, rather than 12.00?

5. Is Avogadro's number a fundamental natural constant?

6. What does "Avogadro's number of entities" mean in connection with a mole of:
   (a) an element such as potassium
   (b) an ionic compound such as KCl
   (c) a molecular solid such as CH_4
   (d) chloride ions

7. How do we convert the quantity of a compound (a) from moles to grams, and (b) from grams to moles?

8. What generally characterizes a chemical compound?

9. What information do we need to calculate:
   (a) the percentage composition by weight
   (b) the molecular weight
   (c) the atomic combining ratio (simplest formula)
   (d) the molecular formula of a compound

10. For a solution, how are molarity, molar mass, mass of solvent, and volume of solution connected? Write this relationship in several different ways.

PROBLEMS

1. Calculate the mass in grams of the following:
   (a) one atom of fluorine
   (b) five atoms of sodium
   (c) one "average" molecule of ethanol, C_2H_5OH

2. Calculate the average atomic weight of the following:
   (a) gallium: 60.4% ^{69}Ga, at.wt 68.9257; 39.6% ^{71}Ga, at.wt 70.9249.
   (b) silver: 51.82% ^{107}Ag, at.wt 106.9041; 48.18% ^{109}Ag, at.wt 108.9047.

3. Natural lithium is a mixture of ^6Li, at.wt 6.0151, and ^7Li, at.wt 7.0160. If the average atomic weight of Li is 6.939, what is the percentage of each isotope present?

4. Calculate the molecular weight of the following:
   (a) $BaCO_3$ (b) $CdBr_2$ (c) $CuCl_2 \cdot 2H_2O$ (d) $H_4P_2O_6$

5. For each of the compounds in Problem 4, calculate the number of moles present in a sample of mass 5.00×10^2 g.

6. Calculate the mass of the following:
   (a) 5.00 mole of HCl   (b) 0.288 mole of $NaClO_4$
   (c) 1.20×10^{-3} mole of CH_4

7. Calculate the molecular weights of the following compounds from the information given.
   (a) 11.00 mole weigh 598 g   (b) 2.30×10^{-2} mole weigh 1.98 g

8. Calculate the percentage composition of the following compounds:
   (a) KI   (b) Na_2O (c) $BaSO_4$ (d) $Ca(NO_3)_2 \cdot 4H_2O$

9. Calculate the simplest formula of each of the following compounds from the data given.
   (a) carbon, 92.3%; hydrogen, 7.7%
   (b) C, H, and O present: carbon, 40.0%; hydrogen, 6.7%
   (c) carbon, 13.6%; fluorine 86.4%
   (d) carbon, 88.9%; hydrogen, 11.1%
   (e) C, H, and S present: carbon, 33.3%; hydrogen, 7.4%

10. Calculate the simplest formula, molecular, formula, and accurate $M.W.$ of the following:
    (a) carbon, 92.3%; hydrogen, 7.7%; $M.W.$ 80 ± 3
    (b) C, H, and O present: carbon, 53.3%; hydrogen, 11.1%; $M.W.$ 95 ± 5
    (c) silicon, 33.0%; fluorine, 67.0%; $M.W.$ 165 ± 10
    (d) C, H, and Cl present: carbon, 49.0%; chlorine, 48.3%; $M.W.$ 145 ± 5
    (e) H, O, and P present: hydrogen, 2.5%; phosphorus, 38.3%; $M.W.$ 160 ± 10

11. Calculate the formula of each of the following:
    (a) a chloride of tin containing 62.6% Sn
    (b) an oxide of vanadium containing 68.0% V
    (c) a bromide of tungsten containing 31.5% W
    (d) a fluoride of manganese containing 50.9% F
    (e) a sulfide of nickel containing 73.3% Ni

12. A nitrate of lead contains 8.5% nitrogen. What is its formula?

13. An acetate of chromium contains 28.2% carbon. What is its formula? (The acetate ion is CH_3COO^-.)

14. Copper oxalate crystallizes as the hydrate, $CuC_2O_4 \cdot xH_2O$. If the crystals contain 39.6% copper, what is the value of x?

15. Sodium chromate forms the crystals $Na_2CrO_4 \cdot xH_2O$. If the chromium content is 15.2%, what is the complete formula?

16. Express the following concentrations in g liter^{-1}.
    (a) 0.348 g dissolved in 75.0 ml
    (b) 10.23 g dissolved in 0.250 liter
    (c) 511 g dissolved in 24.0 liter

17. What mass of solute would be present in the following?
    (a) 4.50 liter, concentration 6.63 g liter^{-1}
    (b) 12.5 ml, concentration 1.18 g liter^{-1}
    (c) 0.150 liter, concentration 16.9 g liter^{-1}

18. What volume of solution would contain 1.50 g of solute if the concentration were
    (a) 21.3 g liter^{-1} (b) 0.688 g liter^{-1}

19. Express the following in terms of molarity.
    (a) 0.226 mole in 0.875 liter
    (b) 1.44×10^{-3} mole in 16.8 ml
    (c) 8.51 mole in 10.5 liter

20. How many moles of solute in the following?
    (a) 2.66 liter of $0.0200M$ solution
    (b) 25.0 ml of $0.0987M$ solution
    (c) 0.125 liter of $0.260M$ solution

21. What volume of a solution which is $0.475M$ would contain
    (a) 1.00 mole   (b) 0.250 mole

22. Calculate the molarity of the following:
    (a) KNO_3, 100 g liter^{-1}
    (b) KCl, 15.0 g liter^{-1}
    (c) $SnSO_4$, 5.11 g in 150 ml solution

23. Express the following concentrations in g liter^{-1}.
    (a) HBr, $1.50M$
    (b) $CaSO_4$, $(1.4 \times 10^{-3})M$
    (c) C_2H_5OH, 3.6×10^{-4} mole in 1.00 ml solution

24. Calculate molecular weights for the solutes in the following:
    (a) a $0.500M$ solution contains 85.6 g liter^{-1}
    (b) a $0.012M$ solution contains 1.5 g in 125 ml
    (c) a $0.481M$ solution contains 2.81 g in 25.0 ml

SOLUTIONS TO PROBLEMS

1. Taking Avogadro's number as 6.02×10^{23} mole^{-1},
   (a) 1 mole of fluorine (F) is 19.0 g; therefore 1 atom mass of fluorine is

$$\frac{19.0 \text{ g mole}^{-1}}{6.02 \times 10^{23} \text{ mole}^{-1}} = 3.16 \times 10^{-23} \text{ g}$$

   (b) 1 mole of sodium (Na) is 23.0 g; therefore 5 atoms mass of sodium is

$$\frac{5 \times 23.0}{6.02 \times 10^{23}} = 1.91 \times 10^{-22} \text{ g}$$

   (c) 1 mole of ethanol (C_2H_5OH) is 46.0 g; therefore 1 molecule mass of ethanol is

$$\frac{46.0}{6.02 \times 10^{23}} = 7.64 \times 10^{-23} \text{ g}$$

2. Take a weighted average:

   (a) $0.604 \times 68.9257 = 41.63$
   $0.396 \times 70.9249 = 28.09$
   average at.wt $= 69.72$

   (b) $0.5182 \times 106.9041 = 55.40$
   $0.4818 \times 108.9047 = 52.47$
   average at.wt $= 107.87$

3. Suppose the mixture of isotopes has a fraction x of ^{6}Li and $(1 - x)$ of ^{7}Li. Then we compute the average atomic weight, knowing it is equal to 6.939.

$$6.0151x + 7.0160(1 - x) = 6.939$$

$$6.0151x - 7.0160x = 6.939 - 7.0160$$

$$-1.001x = -0.0770$$

$$x = \frac{0.0770}{1.001} = 0.077$$

   The element contains 7.7% ^{6}Li and (by difference) 92.3% ^{7}Li.

4. (a) $BaCO_3 = 137.3 + 12.0 + 48.0 = 197.3$
   (b) $CdBr_2 = 112.4 + 159.8 = 272.2$
   (c) $CuCl_2 \cdot 2H_2O = 63.5 + 70.9 + (2 \times 18.0) = 170.4$
   (d) $H_4P_2O_6 = 4.0 + 62.0 + 96.0 = 162.0$

5. In each case, we use

$$\text{number of moles} = \frac{\text{mass in g}}{\text{molar mass}}$$

   (a) $\dfrac{5.00 \times 10^2}{197.3} = 2.53$ mole

   (b) $\dfrac{5.00 \times 10^2}{272.2} = 1.84$ mole

   (c) $\dfrac{5.00 \times 10^2}{170.4} = 2.93$ mole

   (d) $\dfrac{5.00 \times 10^2}{162} = 3.09$ mole

6. (a) HCl has a molar mass of 36.5 g; therefore 5.00 mole HCl is

$$(5.00 \text{ mole}) \times (36.5 \text{ g mole}^{-1}) = 183 \text{ g}$$

(b) $NaClO_4$ has a molar mass of 122.5 g; therefore 0.288 mole is

$$0.288 \times 122.5 = 35.3 \text{ g}$$

(c) CH_4 has a molar mass of 16.0 g;

$$1.20 \times 10^{-3} \times 16.0 = 0.0192 \text{ g}$$

7.   Here we use

$$\text{molar mass} = \frac{\text{mass in g}}{\text{number of moles}}$$

(a) $\text{molar mass} = \dfrac{598 \text{ g}}{11.00 \text{ mole}} = 54.4 \text{ g mole}^{-1}$; $M.W. = 54.4$

(b) $\text{molar mass} = \dfrac{1.98 \text{ g}}{2.30 \times 10^{-2} \text{ mole}} = 86.1 \text{ g mole}^{-1}$; $M.W. = 86.1$

8.   (a)

K	39.1	$(39.1/166.0) \times 100 = 23.6\%$
I	126.9	$(126.9/166.0) \times 100 = 76.4\%$
M.W.	166.0	

(b)

2Na	46.0	$(46.0/62.0) \times 100 = 74.2\%$
O	16.0	$(16.0/62.0) \times 100 = 25.8\%$
M.W.	62.0	

(c)

Ba	137.3	$(137.3/233.4) \times 100 = 58.8\%$
S	32.1	$(32.1/233.4) \times 100 = 13.8\%$
4O	64.0	$(64.0/233.4) \times 100 = 27.4\%$
M.W.	233.4	

(d)

Ca	40.1	$(40.1/236.2) \times 100 = 17.0\%$
2N	28.0	$(28.0/236.2) \times 100 = 11.9\%$
10O	160.0	$(160.0/236.2) \times 100 = 67.7\%$
8H	8.1	$(8.1/236.2) \times 100 = 3.4\%$
M.W.	236.2	

9.   In all cases, we divide by atomic weights to obtain the combining mole ratios of the elements.

(a)

$$\begin{array}{cc} C & H \\ \dfrac{92.3}{12.0} = 7.69 & \dfrac{7.7}{1.01} = 7.62 \end{array}$$

Ratio is 1:1; simplest formula is CH.

(b)   Oxygen (by difference) = $100 - (40.0 + 6.7) = 53.3\%$.

$$\begin{array}{ccc} C & H & O \\ \dfrac{40.0}{12.0} = 3.33 & \dfrac{6.7}{1.01} = 6.63 & \dfrac{53.3}{16.0} = 3.33 \end{array}$$

Ratio is 1:2:1; simplest formula is CH_2O.

(c)

$$\qquad\qquad\text{C}\qquad\qquad\qquad\text{F}$$

$$\frac{13.6}{12.0} = 1.13 \qquad\qquad \frac{86.4}{19.0} = 4.55$$

$$\frac{1.13}{1.13} = 1.00 \qquad\qquad \frac{4.55}{1.13} = 4.03$$

Ratio is 1:4; simplest formula is CF_4.

(d)

$$\qquad\qquad\text{C}\qquad\qquad\qquad\text{H}$$

$$\frac{88.9}{12.0} = 7.41 \qquad\qquad \frac{11.1}{1.01} = 11.0$$

$$\frac{7.41}{7.41} = 1.00 \qquad\qquad \frac{11.0}{7.41} = 1.48$$

Double both to obtain the simplest integral ratio, 2:3; the simplest formula is C_2H_3.

(e) Sulfur (by difference) $= 100 - (33.3 + 7.4) = 59.3\%$.

$$\qquad\text{C}\qquad\qquad\qquad\text{H}\qquad\qquad\qquad\text{S}$$

$$\frac{33.3}{12.0} = 2.78 \qquad \frac{7.4}{1.01} = 7.3 \qquad \frac{59.3}{32.1} = 1.85$$

$$\frac{2.78}{1.85} = 1.50 \qquad \frac{7.3}{1.85} = 4.0 \qquad \frac{1.85}{1.85} = 1.00$$

Double, to obtain integral ratio 3:8:2; simplest formula is $C_3H_8S_2$.

10. (a) Simplest formula (see Solution 9(a)) is CH; molecular formula is $(CH)_n$, *M.W.* $13.0n$:

$$13.0n = 80 \pm 3 \qquad n = \frac{80 \pm 3}{13.0} \approx 6$$

Formula is C_6H_6; *M.W.* 78.1.

(b) Oxygen (by difference) $= 100 - (53.3 + 11.1) = 35.6\%$.

$$\qquad\text{C}\qquad\qquad\qquad\text{H}\qquad\qquad\qquad\text{O}$$

$$\frac{53.3}{12.0} = 4.44 \qquad \frac{11.1}{1.01} = 11.0 \qquad \frac{35.6}{16.0} = 2.23$$

$$\frac{4.44}{2.23} = 2.0 \qquad \frac{11.0}{2.23} = 5.0 \qquad \frac{2.23}{2.23} = 1.0$$

Simplest formula is C_2H_5O; *M.W.* $45.1n$:

$$45.1n = 95 \pm 5 \qquad n = 2$$

Molecular formula is $C_4H_{10}O_2$; *M.W.* 90.1.

(c)

$$\qquad\qquad\text{Si}\qquad\qquad\qquad\text{F}$$

$$\frac{33.0}{28.1} = 1.17 \qquad\qquad \frac{67.0}{19.0} = 3.53$$

Simplest formula is SiF_3; *M.W.* $85.1n$.

$$85.1n = 165 \pm 10 \qquad n = 2$$

Molecular formula is Si_2F_6; $M.W.$ 170.2.

(d)   Hydrogen (by difference) $= 100 - (49.0 + 48.3) = 2.7\%$.

C	H	Cl
$\dfrac{49.0}{12.0} = 4.08$	$\dfrac{2.7}{1.01} = 2.7$	$\dfrac{48.3}{35.5} = 1.36$
$\dfrac{4.08}{1.36} = 3.0$	$\dfrac{2.7}{1.36} = 2.0$	$\dfrac{1.36}{1.36} = 1.0$

Simplest formula is C_3H_2Cl; $M.W.$ 73.5n.

$$73.5n = 145 + 5 \qquad n = 2$$

Molecular formula is $C_6H_4Cl_2$; $M.W.$ 147.0.

(e)   Oxygen (by difference) $= 100 - (2.5 + 38.3) = 59.2\%$.

H	P	O
$\dfrac{2.5}{1.01} = 2.5$	$\dfrac{38.3}{31.0} = 1.24$	$\dfrac{59.2}{16.0} = 3.70$
$\dfrac{2.5}{1.24} = 2.0$	$\dfrac{1.24}{1.24} = 1.0$	$\dfrac{3.7}{1.24} = 3.0$

Simplest formula is H_2PO_3; $M.W.$ 81.0n.

$$81.0n = 160 \pm 10 \qquad n = 2$$

Molecular formula is $H_4P_2O_6$; $M.W.$ 162.0.

11.   In each case, we find the percentage of the second element by difference.

(a)

Sn	Cl
$\dfrac{62.6}{118.7} = 0.527$	$\dfrac{100 - 62.6}{35.5} = 1.05$

Ratio is 1:2; formula is $SnCl_2$.

(b)

V	O
$\dfrac{68.0}{50.9} = 1.34$	$\dfrac{100 - 68.0}{16.0} = 2.00$
$\dfrac{1.34}{1.34} = 1.00$	$\dfrac{2.00}{1.34} = 1.50$

Ratio is 1:1.5 or 2:3; formula is V_2O_3.

(c)

W	Br
$\dfrac{31.5}{183.9} = 0.171$	$\dfrac{100 - 31.5}{79.9} = 0.857$
$\dfrac{0.171}{0.171} = 1.00$	$\dfrac{0.857}{0.171} = 5.0$

Ratio is 1:5, formula is WBr_5.

(d)

	Mn	F

$$\frac{100 - 50.9}{54.9} = 0.894 \qquad \frac{50.9}{19.0} = 2.68$$

$$\frac{0.894}{0.894} = 1.00 \qquad \frac{2.68}{0.894} = 3.0$$

Ratio is 1:3; formula is MnF_3.

(e)

	Ni	S

$$\frac{73.3}{58.7} = 1.25 \qquad \frac{100 - 73.3}{32.1} = 0.831$$

$$\frac{1.25}{0.831} = 1.50 \qquad \frac{0.831}{0.831} = 1.00$$

Ratio is 1.5:1 or 3:2; formula is Ni_3S_2.

12. Let the formula be $Pb(NO_3)_x$:

$$M.W. = 207 + x(14 + 48) = 207 + 62x$$

$$\text{percent nitrogen} = \frac{14x}{207 + 62x} \times 100 = 8.5 \qquad \text{(given)}$$

whence $x = 2$; formula is $Pb(NO_3)_2$.

13. Let the formula be $Cr(CH_3COO)_x$:

$$M.W. = 52 + x(24 + 3 + 32) = 52 + 59x$$

$$\text{percent carbon} = \frac{24x}{52 + 59x} \times 100 = 28.2 \qquad \text{(given)}$$

whence $x = 2$; formula is $Cr(CH_3COO)_2$.

14. Let the formula be $CuC_2O_4 \cdot xH_2O$:

$$M.W. = 63.5 + 24 + 64 + 18x = 152 + 18x$$

$$\text{percent copper} = \frac{63.5}{152 + 18x} \times 100 = 39.6 \qquad \text{(given)}$$

$$6.35 \times 10^3 = (6.02 \times 10^3) + 713x$$

whence $x = 0.5$; formula is $CuC_2O_4 \cdot 1/2H_2O$.

15. Let the formula be $Na_2CrO_4 \cdot xH_2O$:

$$M.W. = 46 + 52 + 64 + 18x = 162 + 18x$$

$$\text{percent chromium} = \frac{52}{162 + 18x} \times 100 = 15.2 \qquad \text{(given)}$$

$$5.2 \times 10^3 = (2.46 \times 10^3) + 274x$$

whence $x = 10$; formula is $Na_2CrO_4 \cdot 10H_2O$.

16. (a) $\dfrac{0.348 \text{ g}}{0.0750 \text{ liter}} = 4.64 \text{ g liter}^{-1}$

    (b) $\dfrac{10.23 \text{ g}}{0.250 \text{ liter}} = 40.9 \text{ g liter}^{-1}$

    (c) $\dfrac{511 \text{ g}}{24.0 \text{ liter}} = 21.3 \text{ g liter}^{-1}$

17. (a) $(4.50 \text{ liter}) \times (6.63 \text{ g liter}^{-1}) = 29.8 \text{ g}$
    (b) $(0.0125 \text{ liter}) \times (1.18 \text{ g liter}^{-1}) = 1.48 \times 10^{-2} \text{ g}$
    (c) $(0.150 \text{ liter}) \times (16.9 \text{ g liter}^{-1}) = 2.54 \text{ g}$

18. (a) $\dfrac{1.50 \text{ g}}{21.3 \text{ g liter}^{-1}} = 7.04 \times 10^{-2} \text{ liter} = 70.4 \text{ ml}$

    (b) $\dfrac{1.50 \text{ g}}{0.688 \text{ g liter}^{-1}} = 2.18 \text{ liter}$

19. (a) $\dfrac{0.226 \text{ mole}}{0.875 \text{ liter}} = 0.258 M$

    (b) $\dfrac{1.44 \times 10^{-3} \text{ mole}}{0.0168 \text{ liter}} = (8.57 \times 10^{-2}) M$

    (c) $\dfrac{8.51 \text{ mole}}{10.5 \text{ liter}} = 0.810 M$

20. (a) $(2.66 \text{ liter}) \times (0.0200 M) = 5.32 \times 10^{-2} \text{ mole}$
    (b) $(0.0250 \text{ liter}) \times (0.0987 M) = 2.47 \times 10^{-3} \text{ mole}$
    (c) $(0.125 \text{ liter}) \times (0.260 M) = 3.25 \times 10^{-2} \text{ mole}$

21. (a) $\dfrac{1.00 \text{ mole}}{0.475 \text{ mole liter}^{-1}} = 2.11 \text{ liter}$

    (b) $\dfrac{0.250 \text{ mole}}{0.475 \text{ mole liter}^{-1}} = 0.526 \text{ liter} = 526 \text{ ml}$

22. (a) KNO_3 has a molar mass of 101 g:

$$\dfrac{100 \text{ g liter}^{-1}}{101 \text{ g mole}^{-1}} = 0.99 M$$

    (b) KCl has a molar mass of 74.6 g:

$$\dfrac{15.0 \text{ g liter}^{-1}}{74.6 \text{ g mole}^{-1}} = 0.201 M$$

    (c) $SnSO_4$ has a molar mass of 215 g:

$$\dfrac{5.11 \text{ g}}{(0.150 \text{ liter}) \times (215 \text{ g mole}^{-1})} = 0.158 M$$

23. (a) HBr has a molar mass of 80.9 g:

$$(1.50 \text{ mole liter}^{-1}) \times (80.9 \text{ g mole}^{-1}) = 121 \text{ g liter}^{-1}$$

(b) $CaSO_4$ has a molar mass of 136 g:

$$(1.4 \times 10^{-3} \text{ mole liter}^{-1}) \times (136 \text{ g mole}^{-1}) = 0.19 \text{ g liter}^{-1}$$

(c) C_2H_5OH has a molar mass of 46.0 g:

$$\frac{(3.6 \times 10^{-4} \text{ mole}) \times (46.0 \text{ g mole}^{-1})}{0.00100 \text{ liter}} = 16.6 \text{ g liter}^{-1}$$

24. (a) $\dfrac{85.6 \text{ g liter}^{-1}}{0.500 \text{ mole liter}^{-1}} = 171 \text{ g mole}^{-1}; \quad M.W. = 171$

(b) $\dfrac{1.5 \text{ g}}{(0.125 \text{ liter}) \times (0.012 \text{ mole liter}^{-1})} = 1.0 \times 10^3 \text{ g}; \quad M.W. = 1.0 \times 10^3$

(c) $\dfrac{2.81}{(0.0250 \text{ liter}) \times (0.481 \text{ mole liter}^{-1})} = 234 \text{ g}; \quad M.W. = 234$

Chapter 3
The Chemical Equation and the Stoichiometry of Reaction

I gave my mind a thorough rest by plunging into a chemical analysis.   THE SIGN OF FOUR

3.1   The Equation

The chemical equation is the basic language of chemistry. It tells us both *qualitatively* what happens when two or more chemical reagents come together, and *quantitatively* how much of one reacts with the other and what amounts of products are formed.

To take a simple example, consider the reaction

$$C + O_2 \longrightarrow CO_2$$

Qualitatively, this tells us that carbon reacts with oxygen to give carbon dioxide. Quantitatively, the equation says that one mole of C reacts with one mole of O_2 to give one mole of CO_2. Knowing the definition of the mole, we see that this relates definite masses of different compounds. However, we may go a little further where gases are involved and relate volumes, using the fact that one mole of any gas will occupy the same volume—called the molar volume—with the value 22.4 liters under conditions of standard temperature and pressure (STP), i.e., 273 K and 1 atmosphere pressure. (The effects on gaseous volumes of changing conditions are discussed fully in Chapter 8.)

Using the equation in conjunction with the mole concept, therefore, we may relate different quantities of reagents and products.

Example 3.1   100 g of carbon is burned to give CO_2. What mass of oxygen is needed? Calculate the mass and volume (at STP) of CO_2 produced.

SOLUTION   Knowing that the molar mass of C is 12.0 g, we convert to molar quantities:

$$100 \text{ g} = \frac{100 \text{ g}}{12.0 \text{ g mole}^{-1}} = 8.33 \text{ mole C}$$

From the equation, $C + O_2 \rightarrow CO_2$, we know that

   1 mole C reacts with 1 mole O_2 so   8.33 mole C reacts with 8.33 mole O_2

Since O_2 has a molar mass of 32.0 g,

$$8.33 \text{ mole } O_2 = 8.33 \times 32.0 = 267 \text{ g } O_2$$

Each mole of C burned produces 1 mole CO_2, therefore,

$$8.33 \text{ mole C give } 8.33 \text{ mole } CO_2$$

We know CO_2 has a molar mass of 44.0 g, so

$$8.33 \text{ mole } CO_2 = 8.33 \times 44.0 = 367 \text{ g } CO_2$$

And, since any gas has a molar volume of 22.4 liter, the volume of 8.33 mole CO_2 is

$$8.33 \times 22.4 = 187 \text{ liter}$$

(*Note*: Obviously, we could also have found the mass of CO_2 by adding together the masses of C and O_2 used.)

Example 3.2 SO_2 is converted to SO_3 by the reaction

$$SO_2 + \frac{1}{2} O_2 \longrightarrow SO_3$$

How much SO_2 and O_2 are needed to make 250 g of SO_3?

   SOLUTION SO_3 has a molar mass of 80.1 g,

$$250 \text{ g } SO_3 = \frac{250}{80.1} = 3.13 \text{ mole}$$

From the equation, we need 1 mole SO_2 to make 1 mole SO_3, so we need 3.13 mole SO_2 (*M.W.* 64.1):

$$3.13 \text{ mole } SO_2 = 3.13 \times 64.1 = 200 \text{ g } SO_2$$

For each mole SO_3, we only need 1/2 mole O_2, so for 3.13 mole we need 3.13/2 mole O_2 (*M.W.* 32.0), which is

$$\frac{3.13}{2} \times 32.0 = 50.0 \text{ g } O_2$$

(*Note*: Here again, the total mass balances up.)

   In this equation, note the use of the coefficient 1/2 for oxygen. It is frequently convenient to use the concept of "half a mole" in an equation, to avoid doubling all the other coefficients. Of course, the doubled equation

$$2SO_2 + O_2 \longrightarrow 2SO_3$$

is perfectly correct and conveys the same information. However, one should avoid excessive use of fractional coefficients, and it would be clumsy to use awkward fractions like 2/3 or 3/4.

Example 3.3 Zinc was dissolved in dilute acid and 530 ml H_2 (STP) was evolved. What was the weight of zinc taken?

SOLUTION The equation is, essentially,

$$Zn + 2H^+ \longrightarrow Zn^{2+} + H_2$$

(*Note*: The identity of the acid does not matter, the reaction concerns only the hydrogen ion.) From the equation, 1 mole H_2 comes from 1 mole zinc. Therefore,

$$\frac{0.530 \text{ liter}}{22.4 \text{ liter mole}^{-1}} = 2.37 \times 10^{-2} \text{ mole } H_2$$

comes from 2.37×10^{-2} mole of Zn (molar mass 65.4 g), which is

$$(2.37 \times 10^{-2} \text{ mole}) \times (65.4 \text{ g mole}^{-1}) = 1.55 \text{ g Zn}$$

3.2 Molecular and Ionic Equations

Having seen what information we can obtain from an equation, let's take a step backward and think how we obtained a balanced equation in the first place. We can roughly divide chemical equations into four groups, according to the type of reaction they represent:

1. *Synthesis*, or formation of compounds from elements:

$$Fe + Cl_2 \longrightarrow FeCl_2$$

2. *Metathesis*, or exchange between compounds:

$$Na_2CO_3 + Ca(NO_3)_2 \longrightarrow CaCO_3(s) + 2NaNO_3$$
$$(CaCO_3 \text{ precipitates from solution})$$

3. *Neutralization*, or acid-base reaction:

$$H_2SO_4 + 2KOH \longrightarrow K_2SO_4 + 2H_2O$$

4. *Redox*, or transfer of electrons:

$$K_2SO_3 + \frac{1}{2}O_2 \longrightarrow K_2SO_4$$

(*Note*: There are many reactions which fall into more than one of the categories, or are difficult to classify.)

Before going any further, we should differentiate between two methods of writing equations: *molecular* and *ionic*. In the former, all species are represented in the equations as neutral molecules, regardless of their actual structures. However, many reactions (including 2, 3, and 4 above) actually involve ionic compounds reacting in solution, so it is simpler (and more correct) to include in the equation only the actual ions taking part in the reaction. Remember that *every ionic salt is regarded as completely ionized in solution*.

Students often have trouble in recognizing ionic compounds and in deciding the manner in which they dissociate into ions. Most ionic compounds contain an electropositive metal (Na^+, K^+, Ca^{2+}, etc.) and a negatively charged anion derived from an acid (e.g., Cl^- from HCl). At the end of this chapter, lists of common positive and negative ions are given as Tables 3.1–3.3. It is recommended you study these.

A glance around any chemical stockroom will show that sodium and potassium salts greatly outnumber the compounds of other elements. This is not because these two elements have an extensive chemistry; quite the contrary, their ions undergo no reactions at all in aqueous solution. Na^+ and K^+ are simply used to provide a positive counterion to a variety of negative ions whose chemistry we wish to study. Once we dissolve the salt in water and free the ions, we can forget about Na^+ or K^+.

Let's rewrite the above equations in ionic form

2. $CO_3^{2-} + Ca^{2+} \longrightarrow CaCO_3(s)$ (Neither Na^+ nor NO_3^- ions are involved.)

3. $2H^+ + 2OH^- \longrightarrow 2H_2O$ (SO_4^{2-} and K^+ remain unaffected throughout.)

4. $SO_3^{2-} + \frac{1}{2}O_2 \longrightarrow SO_4^{2-}$

The ionic equations are simpler and more directly informative about the reaction occurring, so we will use them extensively in this book.

Example 3.4   Rewrite the following equations in ionic form. All reactants and products are in solution, unless otherwise indicated.

(a) $AgNO_3 + KCl \longrightarrow AgCl(s) + KNO_3$
(b) $Ca(OH)_2(s) + 2HNO_3 \longrightarrow Ca(NO_3)_2 + 2H_2O$
(c) $2KMnO_4 + 16HCl \longrightarrow 2KCl + 2MnCl_2 + 5Cl_2(g) + 8H_2O$

SOLUTION   All we have to do is to write out the formulas of all ionic substances as ions, then remove all ions appearing *unchanged* on both sides of the equation.

(a) $Ag^+ + NO_3^- + K^+ + Cl^- \longrightarrow AgCl(s) + K^+ + NO_3^-$
Removing NO_3^- and K^+ gives

$$Ag^+ + Cl^- \longrightarrow AgCl(s)$$

(b) $Ca(OH)_2(s) + 2H^+ + 2NO_3^- \longrightarrow Ca^{2+} + 2NO_3^- + 2H_2O$
Removing $2NO_3^-$ gives

$$Ca(OH)_2(s) + 2H^+ \longrightarrow Ca^{2+} + 2H_2O$$

(c) $2K^+ + 2MnO_4^- + 16H^+ + 16Cl^- \longrightarrow$
$$2K^+ + 2Cl^- + 2Mn^{2+} + 4Cl^- + 5Cl_2(g) + 8H_2O$$

As usual, the K^+ ions can be removed, together with $6Cl^-$ appearing on each side (the remaining $10Cl^-$ at the left are oxidized to Cl_2) to give

$$2MnO_4^- + 16H^+ + 10Cl^- \longrightarrow 2Mn^{2+} + 5Cl_2(g) + 8H_2O$$

A few points should be noted in connection with these problems:

1. Although solids such as $AgCl$, $Ca(OH)_2$, etc. contain ions, we do *not* separate them into ions in our equations because they are held firmly together in the solid. We only include as ions those species which are ionized *in solution*.

2. The strong mineral acids HCl, HNO_3, and H_2SO_4, although they can exist as covalent molecules, are usually regarded as completely ionized in solution (this is a slight oversimplification for H_2SO_4, and we will consider this in more detail later in Chapters 13–14).

3. A common student error is to split a compound into two or more parts which do not balance up to the original. Thus, somebody confused by the presence of Cl_2 in the above

equation might "ionize" $MnCl_2$ into $Mn^{2+} + Cl_2$, not noticing that the charges did not balance. Try to guard against this by familiarizing yourself with the formulas of the common ions, and remember that splitting up a salt must produce positive and negative charges in equal amounts.

3.3 Balancing Equations

All the above equations were presented in a balanced form. Of the four classes of reaction given, equations for the first three may be balanced by common sense and simple inspection. Thus, for a synthesis process, we take the appropriate elements in the required mole ratios according to the compound's formula. For a metathesis (exchange) reaction, we take mole ratios giving the required proportions. For a neutralization reaction, we arrange for the acid to supply just the right number of moles of H^+ that will neutralize the moles of base. These three types are illustrated below:

Example 3.5   Balance the following equations:

(a) $Si + Cl_2 \longrightarrow Si_2Cl_6$
(b) $As + S \longrightarrow As_2S_3$
(c) $PCl_5 + 4H_2O \longrightarrow H_3PO_4 + HCl$
(d) $Ag_2SO_4 + BaCl_2 \longrightarrow AgCl(s) + BaSO_4(s)$
(e) $Ca(OH)_2 + H_3PO_4 \longrightarrow Ca_3(PO_4)_2 + H_2O$
(f) $NH_3 + H_2SO_4 \longrightarrow (NH_4)_2SO_4$

SOLUTION
(a) $2Si + 3Cl_2 \longrightarrow Si_2Cl_6$ (synthesis)
Obviously we need an *atom* ratio of 1:3 (or 2:6) for Si:Cl, so we take 2Si and $3Cl_2$. Both Cl_2 and Si_2Cl_6 are molecular formulas.
(b) $2As + 3S \longrightarrow As_2S_3$ (synthesis)
There is no problem here; we must have a 2:3 ratio of As to S.
(c) This is balanced in phosphorus as it stands. To balance chlorine, we need 5HCl on the right, giving a total of 8 hydrogen on the right. To balance hydrogen, we need $4H_2O$ on the left. Fortunately, this balances the oxygen (4 each side), giving

$$PCl_5 + 4H_2O \longrightarrow H_3PO_4 + 5HCl \quad \text{(exchange)}$$

In an exchange reaction, if we juggle quantities to get most of the elements balanced, the last element (oxygen here) should balance automatically.
(d) To balance chlorine, we need 2AgCl on the right, and this automatically balances silver:

$$Ag_2SO_4 + BaCl_2 \longrightarrow 2AgCl(s) + BaSO_4(s) \quad \text{(exchange)}$$

(e) Each $Ca(OH)_2$ gives $2OH^-$ on ionization, whereas the H_3PO_4 gives $3H^+$. To obtain equivalent quantities for the neutralization, we need to take amounts that will give *six* mole of both OH^- and H^+, that means we need *three* mole of $Ca(OH)_2$ and *two* mole of H_3PO_4 (six is the lowest common multiple of three and two). Result:

$$3Ca(OH)_2 + 2H_3PO_4 \longrightarrow Ca_3(PO_4)_2 + 6H_2O$$

(f) In this reaction, the base is NH_3, which neutralizes an acid by reacting with one H^+ to give NH_4^+. So we need $2NH_3$ for a mole of H_2SO_4, giving

$$2NH_3 + H_2SO_4 \longrightarrow (NH_4)_2SO_4$$

The fourth class of reaction is the electron transfer process, or redox ($=$ REDuction $+$ OXidation) reaction. The detailed description of the processes of oxidation and reduction is given in Chapter 15; for the moment we are only concerned with the method of obtaining a balanced equation to describe the overall process.

In a redox reaction, one of the reactants *loses* electrons while another reactant *gains* electrons. Obviously the number of electrons lost by the first must equal the number gained by the second, so we must arrange the molar ratio of the two in our equation in such a way as to ensure this. To do this, we first divide the redox reaction into two half-reactions. In one of these, electrons will be *lost*, and we call this process *oxidation*. The other half-reaction will involve a reactant undergoing a *gain* of electrons, and we call this process *reduction*. We will therefore refer to the two half-reactions as the "reduction half-reaction" and the "oxidation half-reaction."

The balancing of any redox reaction follows the same basic steps: the separation into two half-reactions (it is not necessary to recognize initially which is the oxidation and which the reduction); the balancing of the two half-reactions separately (using electrons to balance charge); and the combining of the two half-reactions in such a way that the electrons cancel out to give a balanced overall reaction. With reactions involving simple metal ions, this can be done very quickly.

Example 3.6   Balance the equation $Fe^{3+} + Zn \rightarrow Fe^{2+} + Zn^{2+}$ using the half-reaction method.

SOLUTION   The half-reactions are:

(i) $$Fe^{3+} \longrightarrow Fe^{2+}$$

This reaction is balanced in mass, but not in charge. There is a total charge of $3+$ on the left and only $2+$ on the right, so we add one electron to the *left* to make a net total of $2+$ on each side:

$$Fe^{3+} + e^- \longrightarrow Fe^{2+} \qquad \text{(reduction reaction)}$$

(ii) $$Zn \longrightarrow Zn^{2+}$$

This reaction is balanced in mass, but has a charge of $2+$ on the right and no charge on the left, so we add two electrons to the *right* to make a net total charge of zero on each side:

$$Zn \longrightarrow Zn^{2+} + 2e^- \qquad \text{(oxidation reaction)}$$

We must have the electrons canceling out when we add the equations together, so we take twice the reduction reaction and add to the oxidation reaction, so that the $2e^-$ on each side will cancel:

$$2Fe^{3+} + 2e^- \longrightarrow 2Fe^{2+} \qquad \text{(reduction)}$$

$$Zn \longrightarrow Zn^{2+} + 2e^- \qquad \text{(oxidation)}$$

Overall: $$2Fe^{3+} + Zn \longrightarrow 2Fe^{2+} + Zn^{2+}$$

(*Check*: This balances in mass *and* in charge, with a total charge of $6+$ on each side.)

Although this is a very simple example, several generally applicable points should be noted.

(i) It is not necessary to decide which is the oxidation reaction and which the reduction reaction before dividing up the overall process into two. In the above example, it was obvious that Fe^{3+} was gaining electrons and Zn was losing electrons, but in more complex systems, it may not be immediately apparent. In such cases, it is only necessary to divide the process into two half-reactions involving the different reactants. When electrons are added to balance the charge, we can tell at once which half-reaction is which, because in the reduction process the electrons will appear on the *left* of the equation, whereas in the oxidation process they will be on the *right*.

(ii) A half-reaction equation must be balanced in mass first, then balanced in charge by adding electrons (*negative* charge) to the side that has an excess of *positive* charge. It is not necessary for the total charge to add up to zero on both sides, it is only necessary for the charge to be the same on both sides.

(iii) The final equation for the overall reaction should always be checked to see that it balances in mass *and in charge*. If it does balance, you can be sure that it is correct.

Many of the redox reactions we carry out in the laboratory involve more complicated species than simple metal ions. In particular, we use complex ions containing oxygen attached to a metal ion (e.g., permanganate, MnO_4^-; dichromate, $Cr_2O_7^{2-}$). Balancing redox equations involving these species is somewhat more complicated than for simple metal ions, but it may be broken down into a systematic series of steps. Before starting, we must specify one important point; is the reaction occurring in acid solution or in basic solution? The overall equations will be different in the two cases, since acid reactions incorporate H^+ into the equation, while alkaline reactions use OH^-. In general practice, acid reactions are much more common, so we will first outline the steps for acid solution, using as an example the half-reaction in which permanganate goes to Mn^{2+}:

$$MnO_4^- \longrightarrow Mn^{2+}$$

1. We first balance the equation in the element *other* than oxygen. In this case, it is already balanced in Mn.
2. Add sufficient water molecules to balance the oxygen. With four atoms of oxygen on the left, we need $4H_2O$ on the right, giving

$$MnO_4^- \longrightarrow Mn^{2+} + 4H_2O$$

3. Add sufficient H^+ to balance the hydrogen. Clearly we need $8H^+$ on the left to balance the hydrogen in $4H_2O$:

$$MnO_4^- + 8H^+ \longrightarrow Mn^{2+} + 4H_2O$$

4. The equation is now balanced in mass, with one manganese, four oxygen, and eight hydrogen on each side; but it is not balanced in charge. There is a total of $+7$ on the left and $+2$ on the right, so we finally add $5e^-$ on the *left* to produce a total charge of $+2$ on each side.

$$MnO_4^- + 8H^+ + 5e^- \longrightarrow Mn^{2+} + 4H_2O$$

We see that the permanganate ion is being reduced in a five-electron reaction. Note that we made no assumption about what was happening to the ion before starting the balancing process. The addition of $5e^-$ was an automatic result of the balancing procedure, given the starting material (MnO_4^-) and the final product (Mn^{2+}). It is most important to follow the above steps in the order given. It would be wrong, for example, to try to balance the charge first by adding electrons before balancing in mass.

In practice, the above reduction of MnO_4^- must be accompanied by the oxidation of some other species, so we will complete the reaction with the addition of the half-reaction

$$H_2O_2 \longrightarrow O_2 \quad \text{(acid solution)}.$$

This is balanced in oxygen, but not in hydrogen, so we add $2H^+$ (knowing we are in acid solution):

$$H_2O_2 \longrightarrow O_2 + 2H^+$$

This now requires the addition of $2e^-$ to the right side to balance in charge

$$H_2O_2 \longrightarrow O_2 + 2H^+ + 2e^- \quad \text{(oxidation)}$$

This is a two-electron half-reaction, so to combine it with the five-electron reduction half-reaction we have to arrange for the lowest common multiple of two and five (i.e., ten) electrons in each

$$2 \times \text{reduction:} \quad 2MnO_4^- + 16H^+ + 10e^- \longrightarrow 2Mn^{2+} + 8H_2O$$
$$5 \times \text{oxidation:} \quad 5H_2O_2 \longrightarrow 5O_2 + 10H^+ + 10e^-$$

Adding the two together, the electrons cancel out. We can also remove $10H^+$ appearing on each side to give the overall reaction

$$2MnO_4^- + 5H_2O_2 + 6H^+ \longrightarrow 2Mn^{2+} + 5O_2 + 8H_2O$$

which is balanced in mass and in charge ($4+$ on each side).

In alkaline solution, the procedure is very similar, except that OH^- must be used (it would be chemically unrealistic to use H^+, since its concentration is so low in alkaline solution). Let's take as an example the reaction

$$Cr^{3+} + ClO^- \longrightarrow CrO_4^{2-} + Cl^-$$

First, we separate into two half-reactions, one involving Cr and the other Cl. We can't tell yet which is the oxidation and which the reduction.

(i) $$Cr^{3+} \longrightarrow CrO_4^{2-}$$

This is balanced in Cr, but not in oxygen. In alkaline solution, we balance oxygen and hydrogen by using H_2O and OH^-. First add an OH^- on the *opposite* side for each oxygen atom by which the equation is out of balance (4 in this case). This gives

$$Cr^{3+} + 4OH^- \longrightarrow CrO_4^{2-}$$

Now add *one* water molecule on the opposite side for *every* H atom by which the equation is out of balance and, at the same time, add the same number of OH^- to the side that has the excess H. (The difference between OH^- and H_2O is, of course, an H atom, so this balances the equation in hydrogen without disturbing other elements.)

In this reaction, the previous equation had 4H on the left, so we add $4H_2O$ on the right and $4OH^-$ on the left:

$$Cr^{3+} + 8OH^- \longrightarrow CrO_4^{2-} + 4H_2O$$

It only remains to balance this equation in charge. Since the total at the moment is -5 on the left and -2 on the right, we add $3e^-$ to the *right* to make -5 on both sides:

$$Cr^{3+} + 8OH^- \longrightarrow CrO_4^{2-} + 4H_2O + 3e^-$$

We see that this is a three-electron oxidation process.

The other half-reaction is

(ii) $$ClO^- \longrightarrow Cl^-$$

(*Note*: ClO^- represents the hypochlorite ion, an ingredient of common liquid bleach and a powerful oxidizing agent in alkaline solution.)

The equation is balanced in chlorine, but not in oxygen, so we add an hydroxide ion to the side deficient in oxygen:

$$ClO^- \longrightarrow Cl^- + OH^-$$

To balance hydrogen, we now add *one* water molecule on the left and, at the same time, *one* hydroxide on the right, giving:

$$ClO^- + H_2O \longrightarrow Cl^- + 2OH^-$$

This is balanced in mass, and needs the addition of $2e^-$ to the left to bring the total charge to -3 on both sides. It is a reduction process:

$$ClO^- + H_2O + 2e^- \longrightarrow Cl^- + 2OH^-$$

To combine this with the oxidation half-reaction, we must arrange for $6e^-$ in each reaction, 6 being the lowest common multiple of 2 and 3.

$$2 \times \text{oxidation:} \quad 2Cr^{3+} + 16OH^- \longrightarrow 2CrO_4^{2-} + 8H_2O + 6e^-$$
$$3 \times \text{reduction:} \quad 3ClO^- + 3H_2O + 6e^- \longrightarrow 3Cl^- + 6OH^-$$

Adding these together, the electrons cancel out, and we can tidy up H_2O and OH^- to give

$$2Cr^{3+} + 3ClO^- + 10OH^- \longrightarrow 2CrO_4^{2-} + 3Cl^- + 5H_2O$$

This is correctly balanced in mass and in charge (total -7 on each side).

Note: Regarding the nomenclature used in redox reactions—the *oxidizing* agent is the reagent that *carries out the oxidation*. It is itself *reduced*, that is, it takes part in the reduction half-reaction. In the above examples, MnO_4^- and ClO^- were the oxidizing agents. The *reducing* agent is the reagent that *carries out the reduction*. It is itself *oxidized* in the oxidation half-reaction. In the above examples, H_2O_2 and Cr^{3+} were reducing agents.

Let's summarize the procedure for balancing redox equations into the essential steps.

1. Separate the overall reaction into two half-reactions.
2. Balance in elements other than hydrogen and oxygen.
3. Balance in oxygen by adding
   (a) in acid solution: water to balance oxygen, then H^+ to balance hydrogen.
   (b) in basic solution: OH^- to balance oxygen, then H_2O and OH^- to balance hydrogen.
4. Balance in charge by adding the appropriate number of electrons.
5. Multiply the two half-reactions by appropriate factors such that each contains the same number of electrons.
6. Add together the two half-reactions, canceling out the electrons. Cancel out any substances appearing on both sides of the equation (commonly H_2O, H^+, or OH^-).
7. *Check the overall equation* to see that it balances in mass and charge.

Example 3.7   Balance the following equation:

$$Cr_2O_7^{2-} + Fe^{2+} \longrightarrow Cr^{3+} + Fe^{3+} \quad \text{(acid solution)}$$

SOLUTION   Taking the chromium half-reaction first:

$$Cr_2O_7^{2-} \longrightarrow Cr^{3+}$$

We need $2Cr^{3+}$ on the right to balance Cr and $7H_2O$ to balance oxygen:

$$Cr_2O_7^{2-} \longrightarrow 2Cr^{3+} + 7H_2O$$

add $14H^+$ to left to balance hydrogen:

$$Cr_2O_7^{2-} + 14H^+ \longrightarrow 2Cr^{3+} + 7H_2O$$

The mass is balanced, but the charge is $(-2 + 14) = +12$ on left and $2 \times 3 = +6$ on right. Add $6e^-$ to left to balance (making $+6$ on each side):

$$Cr_2O_7^{2-} + 14H^+ + 6e^- \longrightarrow 2Cr^{3+} + 7H_2O$$

This is a reduction.

For the iron reaction, $Fe^{2+} \rightarrow Fe^{3+}$, we need only add an electron on the right to balance in charge:

$$Fe^{2+} \longrightarrow Fe^{3+} + e^- \qquad \text{(oxidation)}$$

Since this is a one-electron process, whereas the reduction half-reaction was a six-electron process, we must add six times the oxidation half-reaction to the reduction reaction to obtain the overall equation

$$Cr_2O_7^{2-} + 6Fe^{2+} + 14H^+ \longrightarrow 2Cr^{3+} + 6Fe^{3+} + 7H_2O$$

(*Check*: This is balanced in mass and charge—a total of $+24$ on each side.)

Example 3.8 Balance the following equation:

$$MnO_2 + O_2 \longrightarrow MnO_4^{2-} \qquad \text{(alkaline solution)}$$

SOLUTION   Taking the manganese half-reaction first:

$$MnO_2 \longrightarrow MnO_4^{2-}$$

This is balanced in Mn, so we add *two* OH^- to the left for *each* of the two additional oxygens needed to compensate for the excess oxygen on the right. At the same time, we add $2H_2O$ to the right to balance hydrogen:

$$MnO_2 + 4OH^- \longrightarrow MnO_4^{2-} + 2H_2O$$

The total charge is -4 on left and -2 on right, so we add $2e^-$ to right to make -4 each side.

$$MnO_2 + 4OH^- \longrightarrow MnO_4^{2-} + 2H_2O + 2e^-$$

This is an oxidation.

The other half-reaction starts with O_2 and must obviously end with OH^- or H_2O, since we are working in alkaline solution.

$$O_2 \longrightarrow \quad ?$$

As usual, we put an OH^- on the opposite side to balance each O atom:

$$O_2 \longrightarrow 2OH^-$$

Now, to balance hydrogen, we put an H_2O on the left for *each* of the two needed H atoms, and, at the same time, two OH^- on the right to preserve the oxygen balance:

$$O_2 + 2H_2O \longrightarrow 4OH^-$$

Finally add $4e^-$ on left to balance in charge:

$$O_2 + 2H_2O + 4e^- \longrightarrow 4OH^-$$

This is the reduction half-reaction and involves $4e^-$, so we add it to twice the two-electron oxidation reaction, tidying up OH^- and H_2O.

$$2MnO_2 + 8OH^- \longrightarrow 2MnO_4^{2-} + 4H_2O + 4e^-$$

Overall: $2MnO_2 + O_2 + 4OH^- \longrightarrow 2MnO_4^{2-} + 2H_2O$

It's worth noting that, although the original equation that we started from ($MnO_2 + O_2 \rightarrow MnO_4^{2-}$) was actually balanced in mass (but not charge), the final, correctly balanced, equation involves a *different* reacting ratio of MnO_2 to O_2.

Example 3.9   The species MnO_4^{2-} referred to in Example 3.8 is the manganate ion, which is bright green and only stable in alkaline solution. On acidification, it reacts to give to MnO_4^- (permanganate) and MnO_2. Give a balanced equation for this process.

SOLUTION   A reaction of this type is called a *disproportionation*. Electron transfer is occurring between the same species, i.e., the same substance is acting as both oxidizing and reducing agent. Our two half-reactions are:

$$MnO_4^{2-} \longrightarrow MnO_4^- \quad \text{and} \quad MnO_4^{2-} \longrightarrow MnO_2$$

The first half-reaction is easily balanced, since it is already balanced in mass and only requires the addition of an electron on the right to balance the charge

$$MnO_4^{2-} \longrightarrow MnO_4^- + e^- \quad \text{(oxidation)}$$

For the second reaction, we have

$$MnO_4^{2-} \longrightarrow MnO_2$$

This is balanced in manganese. Add water to balance the oxygen:

$$MnO_4^{2-} \longrightarrow MnO_2 + 2H_2O$$

and add H^+ to balance hydrogen (we were told that this reaction occurred on acidification):

$$MnO_4^{2-} + 4H^+ \longrightarrow MnO_2 + 2H_2O$$

The total charge is $2+$ on the left and zero on the right, so we add $2e^-$ on the left to balance. This is a reduction:

$$MnO_4^{2-} + 4H^+ + 2e^- \longrightarrow MnO_2 + 2H_2O$$

Add this to twice the reduction half-reaction and the electrons cancel out, giving

$$3MnO_4^{2-} + 4H^+ \longrightarrow 2MnO_4^- + MnO_2 + 2H_2O$$

(*Check*: This is balanced in mass and in charge—a total of -2 on each side.)

Example 3.10   Balance the following equation for alkaline solution:

$$Zn + NO_3^- \longrightarrow ZnO_2^{2-} + NH_3$$

SOLUTION   Take the zinc half-reaction first:

$$Zn \longrightarrow ZnO_2^{2-}$$

Add $2OH^-$ to the left to balance oxygen:

$$Zn + 2OH^- \longrightarrow ZnO_2^{2-}$$

Add $2H_2O$ to the right and another $2OH^-$ to the left to balance hydrogen:

$$Zn + 4OH^- \longrightarrow ZnO_2^{2-} + 2H_2O$$

This is balanced in mass and needs $2e^-$ on the right to balance in charge (an oxidation):

$$Zn + 4OH^- \longrightarrow ZnO_2^{2-} + 2H_2O + 2e^-$$

The other half-reaction involves nitrogen compounds:

$$NO_3^- \longrightarrow NH_3$$

Add $3OH^-$ to the right to balance oxygen:

$$NO_3^- \longrightarrow NH_3 + 3OH^-$$

There is now an excess of 6H on the right, so we add $6H_2O$ to the left and another $6OH^-$ to the right, giving

$$NO_3^- + 6H_2O \longrightarrow NH_3 + 9OH^-$$

All we need is $8e^-$ at the left to balance in charge, giving a half-reaction involving an eight-electron change for each nitrogen atom

$$NO_3^- + 6H_2O + 8e^- \longrightarrow NH_3 + 9OH^-$$

Obviously, we must add 4 times our oxidation half-reaction to this reduction process to give an overall balanced redox process:

$$4Zn + 16OH^- \longrightarrow 4ZnO_2^{2-} + 8H_2O + 8e^-$$

Overall: $4Zn + NO_3^- + 7OH^- \longrightarrow 4ZnO_2^{2-} + NH_3 + 2H_2O$

(We remove those OH^- and H_2O appearing on both sides.)

This last example has been included to show that the balancing method may be easily applied to reactions involving hydrogen compounds, such as NH_3.

3.4 Stoichiometry of Reaction

Once we have obtained a balanced equation, we can use stoichiometry of reaction to solve problems on composition of compounds (indeed, this is how such experiments are carried out in practice.)

Example 3.11 A compound is known to contain only carbon, hydrogen, and oxygen. Complete combustion of a sample weighing 2.15×10^{-2} g gives 4.31×10^{-2} g of CO_2 and 1.75×10^{-2} g of H_2O. What is the simplest formula of the compound? Write an equation for its combustion.

SOLUTION   The first step is to convert the CO_2 and H_2O to molar quantities:

$$CO_2 = \frac{4.31 \times 10^{-2}}{44.0} = 9.78 \times 10^{-4} \text{ mole}$$

$$H_2O = \frac{1.75 \times 10^{-2}}{18.0} = 9.70 \times 10^{-4} \text{ mole}$$

Now, one mole of CO_2 comes from combustion of one mole of combined C, so our sample contained 9.78×10^{-4} mole of combined C. However, it takes *two* moles of combined H to make one mole of H_2O on combustion, so our sample contained

$$2 \times 9.70 \times 10^{-4} = 1.94 \times 10^{-3} \text{ mole of combined H}$$

(*Note*: Clearly, these are in the ratio one mole C to two moles H, so a partial formula is CH_2O_x.)

To find the amount of oxygen, we must work out the masses of the C and H present and then obtain O by difference. First, we multiply by atomic weights:

$$9.78 \times 10^{-4} \text{ mole C weighs } 9.78 \times 10^{-4} \times 12.0 = 1.17 \times 10^{-2} \text{ g}$$
$$1.94 \times 10^{-3} \text{ mole H weighs } 1.94 \times 10^{-3} \times 1.0 = \underline{1.94 \times 10^{-3} \text{ g}}$$
$$\text{total of C and H} = \overline{1.36 \times 10^{-2} \text{ g}}$$

The remaining mass of the original sample was oxygen:

$$\text{mass of O} = (2.15 \times 10^{-2}) - (1.36 \times 10^{-2}) = 0.79 \times 10^{-2} \text{ g}$$

$$\text{moles of combined O} = \frac{0.79 \times 10^{-2}}{16.0} = 4.9 \times 10^{-4} \text{ mole}$$

We now have mole ratios of all three elements:

Carbon	Hydrogen	Oxygen
9.78×10^{-4}	1.94×10^{-3}	4.9×10^{-4} mole

Divide through by the smallest (oxygen):

2.0	4.0	1.0

The simplest formula is therefore C_2H_4O, and the balanced equation for combustion is

$$2C_2H_4O + 5O_2 \longrightarrow 4CO_2 + 4H_2O$$

In this example, we could not write a *complete* balanced equation until we had finished the calculation, because we did not know the relative amounts of the elements present. However, we did not need this to do the calculation, since we knew the stoichiometry of formation of CO_2 from C and of H_2O from H. Whatever the formula of the compound, these do not change.

Example 3.12   A compound contains only carbon, hydrogen, and nitrogen. A sample weighing 7.81×10^{-2} g is burned in such a way that the hydrogen is converted to water, while the nitrogen content ends up as gaseous N_2. If the water produced weighs 4.46×10^{-2} g and the N_2 gas occupies 11.1 ml at STP, calculate the simplest formula of the compound.

SOLUTION   Again, we convert to molar quantities:

$$\text{water:} \quad \frac{4.46 \times 10^{-2}}{18.0} = 2.47 \times 10^{-3} \text{ mole of } H_2O$$

The nitrogen occupied 11.1 ml, (0.0111 liter) so we convert to moles by dividing by 22.4 liter, the volume of 1 mole of gas:

$$\text{nitrogen:} \quad \frac{0.0111}{22.4} = 4.96 \times 10^{-4} \text{ mole of } N_2$$

So in our original compound we have

$$2 \times 2.47 \times 10^{-3} = 4.94 \times 10^{-3} \text{ mole of combined H}$$

$$2 \times 4.96 \times 10^{-4} = 9.92 \times 10^{-4} \text{ mole of combined N}$$

(*Note*: In both cases, we multiply by 2 because a mole of H_2O comes from 2 mole of combined H, and a mole of N_2 comes from 2 mole of combined N.) To obtain the amount of C originally present, we must convert these to masses by multiplying by atomic weights:

$$4.94 \times 10^{-3} \text{ mole H weighs } 4.94 \times 10^{-3} \times 1.01 = 4.99 \times 10^{-3} \text{ g}$$
$$9.92 \times 10^{-4} \text{ mole N weighs } 9.92 \times 10^{-4} \times 14.0 = \underline{1.39 \times 10^{-2} \text{ g}}$$
$$\text{total H} + \text{N} = \overline{1.89 \times 10^{-2} \text{ g}}$$

By difference, the carbon content is

$$(7.81 \times 10^{-2}) - (1.89 \times 10^{-2}) = 5.92 \times 10^{-2} \text{ g}$$

which is

$$\frac{5.92 \times 10^{-2}}{12.0} = 4.93 \times 10^{-3} \text{ mole C}$$

We now have the mole ratios of the three elements:

Carbon	Hydrogen	Nitrogen
4.93×10^{-3}	4.94×10^{-3}	9.92×10^{-4}

Divide through by the smallest (nitrogen):

5.0	5.0	1.0

The simplest formula is therefore C_5H_5N.

Example 3.13   1.48 g of a chloride of vanadium is dissolved in water and all the chlorine present goes to chloride ion. Addition of $AgNO_3$ solution precipitates all the Cl^- as AgCl. If the AgCl precipitate weighs 4.06 g, what is the formula of the chloride?

SOLUTION   We ended up with 4.06 g of AgCl (*M.W.* 143.4), which is

$$\frac{4.06}{143.4} = 2.83 \times 10^{-2} \text{ mole AgCl}$$

The precipitation reaction is

$$Ag^+ + Cl^- \longrightarrow AgCl(s)$$

so the 4.06 g of AgCl came from 2.83×10^{-2} mole of Cl^- (*M.W.* 35.5), which weighed

$$2.83 \times 10^{-2} \times 35.5 = 1.01 \text{ g}$$

This was the amount of chlorine contained in the original sample, so, by difference, the amount of vanadium was

$$1.48 - 1.01 = 0.47 \text{ g V}$$

The atomic weight is 50.9, so we had

$$\frac{0.47}{50.9} = 9.2 \times 10^{-3} \text{ mole V}$$

Now we can calculate the molar combining ratio

Vanadium	Chlorine
9.2×10^{-3} mole	2.83×10^{-2} mole

$$\frac{9.2 \times 10^{-3}}{9.2 \times 10^{-3}} = 1.0 \qquad \frac{2.83 \times 10^{-2}}{9.2 \times 10^{-3}} = 3.0$$

The formula is VCl_3.

Note carefully in the above problems that it is necessary to be very precise about the meaning of "mole" in connection with substances such as hydrogen, nitrogen, or chlorine. It would be very easy to go wrong if we did not each time specify whether we meant molecular N_2, combined N, molecular H_2, etc. For this reason, we write the formula each time (N, N_2, H, H_2, O, etc.) rather than just the name of the element.

3.5 Reaction in Solution: Stoichiometry of Titrations

Many types of reactions occur in solution, and we shall be considering some in detail later in this book. There are precipitation reactions, acid–base reactions, oxidation-reduction reactions, and reactions combining these processes. In this section, we are considering only the stoichiometry of reactions that happen to occur in solution, just as we previously considered the stoichiometry of reactions in general.

As before, we will have to have a balanced equation available to do the calculation. There is one difference, however; if we have one or more reactant in solution, we will often measure out the amount of reactant by measuring the volume of a solution, rather than by weighing the compound directly. This is a very convenient and rapid method for measuring amounts of compounds, and we saw in Chapter 2 how we can make such measurements.

Suppose that we take a measured volume of a solution and add to it some other reagent with which it undergoes a reaction of some type, which we can represent by an equation. If we know the concentration of one solution, we can find the concentration of another solution that it reacts with, provided we make sure that we add the exact quantity of solution necessary for complete reaction. This, of course, is called a *titration*, and is a familiar operation in any chemical laboratory. There are many ways of finding out how much of one solution reacts with another, because the properties of the solution will change when reaction is complete. For example, its color may change (either the color of the reactants, or of an added indicator) or a precipitate may appear, or we may use an electrical detector such as a pH meter to help us. (Some of these methods are discussed in Chapters 13 and 15.)

Alternatively, we may precipitate out some compound derived from our solute, weigh the solid precipitate, and hence find how much was present in solution. Both the above methods are illustrated in the following examples. In every case, of course, we must start with a balanced equation and use the mole concept to find reacting amounts.

Before starting calculations on titrations, we may conveniently consider a simple and useful additional concept, the millimole. As the name implies, this is exactly 10^{-3} mole of a substance, and it is abbreviated to mmole. The convenience of the mmole arises from the fact that it has the same relationship to the milliliter as the mole does to the liter, so that

$$\text{molarity} = \frac{\text{mole of solute}}{\text{liter of solution}} = \frac{\text{mmole of solute}}{\text{ml of solution}}$$

When we carry out titrations, we usually measure volumes in the range 10–50 ml, so the quantity of solute is of the order of a few mmole. Using mmole as the measuring unit makes the arithmetic easier by omitting factors of 1000.

Example 3.14   How many mmole of solute are present in the following?

(a) 12.5 ml of 0.120M solution
(b) 31.8 ml of 0.0760M solution

SOLUTION   We use the relationship

$$\text{mmole of solute} = (\text{volume in ml}) \times (\text{molarity})$$

(a) mmole of solute $= (12.5 \text{ ml}) \times (0.120 \text{ mmole ml}^{-1}) = 1.50 \text{ mmole}$
(b) mmole of solute $= (31.8 \text{ ml}) \times (0.0760 \text{ mmole ml}^{-1}) = 2.42 \text{ mmole}$

Example 3.15   What is the molarity of the following solutions?

(a) 3.03 mmole of solute in 16.6 ml solution
(b) 8.91 mmole of solute in 50.0 ml solution

SOLUTION   Divide the number of mmole by the volume in ml.

(a) $\dfrac{3.03 \text{ mmole}}{16.6 \text{ ml}} = 0.183 M$ (b) $\dfrac{8.91 \text{ mmole}}{50.0 \text{ ml}} = 0.178 M$

Example 3.16   What volume of an 0.0800M solution contains 10.0 mmole of solute?

SOLUTION

$$\text{volume} = \text{ml of solution} = \frac{\text{mmole of solute}}{\text{molarity}} = \frac{10.0 \text{ mmole}}{0.0800 M} = 125 \text{ ml}$$

The above examples are obviously very similar in type to those at the end of Chapter 2, and serve as a review of this important type of calculation before we try titration calculations.

Example 3.17   Excess silver nitrate solution is added to 50.0 ml of 0.110M HCl solution. What weight of insoluble AgCl will precipitate?

SOLUTION   The reaction is

$$AgNO_3 + HCl \longrightarrow AgCl + HNO_3$$

or, including only the essential reactants,

$$Ag^+ + Cl^- \longrightarrow AgCl$$

(The H^+ and NO_3^- do not take part in the reaction.)

The 50.0 ml of 0.110M HCl contain $50.0 \times 0.110 = 5.50$ mmole of HCl. This will produce 5.50 mmole of AgCl precipitate (*M.W.* 143):

$$5.50 \text{ mmole AgCl} = 5.50 \times 10^{-3} \times 143 = 0.787 \text{ g AgCl}$$

Example 3.18   Excess $Ca(NO_3)_2$ solution is added to 150 ml of a solution of KF.

(a) If the precipitated CaF_2 weighs 1.38 g, what is the molarity of the KF solution?
(b) What volume of KF solution would be needed to precipitate 1.00 g of CaF_2?

SOLUTION

(a) The reaction is

$$2F^- + Ca^{2+} \longrightarrow CaF_2(s) \quad (M.W.\ 78.0)$$

Thus,

$$1.38 \text{ g CaF}_2 = \frac{1.38}{78.0} = 1.77 \times 10^{-2} \text{ mole} = 17.7 \text{ mmole CaF}_2$$

From the equation:

1 mole CaF_2 comes from 2 mole F^- (i.e., 2 mole KF)
17.7 mmole CaF_2 comes from $17.7 \times 2 = 35.4$ mmole KF
150 ml of KF solution contain 35.4 mmole KF

$$\text{molarity of KF} = \frac{35.4 \text{ mmole}}{150 \text{ ml}} = 0.236M$$

(b) $1.00 \text{ g CaF}_2 = \dfrac{1.00}{78.0} = 1.28 \times 10^{-2} \text{ mole} = 12.8 \text{ mmole CaF}_2$

From the equation, this comes from

$$12.8 \times 2 = 25.6 \text{ mmole KF}$$

If the solution is $0.236M$, this amount of KF is contained in

$$\frac{25.6}{0.236} = 108 \text{ ml of solution}$$

The same result could be obtained more directly from the data by simple proportion:

1.38 g CaF_2 came from 150 ml KF solution.

1.00 g CaF_2 came from $\dfrac{150 \times 1.00}{1.38} = 108$ ml KF solution.

Note in both the above examples that we added an excess of one of the reagents in order to precipitate *all* of the ion in which we were interested (Cl^- or F^-). We were not concerned about adding the exact, stoichiometric, amount of reagent needed for reaction.

In the following examples, based on titrations, we are careful to have exactly the right amount of each reactant present. In the case of an acid–base reaction, this is known as a neutralization.

Example 3.19   A 25.0 ml portion of $0.110M$ HCl requires 18.7 ml of KOH solution for exact neutralization. What is the molarity of the KOH solution?

SOLUTION   The reaction is

molecular: $HCl + KOH \longrightarrow KCl + H_2O$

ionic: $H^+ + OH^- \longrightarrow H_2O$

In 25.0 ml of $0.110M$ HCl, there are $25.0 \times 0.110 = 2.75$ mmole of HCl. From the equation, 1 mole HCl reacts with 1 mole KOH, so

$$2.75 \text{ mmole HCl reacts with } 2.75 \text{ mmole KOH}$$

This quantity of KOH is contained in 18.7 ml, therefore, molarity of KOH is

$$\frac{2.75}{18.7} = 0.147M$$

Example 3.20   A 10.0 ml portion of H_2SO_4 solution is exactly neutralized by 13.1 ml of $0.100M$ NaOH. What is the molarity of the H_2SO_4 solution?

SOLUTION   The reaction is

$$H_2SO_4 + 2NaOH \longrightarrow Na_2SO_4 + 2H_2O$$

In 13.1 ml of $0.100M$ NaOH there are $13.1 \times 0.100 = 1.31$ mmole of NaOH. From the equation, 2 mole NaOH react with 1 mole H_2SO_4; and 1.31 mmole NaOH react with $1.31/2 = 0.655$ mmole H_2SO_4, which quantity of H_2SO_4 is contained in 10.0 ml. Therefore the molarity of H_2SO_4 is

$$\frac{0.655}{10.0} = 6.55 \times 10^{-2}M$$

Example 3.21   Oxalic acid, $(COOH)_2$, reacts with NaOH according to

$$(COOH)_2 + 2NaOH \longrightarrow (COONa)_2 + 2H_2O$$

If 0.816 g of oxalic acid dihydrate, $(COOH)_2 \cdot 2H_2O$, is dissolved in water and titrated with $0.120M$ NaOH solution, what volume of NaOH will be needed?

SOLUTION   We use the *M.W.* of the dihydrate because that was the form in which the reagent was weighed out. (The water of crystallization takes no part in the reaction and is not included in the equation.) Thus, the *M.W.* of $(COOH)_2 \cdot 2H_2O$ is 126, and

$$0.816 \text{ g} = \frac{0.816}{126} = 6.48 \times 10^{-3} \text{ mole} = 6.48 \text{ mmole}$$

From equation, this will react with $6.48 \times 2 = 12.96$ mmole of NaOH. The NaOH solution is $0.120M$:

$$12.96 \text{ mmole is contained in } \frac{12.96}{0.120} = 108 \text{ ml NaOH}$$

(*Note*: The volume of water used to dissolve the oxalic acid does not matter. As we were titrating *all* of the sample, we only needed the total number of moles, not the molarity of the solution.)

Example 3.22   Potassium hydrogen phthalate, $C_8H_5O_4K$, is often used to standardize basic solutions. It reacts with potassium hydroxide according to

$$C_8H_5O_4K + KOH \longrightarrow C_8H_4O_4K_2 + H_2O$$

If a sample of $C_8H_5O_4K$ weighing 0.248 g is dissolved in water, 15.9 ml of KOH solution is needed for exact neutralization. What is the molarity of the KOH?

SOLUTION   As in the previous example, we do not need to know the volume of the solution. The *M.W.* of $C_8H_5O_4K$ is 204:

$$0.248 \text{ g is } \frac{0.248}{204} = 1.22 \times 10^{-3} \text{ mole} = 1.22 \text{ mmole}$$

From the equation, this reacts with the same number of mole of KOH; and since there are 1.22 mmole of KOH in 15.9 ml of solution, the molarity of the KOH solution is

$$\frac{1.22}{15.9} = 0.0765M$$

Example 3.23   In an experiment similar to Example 3.22, 3.06 g of $C_8H_5O_4K$ is dissolved in water and made up to 250 ml of solution. A 10.0 ml portion of this solution exactly neutralizes 12.1 ml of a solution of NaOH. What is the molarity of the NaOH solution?

SOLUTION   This time we *do* need to calculate the molarity of the solution of $C_8H_5O_4K$, since we are measuring it by volume:

$$3.06 \text{ g of } C_8H_5O_4K = \frac{3.06}{204} = 1.50 \times 10^{-2} \text{ mole} = 15.0 \text{ mmole}$$

This is dissolved in 250 ml of solution, which has a molarity of

$$\frac{15.0}{250} = 0.0600M$$

A 10 ml portion of this solution contains

$$6.00 \times 10^{-2} \times 10.0 = 0.600 \text{ mmole } C_8H_5O_4K$$

This reacts with the same number of moles of NaOH. Thus, 12.1 ml NaOH solution contains 0.600 mmole and the molarity of the NaOH solution is

$$\frac{0.600}{12.1} = 0.0496M$$

Example 3.24   Permanganate ion, MnO_4^- reacts with oxalate ion, $C_2O_4^{2-}$, in acid solution to give Mn^{2+} and CO_2. In the standardization of a solution of permanganate, a sample of sodium oxalate, $Na_2C_2O_4$, weighing 0.188 g, was dissolved in water, acidified, and titrated with $KMnO_4$ solution. If 17.5 ml of $KMnO_4$ solution was required, what was its molarity?

SOLUTION   The partial equation is

$$MnO_4^- + C_2O_4^{2-} \longrightarrow Mn^{2+} + CO_2 \quad \text{(acid solution)}$$

We have considered the MnO_4^- reaction before; its reduction half-reaction is

$$MnO_4^- + 8H^+ + 5e^- \longrightarrow Mn^{2+} + 4H_2O$$

The oxidation of oxalate is easy to balance: the equation,

$$C_2O_4^{2-} \longrightarrow 2CO_2$$

is already balanced in mass, we need only add two electrons:

$$C_2O_4^{2-} \longrightarrow 2CO_2 + 2e^-$$

To balance the overall reaction, we need $10e^-$ in each, coming from $5 \times$ the oxidation reaction and $2 \times$ the reduction reaction:

$$2MnO_4^- + 5C_2O_4^{2-} + 16H^+ \longrightarrow 2Mn^{2+} + 10CO_2 + 8H_2O$$

Although we only consider the oxalate ion, we had to weigh the compound as $Na_2C_2O_4$, so we need the *M.W.* of that compound (134).

$$0.188 \text{ g of } Na_2C_2O_4 = \frac{0.188}{134} = 1.40 \times 10^{-3} \text{ mole} = 1.40 \text{ mmole}$$

From the equation:

5 mole oxalate react with 2 mole MnO_4^{2-}

1 mole oxalate reacts with 2/5 mole MnO_4^-

1.40 mmole oxalate react with $(1.40 \times 2)/5$ mmole $MnO_4^- = 0.560$ mmole MnO_4^-

0.560 mmole MnO_4^- is contained in 17.5 ml of solution

molarity of the $KMnO_4$ solution is $0.560/17.5 = 0.0320M$

The chief hazard in calculations of this type is putting the fraction for the reacting ratio the wrong way up. It is very easy to write 5/2 instead of 2/5 if you are thinking carelessly. This can be avoided, however, by writing down the proportionality steps in a logical order, as was done above.

Example 3.25   What volume of the above $KMnO_4$ solution will be needed to titrate 25.0 ml of $0.100M$ $FeSO_4$ solution? The iron is oxidized to Fe^{3+}, and acidic conditions are used.

SOLUTION   First we balance the equation. The reduction half-reaction is the usual equation for permanganate:

$$MnO_4^- + 8H^+ + 5e^- \longrightarrow Mn^{2+} + 4H_2O$$

while the oxidation half-reaction is a simple change from Fe^{2+} to Fe^{3+}:

$$Fe^{2+} \longrightarrow Fe^{3+} + e^-$$

Multiply this by five and add to the previous half-reaction to balance the overall equation:

$$MnO_4^- + 8H^+ + 5Fe^{2+} \longrightarrow Mn^{2+} + 5Fe^{3+} + 4H_2O$$

The $FeSO_4$ solution contains $25.0 \times 0.100 = 2.50$ mmole Fe^{2+}. And, from the equation,

5 mole Fe^{2+} react with 1 mole MnO_4^-

1 mole Fe^{2+} reacts with 1/5 mole MnO_4^-

2.50 mmole Fe^{2+} reacts with $2.50/5 = 0.500$ mmole MnO_4^-

0.500 mmole of MnO_4^- is contained in

$$\frac{0.500}{0.0320} = 15.6 \text{ ml of } 0.0320M \text{ solution}$$

Example 3.26   Potassium chromate, K_2CrO_4, oxidizes sulfite to sulfate in acid solution according to

$$2CrO_4^{2-} + 3SO_3^{2-} + 10H^+ \longrightarrow 2Cr^{3+} + 3SO_4^{2-} + 5H_2O$$

It is found that a 25.0 ml portion of a solution of sodium sulfite exactly reacts with 28.1 ml of $0.0875M$ K_2CrO_4 solution. What is the molarity of the Na_2SO_3 solution?

SOLUTION   28.1 ml of 0.0875 M K_2CrO_4 contains $28.1 \times 0.0875 = 2.46$ mmole CrO_4^{2-}. CrO_4^{2-}.

From the equation:

2 mole CrO_4^{2-} react with 3 mole SO_3^{2-}
1 mole CrO_4^{2-} reacts with 3/2 mole SO_3^{2-}
2.46 mmole CrO_4^{2-} reacts with $(2.46 \times 3)/2 = 3.69$ mmole SO_3^{2-}
3.69 mmole SO_3^{2-} is contained in 25.0 ml of solution, the molarity of which is

$$\frac{3.69}{25.0} = 0.148M$$

3.6   Limiting Quantity of Reactant

Before leaving the subject of stoichiometry, we should look at one further application of the combining-ratio concept, that of the determination of what happens when a reaction occurs between two reactants which are *not* present in a stoichiometric ratio. The basic principle here is dictated by common sense:

When a reaction occurs between two (or more) reactants which are present in random amounts, the amount of reaction that occurs will be limited by the reactant which is present in the *smaller* quantity, according to the required combining ratio.

Some examples will make this clearer.

Example 3.27   To make one ox cart, I need a body and two wheels. If I have 24 wheels and 18 bodies, how many complete ox carts can I make? What will be left over?

SOLUTION   It is easiest to calculate what would be needed for each set of components:

24 wheels would combine with 12 bodies to give 12 ox carts.
18 bodies would combine with 36 wheels to give 18 ox carts.

We have calculated the *theoretical* yield from each of our given amounts of starting material. We see that the 24 wheels we have would combine with only 12 bodies (which is *less* than the available bodies), whereas the 18 bodies we have would require 36 wheels (which is *more* than the amount of wheels we have)

The number of complete ox carts that I can make is therefore limited by the number of *wheels* which I have available. Although I actually have more wheels than bodies, I have less than the 2:1 ratio that would be needed to use up all the bodies. So I end up with 12 complete ox carts and 6 bodies left over.

If I had started with, say 24 wheels and 10 bodies, then the number of wheels available would have been *greater* than the 2:1 ratio required to use up all the bodies. In other words, it would have been the bodies that limited the number of ox carts I could have made to 10, with 4 wheels left over.

Now we'll try the calculation with chemicals.

Example 3.28   Phosphorus reacts with bromine to give PBr_3. If I allow 50.0 g of phosphorus to react with 200 g bromine, how much PBr_3 will I obtain? What will remain unreacted?

SOLUTION   The equation is

$$2P + 3Br_2 \longrightarrow 2PBr_3$$

We start with the following quantities:

phosphorus: $\dfrac{50.0}{31.0} = 1.61$ mole P

bromine: $\dfrac{200}{160} = 1.25$ mole Br_2

For complete reaction, 2 mole of P needs 3 mole of Br_2, so

$$1.61 \text{ mole P needs } \frac{1.61 \times 3}{2} = 2.41 \text{ mole } Br_2$$

This is *more* than the amount of Br_2 we have, so we cannot react all the phosphorus. The bromine is the limiting reactant:

3 mole Br_2 react with 2 mole P

1.25 mole Br_2 react with $\dfrac{1.25 \times 2}{3} = 0.833$ mole P

This is *less* than the total amount of P which we have, so after *all* the Br_2 has reacted (giving 0.833 mole PBr_3), we have remaining $1.61 - 0.833 = 0.78$ mole P.

Summarizing:
PBr_3 produced: 0.833 mole $= 0.833 \times 271 = 226$ g
unreacted P remaining: 0.78 mole $= 0.78 \times 31.0 = 24$ g

Note that the total mass at the end of the reaction (250 g) is the same as the total mass we started with, giving a check on our arithmetic.

The same type of calculation may be applied to reactions occurring in solution. When we do a titration, we make sure that we have exactly the correct amount of one reactant needed to react with the given quantity of the second reactant. But supposing we do *not* have this stoichiometric ratio, so that one reactant is in excess. What happens then? Obviously we calculate the mole quantities from the volumes and concentrations of the two solutions being mixed, then apply the logic used before.

Example 3.29   15.0 ml of $0.120M$ silver nitrate ($AgNO_3$) solution is mixed with 25.0 ml of $0.080M$ sodium chloride (NaCl) solution. Assuming that the solubility of AgCl is negligibly small, calculate how much AgCl precipitates. Calculate also the concentrations of all ions remaining in solution in appreciable quantity.

SOLUTION   The $AgNO_3$ solution gives

$$15.0 \times 0.120 = 1.80 \text{ mmole each of } Ag^+ \text{ and } NO_3^-$$

The NaCl solution gives

$$25.0 \times 0.080 = 2.00 \text{ mmole each of } Na^+ \text{ and } Cl^-$$

The reaction is

$$Ag^+ + Cl^- \longrightarrow AgCl(s)$$

Obviously, one Ag^+ reacts with one Cl^-, so the amount of AgCl we can obtain is limited by the ion present in the smaller amount, namely Ag^+.

Precipitate: 1.80 mmole AgCl
Solution: $2.00 - 1.80 = 0.20$ mmole Cl^-

We were asked for the actual concentrations of the ions remaining, so we have to take into account the total volume of the solution after mixing (40.0 ml):

$$[Cl^-] = \frac{0.20}{40.0} = 5.0 \times 10^{-3} M$$

$$[Na^+] = \frac{2.00}{40.0} = 5.00 \times 10^{-2} M$$

$$[NO_3^-] = \frac{1.80}{40.0} = 4.50 \times 10^{-2} M$$

(*Note*: Use of *square brackets* in the above equations is a standard abbreviation meaning "the molar concentration of the ion or molecule whose formula is contained within the brackets," and we shall be using this convention frequently in this book.)

Example 3.30   Solution A contains $[Ba^{2+}] = 0.30M$ and solution B contains $[F^-] = 0.40M$. Assuming that BaF_2 has a negligibly small solubility, calculate what happens when equal volumes of solutions A and B are mixed.

SOLUTION   Suppose we mix 1 liter of solution A with 1 liter of solution B. Then we have

$$0.30 \text{ mole of } Ba^{2+} \quad \text{and} \quad 0.40 \text{ mole of } F^-$$

The formation of BaF_2 requires 2 mole F^- for each mole of Ba^{2+}

$$0.30 \text{ mole } Ba^{2+} \quad \text{would need} \quad 2 \times 0.30 = 0.60 \text{ mole } F^-$$

This is *more* than the amount of F^- that we have available, so we cannot precipitate all the Ba^{2+}. The F^- is the limiting reactant, and 0.40 mole F^- reacts with 0.20 mole Ba^{2+} to precipitate BaF_2. This leaves

$$(0.30 - 0.20) = 0.10 \text{ mole } Ba^{2+}$$

Since this is in solution in a total volume of 2 liter, the concentration of Ba^{2+} is

$$\frac{0.10 \text{ mole}}{2 \text{ liter}} = 0.050M$$

The answer is that, on mixing, essentially all the fluoride ion precipitates as BaF_2, leaving the excess barium ion in solution with $[Ba^{2+}] = 0.050M$.

(*Note*: The volume of solution taken did not affect the final answer to this problem. If we had taken some other volume of each solution, the amounts throughout the calculation would have canceled out at the end.)

Calculations of the above type are quite important in practice, as we frequently need to mix solutions together in such a way as to obtain an excess of one reactant. We will find more examples in Chapter 12 (solubility and precipitation) and Chapter 13 (neutralization reactions).

This has been a long chapter, because we have dealt in some detail with these stoichiometric calculations and the mole concept. This is a reflection of their basic importance in all forms of quantitative relationships in chemistry. A thorough understanding of this material is vital to further studies in chemistry.

3.7 Common Ions, Their Names and Formulas

When we are writing out equations in ionic form, we must use the correct formulas, including the values of the charge, for each ion. The three tables following are intended to help you in this translation step by giving the formulas corresponding to the names of the more common cations and anions.

In the ions of Table 3.1, the charge is generally the same as the group of the periodic table in which the element is found (all valence shell electrons have been lost). No other charge is found under usual chemical conditions, so we do not need to specify the charge when naming the ion.

Two very common and important positive ions not mentioned in Table 3.1 are the hydronium ion, H_3O^+, and the ammonium ion, NH_4^+. Only the latter is commonly encountered in stable ionic salts.

TABLE 3.1 Common Positive Ions (Cations) of Elements Having Only One Oxidation State

| Element | | Usual ion or ions | |
Name	Symbol	Name	Symbol
Hydrogen	H	Hydrogen ion (proton)	H^+
Lithium	Li	Lithium ion	Li^+
Sodium	Na	Sodium ion	Na^+
Potassium	K	Potassium ion	K^+
Magnesium	Mg	Magnesium ion	Mg^{2+}
Calcium	Ca	Calcium ion	Ca^{2+}
Strontium	Sr	Strontium ion	Sr^{2+}
Barium	Ba	Barium ion	Ba^{2+}
Zinc	Zn	Zinc ion	Zn^{2+}
Aluminum	Al	Aluminum ion	Al^{3+}

TABLE 3.2   Common Positive Ions (Cations) of Elements Having More Than One Oxidation State

Chromium	Cr	Chromium(III)	(chromic)	Cr^{3+}
Manganese	Mn	Manganese(II)	(manganous)	Mn^{2+}
Iron	Fe	Iron(II)	(ferrous)	Fe^{2+}
		Iron(III)	(ferric)	Fe^{3+}
Cobalt	Co	Cobalt(II)	(cobaltous)	Co^{2+}
Nickel	Ni	Nickel(II)		Ni^{2+}
Copper	Cu	Copper(I)	(cuprous)	Cu^{+}
		Copper(II)	(cupric)	Cu^{2+}
Silver	Ag	Silver(I)	(argentous)	Ag^{+}
Tin	Sn	Tin(II)	(stannous)	Sn^{2+}
		Tin(IV)	(stannic)	Sn^{4+}
Mercury	Hg	Mercury(I)	(mercurous)	Hg_2^{2+}
		Mercury(II)	(mercuric)	Hg^{2+}
Lead	Pb	Lead(II)	(plumbous)	Pb^{2+}

TABLE 3.3   Common Negative Ions (Anions)

Anion	Formula	Parent acid	Formula	Anhydride	Formula
Hydride*	H^-	Hydrogen	H_2		
Oxide	O^{2-}	Water	H_2O		
Hydroxide	OH^-	Water	H_2O		
Peroxide	O_2^{2-}	Hydrogen Peroxide	H_2O_2		
Fluoride	F^-	Hydrofluoric	HF		
Chloride	Cl^-	Hydrochloric	HCl		
Bromide	Br^-	Hydrobromic	HBr		
Iodide	I^-	Hydriodic	HI		
Nitrate	NO_3^-	Nitric	HNO_3	Nitrogen(V) Oxide	N_2O_5
Nitrite	NO_2^-	Nitrous	HNO_2	Nitrogen(III) Oxide	N_2O_3
Nitride	N^{3-}	Ammonia	H_3N		
Sulfate	SO_4^{2-}	Sulfuric	H_2SO_4	Sulfur Trioxide	SO_3
Sulfite	SO_3^{2-}	Sulfurous	H_2SO_3	Sulfur Dioxide	SO_2
Sulfide	S^{2-}	Hydrosulfurous	H_2S		
Carbonate	CO_3^{2-}	Carbonic	H_2CO_3	Carbon Dioxide	CO_2
Acetate	CH_3COO^-	Acetic	CH_3COOH		
Oxalate	$C_2O_4^{2-}$	Oxalic	$H_2C_2O_4$		
Cyanide	CN^-	Hydrocyanic	HCN		
Phosphate	PO_4^{3-}	Phosphoric	H_3PO_4	Phosphorus(V) Oxide	P_2O_5
Chromate	CrO_4^{2-}	Chromic	H_2CrO_4	Chromium(VI) Oxide	CrO_3
Dichromate	$Cr_2O_7^{2-}$	(Chromic + CrO_3)			
Permanganate	MnO_4^-	Permanganic	$HMnO_4$	Manganese(VII) Oxide	Mn_2O_7

*Hydride is immediately decomposed by water:

$$H^- + H_2O \longrightarrow H_2 + OH^-$$

It only exists in a few compounds such as lithium hydride, LiH (i.e., Li^+ and H^-).

For a number of metals, particularly transition elements, there is more than one common oxidation state, so we have to specify the charge when naming the ion. Roman numerals, in brackets, are the correct way of doing this, but the older "-ous" and "-ic" terminations are still used enough that you should be familiar with them. Only the most commonly occurring ions are given in Table 3.2.

The negative ions may all be regarded as derived from acids by the loss of one or more hydrogen ions. The acid, in turn, may sometimes (not always) be the result of reacting water with a substance (usually the oxide of a nonmetal) called an acid anhydride. In Table 3.3, we give the anion; the parent acid; and, where applicable, the acid anhydride.

Many ions in Table 3.3 have a charge of 2−, and are derived from acids that lose two hydrogen ions (diprotic acids). Usually, an intermediate ion may be formed in which only one hydrogen ion has been lost. We denote this by putting "bi" before the name of the ion, or, better, by inserting "hydrogen" into the name of the compound. For example,

CO_3^{2-} carbonate ion

HCO_3^- bicarbonate ion

$NaHCO_3$ sodium bicarbonate (sodium hydrogen carbonate)

Similarly, HSO_4^- (bisulfate); HSO_3^- (bisulfite).

Of the acids in Table 3.3, the ones we usually regard as "strong" acids, which are completely ionized in aqueous solution, are: HCl, HBr, HI, HNO_3, H_2SO_4, H_2CrO_4 and $HMnO_4$. In writing an equation in ionic form, always represent these acids as being completely in the ionic form. (Chapter 13 has a detailed discussion of this problem.)

KEY WORDS

chemical equation	millimole
oxidation	stoichiometry of reaction
reduction	half-reaction
titration	neutralization

STUDY QUESTIONS

1. What does a chemical equation tells us?

2. How do we convert from stoichiometric reacting quantities in moles to reacting quantities in grams?

3. Look at the samples of "synthesis" reactions given in this chapter. What other class of reaction type do they belong to?

4. What characterizes a redox reaction?

5. Why do we divide a redox reaction into two half-reactions? How do we differentiate between them?

6. Why do oxidation and reduction occur together?

7. Why do we introduce the millimole? How is it related to solution volume and concentration? Write this relationship in several different ways.

8. Why are titrations used in chemical analysis? Apart from titration, what other methods are available to determine the solute concentration in an unknown solution?

9. What do we mean by "limiting quantity of reactant"? What information do we need to determine which is the limiting reactant?

PROBLEMS

1. Reaction of FeO with oxygen gives Fe_2O_3.
   (a) Write a balanced equation for this process.
   (b) What weight of FeO is needed to make 50.0 g of Fe_2O_3?

2. Zinc reacts with sulfur to give zinc sulfide, ZnS.
   (a) How much sulfur is needed to react with 48.0 g of zinc?
   (b) How much ZnS is produced?

3. Sodium reacts with water according to

$$Na + H_2O \longrightarrow NaOH + \frac{1}{2} H_2(g)$$

   What volume of hydrogen is produced when 1.00 g Na reacts with water?

4. Ammonium dichromate decomposes on heating according to

$$(NH_4)_2Cr_2O_7 \longrightarrow Cr_2O_3 + N_2 + 4H_2O$$

   If 0.152 g of Cr_2O_3 remained after such a reaction,
   (a) What weight of $(NH_4)_2Cr_2O_7$ was taken?
   (b) What volume of nitrogen was evolved?

5. A compound is known to contain C, H, and O. 2.81 g burned in oxygen gave 5.75 g CO_2 and 1.76 g H_2O. Calculate the simplest formula of the compound.

6. A compound contains only C and H. On combustion, 0.588 g gives 1.73 g CO_2 and 1.06 g H_2O. The *M.W.* is 32 ± 5. Calculate
   (a) The simplest formula
   (b) The molecular formula

7. A compound contains C, H, and S. On burning, a sample weighing 0.0116 g gave 0.0226 g CO_2. In another reaction, 0.223 g of the compound yielded 0.576 g of $BaSO_4$ corresponding to its sulfur content. Calculate the simplest formula.

8. A chloride of manganese was subjected to reactions that converted all the manganese to MnO_2. If 0.613 g of the chloride yielded 0.330 g of MnO_2, what was the formula of the chloride?

9. Write balanced equations for the following reactions:
    (a) Synthesis of PCl_3 from yellow phosphorus (P_4) and chlorine (Cl_2).
    (b) Synthesis of VOF_3 from vanadium metal, oxygen (O_2), and fluorine (F_2).

10. Write balanced molecular equations for the following reactions:
    (a) Chromium reacts with oxygen to give chromium(III) oxide, Cr_2O_3.
    (b) Chromium(III) oxide dissolves in acid.
    (c) A solution of Cr^{3+} ions is treated with base, precipitating chromium(III) hydroxide.
    (d) Octane thiol, $C_8H_{17}SH$, burns to give carbon dioxide, water, and sulfur dioxide.
    (e) Octane thiol burns in fluorine to give carbon tetrafluoride, hydrogen fluoride, and sulfur hexafluoride.
    (f) Barium carbonate dissolves in nitric acid to give carbon dioxide and aqueous barium nitrate.
    (g) Aqueous barium nitrate reacts with aqueous lithium sulfate to precipitate solid barium sulfate.
    (h) Solutions of silver sulfate and barium chloride are mixed, precipitating silver chloride and barium sulfate.
    (i) Lead(II) acetate solution is mixed with hydrochloric acid, producing insoluble lead(II) chloride.
    (j) Gaseous ammonia reacts with pure nitric acid to give solid ammonium nitrate.
    (k) Solid ammonium nitrate decomposes on heating to give nitrous oxide (N_2O) and one other product.
    (l) Calcium carbonate (solid) dissolves in water containing dissolved carbon dioxide. The solution contains bicarbonate ions.
    (m) Liquid germanium tetrachloride (pure, not a solution) reacts with aqueous silver nitrate to precipitate insoluble silver chloride and insoluble germanium dioxide. (*Hint*: water may be a reactant!)

11. Rewrite the equations of Example 10 in ionic form, for reactions involving ionic substances.

12. Complete and balance the following equations for reactions occurring in acid solution. In each case, identify the oxidizing agent and the reducing agent.
    (a) $VO_3^- + Al \longrightarrow VO^{2+} + Al^{3+}$
    (b) $Cu + NO_3^- \longrightarrow Cu^{2+} + NO_2(g)$
    (c) $Cr_2O_7^{2-} + SO_3^{2-} \longrightarrow Cr^{3+} + SO_4^{2-}$
    (d) $ClO_2^- + I^- \longrightarrow Cl^- + I_2$
    (e) $Ce^{3+} + O_3 \longrightarrow Ce^{4+} + O_2$
    (f) $Ce^{4+} + H_2O_2 \longrightarrow Ce^{3+} + O_2$
    (g) $TcO_4^- + Ti \longrightarrow Tc^{2+} + Ti^{3+}$
    (h) Copper(I) sulfate (solid) reacts with water to give aqueous copper(II) sulfate and copper metal.
    (i) Metallic zinc dissolves in acidified permanganate solution, giving Mn^{2+} ions and Zn^{2+} ions.
    (j) Potassium bromide (solid) and potassium dichromate (solid) are treated with sulfuric acid, evolving gaseous bromine and producing Cr^{3+}.

13. Complete and balance the following equations for reactions occurring in alkaline solution. In each case, identify the oxidizing agent and the reducing agent.
    (a) $AsO_2^- + Br_2 \longrightarrow AsO_4^{3-} + Br^-$
    (b) $O_2 + Mn^{2+} \longrightarrow MnO_2$
    (c) $Fe^{3+} + Cl_2 \longrightarrow FeO_4^{2-} + Cl^-$
    (d) $Ag_2O(s) + HPO_3^{2-} \longrightarrow Ag(s) + PO_4^{3-}$
    (e) Chromium(III) hydroxide is treated with alkaline hydrogen peroxide, giving chromate ion.

14. What is the molarity of the following solutions?
    (a) 1.28 mmole solute in 17.5 ml solution
    (b) 21.1 mmole solute in 75.3 ml solution

15. How many mmole of solute are present in the following?
    (a) 12.2 ml of $0.0115M$ solution
    (b) 27.5 ml of $0.223M$ solution

16. What volume of $0.375M$ solution will contain 28.9 mmole solute?

17. Excess $AgNO_3$ solution is added to 20.0 ml of $0.210M$ HBr. What weight of insoluble AgBr is precipitated?

18. Excess $BaCl_2$ solution is added to 15.0 ml of $3.00M$ H_2SO_4. What weight of insoluble $BaSO_4$ will be precipitated?

19. Gaseous H_2S was passed through 125 ml of a solution containing Cu^{2+} until all the copper had precipitated as CuS. If 7.81 g CuS was formed, what was the molarity of the Cu^{2+} in the solution?

20. From the equation $H_2SO_4 + 2KOH \rightarrow K_2SO_4 + 2H_2O$, find what volume of $0.0800M$ KOH reacts with 25.0 ml of $0.112M$ H_2SO_4.

21. Given, the reaction $2HNO_3 + Ca(OH)_2 \rightarrow Ca(NO_3)_2 + 2H_2O$: If 10.0 ml of $Ca(OH)_2$ solution react with 17.6 ml of $0.100M$ HNO_3, what is the molarity of the $Ca(OH)_2$ solution?

22. From the equation $2H_3PO_4 + 3Ba(OH)_2 \rightarrow Ba_3(PO_4)_2 + 3H_2O$, find the molarity of a solution of H_3PO_4, if 27.1 ml of it react with 20.0 ml of $0.098M$ $Ba(OH)_2$.

23. Given, the reaction $CaCO_3 + 2HNO_3 \rightarrow Ca(NO_3)_2 + H_2O + CO_2$: If 18.3 ml HNO_3 solution exactly react with 0.250 g of $CaCO_3$, what is the molarity of the acid?

24. Returning to Problem 12, calculate in each case—except (e), (h), and (j), where the reagents are not in solution—the volume of an $0.150M$ solution of the oxidizing agent that would react with 5.00 mmole of the reducing agent.

25. Returning to Problem 13, calculate in each case—except (e)—the volume of an $0.0800M$ solution of the reducing agent that would react with 2.00 mmole of the oxidizing agent.

26. (a) Complete and balance the following for acid solution:

$$MnO_4^- + HCOO^- \longrightarrow Mn^{2+} + CO_2$$

   (b) If a sample of potassium formate, HCOOK, weighing 0.7064 g, is dissolved in dilute acid and requires 28.0 ml of $KMnO_4$ solution for titration, what is the molarity of the latter?

27. (a) Complete and balance the following for acid solution:

$$MnO_4^- + H_2O_2 \longrightarrow Mn^{2+} + O_2$$

   (b) 10.0 ml of H_2O_2 solution need 18.6 ml of $0.0200M$ $KMnO_4$ for complete reaction. What is the molarity of the H_2O_2?
   (c) What volume of oxygen is evolved in this titration?

28. (a) Complete and balance the following for acid solution:

$$Cr_2O_7^{2-} + Sn^{2+} \longrightarrow Cr^{3+} + Sn^{4+}$$

   (b) A sample is known to be a mixture of SnO and SnO_2. A portion of the mixture weighing 2.00 g is dissolved in dilute acid and titrated with $0.150M$ $K_2Cr_2O_7$ solution, 18.5 ml being required. What percentage of the mixture is SnO? (*Hint*: Write equations for the reactions of SnO and SnO_2 with H^+.)

29. A solution of sodium thiosulfate, $Na_2S_2O_3$, is to be standardized by titration against iodine liberated from standard KIO_3 solution. The latter is made up by dissolving 2.03 g of KIO_3 in water and making up to 250 ml. 10.0 ml of this solution are then mixed with excess KI solution and the following reaction occurs:

$$IO_3^- + 5I^- + 6H^+ \longrightarrow 3I_2 + 3H_2O$$

   The resulting iodine is titrated with the $Na_2S_2O_3$ solution according to:

$$I_2 + 2S_2O_3^{2-} \longrightarrow 2I^- + S_4O_6^{2-}$$

   and it is found that 24.4 ml are needed. What is the molarity of the $Na_2S_2O_3$ solution?

30. If 10.0 g of iron is heated with 10.0 g of sulfur, how much FeS will be produced? What will remain unreacted?

31. 5.0 g of gaseous hydrogen is mixed with 10.0 g of gaseous oxygen and the mixture ignited.
   (a) How much water is produced?
   (b) What quantity of which gas remains after the explosion?

32. Thionyl chloride, $SOCl_2$, reacts with water according to

$$SOCl_2 + H_2O \longrightarrow SO_2 + 2HCl$$

   If 5.0 g $SOCl_2$ is mixed with 1.0 g H_2O, how much HCl will be produced?

33. Solution A contains $[Ba^{2+}] = 0.25M$ and solution B contains $[SO_4^{2-}] = 0.40M$. Assuming that barium sulfate has a negligible solubility in water, calculate what happens when 20 ml of solution A is mixed with 10 ml of solution B. What remains in solution?

34. Given that copper(II) hydroxide is insoluble, calculate the amount of precipitate obtained by mixing 150 ml of $0.120M$ copper(II) sulfate solution with 200 ml of $0.150M$ sodium hydroxide solution.

SOLUTIONS TO PROBLEMS

1. (a) $2FeO + \frac{1}{2}O_2 \longrightarrow Fe_2O_3$

   (b) $50.0 \text{ g } Fe_2O_3 = \frac{50.0}{160} = 0.313 \text{ mole}$

   $2 \times 0.313 \text{ mole } FeO = 2 \times 0.313 \times 71.8 = 44.9 \text{ g } FeO$

2. (a) $Zn + S \longrightarrow ZnS$

   $48.0 \text{ g } Zn = \frac{48.0}{65.4} = 0.734 \text{ mole } Zn \text{ reacts with}$

   $0.734 \text{ mole } S = 0.734 \times 32.0 = 23.5 \text{ g } S$
   (b) product: $0.734 \times 97.4 = 71.5 \text{ g } ZnS$

3. $1.00 \text{ g } Na = \frac{1.00}{23.0} = 0.0435 \text{ mole } Na \text{ gives}$

   $\frac{0.0435}{2} \text{ mole } H_2 = \frac{0.0435 \times 22.4}{2} = 0.487 \text{ liter } H_2$

4. $0.152 \text{ g of } Cr_2O_3 = \frac{0.152}{152} = 1.00 \times 10^{-3} \text{ mole } Cr_2O_3 \text{ came from}$

   $1.00 \times 10^{-3} \text{ mole } (NH_4)_2Cr_2O_7 \ (M.W. \ 252)$
   (a) mass of $(NH_4)_2Cr_2O_7$: $252 \times 1.00 \times 10^{-3} = 0.252 \text{ g}$
   (b) produced: $1.00 \times 10^{-3} \text{ mole } N_2 = 1.00 \times 10^{-3} \times 22.4 \text{ liter} = 22.4 \text{ ml } N_2$

5. $5.75 \text{ g } CO_2 = \frac{5.75}{44.0} = 0.131 \text{ mole } CO_2, \text{ from } 0.131 \text{ mole } C \text{ in compound}$

   mass of C: $0.131 \times 12.0 = 1.57 \text{ g}$

   $1.76 \text{ g } H_2O = \frac{1.76}{18.0} = 0.0978 \text{ mole } H_2O, \text{ from } 0.0978 \times 2 \text{ mole } H \text{ in compound}$

   mass of H: $0.0978 \times 2 \times 1.01 = 0.198 \text{ g } H$
   mass of O in compound is, by difference,

   $$2.81 - (1.57 + 0.198) = 1.04 \text{ g } O$$

   $1.04 \text{ g } O = \frac{1.04}{16.0} = 0.0651 \text{ mole}$

molar ratio:	C	H	O
	0.131	0.196	0.0651

$$\frac{0.131}{0.0651} = 2.0 \qquad \frac{0.196}{0.0651} = 3.0 \qquad \frac{0.0651}{0.0651} = 1.0$$

The simplest formula is C_2H_3O.

6. $1.73 \text{ g } CO_2 = \dfrac{1.73}{44.0} = 0.0393$ mole

There was 0.0393 mole C in the sample.

$1.06 \text{ g } H_2O = \dfrac{1.06}{18.0} = 0.0589$ mole

There was $0.0589 \times 2 = 0.118$ mole H in the sample.

molar ratio: $\dfrac{H}{C} = \dfrac{0.118}{0.0393} = 3.00$

(a)   The simplest formula is CH_3, *M.W.* $15.0n$
(b) $15.0n = 32 \pm 5$, $n = 2$, molecular formula: C_2H_6

7.   Note that there are two different experiments here.

(i) $0.0226 \text{ g } CO_2 = \dfrac{0.0226}{44.0} = 5.14 \times 10^{-4}$ mole CO_2 from 5.14×10^{-4} mole C (at wt. 12.0)

mass of C: $5.14 \times 10^{-4} \times 12.0 = 6.17 \times 10^{-3}$ g

percent C in original sample: $\dfrac{6.17 \times 10^{-3}}{0.0116} \times 100 = 53.2\%$

(ii) $0.576 \text{ g of } BaSO_4 = \dfrac{0.576}{233} = 2.47 \times 10^{-3}$ mole $BaSO_4$ from 2.47×10^{-3} mole S

mass of S: $2.47 \times 10^{-3} \times 32.1 = 7.94 \times 10^{-2}$ g

percent S in original sample: $\dfrac{7.94 \times 10^{-2}}{0.223} \times 100 = 35.6\%$

Having expressed the C and S contents on a common basis (as percent), we may combine them to find the hydrogen content
percent H: $100 - (53.2 + 35.6) = 11.2\%$

molar ratio:	C	H	S
	$\dfrac{53.2}{12.0} = 4.43$	$\dfrac{11.2}{1.01} = 11.1$	$\dfrac{35.6}{32.1} = 1.11$

The ratio is 4:10:1; the simplest formula is $C_4H_{10}S$.

8. $0.330 \text{ g } MnO_2 = \dfrac{0.330}{86.9} = 3.80 \times 10^{-3}$ mole MnO_2, from 3.80×10^{-3} mole Mn

mass of Mn $= 3.80 \times 10^{-3} \times 54.9 = 0.209$ g
mass of chlorine in sample $= 0.613 - 0.209 = 0.404$ g

mole of chlorine $= \dfrac{0.404}{35.5} = 1.14 \times 10^{-2}$ mole

$$\text{combining ratio} = \frac{\text{Cl}}{\text{Mn}} = \frac{1.14 \times 10^{-2} \text{ mole}}{3.80 \times 10^{-3} \text{ mole}} = 3.0$$

The formula is $MnCl_3$.

9. (a) $P_4 + 6Cl_2 \longrightarrow 4PCl_3$ (b) $2V + O_2 + 3F_2 \longrightarrow 2VOF_3$

10. (a) $2Cr + 3/2\,O_2 \longrightarrow Cr_2O_3(s)$
    (b) $Cr_2O_3(s) + 6H^+ \longrightarrow 2Cr^{3+} + 3H_2O$
    (c) $Cr^{3+} + 3OH^- \longrightarrow Cr(OH)_3(s)$
    (d) $C_8H_{17}SH + 27/2\,O_2 \longrightarrow 8CO_2 + 9H_2O + SO_2$
    (e) $C_8H_{17}SH + 28F_2 \longrightarrow 8CF_4 + 18HF + SF_6$
    (f) $BaCO_3(s) + 2HNO_3 \longrightarrow Ba(NO_3)_2 + H_2O + CO_2(g)$
    (g) $Ba(NO_3)_2 + Li_2SO_4 \longrightarrow BaSO_4(s) + 2LiNO_3$
    (h) $Ag_2SO_4 + BaCl_2 \longrightarrow 2AgCl(s) + BaSO_4(s)$
    (i) $Pb(CH_3COO)_2 + 2HCl \longrightarrow PbCl_2(s) + 2CH_3COOH$
    (j) $NH_3(g) + HNO_3 \longrightarrow NH_4NO_3(s)$
    (k) $NH_4NO_3(s) \longrightarrow N_2O + 2H_2O$
        (Obviously, the "other product" must be water.)
    (l) $CaCO_3(s) + H_2O + CO_2 \longrightarrow Ca(HCO_3)_2$
    (m) $GeCl_4(l) + 4AgNO_3 + 2H_2O \longrightarrow 4AgCl(s) + GeO_2(s) + 4HNO_3$

11. (f) $BaCO_3(s) + 2H^+ \longrightarrow Ba^{2+} + H_2O + CO_2$
    (g) $Ba^{2+} + SO_4^{2-} \longrightarrow BaSO_4(s)$
    (h) $2Ag^+ + SO_4^{2-} + Ba^{2+} + 2Cl^- \longrightarrow 2AgCl(s) + BaSO_4(s)$
    (i) $Pb^{2+} + 2CH_3COO^- + 2H^+ + 2Cl^- \longrightarrow PbCl_2(s) + 2CH_3COOH$
        (We include CH_3COOH because this weak acid is largely undissociated.)
    (l) $CaCO_3(s) + H_2O + CO_2 \longrightarrow Ca^{2+} + 2HCO_3^-$
    (m) $GeCl_4(l) + 4Ag^+ + 2H_2O \longrightarrow 4AgCl(s) + GeO_2(s) + 4H^+$
    The other equations of Problem 10 are either in ionic form already, or do not involve ionic substances.

12. In each case, the reduction half-reaction is given first.

    (a) $VO_3^- + 4H^+ + e^- \longrightarrow VO^{2+} + 2H_2O$ (ox. ag. VO_3^-)
    $ Al \longrightarrow Al^{3+} + 3e^-$ (red. ag. Al)
    $\overline{}$
    $3VO_3^- + Al + 12H^+ \longrightarrow 3VO^{2+} + Al^{3+} + 6H_2O$

    (b) $NO_3^- + 2H^+ + e^- \longrightarrow NO_2 + H_2O$ (ox. ag. NO_3^-)
    $ Cu \longrightarrow Cu^{2+} + 2e^-$ (red. ag. Cu)
    $\overline{}$
    $Cu + 2NO_3^- + 4H^+ \longrightarrow Cu^{2+} + 2NO_2 + 2H_2O$

    (c) $Cr_2O_7^{2-} + 14H^+ + 6e^- \longrightarrow 2Cr^{3+} + 7H_2O$ (ox. ag. $Cr_2O_7^{2-}$)
    $ SO_3^{2-} + H_2O \longrightarrow SO_4^{2-} + 2H^+ + 2e^-$ (red. ag. SO_3^{2-})
    $\overline{}$
    $Cr_2O_7^{2-} + 3SO_3^{2-} + 8H^+ \longrightarrow 2Cr^{3+} + 3SO_4^{2-} + 4H_2O$

(d) $ClO_2^- + 4H^+ + 4e^- \longrightarrow Cl^- + 2H_2O$ (ox. ag. ClO_2^-)
$\qquad\qquad\qquad\quad 2I^- \longrightarrow I_2 + 2e^-$ (red. ag. I^-)
───
$ClO_2^- + 4I^- + 4H^+ \longrightarrow Cl^- + 2I_2 + 2H_2O$

(e) $O_3 + 2H^+ + 2e^- \longrightarrow O_2 + H_2O$ (ox. ag. O_3, ozone)
$\qquad\qquad\quad Ce^{3+} \longrightarrow Ce^{4+} + e^-$ (red. ag. Ce^{3+})
───
$2Ce^{3+} + O_3 + 2H^+ \longrightarrow 2Ce^{4+} + O_2 + H_2O$

(f) $Ce^{4+} + e^- \longrightarrow Ce^{3+}$ (ox. ag. Ce^{4+})
$\qquad\quad H_2O_2 \longrightarrow O_2 + 2H^+ + 2e^-$ (red. ag. H_2O_2)
───
$2Ce^{4+} + H_2O_2 \longrightarrow 2Ce^{3+} + O_2 + 2H^+$

(g) $TcO_4^- + 8H^+ + 5e^- \longrightarrow Tc^{2+} + 4H_2O$ (ox. ag. TcO_4^-, pertechnetate)
$\qquad\qquad\qquad\quad Ti \longrightarrow Ti^{3+} + 3e^-$ (red. ag. Ti)
───
$3TcO_4^- + 5Ti + 24H^+ \longrightarrow 3Tc^{2+} + 5Ti^{3+} + 12H_2O$

(h) $Cu_2SO_4(s) + 2e^- \longrightarrow 2Cu(s) + SO_4^{2-}$ (ox. ag. Cu^+ in Cu_2SO_4)
$\qquad\quad Cu_2SO_4(s) \longrightarrow 2Cu^{2+} + SO_4^{2-} + 2e^-$ (red. ag. Cu^+ in Cu_2SO_4)
───
$Cu_2SO_4(s) \longrightarrow Cu^{2+} + Cu(s) + SO_4^{2-}$ (disproportionation of Cu^+)

(i) $MnO_4^- + 8H^+ + 5e^- \longrightarrow Mn^{2+} + 4H_2O$ (ox. ag. MnO_4^-)
$\qquad\qquad\qquad Zn(s) \longrightarrow Zn^{2+} + 2e^-$ (red. ag. Zn)
───
$2MnO_4^- + 5Zn(s) + 16H^+ \longrightarrow 2Mn^{2+} + 5Zn^{2+} + 8H_2O$

(j) $K_2Cr_2O_7(s) + 14H^+ + 6e^- \longrightarrow 2Cr^{3+} + 2K^+ + 7H_2O$ (ox. ag. $K_2Cr_2O_7$)
$\qquad\qquad\qquad 2KBr(s) \longrightarrow 2K^+ + Br_2 + 2e^-$ (red. ag. KBr)
───
$K_2Cr_2O_7(s) + 6KBr + 14H^+ \longrightarrow 2Cr^{3+} + 3Br_2 + 8K^+ + 7H_2O$

or, alternatively, solid K_2SO_4 and $Cr_2(SO_4)_3$ could be products.

13. (a) $Br_2 + 2e^- \longrightarrow 2Br^-$ (ox. ag. Br_2)
$\qquad AsO_2^- + 4OH^- \longrightarrow AsO_4^{3-} + 2H_2O + 2e^-$ (red. ag. AsO_2^-)
───
$Br_2 + AsO_2^- + 4OH^- \longrightarrow 2Br^- + AsO_4^{3-} + 2H_2O$

(b) $O_2 + 2H_2O + 4e^- \longrightarrow 4OH^-$ (ox. ag. O_2)
$\qquad Mn^{2+} + 4OH^- \longrightarrow MnO_2 + 2H_2O + 2e^-$ (red. ag. Mn^{2+})
───
$O_2 + 2Mn^{2+} + 4OH^- \longrightarrow 2MnO_2 + 2H_2O$

(c) $Cl_2 + 2e^- \longrightarrow 2Cl^-$ (ox. ag. Cl_2)
$\qquad Fe^{3+} + 8OH^- \longrightarrow FeO_4^{2-} + 4H_2O + 3e^-$ (red. ag. Fe^{3+})
───
$3Cl_2 + 2Fe^{3+} + 16OH^- \longrightarrow 2FeO_4^{2-} + 6Cl^- + 8H_2O$

(d) $Ag_2O(s) + H_2O + 2e^- \longrightarrow 2Ag(s) + 2OH^-$ (ox. ag. Ag_2O)
$\qquad HPO_3^{2-} + 3OH^- \longrightarrow PO_4^{3-} + 2H_2O + 2e^-$ (red. ag. HPO_3^{2-})
───
$Ag_2O + HPO_3^{2-} + OH^- \longrightarrow 2Ag(s) + PO_4^{3-} + H_2O$

(e)

$$H_2O_2 + 2e^- \longrightarrow 2OH^- \qquad \text{(ox. ag. } H_2O_2)$$
$$Cr(OH)_3(s) + 5OH^- \longrightarrow CrO_4^{2-} + 4H_2O + 3e^- \qquad \text{(red. ag. } Cr(OH)_3)$$

$$2Cr(OH)_3(s) + 3H_2O_2 + 4OH^- \longrightarrow 2CrO_4^{2-} + 8H_2O$$

14. (a) $\dfrac{1.28 \text{ mmole}}{17.5 \text{ ml}} = 0.0731M$ (b) $\dfrac{21.1 \text{ mmole}}{75.3 \text{ ml}} = 0.280M$

15. (a) $(12.2 \text{ ml}) \times (0.0115 \text{ mmole ml}^{-1}) = 0.140 \text{ mmole}$
    (b) $(27.5 \text{ ml}) \times (0.223 \text{ mmole ml}^{-1}) = 6.13 \text{ mmole}$

16. ml of solution $= \dfrac{28.9 \text{ mmole}}{0.375 \text{ mmole ml}^{-1}} = 77.1 \text{ ml}$

17. 20.0 ml of $0.210M$ HBr contain $20.0 \times 0.210 = 4.20$ mmole Br^-, which gives 4.20 mmole AgBr.
    mass: $(4.20 \times 10^{-3}) \times 187.8 = 0.789$ g AgBr

18. 15.0 ml of $3.00M$ H_2SO_4 contain $15.0 \times 3.00 = 45.0$ mmole SO_4^{2-}, which gives 45.0 mmole $BaSO_4$.
    mass: $(45.0 \times 10^{-3}) \times 233 = 10.5$ g $BaSO_4$

19. 7.81 g of CuS $= \dfrac{7.81}{95.5} = 0.0818$ mole $= 81.8$ mmole

    125 ml solution contained 81.8 mmole Cu^{2+}

    molarity of Cu^{2+}: $\dfrac{81.8}{125} = 0.654M$

20. 25.0 ml of $0.112M$ H_2SO_4 contain $25.0 \times 0.112 = 2.80$ mmole. From the equation, this reacts with $2 \times 2.80 = 5.60$ mmole KOH, which is contained in $5.60/0.0800 = 70.0$ ml of $0.0800M$ solution.

21. 17.6 ml of $0.100M$ HNO_3 contain $17.6 \times 0.100 = 1.76$ mmole. From the equation, this reacts with $1.76/2 = 0.880$ mmole $Ca(OH)_2$, which is contained in 10.0 ml solution.

    molarity of $Ca(OH)_2$: $\dfrac{0.880}{10.0} = 0.0880M$

22. 20.0 ml of $0.098M$ $Ba(OH)_2$ contain $20.0 \times 0.098 = 1.96$ mmole
    3 mole $Ba(OH)_2$ react with 2 mole H_3PO_4
    1 mole $Ba(OH)_2$ reacts 2/3 mole H_3PO_4
    1.96 mmole $Ba(OH)_2$ reacts $(1.96 \times 2)/3 = 1.31$ mmole H_3PO_4

    molarity of H_3PO_4: $\dfrac{1.31 \text{ mmole}}{27.1 \text{ ml}} = 0.0482M$

23. 0.250 g $CaCO_3 = \dfrac{0.250}{100} = 2.50 \times 10^{-3}$ mole $= 2.50$ mmole, which reacts with

    $2.50 \times 2 = 5.00$ mmole HNO_3

    molarity of HNO_3: $\dfrac{5.00 \text{ mmole}}{18.3 \text{ ml}} - 0.273M$

24. Suppose, in the balanced equation, that x mole oxidizing agent react with y mole reducing agent. Then x/y mole oxidizing agent react with 1 mole reducing agent. In these calculations, we are using 5.00 mmole of reducing agent, so we need $5x/y$ mmole oxidizing agent. If the concentration of oxidizing agent is $0.150M$, this is contained in $5x/0.150y$ ml.

(a) $x = 3, y = 1$ $\dfrac{5x}{0.150y} = 100$ ml

(b) $x = 2, y = 1$ $\dfrac{5x}{0.150y} = 66.7$ ml

(c) $x = 1, y = 3$ $\dfrac{5x}{0.150y} = 11.1$ ml

(d) $x = 1, y = 4$ $\dfrac{5x}{0.150y} = 8.33$ ml

(f) $x = 2, y = 1$ $\dfrac{5x}{0.150y} = 66.7$ ml

(g) $x = 3, y = 5$ $\dfrac{5x}{0.150y} = 20.0$ ml

(i) $x = 2, y = 5$ $\dfrac{5x}{0.150y} = 13.3$ ml

25. Again, suppose x mole of oxidizing agent reacts with y mole of reducing agent, then y/x mole of reducing agent reacts with 1 mole of oxidizing agent. We have, in each case, 2.00 mmole of oxidizing agent, which would react with $2.00y/x$ mmole of reducing agent. If the concentration of the latter is $0.0800M$, the volume of solution of reducing agent will be $2.00y/0.0800x$.

(a) $x = 1, y = 1$ $\dfrac{2.00y}{0.0800x} = 25.0$ ml

(b) $x = 1, y = 2$ $\dfrac{2.00y}{0.0800x} = 50.0$ ml

(c) $x = 3, y = 2$ $\dfrac{2.00y}{0.0800x} = 16.7$ ml

(d) $x = 1, y = 1$ $\dfrac{2.00y}{0.0800x} = 25.0$ ml

26. (a)
$$MnO_4^- + 8H^+ + 5e^- \longrightarrow Mn^{2+} + 4H_2O$$
$$HCOO^- \longrightarrow CO_2 + H^+ + 2e^-$$

$$2MnO_4^- + 5HCOO^- + 11H^+ \longrightarrow 2Mn^{2+} + 5CO_2 + 8H_2O$$

(b) 0.7064 g HCOOK $= \dfrac{0.7064}{84.11} = 8.399 \times 10^{-3}$ mole $= 8.399$ mmole

from the equation, 5 mole $HCOO^-$ react with 2 mole MnO_4^- so
1 mole $HCOO^-$ reacts with 2/5 mole MnO_4^-, and
8.399 mmole reacts with $2/5 \times 8.399 = 3.359$ mmole MnO_4^-

molarity of $KMnO_4$: $\dfrac{3.359 \text{ mmole}}{28.0 \text{ ml}} = 0.120M$

(*Note*: Both $KMnO_4$ and HCOOK are, of course, completely ionized, the K^+ ions taking no part in the reaction.)

27. (a)
$$MnO_4^- + 8H^+ + 5e^- \longrightarrow Mn^{2+} + 4H_2O$$
$$H_2O_2 \longrightarrow O_2 + 2H^+ + 2e^-$$

$$2MnO_4^- + 5H_2O_2 + 6H^+ \longrightarrow 2Mn^{2+} + 5O_2 + 8H_2O$$

(b) 18.6 ml of $0.0200M$ $KMnO_4$ contain $18.6 \times 0.0200 = 0.372$ mmole MnO_4^-
from the equation, 2 mole MnO_4^- react with 5 mole H_2O_2, so
1 mole MnO_4^- reacts with 5/2 mole H_2O_2
0.372 mmole MnO_4^- reacts with $5/2 \times 0.372 = 0.930$ mmole H_2O_2

$$\text{molarity of } H_2O_2: \quad \frac{0.930}{10.0} = 0.0930M$$

(c) 0.930 mmole O_2 is evolved
(Obviously, 1 mmole gas has a volume 22.4 ml.)
volume of O_2: $0.930 \times 22.4 = 20.8$ ml (at STP)

28. (a)
$$Cr_2O_7^{2-} + 14H^+ + 6e^- \longrightarrow 2Cr^{3+} + 7H_2O$$
$$Sn^{2+} \longrightarrow Sn^{4+} + 2e^-$$

$$Cr_2O_7^{2-} + 3Sn^{2+} + 14H^+ \longrightarrow 2Cr^{3+} + 3Sn^{4+} + 7H_2O$$

(b) 18.5 ml $0.150M$ $K_2Cr_2O_7$ contain $18.5 \times 0.150 = 2.78$ mmole $Cr_2O_7^{2-}$
from the equation, this reacts with $2.78 \times 3 = 8.34$ mmole Sn^{2+}
The reaction of SnO with acid is

$$SnO + 2H^+ \longrightarrow Sn^{2+} + H_2O$$

8.34 mmole Sn^{2+} come from 8.34 mmole SnO with mass of $(8.34 \times 10^{-3}) \times 135 = 1.12$ g
This represents $(1.12/2.00) \times 100 = 56.0\%$ of original sample of 2.00 g of mixed oxides.
(The SnO_2 dissolves to give Sn^{4+}, which does not react with $Cr_2O_7^{2-}$.)

29. The KIO_3 solution is 2.03 g in 250 ml, so its molarity is

$$\frac{2.03 \text{ g}}{(0.250 \text{ liter}) \times (214 \text{ g mole}^{-1})} = 0.0379M$$

10.0 ml of this solution contain $10.0 \times 0.0379 = 0.379$ mmole IO_3^-
From first equation, this gives $0.379 \times 3 = 1.14$ mmole I_2.
From second equation, this reacts with $1.14 \times 2 = 2.28$ mmole $S_2O_3^{2-}$.

$$\text{molarity of } S_2O_3^{2-}: \quad \frac{2.28}{24.4} = 0.0934M$$

30. 10.0 g Fe is $\dfrac{10.0}{55.8} = 0.179$ mole   10.0 g S is $\dfrac{10.0}{32.1} = 0.313$ mole

They react in a 1:1 molar ratio: $Fe + S \longrightarrow FeS$
Iron is limiting.
produced: 0.179 mole FeS with mass $0.179 \times 87.8 = 15.7$ g
remaining: $0.313 - 0.179 = 0.134$ mole S, with mass $0.134 \times 32.1 = 4.30$ g

31. (a) 5.0 g of $H_2 = \dfrac{5.0}{2.02} = 2.48$ mole   10.0 g of $O_2 = \dfrac{10.0}{32.0} = 0.313$ mole

2 mole H_2 react with 1 mole O_2
H_2 present is more than twice O_2
O_2 is limiting; 1 mole O_2 gives 2 mole H_2O
0.313 mole O_2 gives $0.313 \times 2 = 0.626$ mole H_2O
     mass of H_2O: $0.626 \times 18.0 = 11.3$ g

(b) H_2 remaining: $2.48 - (2 \times 0.313) = 1.85$ mole
     mass of H_2 remaining: $1.85 \times 2.02 = 3.8$ g

32. 5.0 g $SOCl_2 = \dfrac{5.0}{119} = 0.042$ mole 1.0 g $H_2O = \dfrac{1.0}{18} = 0.056$ mole

They react in a 1:1 molar ratio.
$SOCl_2$ is limiting.
     produced: $0.042 \times 2 = 0.084$ mole HCl, with mass $0.084 \times 36.5 = 3.1$ g

33.   20 ml of a solution with $[Ba^{2+}] = 0.25M$
mmole of $Ba^{2+} = 20 \times 0.25 = 5.0$ mmole
10 ml of a solution with $[SO_4^{2-}] = 0.40M$
mmole of $SO_4^{2-} = 10 \times 0.40 = 4.0$ mmole
They react in a 1:1 molar ratio.
SO_4^{2-} is limiting; 4.0 mmole $BaSO_4$ is produced
$5.0 - 4.0 = 1.0$ mmole Ba^{2+} remains
total solution volume $= 20 + 10 = 30$ ml

$$[Ba^{2+}] = \dfrac{1.0 \text{ mmole}}{30 \text{ ml}} = 0.033M$$

34.   150 ml $0.120M$ $CuSO_4$ contain $150 \times 0.120 = 18.0$ mmole Cu^{2+}
200 ml $0.150M$ NaOH contain $200 \times 0.150 = 30.0$ mmole OH^-

$$Cu^{2+} + 2OH^- \longrightarrow Cu(OH)_2(s)$$

They react in a 1:2 ratio, but we have *less* than twice as much OH^- as Cu^{2+}.
OH^- is limiting
$30.0/2 = 15.0$ mmole $Cu(OH)_2$ precipitates
$18.0 - 15.0 = 3.0$ mmole Cu^{2+} remains
total solution volume $= 150 + 200 = 350$ ml

$$[Cu^{2+}] = \dfrac{3.0 \text{ mmole}}{350 \text{ ml}} = 0.0086M$$

Chapter 4
Enthalpy and Heat of Reaction

When you have eliminated the impossible, whatever remains, however improbable, *must be the truth.* THE SIGN OF FOUR

In the last chapter, we considered several aspects of a chemical reaction, and, in particular, the information that can be obtained from a balanced chemical equation. In this chapter, we shall extend this study to consider the energy change involved in a chemical reaction. We must start, however, by defining the units in which energy changes are measured.

4.1 Units

For many years chemists used the calorie as a unit for measuring energy. This was convenient, because a calorie was originally defined as "the heat absorbed in raising the temperature of one gram of water through one celsius degree." Since the energy evolved in a reaction often appears in the form of heat, which may conveniently be measured by its effect on the temperature of a known mass of water, the calculation of heat of reaction was straightforward. However, the calorie has the disadvantage that it must be related to other energy units by conversion factors that are not exact powers of ten. For example, by definition,

$$1 \text{ calorie} = 4.184 \text{ joule}$$

With the introduction of SI units, the joule will eventually displace the calorie as a measure of energy for chemists, because the joule may be easily related to other units of energy, force, acceleration, etc., without the use of clumsy conversion factors which have to be experimentally determined or defined.

$$1 \text{ joule} = 1 \text{ newton} \times 1 \text{ meter}$$

$$\begin{matrix} \text{(unit of} & \times & \text{(unit of} \\ \text{force)} & & \text{length)} \end{matrix}$$

$$1 \text{ newton} = 1 \text{ kilogram} \times 1 \text{ meter} \times (1 \text{ second})^{-2}$$

and, going from SI units to cgs units we have

$$1 \text{ joule} = 10^7 \text{ erg}$$

All these are exact relationships, and the transformations and calculations based on them are very straightforward.

We will therefore use the joule (abbreviation J) as the unit for measuring energy throughout this book. For chemical reactions on the mole scale, the joule is rather too small a unit, so we often use the kilojoule (kJ), which is, of course, 1000 joule.

Since the calorie will remain in wide use among chemists for some years to come, additional problems involving conversion between calories and joules are included, so that students will become familiar with both sets of units. A convenient concept to start with is that of heat capacity (otherwise known as specific heat), which is a measure of the amount of heat required to change the temperature of a substance. This will obviously depend on the quantity of substance that is present, which we may express on a 1 g or a 1 mole basis, and on the temperature interval through which it is heated or cooled.

Using SI units, temperature intervals are measured in kelvin (K). The kelvin, or absolute, scale of temperature is related to the Celsius scale (°C), but starts at the absolute zero of temperature (-273.15°C). It follows that 0°C is 273.15 K and the conversion is

$$(\text{temperature in K}) = (\text{temperature in °C}) + 273.15$$

For most purposes, we can forget the last two figures and remember the difference as 273. (*Note*: We will speak of temperature in "kelvin" rather than "degrees kelvin.")

It should be clear that a *difference* in temperature will be the same in Celsius or in kelvin. Thus, the difference between 20°C and 30°C is 10 Celsius degrees. If we add 273 to each, we have 293 K and 303 K, but the difference is *numerically* the same at 10 K. We can therefore measure differences in temperature in kelvin by subtracting temperatures measured in °C.

The molar heat capacity of a substance is the heat necessary to raise the temperature of one mole of that substance through one kelvin.

If we measure the heat energy in joules, the units of molar heat capacity will be J K^{-1} mole^{-1}. From this value, we can find the amount of heat necessary to raise other amounts through various temperature differences by simple proportion.

Example 4.1 The heat capacity of water is 1.00 cal K^{-1} g^{-1}.

(a) What is its molar heat capacity in J K^{-1} mole^{-1}?
(b) How much heat energy would have to be supplied to raise the temperature of 100 g of water from 25°C to 100°C?

SOLUTION

(a) Knowing that 1 cal = 4.184 J, the heat capacity is

$$(1.00 \text{ cal K}^{-1} \text{ g}^{-1}) \times (4.18 \text{ J cal}^{-1}) = 4.18 \text{ J K}^{-1} \text{ g}^{-1}$$

The molar mass is 18.0 g, so, on a molar basis, the heat capacity is

$$(4.18 \text{ J K}^{-1} \text{ g}^{-1}) \times (18.0 \text{ g mole}^{-1}) = 75.2 \text{ J K}^{-1} \text{ mole}^{-1}$$

(b) 100 g of water is $100/18.0 = 5.56$ mole, so the heat necessary to raise its temperature through 75 K ($= 100 - 25$) will be

$$(5.56 \text{ mole}) \times (75 \text{ K}) \times (75.2 \text{ J K}^{-1} \text{ mole}^{-1}) = 3.14 \times 10^4 \text{ J} = 31.4 \text{ kJ}$$

4.2   Enthalpy Changes

When a chemical change occurs, there are several ways of evaluating the energy change accompanying the process. In this chapter, we are going to be concerned only with the change in the *enthalpy* of the system. Enthalpy is a difficult quantity to define in a few words, but it can be described as the "heat content" of the system. As a change occurs, heat energy goes into the system or flows out of it, and we define the enthalpy change as follows:

> The enthalpy change accompanying some process occurring in a system is equal to the amount of heat added to the system when that process occurs at constant volume.

We give the symbol H to enthalpy, and a change in enthalpy is called ΔH. The enthalpy change, ΔH, accompanying any process may be positive or negative, the significance of a negative value being that heat energy comes *out* of the system when the change occurs.

A point that often confuses students is the sign attached to ΔH values to signify increase or decrease in enthalpy content. Remember, we are always looking at the change from the viewpoint of the system, *not* from the viewpoint of the observer. When a substance burns, for example, much energy is evolved as heat, and an observer will see this energy heating up the surroundings, the observer's hand, etc. But this heat evolved has been *lost* by the system, so the enthalpy content of the system has *decreased*, and the value of ΔH will be *negative*.

Conversely, in a process such as melting, we have to put heat energy into the system. The enthalpy content of the system will increase and ΔH will be *positive*.

A process in which ΔH is negative is said to be *exothermic* (heat going out), while a process in which ΔH is positive is said to be *endothermic* (heat going in). ΔH is conveniently measured on a molar basis.

Example 4.2   The heat of fusion of ice is 79.7 cal g^{-1} (i.e., this much heat must be supplied to melt 1 g of ice). Calculate ΔH in going from (a) ice to water (b) water to ice, at 0°C throughout. Express your answer in units of both kcal mole^{-1} and kJ mole^{-1}.

SOLUTION   First we put the heat of fusion on a molar basis by multiplying by the molar mass.

$$(79.7 \text{ cal g}^{-1}) \times (18.0 \text{ g mole}^{-1}) = 1.44 \times 10^3 \text{ cal mole}^{-1}$$

(a) When we go from ice to water, this amount of energy must be supplied, so the enthalpy content of the system *increases* and ΔH is positive

$$\Delta H = + 1.44 \times 10^3 \text{ cal mole}^{-1} = 1.44 \text{ kcal mole}^{-1}$$
$$= 1.44 \times 4.18 = 6.00 \text{ kJ mole}^{-1}$$

(b) When water freezes to form ice, the quantity of energy is the same, but heat is now being lost by the system and ΔH is negative:

$$\Delta H = - 1.44 \text{ kcal mole}^{-1} = -6.00 \text{ kJ mole}^{-1}$$

This example illustrates a few important points. First, we notice that the changes from water to ice were occurring at a constant temperature (0°C). This ensures that all the heat gained or lost was associated with the change from liquid to solid state, rather than being used in changing the temperature of the ice or the water. When we consider the enthalpy changes accompanying a chemical reaction, we shall find it very useful to treat these as occurring at constant temperature, so that the ΔH value corresponds to the chemical change, rather than to the heating up of a reactant or product.

The second point is something that we should expect on commonsense principles, but it deserves to be stated explicitly.

For any physical or chemical process where there is an enthalpy change, the reverse process will be accompanied by the reverse enthalpy change.

Turning to a chemical reaction to illustrate this, we find

$$C + O_2 \longrightarrow CO_2 \qquad \Delta H = -394 \text{ kJ}$$

$$CO_2 \longrightarrow C + O_2 \qquad \Delta H = +394 \text{ kJ}$$

Note that we simply write "kJ" or "kcal" for enthalpy quantities that are written beside a specific equation. There is no need to specify "per mole," because the quantity of energy is associated with the specific molar quantities referred to in the equation. These will depend on the coefficients of the various reactants in the equation, which will often differ. For example, if we write

$$2Cr + \frac{3}{2}O_2 \longrightarrow Cr_2O_3 \qquad \Delta H = -1128 \text{ kJ}$$

the quantities corresponding to this ΔH value are 2 mole Cr, 3/2 mole O_2, and 1 mole Cr_2O_3, so we could not express ΔH in "kJ per mole" without specifying which substance was used as a basis for counting moles.

We often refer to a ΔH value as "enthalpy of reaction" or "heat of reaction," defined as follows:

Enthalpy of reaction is the enthalpy change accompanying a reaction represented by a balanced chemical equation.

The equation enables us to relate ΔH to the number of moles of each reactant. If we should multiply all the coefficients in the equation by 2, 3, ..., then the ΔH value is multiplied by the same factor, e.g.,

$$4Cr + 3O_2 \longrightarrow 2Cr_2O_3 \qquad \Delta H = -2256 \text{ kJ}$$

It is important to specify the phase (solid, liquid, or gaseous) of each substance taking part in the reaction. The above equation, for example, should be written

$$2Cr(s) + \frac{3}{2}O_2(g) \longrightarrow Cr_2O_3(s) \qquad \Delta H = -1128 \text{ kJ}$$

indicating that we are burning solid Cr in gaseous O_2 to give solid Cr_2O_3. For carbon, where there are two solid forms, we should be even more specific and write

$$C(graphite) + O_2(g) \longrightarrow CO_2(g) \qquad \Delta H = -394 \text{ kJ}$$

indicating that we are measuring the enthalpy change when solid carbon (graphite) burns in gaseous oxygen to give gaseous carbon dioxide. These are the forms in which we would normally encounter these substances (unless we're lucky enough to come on the carbon in the form of diamonds, in which case we'd hardly want to burn it!). For other substances, we have to be more careful, since more than one phase may be common. Take water, for example:

$$H_2(g) + \frac{1}{2}O_2(g) \longrightarrow H_2O(l) \qquad \Delta H = -286 \text{ kJ}$$

$$H_2(g) + \frac{1}{2}O_2(g) \longrightarrow H_2O(g) \qquad \Delta H = -242 \text{ kJ}$$

The enthalpy change for the combustion of hydrogen will be different, depending on whether we end up with liquid water or gaseous water. What does the difference between the two ΔH values (44 kJ) represent? Clearly, it should be the enthalpy difference between liquid and gaseous water; in other words, the heat of vaporization of water. Experiment verifies that this is the case. In other words, if we subtract one equation from the other, the ΔH value for the resulting equation will be found by subtracting the corresponding ΔH values

$$H_2O(g) \longrightarrow H_2O(l) \qquad \Delta H = -286 - (-242) = -44 \text{ kJ}$$

Similarly, if we add two or more chemical equations, ΔH for the overall process will be the sum of the ΔH values for each component equation.

This is a completely general result, which is embodied in *Hess' law of heat summation*:

The enthalpy change for a given reaction is a constant (under given conditions of temperature and pressure), and is independent of the number of steps by which the reaction occurs.

Common sense tells us that this must be so. If, for any reaction, we could find two routes with different ΔH values, we could get an endless supply of energy out by repeatedly going "up" the smaller ΔH route and coming "down" the larger ΔH route. Obviously, this is an impossibility.

Hess' law is very useful in relating the enthalpy changes occurring in a sequence of reactions.

Example 4.3   Using the following data

			ΔH, kJ
(i)	$Sn(s) + Cl_2(g) \longrightarrow SnCl_2(s)$		-350
(ii)	$SnCl_2(s) + Cl_2(g) \longrightarrow SnCl_4(l)$		-195

evaluate ΔH for the reaction

(iii)	$Sn(s) + 2Cl_2(g) \longrightarrow SnCl_4(l)$

SOLUTION   If we add together equations (i) and (ii), canceling out the $SnCl_2$ which appears on both sides, the result is equation (iii). The ΔH value for (iii) is therefore the sum of (i) and (ii), namely $-350 - 195 = -545$ kJ.

Example 4.4   By considering the following reactions

ΔH, kJ

(i) $\frac{1}{2}N_2(g) + \frac{1}{2}O_2(g) \longrightarrow NO(g)$ $+90$

(ii) $NO(g) + \frac{1}{2}O_2(g) \longrightarrow NO_2(g)$ -56

(iii) $2NO_2(g) \longrightarrow N_2O_4(g)$ -58

evaluate ΔH for the reaction

(iv) $N_2(g) + 2O_2(g) \longrightarrow N_2O_4(g)$

SOLUTION   Our reaction sequence proceeds through the oxides NO and NO_2 to the final product N_2O_4. Since we need to have two N atoms in the final molecule, we must double equations (i) and (ii) throughout, including their ΔH values.

ΔH, kJ

$2 \times$ (i) $N_2(g) + O_2(g) \longrightarrow 2NO(g)$ $+180$

$2 \times$ (ii) $2NO(g) + O_2(g) \longrightarrow 2NO_2(g)$ -112

(iii) $2NO_2(g) \longrightarrow N_2O_4(g)$ -58

When we add the three equations, the intermediate oxides NO and NO_2 cancel out to give equation (iv), for which we know that

$$\Delta H = +180 - 112 - 58 = +10 \text{ kJ}$$

Example 4.5   Carbon burns in oxygen to give two oxides:

ΔH, kJ

(i) $C(s) + \frac{1}{2}O_2(g) \longrightarrow CO(g)$ -111

(ii) $C(s) + O_2(g) \longrightarrow CO_2(g)$ -394

What will be the value of ΔH for the reaction

(iii) $CO(g) + \frac{1}{2}O_2(g) \longrightarrow CO_2(g)$

SOLUTION   Equation (iii) is the result of subtracting (i) from (ii), in other words *reversing* (i), including the sign of ΔH, and adding it to (ii).

ΔH, kJ

(i) reversed: $CO(g) \longrightarrow C(s) + \frac{1}{2}O_2(g)$ $+111$

(ii) $C(s) + O_2(g) \longrightarrow CO_2(g)$ -394

Adding these gives (iii), so the value of ΔH for (iii) is $+111 - 394 = -283$ kJ.

4.3   Measurement of ΔH: The Calorimeter

It is often convenient to measure ΔH experimentally in an apparatus called a calorimeter. This contains a known amount of water, which absorbs the heat evolved in the reaction process. Knowing the specific heat of water (for our purposes, we may take this as 1.00 cal g^{-1} K^{-1}) we calculate the heat evolved from the rise in temperature. However, some of the heat will go toward warming up the calorimeter itself, so we have to know how much to allow for this. This is done by measuring the heat capacity of the calorimeter, that is, the amount of heat absorbed by the calorimeter per unit of temperature change. In practice, certain other corrections have to be made, particularly in reactions where there is a change in gaseous volumes, but an example will make the principle clear.

Example 4.6   In an experiment to determine ΔH for the reaction

$$S(s) + O_2(g) \longrightarrow SO_2(g)$$

4.80 g of sulfur is put into a calorimeter with an excess of gaseous oxygen. The calorimeter has a heat capacity of 190 cal K^{-1} and contains 924 g of water. During the reaction, the temperature rises from 21.80°C to 31.35°C. Calculate ΔH, in units of kcal and kJ.

SOLUTION   The heat absorbed by the system is divided between that absorbed by the water and that absorbed by the calorimeter.

the temperature rise is $31.35 - 21.80 = 9.55$ K

heat absorbed by water $= 924 \times 1.00 \times 9.55 = 8.82 \times 10^3$ cal

(the heat capacity of water being 1.00 cal g^{-1} K^{-1})

heat absorbed by calorimeter $= 190 \times 9.55 = 1.81 \times 10^3$ cal

total heat absorbed $= (8.82 \times 10^3) + (1.81 \times 10^3) = 1.06 \times 10^4$ cal.

This is equal to the total heat given out in the reaction. Converting to a mole basis:

$$4.80 \text{ g S} = \frac{4.80}{32.0} = 0.150 \text{ mole}$$

0.150 mole S evolves 1.06×10^4 cal

$$1 \text{ mole S evolves } \frac{1.06 \times 10^4}{0.150} = 7.07 \times 10^4 \text{ cal} = 70.7 \text{ kcal}$$

This heat was evolved during the reaction (remember that the temperature of the water *increased*), so ΔH is negative:

$$\Delta H = -70.7 \text{ kcal} = -70.7 \times 4.184 = -296 \text{ kJ}$$

4.4   Heat of Formation and Heat of Combustion

For any compound, there is one quantity of fundamental importance; the enthalpy change when it is formed from its component elements. We usually call this the heat of formation of a compound, defined as follows:

The standard heat (enthalpy) of formation of a compound is the enthalpy change when a mole of the compound is formed from the required quantities of its component elements *in their standard states.*

Standard states for elements and compounds are simply the states in which they exist under the standard conditions of 298 K and one atmosphere pressure. For most elements, this means as solids; but two are liquid (Hg and Br_2) and a few are found as diatomic gaseous molecules (H_2, N_2, O_2, F_2, Cl_2) or monatomic gases (He, Ne, Ar, Kr, Xe, Rn).

The standard heat of formation of an element *in its standard state* is, by definition, zero.

We indicate enthalpy changes occurring between substances in their standard states with the symbol ΔH^0, while standard heats of formation are denoted as ΔH_f^0. We have already met several ΔH_f^0 values in this chapter, i.e.,

$$CO_2(g): \quad \Delta H_f^0 = -394 \text{ kJ mole}^{-1} \qquad SO_2(g): \quad \Delta H_f^0 = -296 \text{ kJ mole}^{-1}$$

$$H_2O(l): \quad \Delta H_f^0 = -286 \text{ kJ mole}^{-1} \qquad H_2O(g): \quad \Delta H_f^0 = -242 \text{ kJ mole}^{-1}$$

All of these compounds may be easily prepared by burning the appropriate element in oxygen. This combustion process is often of great importance in determining ΔH values, and we define its accompanying ΔH value as follows:

The heat (enthalpy) of combustion of an element or compound is the enthalpy change when a mole of that compound reacts with oxygen to form products specified by a balanced equation.

The last phrase is an important qualification, because we must specify what is produced by burning the substance. Carbon, for example, can burn to give either CO or CO_2, while hydrogen-containing substances can give H_2O in either gaseous or liquid form. Both the nature and phase of the products will affect the value of ΔH for the combustion process.

We see that, for compounds such as CO_2, SO_2, and H_2O, the heat of formation will be the same as the heat of combustion of the appropriate element. By simply putting the element in our calorimeter and burning it in oxygen, we may find ΔH_f^0 values for many oxides.

But how are we to find the heats of formation of compounds other than oxides? We might fill our calorimeter with chlorine or fluorine instead of oxygen, put in the appropriate element, and let it burn to give the chloride or fluoride; but what about a compound such as methane, CH_4, which cannot be made directly from the elements carbon and hydrogen?

We can get around this problem quite easily for methane by measuring its heat of combustion and making use of Hess' law. We find by experiment that

$$CH_4(g) + 2O_2(g) \longrightarrow CO_2(g) + 2H_2O(l) \qquad \Delta H^0 = -890 \text{ kJ}$$

From the results of other experiments, we know that

$$C(s) + O_2(g) \longrightarrow CO_2(g) \qquad \Delta H^0 = -394 \text{ kJ}$$

$$2H_2(g) + O_2(g) \longrightarrow 2H_2O(l) \qquad \Delta H^0 = -(2 \times 286) = -572 \text{ kJ}$$

So we know the heats of formation of three out of the four substances involved in the combustion of CH_4 (ΔH_f^0 for O_2 is zero). If we had started with the three elements, C, H_2, and O_2, we could have gone to CO_2 and $2H_2O$ in two ways, either through CH_4 or by direct combustion. Let's illustrate this diagrammatically:

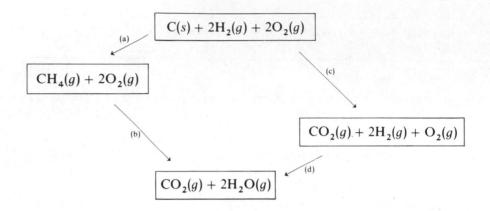

The highest enthalpy state of the system is at the top of the diagram, and the lowest enthalpy state at the bottom.

Hess' law tells us that the overall ΔH value going from the elements to the final products must be the same, whether we take the route $A + B$ or the route $C + D$. We can write

$$\Delta H_A + \Delta H_B = \Delta H_C + \Delta H_D$$

or

$$\Delta H_B = (\Delta H_C + \Delta H_D) - \Delta H_A$$

Looking at the significance of each ΔH value, we see that:

ΔH_B is ΔH^0 for the combustion of 1 mole CH_4 (-890 kJ)
ΔH_D is ΔH_f^0 for 2 mole $H_2O(g)$ (-572 kJ)
ΔH_C is ΔH_f^0 for 1 mole $CO_2(g)$ (-394 kJ)
ΔH_A is ΔH_f^0 for 1 mole CH_4 (unknown)

So, with respect to the equation

$$CH_4(g) + 2O_2(g) \longrightarrow CO_2(g) + 2H_2O(g)$$

we have shown that the heat of reaction, ΔH_B, is equal to the total heat of formation of the products (CO_2 and H_2O) *less* the total heat of formation of the reactants (CH_4 and O_2). This is a general result which is very useful to us.

For any reaction, the heat (enthalpy) of the reaction as written is equal to the total heat of formation of the products *minus* the total heat of formation of the reactants.

As usual with a relationship, we can use this to find any one quantity if all the others are known. Thus, we could find a heat of combustion from the various heats of formation, or the other way round.

Returning to the CH_4 combustion reaction, we have

$$\Delta H \text{ comb.}(CH_4) = \Delta H_f^0(CO_2) + 2\Delta H_f^0(H_2O) - \Delta H_f^0(CH_4)$$

$$-890 = -572 - 394 - \Delta H_f^0(CH_4)$$

$$\Delta H_f^0(CH_4) = 890 - 572 - 394 = -76 \text{ kJ}$$

ΔH_f^0 for $CH_4(g)$ is -76 kJ mole^{-1}, and the formation reaction is exothermic.

This principle will be illustrated by some similar examples. Always be careful in these calculations to get the signs right, especially when subtracting negative quantities.

Example 4.7 Calculate ΔH^0 for the reaction

$$2FeO(s) + \frac{1}{2}O_2(g) \longrightarrow Fe_2O_3(s)$$

given that the ΔH_f^0 values are: for FeO, -266 kJ mole^{-1}; for Fe_2O_3, -821 kJ mole^{-1}. State whether the reaction is exothermic or endothermic.

SOLUTION   Remember that ΔH_f^0 is zero for $O_2(g)$.

for the product, Fe_2O_3: $\Delta H_f^0 = -821$ kJ
for the reactants: $\Delta H_f^0 = -(2 \times 266) = -532$ kJ
for the reaction as written: $\Delta H^0 = -821 - (-532) = -289$ kJ

This is evolved as heat; the reaction is exothermic.

Example 4.8 When one mole of benzene, $C_6H_6(l)$, is burned to $CO_2(g)$ and $H_2O(l)$, 782 kcal of heat is evolved. Calculate ΔH_f^0 of benzene in units of kJ mole^{-1} and kcal mole^{-1}, given ΔH_f^0 for CO_2 and H_2O to be -394 and -286 kJ mole^{-1}, respectively. Comment on your answer.

SOLUTION   First write a balanced equation:

$$C_6H_6(l) + \frac{15}{2}O_2(g) \longrightarrow 6CO_2(g) + 3H_2O(l)$$

We know that 782 kcal is evolved in this process (exothermic) so

$$\Delta H^0 = -782 \text{ kcal} = -(782 \times 4.18) = -3.27 \times 10^3 \text{ kJ}$$

Total ΔH_f^0 of the products is

$$-(6 \times 394) - (3 \times 286) = -2364 - 858 = -3.22 \times 10^3 \text{ kJ}$$

if $\Delta \dot{H}_f^0$ for benzene is x kJ mole^{-1}, we can write

$$\text{enthalpy of reaction} = -3.27 \times 10^3 = -(3.22 \times 10^3) - x$$

$$x = (3.27 \times 10^3) - (3.22 \times 10^3) = +50 \text{ kJ mole}^{-1}$$

in kcal, $\Delta H_f^0 = \dfrac{50}{4.18} = +12$ kcal mole^{-1}

We note that, for benzene, ΔH_f^0 is positive, while for methane the enthalpy of formation is negative. This means that the formation reaction for making benzene from its elements is endothermic. Benzene has a higher enthalpy content than carbon and hydrogen in the elementary state. (Such compounds are sometimes said to be "endothermic" compounds.)

Example 4.9   The values for the heat of combustion of the two elementary forms of carbon (giving CO_2) are found to be

$$\text{graphite: } -7.845 \text{ kcal g}^{-1} \qquad \text{diamond: } -7.881 \text{ kcal g}^{-1}$$

Assuming that graphite is the standard state of carbon, what is ΔH_f^0 for diamond? Which form of solid carbon has the lower enthalpy content?

SOLUTION   Converting the heats of combustion to a molar basis:

graphite: $(-7.845 \text{ kcal g}^{-1}) \times (4.184 \text{ J cal}^{-1}) \times (12.01 \text{ g mole}^{-1}) = -394.2 \text{ kJ mole}^{-1}$
diamond: $(-7.881) \times 4.184 \times 12.01 = -396.0 \text{ kJ mole}^{-1}$

So we can write ΔH for the combustion reactions.

$$\text{C (graphite)} + O_2 \longrightarrow CO_2(g) \qquad \Delta H^0 = -394.2 \text{ kJ}$$

$$\text{C (diamond)} + O_2 \longrightarrow CO_2(g) \qquad \Delta H^0 = -396.0 \text{ kJ}$$

If we reverse the second equation and add it to the first, CO_2 and O_2 cancel out, giving

$$\text{C (graphite)} \longrightarrow \text{C (diamond)}$$

$$\Delta H^0 = -394.2 + 396.0 = +1.8 \text{ kJ}$$

Taking ΔH_f^0 for graphite (the standard state) as zero, this means that ΔH_f^0 for diamond is $+1.8$ kJ mole^{-1}.

Graphite has the lower enthalpy content. Whenever an element exists in more than one form (a phenomenon known as *allotropy*), that with the lowest enthalpy content is usually taken as the standard state for that element.

4.5   Bond Energy

One of the simplest of all chemical reactions is the formation of a diatomic molecule from two atoms, for example,

$$\text{H}(g) + \text{H}(g) \longrightarrow \text{H}_2(g)$$

This is a highly exothermic reaction, with $\Delta H^0 = -432$ kJ. The reverse reaction, in which a mole of gaseous molecular hydrogen splits up into two moles of gaseous atomic hydrogen, will obviously be endothermic.

$$\text{H}_2(g) \longrightarrow 2\text{H}(g) \qquad \Delta H = +432 \text{ kJ}$$

The energy necessary to break the bond in the H_2 molecule is called the *bond dissociation energy*, usually shortened to bond energy, for this molecule. It is a generally useful and important concept. Obviously there is only one covalent bond in each H_2 molecule, but other molecules may have several bonds, each with an associated bond energy.

The bond energy for a given bond in a molecule is ΔH for the process in which that bond is broken in the molecule, the process occurring in the gas phase, and the resulting fragments separated by an infinite distance.

Obviously, in the case of a diatomic molecule, the fragments would be two atoms. Thus, we speak of the bond energies of H_2, N_2, F_2, etc., as 432 kJ, 940 kJ, 155 kJ, respectively, where this is the energy to separate a mole of molecules into two moles of atoms. However, the fragments could be stable molecules, as in the dissociation of N_2O_4 to give $2NO_2$

$$N_2O_4(g) \longrightarrow 2NO_2(g) \qquad \Delta H = +58 \text{ kJ}$$

For this reaction, ΔH represents the bond energy of the N—N bond present in N_2O_4.

We must be very careful to distinguish between bond energies and standard heats of formation. Bond energies represent the difference in enthalpy between a molecule and the *atoms* formed by breaking one or more bonds. Heats of formation, on the other hand, refer to the enthalpy difference between the compound and the elements in their standard states, so a bond energy is *not* the negative of ΔH_f^0. In HCl, for example, the two processes are:

bond breaking: $\quad HCl(g) \longrightarrow H(g) + Cl(g)$

$\Delta H_f^0: \quad \dfrac{1}{2} H_2(g) + \dfrac{1}{2} Cl_2(g) \longrightarrow HCl(g)$

The difference between these is brought out in the following example, in which we apply Hess' law to find the bond energy.

Example 4.10 Calculate the bond energy in the HCl molecule from the following information:

ΔH_f^0 for $HCl(g)$: $\quad -92.4 \text{ kJ mole}^{-1}$
bond energies: $\quad H_2$, 432 kJ mole^{-1}; Cl_2, 238 kJ mole^{-1}

SOLUTION   The bond energies enable us to work out ΔH_f^0 values for H atoms and Cl atoms:

ΔH_f^0 for H: $\quad \dfrac{432}{2} = 216 \text{ kJ mole}^{-1}$

ΔH_f^0 for Cl: $\quad \dfrac{238}{2} = 119 \text{ kJ mole}^{-1}$

The bond energy of HCl is ΔH for the reaction

$$HCl(g) \longrightarrow H(g) + Cl(g)$$

Knowing ΔH_f^0 for each substance here, we can find ΔH for the reaction:

$$\Delta H = (216 + 119) - (-92.4) = 427 \text{ kJ mole}^{-1}$$

The bond energy in HCl is 427 kJ mole^{-1}

Note carefully the relationship between the heat of formation of an *atom* and the bond energy of a diatomic *molecule*. When we split up a mole of H_2, Cl_2, O_2, etc. we make *two* moles of atoms:

$$H_2(g) \longrightarrow 2H(g) \qquad \Delta H = 432 \text{ kJ}$$

Since $H_2(g)$ represents the standard state for the element, we have made H (*atomic* hydrogen) from the element in its standard state, so this represents ΔH_f^0 for atomic hydrogen.

However, we made *two* moles of atoms, so the value of ΔH^0_f per mole of H atoms will be 1/2 this value, or 216 kJ mole^{-1}. This is generally true:

$$\Delta H^0_f \text{ per mole atoms} = \frac{\text{bond energy per mole of diatomic molecules}}{2}$$

You should also realize that ΔH^0_f cannot be zero for atomic hydrogen, because the element is not in its standard state. The same applies to most elements when they are in a monatomic, gaseous, form. For example, the standard state of sodium, Na(s), is the solid metal, so ΔH^0_f for Na(g) corresponds to the process

$$\text{Na}(s) \longrightarrow \text{Na}(g) \qquad \Delta H = +109 \text{ kJ}$$

We call this the heat of sublimation of sodium, since it is the transformation of the solid to the gas. It is very similar to a bond energy, since the attractive forces holding together sodium atoms in the solid have to be overcome.

In the case of a compound with more than two atoms, we often speak of *average* bond energy for the bonds present. Thus, in talking of the energy of the C—H bonds in CH_4, we would calculate 1/4 of the energy necessary to break all four of the bonds to give C + 4H. If we broke the C—H bonds one at a time, we would find different values for the four steps, since each corresponds to different reactions as the H atom is removed in succession from CH_4, CH_3, CH_2, and CH.

Example 4.11   Calculate the average bond energy in methane, CH_4, from the following data:

	kJ mole^{-1}
heat of formation of $CH_4(g)$	−76
heat of formation of C(g)	+736
bond energy of $H_2(g)$	+432

SOLUTION   The bond-breaking reaction is

$$CH_4(g) \longrightarrow C(g) + 4H(g)$$

To give 4 mole H(g), we must break up 2 mole H_2:

$$2H_2(g) \longrightarrow 4H(g) \qquad \Delta H = 2 \times 432 = 864 \text{ kJ}$$

We have now all the ΔH^0_f values needed, so we can calculate for the bond-breaking reaction

$$\Delta H = 736 + 864 - (-76) = 1676 \text{ kJ}$$

This corresponds to the breaking of *four* bonds, so the average C—H bond energy is

$$\frac{1676}{4} = 419 \text{ kJ mole}^{-1} \quad \text{per bond}$$

Example 4.12   Calculate the average bond energy in $SnCl_4(l)$, for which $\Delta H^0_f = -534$ kJ mole^{-1}. The bond energy of Cl_2 is 238 kJ mole^{-1} and the heats of vaporization are Sn(s), 301 kJ mole^{-1}; and $SnCl_4(l)$, 133 kJ mole^{-1}.

SOLUTION   The key reaction in this problem is

$$SnCl_4(g) \longrightarrow Sn(g) + 4Cl(g)$$

We can calculate ΔH_f^0 for each of the three substances in this equation from the given data:

$$SnCl_4(l) \longrightarrow SnCl_4(g) \qquad \Delta H = +133 \text{ kJ} \quad \text{(vaporization)}$$

Knowing ΔH_f^0 for $SnCl_4(l)$ is -534 kJ mole^{-1}, ΔH_f^0 for $SnCl_4(g)$ must be $+133 - 534 = -401$ kJ mole^{-1}. Similarly for gaseous Sn,

$$Sn(s) \longrightarrow Sn(g) \qquad \Delta H = +301 \text{ kJ}$$

Since ΔH_f^0 for $Sn(s)$ is, by definition, zero (standard state), ΔH_f^0 for $Sn(g)$ is $+301$ kJ mole^{-1}.

$$2Cl_2(g) \longrightarrow 4Cl(g) \qquad \Delta H = 2 \times 238 = 476 \text{ kJ}$$

(This is the heat of formation of 4 mole atomic chlorine.) Combining the three ΔH_f values gives ΔH for the first equation

$$\Delta H = 301 + 476 - (-401) = 1178 \text{ kJ}$$

This is the energy required to break *four* Sn—Cl bonds, so the average bond energy is

$$\frac{1178}{4} = 295 \text{ kJ mole}^{-1} \quad \text{per bond}$$

The three examples above demonstrate the general approach to finding bond energies. The important starting point is to write down the equation corresponding to the bond-breaking reaction in question. Then work out heats of formation for *all* the reactants and products in the equation, so that you can find ΔH for the bond-breaking reaction from the difference between the total ΔH_f of the products and the total ΔH_f of the reactants. Be careful with the signs!

4.6 Binding Energy

Finally, in this chapter, we consider a slightly different topic; the energy change accompanying the formation of an atomic nucleus from its component protons and neutrons. We use the term "binding energy" to describe this (be careful not to confuse this with "bond energy," discussed in the previous section!).

The molar binding energy of an isotope is the energy evolved when a mole of that isotope is formed from its component protons, neutrons, and electrons.

This formation process can be regarded as taking place in two stages, the formation of the nucleus from protons and neutrons, then the addition of electrons to the nucleus to form a neutral atom. Both of these processes are exothermic; but the energy change in the first step is by far the larger, so we can, without appreciable error, regard the binding energy as corresponding solely to the formation of the nucleus.

We cannot carry out a nuclear formation reaction in the laboratory, but we can work out the binding energy by an indirect route. If we add up the total mass of the component parts needed to make a mole of a particular isotope, then compare this with the molar mass of the actual isotope, we shall find a small discrepancy. The molar mass of the isotope is always a little *less* than the total of the components, and the difference (called the *mass defect*) has been converted into the binding energy holding the nucleus together. We know the equivalence of mass and energy by Einstein's equation, so we can easily find the binding energy.

The helium isotope $_2^4$He forms a simple example. Each atom contains 2 protons, 2 neutrons, and 2 electrons, so to make a mole of ^{4}He we need 2 moles of each of these fundamental particles, whose atomic weights are all accurately known.

$$2 \text{ mole protons, atomic weight } 1.007277 = 2.014554 \text{ g}$$
$$2 \text{ mole neutrons, atomic weight } 1.008665 = 2.017330$$
$$2 \text{ mole electrons, atomic weight } 0.000549 = 0.001098$$

$$\text{total mass} = 4.032982 \text{ g}$$

(*Note*: Although the mass of the electrons is very small, it is included in the above calculation because we are carrying 7 significant figures.)

Now let's compare this with the experimentally determined molar mass of $_2^4$He. We find this to be 4.00260 g (very close to that of natural helium, which is essentially all ^{4}He). The difference between the total mass of the components and the molar mass of ^{4}He is obviously appreciable, amounting to 0.75%.

$$\text{mass defect} = 4.032982 - 4.00260 = 0.03038 \text{ g mole}^{-1}$$

To work out the amount of binding energy equivalent to this mass, we use Einstein's well-known equation $E = mc^2$. If m is in kg and c, the velocity of light, is in m s^{-1}, then E comes out in joules. In this case $m = 3.04 \times 10^{-5}$ kg and $c = 3.00 \times 10^8$ m s^{-1}.

$$E = (3.04 \times 10^{-5} \text{ kg mole}^{-1}) \times (3.00 \times 10^8 \text{ m s}^{-1})^2$$
$$= 2.74 \times 10^{12} \text{ J mole}^{-1} = 2.74 \times 10^9 \text{ kJ mole}^{-1}$$

This is sufficient energy to heat 6.5 million kg of water from 0° to 100°C!

Comparing this with the energies encountered in ordinary chemical process we see the immensely greater magnitude of enthalpy changes associated with nuclear changes.

The reaction responsible for this energy is solely the formation of the helium nucleus. Although the electrons' mass was included for accuracy in the calculation, the energy liberated by addition of two electrons to the nucleus to form the atom was only a few hundred kJ, quite negligible here.

We may calculate one further quantity in this calculation, the binding energy per nucleon. The word "nucleon" is a term including all particles in the nucleus, both protons and neutrons, so the total number of nucleons present in a nucleus will obviously be equal to the mass number of the isotope in question. The binding energy on a per nucleon basis gives a better indication of overall nuclear stability than the total binding energy for an atom.

For $_2^4$He, there are in total four nucleons in the nucleus, so the binding energy per nucleon is

$$\frac{2.74 \times 10^9}{4} = 6.9 \times 10^8 \text{ kJ mole}^{-1} \text{ per nucleon}$$

Example 4.13 Calculate the mass defect, total binding energy, and binding energy per nucleon for $_{29}^{63}$Cu, which has atomic weight 62.9298.

SOLUTION   Each nucleus in this isotope contains 29 protons and $(63 - 29) = 34$ neutrons, so we work out the total mass of the particles taken separately:

$$
\begin{array}{rl}
\text{29 mole of protons at } 1.007277 = & 29.21103 \text{ g} \\
\text{34 mole of neutrons at } 1.008665 = & 34.29461 \\
\text{29 mole of electrons at } 0.000549 = & 0.01592 \\
\hline
\text{total mass} = & 63.52156 \text{ g} \\
\text{subtract molar mass of } ^{63}\text{Cu} & -62.9298 \\
\hline
\text{mass defect} = & 0.5918 \text{ g} \\
= & 5.92 \times 10^{-4} \text{ kg mole}^{-1}
\end{array}
$$

The binding energy is therefore

$$5.92 \times 10^{-4} \times (3.00 \times 10^8)^2 = 5.33 \times 10^{13} \text{ J mole}^{-1}$$

and the binding energy per nucleon is

$$\frac{5.33 \times 10^{13}}{63} = 8.45 \times 10^{11} \text{ joule} = 8.45 \times 10^8 \text{ kJ mole}^{-1} \text{ per nucleon}$$

The binding energy *per nucleon* is similar in magnitude to that found previously for the helium nucleus.

KEY WORDS

joule, kilojoule	calorie
enthalpy, ΔH	Hess' law
calorimeter	heat of formation
heat of combustion	bond energy
mass defect	binding energy

STUDY QUESTIONS

1. What is the connection between the joule and the calorie? Which is the larger quantity of energy?

2. How is ΔH for a reaction related to ΔH for the reverse reaction?

3. What do "endothermic" and "exothermic" mean, in terms of ΔH values for reaction?

4. How is ΔH related to the heats of formation of the various substances in a reaction?

5. What is ΔH_f^0 for an element in its standard state?

6. How is bond energy related to ΔH_f^0 for atoms of elements commonly found as diatomic molecules?

7. Why do we talk of "average" bond energy when a molecule contains more than one identical bond?

8. Does the magnitude of nuclear binding energies tell us anything about the energy needed to break up a nucleus? Why does a nucleus not break up in an ordinary chemical reaction?

9. Why did we take helium as a simple example of a mass defect and binding energy calculation, rather than hydrogen?

10. Could we use mass-defect calculations to measure ΔH in chemical reactions? What practical difficulties would be involved?

PROBLEMS

Take 1 calorie = 4.18 joule.

1. The heats of fusion of the following substances are given in the units of cal g^{-1}. In each case, work out ΔH for the process solid $\rightarrow$ liquid in units of kcal mole^{-1} and kJ mole^{-1}
   (a) Al_2Cl_6, 63.4 cal g^{-1} (b) NH_4NO_3, 16.2 cal g^{-1}

2. Given the following ΔH^0 values,

		ΔH, kJ
(i)	$2Cu(s) + \frac{1}{2}O_2(g) \longrightarrow Cu_2O(s)$	-167
(ii)	$Cu(s) + \frac{1}{2}O_2(g) \longrightarrow CuO(s)$	-155

   calculate ΔH^0 for the reaction

   (iii) $Cu_2O(s) + \frac{1}{2}O_2(g) \longrightarrow 2CuO(s)$

3. A calorimeter has a heat capacity of 350 cal K^{-1} and contains 880 g of water. 3.00 g of acetic acid, $CH_3COOH(l)$, is burned in excess oxygen in the calorimeter to give $CO_2(g)$ and $H_2O(l)$, and the temperature is observed to rise from 20.1°C to 28.6°C. Calculate the molar heat of combustion of acetic acid in kcal and kJ.

4. A calorimeter has a heat capacity of 320 cal K^{-1} and contains 718 g of water. Initially the apparatus is at 22.0°C and contains 1.50 g of ethyl alcohol, C_2H_5OH, in oxygen. If the heat of combustion of ethyl alcohol is -328 kcal mole^{-1}, what will be the temperature of the calorimeter when combustion is complete?

5. A calorimeter has a heat capacity of 290 cal K^{-1} and contains 600 g of water. A strip of solid magnesium of mass 1.60 g, in excess oxygen inside the calorimeter, is ignited electrically and burns completely to MgO. If the temperature increases from 21.8°C to 32.4°C, what is ΔH_f^0 for MgO? Give your answer in kJ mole^{-1}.

6. Calculate ΔH^0 for the reaction

$$2Mn_3O_4(s) + \frac{1}{2}O_2(g) \longrightarrow 3Mn_2O_3(s)$$

given that the ΔH_f^0 values are Mn_3O_4, -1385; Mn_2O_3, -970 kJ mole^{-1}.

7. Calculate ΔH_f^0 for $SO_3(g)$, given that for the reaction

$$SO_2(g) + \frac{1}{2}O_2(g) \longrightarrow SO_3(g)$$

ΔH^0 is -100 kJ. The ΔH_f^0 for $SO_2(g)$ is -296 kJ mole^{-1}.

8. Calculate ΔH^0 for the reaction

$$TiF_4(s) + 2H_2O(l) \longrightarrow TiO_2(s) + 4HF(l)$$

given that the ΔH_f^0 values are $TiF_4(s)$, -1550; $TiO_2(s)$, -890; $H_2O(l)$, -286; $HF(l)$, -298 kJ mole^{-1}.

In problems 9–11, take ΔH_f^0 for $CO_2(g)$ as -394 kJ mole^{-1} and ΔH_f^0 for $H_2O(l)$ as -286 kJ mole^{-1}.

9. Use the result of problem 3 to calculate ΔH_f^0 for acetic acid in kJ mole^{-1}.

10. From the following values of heats of combustion of organic molecules (to $CO_2(g)$, and $H_2O(l)$) calculate their heats of formation in kJ mole^{-1}. State in each case whether formation of the compound is an endothermic or exothermic process.
    (a) ethylene, $C_2H_4(g)$, -1410 kJ mole^{-1}
    (b) propane, $C_3H_8(g)$, -2220 kJ mole^{-1}
    (c) oxalic acid, $(COOH)_2(s)$, -252 kJ mole^{-1}

11. Given the following heats of formation, calculate the heat of combustion of each compound in kJ mole^{-1}. The products in each case will be $CO_2(g)$ and $H_2O(l)$.
    (a) ethane, $C_2H_6(g)$, -107 kJ mole^{-1}
    (b) acetylene, $C_2H_2(g)$, $+227$ kJ mole^{-1}
    (c) benzoic acid, $C_7H_6O_2(s)$, -390 kJ mole^{-1}

12. Phosphorus exists in two solid forms known as red phosphorus and white phosphorus. Reaction of either of these with gaseous chlorine gives PCl_3. Calculate ΔH for the process

$$P(red) \longrightarrow P(white)$$

given that ΔH_f^0 for PCl_3 is -306 kJ mole^{-1} from P(white) but -288 kJ mole^{-1} from P(red).

13. Calculate the bond energy in $HF(g)$ from the following data:
    heat of formation of $HF(g)$ -268 kJ mole^{-1}
    bond energy of hydrogen 432 kJ mole^{-1}
    bond energy of fluorine 155 kJ mole^{-1}

14. Calculate the average bond energy in $H_2O(g)$ from the following data:

    heat of formation of $H_2O(g)$ -242 kJ mole^{-1}
    bond energy of hydrogen 432 kJ mole^{-1}
    bond energy of oxygen 493 kJ mole^{-1}

15. Calculate the average bond energy in $CCl_4(g)$ from the following data:

    heat of formation of $CCl_4(g)$ -106 kJ mole^{-1}
    heat of formation of $C(g)$ $+736$ kJ mole^{-1}
    bond energy of chlorine 238 kJ mole^{-1}

16. Calculate the bond energy in $NaCl(g)$ from the following data:

    heat of formation of $NaCl(g)$ -182 kJ mole^{-1}
    heat of formation of $Na(g)$ $+109$ kJ mole^{-1}
    bond energy of chlorine 238 kJ mole^{-1}

17. Calculate on a molar basis the mass defect, total binding energy, and binding energy per nucleon, for the following:

    (a) $^{12}_{6}C$ (b) $^{35}_{17}Cl$, at. wt 34.9689   (c) $^{113}_{49}In$, at. wt 112.9043

    Atomic weights of the fundamental particles are proton, 1.007277; neutron, 1.008665; and electron, 0.000549.

18. (a) The atomic weight of ^1H is 1.007825. How does this compare with the sum of the atomic weights of its two component particles?

    (b) Given that ΔH for the ionization of hydrogen, $H(g) \rightarrow H^+(g) + e^-(g)$, is $+1305$ kJ mole^{-1}, use Einstein's equation to calculate the change in mass accompanying this process. Express your answer in g mole^{-1} and parts per million. Would you expect this to be detectable by experiment?

SOLUTIONS TO PROBLEMS

1. (a) Al_2Cl_6 has a molar mass of 267g

    $$\begin{aligned}\Delta H &= (63.4 \text{ cal g}^{-1}) \times (267 \text{ g mole}^{-1}) \\ &= 1.69 \times 10^4 \text{ cal mole}^{-1} = 16.9 \text{ kcal mole}^{-1} \\ &= (16.9 \text{ kcal mole}^{-1}) \times (4.18 \text{ J cal}^{-1}) = 70.8 \text{ kJ mole}^{-1}\end{aligned}$$

    (b) NH_4NO_3 has a molar mass of 80.0 g

    $$\begin{aligned}\Delta H &= (16.2 \text{ cal g}^{-1}) \times (80.0 \text{ g mole}^{-1}) \\ &= 1.30 \times 10^3 \text{ cal mole}^{-1} = 1.30 \text{ kcal mole}^{-1} \\ &= (1.30 \text{ kcal mole}^{-1}) \times (4.18 \text{ J cal}^{-1}) = 5.42 \text{ kJ mole}^{-1}\end{aligned}$$

2. To obtain equation (iii), we have to add together the reverse of (i) and twice (ii):

    $$\Delta H^0, \text{ kJ}$$

    reverse (i): $Cu_2O \longrightarrow 2Cu + \frac{1}{2}O_2$ $+167$

    twice (ii): $2Cu + O_2 \longrightarrow 2CuO$ -310

    add, canceling out 2Cu and $1/2O_2$, to get ΔH^0 for (iii): $+167 - 310 = -143$ kJ

3. temperature rise: $28.6 - 20.1 = 8.5$ K
    heat absorbed by calorimeter + water: $8.5(350 + 880) = 1.05 \times 10^4$ cal
    molecular weight of acetic acid is 60.0
    3.00 g $CH_3COOH = 3.00/60.0 = 5.00 \times 10^{-2}$ mole
    5.00×10^{-2} mole gives 1.05×10^4 cal on combustion
    1 mole gives $(1.05 \times 10^4)/(5.00 \times 10^{-2}) = 2.1 \times 10^5$ cal

    $$\text{heat of combustion} = -2.1 \times 10^2 \text{ kcal mole}^{-1} \text{ or}$$
    $$= (-2.1 \times 10^2) \times 4.18 = -8.8 \times 10^2 \text{ kJ mole}^{-1}$$

    (negative sign because heat is evolved)

4. 1.50 g ethyl alcohol $= 1.50/46.0 = 3.26 \times 10^{-2}$ mole
    heat evolved on combustion will be $3.26 \times 10^{-2} \times 328 = 10.7$ kcal
    If the rise in temperature is ΔT K, then the heat absorbed by calorimeter + water is

    $$\Delta T(320 + 718) \text{ cal}$$

    which must equal the heat evolved in reaction:

    $$\Delta T(320 + 718) = 10.7 \times 10^3$$

    $$\Delta T = \frac{10.7 \times 10^3}{1038} = 10.3 \text{ K}$$

    final temperature: $22.0 + 10.3 = 32.3°C$

5. temperature rise: $32.4 - 21.8 = 10.6$ K
    heat absorbed: $10.6 (290 + 600) = 9.43 \times 10^3$ cal
    1.60 g Mg $= 1.60/24.3 = 6.58 \times 10^{-2}$ mole

    $$Mg(s) + \frac{1}{2} O_2(g) \longrightarrow MgO(s)$$

    we produced 6.58×10^{-2} mole MgO

    $$\Delta H_f^0(MgO): \quad \frac{-9.43 \text{ kcal}}{6.58 \times 10^{-2} \text{ mole}} = -1.43 \times 10^2 \text{ kcal mole}^{-1}$$

    $$= (-1.43 \times 10^2) \times 4.18 = -599 \text{ kJ mole}^{-1}$$

    (Again, ΔH is negative because heat is evolved.)

In the following examples, we use the same approach in each case; make up a balanced equation, and use $\Delta H^0(\text{reaction}) = \text{total } \Delta H_f^0(\text{products}) - \text{total } \Delta H_f^0(\text{reactants})$.

6. $\Delta H^0 = 3 \times (-970) - 2 \times (-1385) = -140$ kJ

7. $-100 = \Delta H_f^0(SO_3) - (-296)$ so $\Delta H_f^0(SO_3) = -100 - 296 = -396$ kJ mole^{-1}

8. $\Delta H^0 = -890 + 4 \times (-298) - [-1550 + 2 \times (-286)] = +40$ kJ

9. The combustion reaction is $CH_3COOH + 2O_2 \longrightarrow 2CO_2 + 2H_2O$.
    Knowing ΔH^0 for this is -880 kJ, gives $-880 = 2 \times (-394) + 2 \times (-286) - \Delta H_f^0(CH_3COOH)$.

    $$\Delta H_f^0(CH_3COOH) = 880 - 788 - 572 = -480 \text{ kJ mole}^{-1}$$

10. (a) $C_2H_4 + 3O_2 \longrightarrow 2CO_2 + 2H_2O$

        $-1410 = 2 \times (-394) + 2 \times (-286) - \Delta H_f^0(C_2H_4)$

           $\Delta H_f^0(C_2H_4) = 1410 - 788 - 572 = +50$ kJ mole^{-1} (endothermic process)

  (b) $C_3H_8 + 5O_2 \longrightarrow 3CO_2 + 4H_2O$

        $-2220 = 3 \times (-394) + 4 \times (-286) - \Delta H_f^0(C_3H_8)$

           $\Delta H_f^0(C_3H_8) = 2220 - 1182 - 1144 = -106$ kJ mole^{-1} (exothermic process)

  (c) $(COOH)_2 + 1/2O_2 \longrightarrow 2CO_2 + H_2O$

        $-252 = 2 \times (-394) - 286 - \Delta H_f^0[(COOH)_2]$

           $\Delta H_f^0[(COOH)_2] = 252 - 788 - 286 = -822$ kJ mole^{-1} (exothermic process)

11. (a) $C_2H_6 + 7/2O_2 \longrightarrow 2CO_2 + 3H_2O$

        ΔH^0: $2 \times (-394) + 3 \times (-286) - (-107) = -1.54 \times 10^3$ kJ

  (b) $C_2H_2 + 5/2O_2 \longrightarrow 2CO_2 + H_2O$

        ΔH^0: $2 \times (-394) - 286 - 227 = -1.30 \times 10^3$ kJ

  (c) $C_7H_6O_2 + 15/2O_2 \longrightarrow 7CO_2 + 3H_2O$

        ΔH^0: $7 \times (-394) + 3 \times (-286) - (-390) = -3.23 \times 10^3$ kJ

12. Write the equation for each reaction:

                                                                ΔH^0, kJ

$$P(\text{red}) + 3/2Cl_2(g) \longrightarrow PCl_3 \qquad -288$$

$$P(\text{white}) + 3/2Cl_2(g) \longrightarrow PCl_3 \qquad -306$$

Reversing the second equation and adding, when Cl_2 and PCl_3 cancel out, gives

$$P(\text{red}) \longrightarrow P(\text{white})$$

  ΔH^0: $-288 + 306 = +18$ kJ

(White phosphorus has the higher enthalpy content.)

13. The bond-breaking reaction is $HF(g) \longrightarrow H(g) + F(g)$.

$$\Delta H = \frac{432}{2} + \frac{155}{2} - (-268) = 562 \text{ kJ}$$

bond energy in HF is 562 kJ mole^{-1}

14. The bond-breaking reaction is $H_2O(g) \longrightarrow 2H(g) + O(g)$.

$$\Delta H = 432 + \frac{493}{2} - (-242) = 920 \text{ kJ}$$

which corresponds to breaking *two* O—H bonds

    average bond energy: $\dfrac{920}{2} = 460$ kJ mole^{-1}

(Note that the bond-breaking gives 2H (atomic H) rather than H_2.)

15.   The bond-breaking reaction is $CCl_4(g) \longrightarrow C(g) + 4Cl(g)$.

$\Delta H = 736 + 2 \times 238 - (-106) = 1318$ kJ

which corresponds to breaking *four* C—Cl bonds

average bond energy: $\dfrac{1318}{4} = 330$ kJ mole^{-1}

(Note that we have to break up two moles of Cl_2 to give the required 4Cl(atomic).)

16.   The bond-breaking reaction is $NaCl(g) \longrightarrow Na(g) + Cl(g)$.

$\Delta H = 109 + \dfrac{238}{2} - (-182) = 410$ kJ

bond energy in NaCl(g):   410 kJ mole^{-1}

17.   (a)   Particles going to make up a mole of $^{12}_{6}C$:
6 mole protons at 1.007277 = 6.043662 g
6 mole neutrons at 1.008665 = 6.051990
6 mole electrons at 0.000549 = 0.003294
————————
total = 12.098946

Subtract the molar mass of $^{12}_{6}C$. (What's that you say? It's not given in the problem data? Perhaps that's because there's something a bit special about $^{12}_{6}C$ and its molar mass of 12 g exactly, by definition. Remember now?)
mass defect: $12.098946 - 12 = 0.098946$ g mole$^{-1} = 9.89 \times 10^{-5}$ kg mole^{-1}
binding energy: $(9.89 \times 10^{-5}) \times (3.00 \times 10^8)^2 = 8.91 \times 10^{12}$ J mole^{-1}
$= 8.91 \times 10^9$ kJ mole

the nucleus contains 12 nucleons

binding energy per nucleon: $\dfrac{8.91 \times 10^9}{12} = 7.42 \times 10^8$ kJ mole^{-1}

(b)   Particles making up a mole of $^{35}_{17}Cl$:

17 mole protons at 1.007277   =   17.123709 g
18 mole neutrons at 1.008665 =   18.155970
17 mole electrons at 0.000549 =   0.009333
————————
total =   35.289012
subtract molar mass of ^{35}Cl -34.9689
————————
mass defect: 0.3201 g mole$^{-1} = 3.20 \times 10^{-4}$ kg mole^{-1}

binding energy: $(3.20 \times 10^{-4}) \times (3.00 \times 10^8)^2 = 2.88 \times 10^{13}$ J mole^{-1}
$= 2.88 \times 10^{10}$ kJ mole^{-1}

binding energy per nucleon: $\dfrac{2.88 \times 10^{10}}{35} = 8.23 \times 10^8$ kJ mole^{-1}

(c) Particles making up a mole of $^{113}_{49}In$:

$$
\begin{array}{lcr}
\text{49 mole protons at } 1.007277 & = & 49.356573 \text{ g} \\
\text{64 mole neutrons at } 1.008665 & = & 64.554560 \\
\text{49 mole electrons at } 0.000549 & = & 0.026901 \\
\end{array}
$$

$$
\begin{array}{lr}
\text{total} = & 113.938034 \\
\text{subtract molar mass of } ^{113}_{49}In & -112.9043 \\
\end{array}
$$

mass defect: 1.0337 g. $mole^{-1} = 1.03 \times 10^{-3}$ kg $mole^{-1}$

binding energy: $(1.03 \times 10^{-3}) \times (3.00 \times 10^8)^2 = 9.30 \times 10^{13}$ J $mole^{-1}$

$$= 9.30 \times 10^{10} \text{ kJ mole}^{-1}$$

binding energy per nucleon: $\dfrac{9.30 \times 10^{10}}{113} = 8.23 \times 10^8$ kJ $mole^{-1}$

18. (a) Total mass of 1 mole protons + 1 mole electrons would be $1.007277 + 0.000549 = 1.007826$ g. Within the accuracy of the given data, this is not significantly different from the mass of 1 mole atomic hydrogen, ^1H

(b) According to Einstein's equation, the mass change corresponding to ΔH of 1305 kJ is

$$\frac{1305 \times 10^3}{(3.00 \times 10^8)^2} = 1.45 \times 10^{-11} \text{ kg} = 1.45 \times 10^{-8} \text{ g}$$

In a mass of about 1 g, this amounts to about 0.015 parts per million, whereas the masses of ^1H and the fundamental particles were only given above to about 1 ppm. Thus, we would not expect the mass change for the given reaction to be detectable by experiment unless we could considerably improve the accuracy of our mass determinations.

Chapter 5

The Electronic Structure of Atoms: The Periodic Table

How simple the explanation may be of an affair which appears at first sight to be almost inexplicable. THE ADVENTURE OF THE NOBLE BACHELOR

The temptation to form premature theories based on insufficient data is the bane of our profession. THE VALLEY OF FEAR

In this chapter, we consider the ways in which electrons are arranged in atoms. Once we have established these arrangements, we will see certain patterns and similarities occurring between groups of atoms, and it is on the basis of these resemblances that we put groups of atoms together to form the periodic table. Looking at this the other way round, we know that the electronic arrangement of an atom, which largely determines its chemical properties, may be deduced very easily from its position in the periodic table.

It is essential, therefore, to appreciate the underlying rules that govern the way in which electrons are fitted into atoms. Once we have this knowledge, we can build up the various *shells* of electrons as they are found in any atom. As a start we will define basic terms.

A neutral atom is an atom in which the number of electrons around the nucleus is equal to the total positive charge on the nucleus. It therefore has no net charge. Since the positive charge on the nucleus is equal to the atomic number of the element, we have only to look at the atomic number of the element to know how many electrons we have to fit into the neutral atom.

A positive ion is an atom which has *lost* one or more electrons. If the charge on the ion is $n+$, obviously n electrons have been lost, so the total number of electrons remaining is (atomic number) $- n$. A positive ion is often called a *cation*, because it is attracted to the *cathode* in electrolysis. Examples: Na^+, K^+, Ba^{2+}, Al^{3+}.

A negative ion is an atom that has *gained* one or more electrons. For a charge of $n-$, n electrons have been gained, so the total number of electrons present is (atomic number) $+ n$. Such an ion is called an *anion*, because it is attracted to the *anode* in electrolysis. Examples: Cl^-, I^-, O^{2-}, S^{2-}.

Ground state is the lowest possible energy state for any atom, ion, or molecule. There are an indefinitely large number of different ways in which any group of electrons may be arranged, but *one* of these arrangements (or sometimes two or three equivalent arrangements) will be of *lower* energy than all the rest. This arrangement is the ground state.

It is most important to realize that the ground state is *not* the only possible way in which the electrons may be arranged. It is simply the lowest energy condition of many possible arrangements. Tables of "electronic configurations of the elements," included in any textbook, are always ground-state configuration; and when we speak of a "configuration" without any qualification, it is the ground-state arrangement which we mean.

Excited state is any possible arrangement of electrons *other than* the ground-state configuration. Obviously, an excited-state configuration must be of higher energy than the ground state.

Note particularly that the concepts of ground state and excited state may be applied equally well to an *ion* as to a neutral atom. There is a well-established ground-state configuration of Na^+, F^-, etc., which may be considered quite independently of the configuration of the corresponding neutral atom.

Although Na^+ is in a higher energy state (higher enthalpy) than Na, we do not regard Na^+ as an excited state of Na, because they have different numbers of electrons. Every different chemical species has its own ground state and corresponding set of excited states.

5.1   Quantum Numbers

To describe any and each electron in an atom or ion, we use four *quantum numbers*. When we have specified the value of each quantum number, we have completely described an electron. However, we will not encounter the same set of numbers twice in one atom, because *Pauli's exclusion principle* states that, in any atom, no two electrons may have the same set of values for all four quantum numbers.

As we build up atoms (i.e., as we move through the periodic table), we must bear Pauli's principle in mind, because it will determine the set of quantum numbers we give to each successive electron as it is added.

The four quantum numbers have the following names and values:

1. *The principal quantum number*, symbol n, may have any positive, integral value, *not* including zero. That is, it may be 1, 2, 3, 4, 5, ..., etc. In practice, the value of n never goes above 7 in the ground-state configuration of any known atom.

We use the word "shell" to designate a group of energy levels having the same value of n. It is very important to realize that this is simply a bookkeeping device to keep track of the electrons. It does *not* imply that all atoms in a shell are in the same place, or have the same energy, etc.

We use capital letters to denote these groups of electrons in an atom:

The K shell means "those electrons having $n = 1$"
The L shell means "those electrons having $n = 2$"
The M shell means "those electrons having $n = 3$"
The N shell means "those electrons having $n = 4$"

Once we have fixed the value of n for an electron, we have placed limitations on the values that the other quantum numbers may have, according to the rules that follow.

2. *The orbital (azimuthal) quantum number*, symbol ℓ, may have a positive integral value,

including zero, up to a *maximum* value of $n - 1$. This at once limits the values of ℓ very much. As you may easily see:

If $n = 1$, ℓ may have the value 0 only.
If $n = 2$, ℓ may be 0 or 1 only.
If $n = 3$, ℓ may be 0, 1, or 2 only.
If $n = 4$, ℓ may be 0, 1, 2, or 3 only.

We have "names" for an electron or a group of electrons having a particular value for the orbital quantum number. They are the letters s, p, d, and f, which derive from ancient spectroscopic terms.

If $\ell = 0$, we call them s electrons.
If $\ell = 1$, we call them p electrons.
If $\ell = 2$, we call them d electrons.
If $\ell = 3$, we call them f electrons.

(*Note*: In practice, values of ℓ greater than 3 are of no importance.) The different values of the orbital quantum number enable us to divide up each shell into a number of *subshells* (or *orbitals*). The number of subshells in each shell increases as the value of n for the shell increases; in fact, it is equal to the value of n.

We denote these subshells by combining the *numerical* value of n with the *letter* denoting the value of ℓ. Thus, $2p$ denotes a subshell with $n = 2$ and $\ell = 1$; $4d$ denotes a subshell with $n = 4$ and $\ell = 2$; etc. Let's list the subshells present in the first four shells:

K shell ($n = 1$) contains the $1s$ subshell only.
L shell ($n = 2$) contains the $2s$ and $2p$ subshells.
M shell ($n = 3$) contains the $3s$, $3p$, and $3d$ subshells.
N shell ($n = 4$) contains the $4s$, $4p$, $4d$, and $4f$ subshells.

As you can see, there are some combinations that are not allowed by the rules we have introduced. Thus, we could not talk of a $1p$, $2d$, $3f$ subshell. Such a combination is *impossible*.

3. *The magnetic quantum number*, symbol m_ℓ, has a value determined by the value of the previous quantum number, ℓ. It may have any integral value (including zero) between $-\ell$ and $+\ell$. Since the values of ℓ found in practice are 0, 1, 2, and 3, the practical possible values of m_ℓ are:

If $\ell = 0$ (s electrons), m_ℓ may be 0 only.
If $\ell = 1$ (p electrons), m_ℓ may be -1, 0, or $+1$.
If $\ell = 2$ (d electrons), m_ℓ may be -2, -1, 0, $+1$, or $+2$.
If $\ell = 3$ (f electrons), m_ℓ may be -3, -2, -1, 0, $+1$, $+2$ or $+3$.

We see that the number of different possible values of m_ℓ in each case is $(2\ell + 1)$.

4. *The spin quantum number*, symbol m_s, has a value independent of the values that an electron may have for its other three quantum numbers: m_s may be either $+1/2$ or $-1/2$.

Having set out the rules for the possible values of the quantum numbers, we can work out the maximum number that can be accommodated in each shell and subshell in an atom. All we have to do is work out the number of *different* allowable combinations.

Let's start with the subshells. In an s-type subshell, we know that $\ell = 0$, so the only possible value of m_ℓ is 0 also. However, m_s may be $+1/2$ or $-1/2$, so we can fit in two electrons only, with the combinations:

$$\ell = 0, \; m_\ell = 0, \; m_s = +1/2 \quad \text{and} \quad \ell = 0, \; m_\ell = 0, \; m_s = -1/2$$

Any s-type subshell (1s, 2s, 3s, 4s, etc.) may contain a maximum of *two* electrons. In a p-type subshell, we have a little more freedom of choice. When $\ell = 1$, there are three possible values of m_ℓ, namely -1, 0, and $+1$.

For *each* of these, m_s may be $+1/2$ or $-1/2$, so we have a total of $3 \times 2 = 6$ possible combinations of quantum numbers:

$$\ell = 1, \; m_\ell = -1, \; m_s = +1/2 \qquad \ell = 1, \; m_\ell = -1, \; m_s = -1/2$$

$$\ell = 1, \; m_\ell = 0, \; m_s = +1/2 \qquad \ell = 1, \; m_\ell = 0, \; m_s = -1/2$$

$$\ell = 1, \; m_\ell = +1, \; m_s = +1/2 \qquad \ell = 1, \; m_\ell = +1, \; m_s = -1/2$$

To make up the complete set of four quantum numbers for each electron, we combine these with possible values of n to get our various p-type subshells, each one of which may hold a maximum of *six* electrons. So we have the 2p, 3p, 4p, 5p, ..., subshells.

Moving on to the d-type subshells, where $\ell = 2$, we have five values of m_ℓ and, for *each* of these, two values of m_s so there are *ten* possible combinations:

$$\ell = 2, \; m_\ell = -2, \; m_s = +1/2 \qquad \ell = 2, \; m_\ell = -2, \; m_s = -1/2$$

$$\ell = 2, \; m_\ell = -1, \; m_s = +1/2 \qquad \ell = 2, \; m_\ell = -1, \; m_s = -1/2$$

$$\ell = 2, \; m_\ell = 0, \quad m_s = +1/2 \qquad \ell = 2, \; m_\ell = 0, \quad m_s = -1/2$$

$$\ell = 2, \; m_\ell = +1, \; m_s = +1/2 \qquad \ell = 2, \; m_\ell = +1, \; m_s = -1/2$$

$$\ell = 2, \; m_\ell = +2, \; m_s = +1/2 \qquad \ell = 2, \; m_\ell = +2, \; m_s = -1/2$$

The various d-type subshells, with different values of n, will each be able to accommodate a maximum of *ten* electrons. So we have the 3d, 4d, 5d, 6d, ...

Finally (as far as chemically significant orbitals are concerned), we have the f subshells where $\ell = 3$. Without writing out the full combinations of quantum numbers, we can see that we have *seven* possible values of m_ℓ and, for each of these, two values of m_s, so a maximum of *fourteen* electrons may be accommodated:

in an f subshell, $\ell = 3$

m_ℓ may be -3, -2, -1, 0, $+1$, $+2$, or $+3$

for each of these, $m_s = +1/2$ or $-1/2$

total possible combinations $7 \times 2 = 14$

This applies to the 4f and 5f subshells (the 6f, etc., are not important to us).

Summarizing, we see that the maximum number of electrons that may be accommodated in a particular subshell depends only on its orbital quantum number, ℓ, and not on the value of n. Any s subshell has 2 electrons; any p subshell, 6 electrons; any d subshell, 10 electrons; any f subshell, 14 electrons.

We can now compute the total number of electrons the shells can accommodate:

$n = 1$ (K shell):   1s subshell only,   2 electrons

$n = 2$ (L shell):   2s subshell,   2 electrons
                        2p subshell,   6 electrons
                        _____
                        8 electrons

$n = 3$ (M shell):   3s subshell,   2 electrons
                        3p subshell,   6 electrons
                        3d subshell,   10 electrons
                        _____
                        18 electrons

$n = 4$ (N shell):   4s subshell,   2 electrons
                        4p subshell,   6 electrons
                        4d subshell,   10 electrons
                        4f subshell,   14 electrons
                        _____
                        32 electrons

This could be extended for any value of n, with the maximum number of electrons accommodated being $2n^2$ in each case.

5.2   Electronic Configurations of the Elements

As we go through the periodic table, progressively filling with electrons the various shells and subshells in the atom, we have to place each successive electron in an available orbital in such a way that the atom has the *lowest* energy state (so that it is a ground-state configuration). Table 5.1 shows the electronic arrangements of the first 92 elements, up to uranium. (The further elements appear to continue filling the 5f orbital, but they are all unstable radioactive isotopes of limited chemical significance.)

We might expect that the various shells would be filled in a regular progression, i.e., first the K, then the L, the M, the N, etc., but a glance at the table shows that this is not exactly the case. As we go through the first 18 elements, we fill the 1s, 2s, 2p, 3s, and 3p orbitals in regular order, but with the next two elements (potassium and calcium) the next two electrons go into the 4s, rather than the 3d, orbital.

This is because, *for the elements K and Ca*, the 4s orbital is of lower energy than the 3d orbital. In other words, for potassium, the configuration $1s^2 2s^2 2p^6 3s^2 3p^6 4s^1$ is of *lower* energy than the configuration $1s^2 2s^2 2p^6 3s^2 3p^6 3d^1$, and the former is therefore the ground state.

A very similar order of filling is found with the elements Rb and Sr, where the 5s orbital is filled before the 4d, and with Cs and Ba, where the 6s orbital is filled before the 5d or the 4f orbitals.

A mnemonic for remembering the order of filling the orbitals is shown as Figure 5.1. It is very important to realize that this order is *not* always the same as the order of the energy levels of the various orbitals. As we increase the atomic number (charge on the nucleus), the order of the energy levels of the orbitals changes very much, and this is indicated roughly in Figure 5.2 (which is not intended to be a scale diagram).

TABLE 5.1   Electronic Configurations of the Elements

	H	He	Li	Be	B	C	N	O	F	Ne	Na	Mg	Al	Si	P	S	Cl	Ar
Z	1	2	3	4	5	6	7	8	9	10	11	12	13	14	15	16	17	18
$1s$	1	2	2	2	2	2	2	2	2	2	2	2	2	2	2	2	2	2
$2s$		1	2	2	2	2	2	2	2	2	2	2	2	2	2	2	2	2
$2p$			1	2	3	4	5	6	6	6	6	6	6	6	6	6	6	6
$3s$											1	2	2	2	2	2	2	2
$3p$													1	2	3	4	5	6

The K and L shells are now full, plus $3s^2 3p^6$, and we add:

	K	Ca	Sc	Ti	V	Cr	Mn	Fe	Co	Ni	Cu	Zn	Ga	Ge	As	Se	Br	Kr
Z	19	20	21	22	23	24	25	26	27	28	29	30	31	32	33	34	35	36
$3d$			1	2	3	5	5	6	7	8	10	10	10	10	10	10	10	10
$4s$	1	2	2	2	2	1	2	2	2	2	1	2	2	2	2	2	2	2
$4p$													1	2	3	4	5	6

The K, L, and M shells are now full, plus $4s^2 4p^6$, and we add:

	Rb	Sr	Y	Zr	Nb	Mo	Tc	Ru	Rh	Pd	Ag	Cd	In	Sn	Sb	Te	I	Xe
Z	37	38	39	40	41	42	43	44	45	46	47	48	49	50	51	52	53	54
$4d$			1	2	4	5	6	7	8	10	10	10	10	10	10	10	10	10
$5s$	1	2	2	2	1	1	1	1	1		1	2	2	2	2	2	2	2
$5p$													1	2	3	4	5	6

The K, L, and M shells are now full, plus $4s^2 4p^6 4d^{10} 5s^2 5p^6$, and we add:

	Cs	Ba	La	Ce	Pr	Nd	Pr	Sm	Eu	Gd	Tb	Dy	Ho	Er	Tm	Yb	Lu
Z	55	56	57	58	59	60	61	62	63	64	65	66	67	68	69	70	71
$4f$				2	3	4	5	6	7	7	9	10	11	12	13	14	14
$5d$			1							1							1
$6s$	1	2	2	2	2	2	2	2	2	2	2	2	2	2	2	2	2

The K, L, M and N shells are now full, plus $5s^2 5p^6$, and we add:

	Hf	Ta	W	Re	Os	Ir	Pt	Au	Hg	Tl	Pb	Bi	Po	At	Rn	Fr	Ra	Ac	Th	Pa	U
Z	72	73	74	75	76	77	78	79	80	81	82	83	84	85	86	87	88	89	90	91	92
$5d$	2	3	4	5	6	7	9	10	10	10	10	10	10	10	10	10	10	10	10	10	10
$5f$																				2	3
$6s$	2	2	2	2	2	2	1	1	2	2	2	2	2	2	2	2	2	2	2	2	2
$6p$										1	2	3	4	5	6	6	6	6	6	6	6
$6d$																		1	2	1	1
$7s$																1	2	2	2	2	2

$$\to 1s \to$$
$$\to 2s \to$$
$$\to 2p \to 3s \to$$
$$\to 3p \to 4s \to$$
$$\to 3d \to 4p \to 5s \to$$
$$\to 4d \to 5p \to 6s \to$$
$$\to 4f \to 5d \to 6p \to 7s \to$$
$$\to 5f \to 6d \to \ldots$$

Figure 5.1 The order of filling the orbitals. The elements lanthanum (^{57}La), $5d^1$, and actinium (^{89}Ac), $6d^1$, are exceptions to the above sequence.

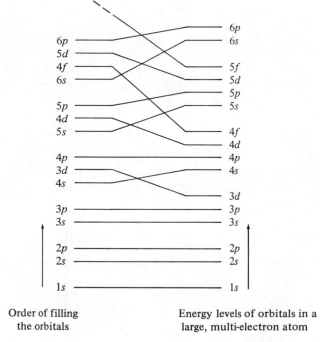

Order of filling
the orbitals

Energy levels of orbitals in a
large, multi-electron atom

Figure 5.2 Energy levels of the orbitals. (*Note:* Figure shows the orders only and is *not* a scale drawing.)

We divide the elements into three classes, according to the orbitals that are being filled.

1. Elements in which s and p orbitals are being filled are called *main group* elements (discussed below in more detail).
2. Elements in which d orbitals are being filled are called *transition* elements. They are in three groups: the first transition series (^{21}Sc to ^{30}Zn), in which the $3d$ orbital is filled; the second transition series (^{39}Y to ^{48}Cd), in which the $4d$ orbital is filled; and the third transition series (^{57}La plus ^{72}Hf to ^{80}Hg), in which the $4d$ orbital is filled.

3. Elements in which f orbitals are being filled are called *inner transition* elements. There are two groups of these: the lanthanide series (^{58}Ce to ^{71}Lu) and the actinide series (^{90}Th to ^{103}Lw), in which the $5f$ shell is filled.

It is important to note that the relative energy levels of the orbitals change when an ion is formed. As a result, when a transition element forms a positive ion, it is the s electrons that are first lost, rather than the d electrons, for example,

Fe   is (K, L) $3s^2 3p^6 3d^6 4s^2$

Fe^{2+} is (K, L) $3s^2 3p^6 3d^6$

Fe^{3+} is (K, L) $3s^2 3p^6 3d^5$

Note that we use two methods of representing an electronic configuration. In a *full* configuration, we put in all the orbitals: $1s^2 2s^2 2p^6 \ldots$. In a *partial* configuration, we abbreviate the *completely filled* shells to the appropriate capital letter. Thus (K) means $1s$; (K, L) means $1s^2 2s^2 2p^6$; (K, L, M) means $1s^2 2s^2 2p^6 3s^2 3p^6 3d^{10}$. Obviously this method of abbreviation saves a lot of time when we write the configuration of a large atom, while conveying exactly the same information:

Hg, full: $1s^2 2s^2 2p^6 3s^2 3p^6 3d^{10} 4s^2 4p^6 4d^{10} 4f^{14} 5s^2 5p^6 5d^{10} 6s^2$

Hg, partial: (K, L, M, N) $5s^2 5p^6 5d^{10} 6s^2$

An alternative way of writing a partial configuration is to write the symbol for the preceding noble gas, followed by the electrons present above this core, for example,

Na could be written [Ne] $3s^1$

Fe could be written [Ar] $3d^6 4s^2$

Hg could be written [Xe] $5d^{10} 6s^2$

In writing down configurations, the order in which we put the various orbitals is of no great significance. In particular, it doesn't make any difference whether the s precedes or follows the d, for example,

Fe: (K, L) $3s^2 3p^6 3d^6 4s^2$

Fe: (K, L) $3s^2 3p^6 4s^2 3d^6$

Both of these convey exactly the same information and are equally correct. The first one is preferable, because it keeps together in sequence the orbitals with the same value of n, and reminds us which electrons are lost when the ion forms.

5.3   Anomalies in Electron Distribution

A close investigation of Table 5.1 will show a number of small irregularities as the various orbitals are filled. Take the following sequence:

$_{23}$V is (K, L) $3s^2 3p^6 3d^3 4s^2$
$_{24}$Cr is (K, L) $3s^2 3p^6 3d^5 4s^1$
$_{25}$Mn is (K, L) $3s^2 3p^6 3d^5 4s^2$

The "expected" arrangement of the last two orbitals in chromium would be $3d^4 4s^2$, to give a regular sequence with its neighbors. However, five electrons in the $3d$ orbital is exactly half of its maximum capacity (ten), and this "half-filled" orbital has a slightly increased stability. This makes the ground state $3d^5 4s^1$, where the $3d$ orbital has borrowed an electron from the $4s$ to make up a total of five $3d$ electrons.

A similar effect is seen with copper, where the configuration is $3d^{10} 4s^1$, rather than $3d^9 4s^2$, because ten electrons is the completely filled arrangement for a d orbital and this, too, has a slight extra degree of stability.

The same effect is seen in the corresponding elements in the second transition series (Mo and Ag), with some other irregularities. Palladium, for example, manages to complete its $4d^{10}$ orbital by borrowing *both* $5s$ electrons. Details of the irregularities in the second and third transition series are not important to us now.

5.4 The Valence Shell

In any atom larger than helium, there will be electrons present with more than one value of n. The electrons that primarily determine the chemical properties of a main-group element are those with the *highest* value of n. We call these the *valence electrons* and the shell containing them is the *valence shell*. A few examples will make this clear:

Na $1s^2 2s^2 2p^6 3s^1$ valence shell: $3s^1$
Br $(K, L, M)4s^2 4p^5$ valence shell: $4s^2 4p^5$
Hg $(K, L, M, N)5s^2 5p^6 5d^{10} 6s^2$ valence shell: $6s^2$

In an atom of a main-group element, the valence shell always contains electrons in s and p orbitals only, and has a maximum population of eight electrons. (In transition elements, where electrons in d orbitals are involved in bonding, the concept of the valence shell has to be widened somewhat, and that topic is outside the scope of this book.)

For the main group elements, the number of electrons in the valence shell is easily worked out from the position of the element in the periodic table, because it is equal to the number of the group containing the element. We can thus summarize the eight possible configurations of the valence shell (in each case, n denotes the values of the principal quantum number):

Group I. One electron. Configuration ns^1 (H, Li, Na, K, Rb, Cs, Fr) $n = 1, 2, 3, 4, 5, 6, 7$.
Group II. Two electrons. Configuration ns^2 (Be, Mg, Ca, Sr, Ba, Ra) $n = 2, 3, 4, 5, 6, 7$. (He, configuration $1s^2$, is usually included in group VIII because of its chemical resemblance to the other inert gases.)
Group III. Three electrons. Configuration $ns^2 np^1$ (B, Al, Ga, In, Tl) $n = 2, 3, 4, 5, 6$.

Group IV. Four electrons. Configuration $ns^2 np^2$ (C, Si, Ge, Sn, Pb) $n = 2, 3, 4, 5, 6$.
Group V. Five electrons. Configuration $ns^2 np^3$ (N, P, As, Sb, Bi) $n = 2, 3, 4, 5, 6$.
Group VI. Six electrons. Configuration $ns^2 np^4$ (O, S, Se, Te, Po) $n = 2, 3, 4, 5, 6$.
Group VII. Seven electrons. Configuration $ns^2 np^5$ (F, Cl, Br, I, At) $n = 2, 3, 4, 5, 6$.
Group VIII. Eight electrons. Configuration $ns^2 np^6$ (Ne, Ar, Kr, Xe, Rn) $n = 2, 3, 4, 5, 6$.

Obviously it's much easier to work out the valence-shell configuration of an element from a glance at the periodic table (always available to you), rather than by writing out the complete electronic configuration. Remember that the number of electrons in the valence shell is given by the group (vertical column) containing the element, while the value of n is given by the horizontal row containing the element (count down, starting with H).

It's often convenient to represent the valence-shell configuration visually by using the "little boxes" system. We draw out the orbitals as a set of square boxes, each of which may contain a maximum of two electrons (of opposite spin). Any s orbital has one box, any p orbital three boxes, any d orbital five boxes. Let's use this for the first row elements:

Note the following points:

1. A pair of electrons in the same box is represented by a pair of arrows, one up and one down.
2. If a box only contains one electron, we conventionally point its arrow upward.
3. The $2s$ orbital is filled *before* the $2p$, because it is of lower energy.
4. *Within* the $2p$ orbital, the three boxes (which correspond to different values of m_ℓ, the magnetic quantum number) are of equal energy. Such orbitals are said to be *degenerate*. As we put electrons into these orbitals, we must remember the rule of maximum multiplicity (*Hund's rule*), which states that

Each one of a set of orbitals of equal energy will be singly occupied before any one of them is doubly occupied

In other words, one arrow in each box before we put two into one box. This means that, for C, N, and O, we have a preferred (lower energy) configuration of the $2p$ orbital:

C is [↑ | ↑ |]   rather than   [↑↓ | |]

N is [↑ | ↑ | ↑]   rather than   [↑↓ | ↑ |]

O is [↑↓ | ↑ | ↑]   rather than   [↑↓ | ↑↓ |]

5. It's hardly necessary to point out that the little boxes are just a bookkeeping device, and not intended to represent the spatial arrangement of the electrons.

In the preceding pages, we have seen that the arrangement of elements in the periodic table may be related to their electronic configurations. Although the rules governing the permitted values of the quantum numbers were presented in a somewhat arbitrary manner, their success in predicting the structure of the table (confirmed, of course, by the similarities found in the chemical properties of the elements in the same group) gives us confidence that our theories are correct.

KEY WORDS

Pauli's exclusion principle   shell
quantum number   orbital (subshell)
valence shell   Hund's rule
ground state   excited state

STUDY QUESTIONS

1. What are the names of the four quantum numbers of an electron? What are the permitted values for each of them?

2. What is the significance of the division of a shell into orbitals (subshells)? How many orbitals are there in each shell?

3. How many electrons may be accommodated in each shell? In each orbital? Why?

4. Why do we always tabulate ground-state electronic configurations, rather than excited states?

5. What *chemical* evidence supports the suggestion that the valence shell is $4s^1$ in potassium and $5s^1$ in rubidium?

6. Are the energy levels of the various orbitals always in the same order, regardless of which atom is being considered?

7. What is a "valence shell" in an atom? Why is it important?

8. What are the eight types of configuration found for the eight main groups of the periodic table?

9. Why are the electron configurations of Cr and Cu anomalous?

10. What type of electron configuration do very many ions of main group elements have?

11. What is meant by Hund's rule? Can you suggest a simple physical explanation for the fact that the three $2p$ electrons in a nitrogen atom prefer to have different values of their magnetic quantum numbers?

PROBLEMS

In answering the following questions, you may refer to a periodic table (e.g., Appendix D) but you should try to avoid copying the answers from the body of this chapter.

1. (a) How many electrons may be accommodated in the shell with $n = 3$?
   (b) Write out their various combinations of quantum numbers.

2. Give the maximum number of electrons that may be accommodated in the following orbitals: $4p$, $5f$, $6d$, $7s$.

3. Which is "odd man out" in the group $3p$, $2d$, $4s$, $6f$?

4. Give the full configuration (ground state) of the following atoms: Si, Co, Br, Sr, As, V.

5. Give the full configuration (ground state) of the following ions: S^{2-}, Rb^+, N^{3-}, Mg^{2+}, Ti^{4+}, Cl^-.

6. How many electrons in a ground-state zinc atom have
   (a) orbital quantum number (ℓ) equal to $+1$
   (b) magnetic quantum number (m_ℓ) equal to -1
   (c) $\ell = 2$ and $m_\ell = 0$
   (d) $\ell = 3$ and $m_\ell = +1$
   (e) $m_s = -1/2$

7. Give partial configurations of the following: Sn, Tl, Ba, La, I, Cs^+, Se^{2-}, Ag^{2+}.

8. Using the "boxes" notation, give the valence-shell configuration of the following, naming the subshells involved: P^{3-}, Ge, Xe, Be, Ba, Mg^+, Al.

9. The following are excited-state configurations for some neutral atoms. Give the corresponding ground-state configuration of each, and identify the atoms:
   (a) $(K, L, M) 4p^3$ (b) $(K, L, M) 4s^2 4p^6 4d^4$
   (c) $1s^2 2s^2 3s^2$ (d) $(K, L) 3s^2 3p^6 3d^4 4s^2$

10. Consider the following group: Ne, F^-, O^{2-}, N^{3-}, Na^+, Mg^{2+}, Al^{3+}.
    (a) What do they all have in common?
    (b) How do they differ?
    (c) Arrange them in order of decreasing size, i.e., starting with the largest. Give a reason for your choice of order. (*Note:* this is an easy question, and is closely related to the material of this chapter.)

SOLUTIONS TO PROBLEMS

1. (a) 18 electrons.
   (b) two electrons in the $3s$, six in the $3p$, and ten in the $3d$. For all of them $n = 3$, and the other quantum numbers have the values given on pp. 100–101.

2. $4p$, six electrons; $5f$, fourteen; $6d$, ten; $7s$, two.

3. The $2d$ does not exist.

4. Si: $1s^2 2s^2 2p^6 3s^2 3p^2$
   Co: $1s^2 2s^2 2p^6 3s^2 3p^6 3d^7 4s^2$
   Br: $1s^2 2s^2 2p^6 3s^2 3p^6 3d^{10} 4s^2 4p^5$
   Sr: $1s^2 2s^2 2p^6 3s^2 3p^6 3d^{10} 4s^2 4p^6 5s^2$
   As: $1s^2 2s^2 2p^6 3s^2 3p^6 3d^{10} 4s^2 4p^3$
   V: $1s^2 2s^2 2p^6 3s^2 3p^6 3d^3 4s^2$

5. S^{2-}: $1s^2 2s^2 2p^6 3s^2 3p^6$
   Rb^+: $1s^2 2s^2 2p^6 3s^2 3p^6 3d^{10} 4s^2 4p^6$
   N^{3-}: $1s^2 2s^2 2p^6$
   Mg^{2+}: $1s^2 2s^2 2p^6$
   Ti^{4+}: $1s^2 2s^2 2p^6 3s^2 3p^6$
   Cl^-: $1s^2 2s^2 2p^6 3s^2 3p^6$

6. (a) $\ell = 1$ means p electrons. A zinc atom has just two filled p orbitals, the $2p$ and the $3p$. With six electrons in each, this gives a total of twelve.
   (b) If $m_\ell = -1$, ℓ must be 1 or more, so we will only find this value in p, d, or f orbitals. In zinc, we are restricted to the filled $2p$, $3p$, and $3d$ orbitals, each containing a pair of electrons differing in m_s. Total: 3 pairs, or six electrons.
   (c) $\ell = 2$ means a d orbital. Only one d orbital is occupied in zinc, the $3d$, which contains a pair of electrons with $m_\ell = 0$. Answer: two electrons.
   (d) $\ell = 3$ means an f orbital. No f orbital is occupied in zinc, so the answer is zero.
   (e) Zinc contains equal numbers of electrons with $m_s = +1/2$ and $m_s = -1/2$, so one-half of the total 30 electrons have $m_s = -1/2$. *Answer:* 15 electrons.

7.   Sn: $(K, L, M) 4s^2 4p^6 4d^{10} 5s^2 5p^2$
      Tl: $(K, L, M, N) 5s^2 5p^6 5d^{10} 6s^2 6p^1$
      Ba: $(K, L, M) 4s^2 4p^6 4d^{10} 5s^2 5p^6 6s^2$
      La: $(K, L, M) 4s^2 4p^6 4d^{10} 5s^2 5p^6 5d^1 6s^2$
       I: $(K, L, M) 4s^2 4p^6 4d^{10} 5s^2 5p^5$
     Cs^+: $(K, L, M) 4s^2 4p^6 4d^{10} 5s^2 5p^6$
    Se^{2-}: $(K, L, M) 4s^2 4p^6$
    Ag^{2+}: $(K, L, M) 4s^2 4p^6 4d^9$

8.

P^{3-}	$3s$ [↑↓]	$3p$	[↑↓] [↑↓] [↑↓]	
Ge	$4s$ [↑↓]	$4p$	[↑] [↑] [ ]	
Xe	$5s$ [↑↓]	$5p$	[↑↓] [↑↓] [↑↓]	

Be	$2s$ [↑↓]	
Ba	$6s$ [↑↓]	
Mg^+	$3s$ [↑]	
Al	$3s$ [↑↓]	$3p$ [↑] [ ] [ ]

9. (a) Ga: $(K, L, M) 4s^2 4p^1$ (b) Zr: $(K, L, M) 4s^2 4p^6 4d^2 5s^2$
    (c) C: $1s^2 2s^2 2p^2$ (d) Cr: $(K, L) 3s^2 3p^6 3d^5 4s^1$

10. (a) They all have the same electronic arrangement (isoelectronic) of 10 electrons.
     (b) They differ in nuclear charge.
     (c) N^{3-}, O^{2-}, F^-, Ne, Na^+, Mg^{2+}, Al^{3+}. This is the order of *increasing* nuclear charge. Since the number of electrons is constant, the increasing nuclear charge will pull in the electrons with a greater electrostatic attraction, thus making the ion smaller as the atomic number increases.

Chapter 6

Crystal Structures and the Ionic Bond

"I've found it! I've found it!" he shouted, running towards us with a test-tube in his hand.   A STUDY IN SCARLET

Most solids are crystalline in nature. In particular, all solids that have an ionic structure are crystalline, and the reason for this is not hard to see. In an ionic solid, where there are strong electrostatic forces of attraction between ions of opposite charge, the potential energy of the system will be at a minimum when the ions are as close as possible, and when as many ions of one charge as possible are surrounding each ion of opposite charge. To achieve this requires a regular arrangement of the ions, which forms a regular, crystalline, lattice arrangement by continued repetition.

6.1   The Born–Haber Cycle and Lattice Energy

Before describing any of the actual arrangements of ions in a crystal lattice, we should look at the enthalpy changes that accompany the formation of an ionic compound. There are several steps involved, which may conveniently be fitted together in a form known as a *Born–Haber cycle*.

Let's consider the formation of sodium chloride from its elements. If we react solid sodium with gaseous chlorine, we can measure ΔH_f^0 for solid sodium chloride. The reaction is highly exothermic:

$$\text{Na}(s) + \frac{1}{2}\text{Cl}_2(g) \longrightarrow \text{NaCl}(s) \qquad \Delta H_f^0 = -411 \text{ kJ} \qquad \text{(a)}$$

Now we should reconsider the overall formation reaction by splitting it up into several small steps, each with a ΔH value. We start by vaporizing the Na:

$$\text{Na}(s) \longrightarrow \text{Na}(g) \qquad \Delta H = +109 \text{ kJ} \qquad \text{(b)}$$

111

This is simply the heat of vaporization of sodium. Now we have to convert the gaseous atom into an ion:

$$Na(g) \longrightarrow Na^+(g) + e^- \qquad \Delta H = +495 \text{ kJ} \tag{c}$$

This amount of energy is called the *ionization energy* (sometimes ionization potential) of sodium. By definition:

Ionization energy is the energy required to remove an electron completely from an isolated (gaseous) atom.

Ionization energy is always positive, i.e., the process is endothermic. We speak of the *first* ionization energy when we remove an electron from a neutral atom, giving an ion with a 1+ charge, then the second and third ionization energies, as further electrons are removed to give ions with charges of 2+, 3+, etc.

Starting with gaseous chlorine, we have to go through two similar steps to produce a chloride ion, first by forming the gaseous atoms by supplying the bond dissociation energy:

$$\frac{1}{2}Cl_2(g) \longrightarrow Cl(g) \qquad \Delta H = +\frac{242}{2} = +121 \text{ kJ} \tag{d}$$

Note that we take half of the molar bond energy of Cl_2, because we started with only 1/2 mole. The second step, the formation of gaseous Cl^- ion from the Cl atom, is exothermic:

$$Cl(g) + e^- \longrightarrow Cl^-(g) \qquad \Delta H = -347 \text{ kJ} \tag{e}$$

This amount of energy is called the *electron affinity* of chlorine. By definition:

Electron affinity is the energy change accompanying the process in which an electron is added to an isolated (gaseous) atom.

Electron affinity is only important for reactive nonmetals that form negative ions, such as F^-, Cl^-, etc. It is usually negative (as with Cl^- above), i.e., the process is exothermic, but it can be positive (as with O^{2-}, see later).

If we add together the last four equations, (b), (c), (d), and (e), we obtain the equation

$$Na(s) + \frac{1}{2}Cl_2(g) \longrightarrow Na^+(g) + Cl^-(g)$$

and the overall ΔH for this process is the sum of the ΔH values for the four steps:

$$\Delta H = +109 + 495 + 121 + (-347) = +378 \text{ kJ}$$

We see that the formation of the gaseous ions is a highly endothermic process, absorbing 378 kJ, whereas the formation of the ionic *solid* from the same starting materials was highly exothermic according to equation (a). So the last step, the condensation of the gaseous ions to form the solid, must be highly exothermic:

$$Na^+(g) + Cl^-(g) \longrightarrow NaCl(s) \qquad \Delta H = -789 \text{ kJ} \tag{f}$$

This amount of energy is called the *lattice energy* of sodium chloride. It is defined as the enthalpy change when a mole of solid forms from gaseous ions at infinite separation. It is always negative.

If we add this equation to the previous one, we will produce the equation for ΔH_f^0 of NaCl:

$$Na(s) + \frac{1}{2} Cl_2(g) \longrightarrow NaCl(s) \qquad \Delta H = +378 - 789 = -411 \text{ kJ}$$

and we see that the ΔH value agrees with that we obtained previously by the direct reaction. Of course, we were expecting this from Hess' law.

These various steps and their interrelationship are conveniently summarized in a diagram called a *Born–Haber cycle*.

A Born–Haber cycle is simply a summary of the various steps that occur in the formation of an ionic compound from its elements. At top left, we have the stoichiometric amounts of the elements (in their standard states) and at top right, the ionic solid. The remaining part of the cycle is filled by the intermediates in the formation process, and we put ΔH values beside each step.

Note that we must have the correct amounts of substances throughout the cycle to produce 1 mole of ionic solid, and we must include each of the intermediate steps.

When we have set out the cycle, we can use Hess' law to relate the various ΔH quantities. If we go from the elements to the ionic solid, we can do it in one step, across the top of the diagram, and the value of ΔH will be (a), the standard heat of formation. By going through the intermediate atoms and ions, we can make the same overall change from elements to ionic solid, and the ΔH value will be the sum of the ΔH values of the individual steps, in this case (b) + (d) + (c) + (e) + (f).

Since the overall ΔH change must be the same by either route, we can write

$$(a) = (b) + (c) + (d) + (e) + (f)$$

(NOTE: As in any other thermochemical calculation, the ΔH values must be added up with careful attention to their respective signs.)

This relationship can be used to find any one of the various ΔH values if all the others are known. It might seem that the lattice energy would be the "unknown" that we would normally want to find, because this quantity cannot be found by direct experiment. In practice, however, this is not always the case. Although we cannot measure lattice energy by any single experiment, we can obtain quite an accurate value of it by calculation alone, provided that we know the structure of the solid. We can therefore use a Born–Haber cycle to calculate some other quantity, in particular, the electron affinity, which is often a difficult quantity to find experimentally. Both of these processes will be illustrated by examples.

Although we could do these problems by adding or subtracting a series of balanced equations (each one with its associated ΔH value), as we did in Chapter 4, the use of the Born–Haber cycle enables us to see at a glance how the various quantities are related.

Example 6.1   Calculate the lattice energy of potassium bromide from the following data:

standard heat of formation of KBr(s)	-392 kJ mole^{-1}
heat of sublimation of K(s)	$+90$ kJ mole^{-1}
dissociation energy of Br$_2$(g)	$+190$ kJ mole^{-1}
heat of vaporization of Br$_2$(l)	$+31$ kJ mole^{-1}
ionization energy of K(g)	$+418$ kJ mole^{-1}
electron affinity of Br(g)	-323 kJ mole^{-1}

SOLUTION   The best way to tackle this problem is to put all the reactants down in the form of a Born–Haber cycle. Comparing this with the previous example (NaCl), we see there is one extra step involved. Since the standard state of bromine is liquid, we have to include the heat of vaporization of Br$_2$ when going through the cycle. Note also that the above figures on heat of vaporization and dissociation energy are given per mole of Br$_2$. Since we start with 1/2 mole of Br$_2$, we must halve the given values.

$$
\begin{array}{ccc}
\text{K}(s) + \dfrac{1}{2}\,\text{Br}_2(l) & \xrightarrow{\ -392\ } & \text{KBr}(s) \\
\Big\downarrow +90 & \dfrac{+31}{2}=+16 & \Big\uparrow x\ (\text{lattice energy}) \\
& \dfrac{1}{2}\,\text{Br}_2(g) & \\
& \dfrac{+190}{2}=+95 & \\
\text{K}(g) + \text{Br}(g) & \xrightarrow{\ +418-323\ } & \text{K}^+(g) + \text{Br}^-(g)
\end{array}
$$

Using Hess' law, we can write

$$-392 = +90 + 16 + 95 + 418 - 323 + x$$

$$x = -688 \text{ kJ mole}^{-1}$$

These calculations are very straightforward provided we can write down exactly what chemical process is going on for each ΔH step around the cycle. Obviously we must preserve a simple mass balance as we go; we must end up with the same amounts of elements that we started with. A very common source of student error is to forget that, for a diatomic molecule such as Br$_2$, Cl$_2$, etc., we only need 1/2 mole to give one mole of the negative ion.

Example 6.2 Calculate the electron affinity of the oxygen atom in gaining two electrons to give the oxide ion, O^{2-}, from the following data:

standard heat of formation of MgO(s)	-600 kJ mole^{-1}
lattice energy of MgO(s)	-3860 kJ mole^{-1}
ionization energy of Mg(g) to give $Mg^{2+}(g)$	$+2170$ kJ mole^{-1}
dissociation energy of $O_2(g)$	$+494$ kJ mole^{-1}
heat of sublimation of Mg(s)	$+150$ kJ mole^{-1}

SOLUTION   Again, we write a Born–Haber cycle:

$$Mg(s) + \frac{1}{2} O_2(g) \xrightarrow{\ -600\ } MgO(s)$$

$$\downarrow +150 \qquad \downarrow \frac{+494}{2} = +247 \qquad \nearrow -3860$$

$$Mg(g) + O(g) \xrightarrow{\ +2170 + x\ } Mg^{2+}(g) + O^{2-}(g)$$

where x denotes the electron affinity of oxygen. Our summation equation for the various steps gives

$$-600 = 150 + 247 + 2170 + x - 3860$$

$$x = +693 \text{ kJ mole}^{-1}$$

(i.e., the formation process is endothermic). The very large lattice energy for MgO, nearly five times that of NaCl, is a result of the double charge that the Mg^{2+} and O^{2-} ions carry. They attract each other more strongly than Na^+ and Cl^-, so more energy is liberated when they come together.

6.2 Crystal Structures: Metals

It is not desirable or possible in a book of this size to discuss in any detail the many different shapes and classes of crystals. Instead, we shall look at some of the very simplest structures of the cubic type, to see what information can be gained about the size and arrangement of the ions or atoms that are present.

It is convenient to start by looking at the structures found in some metals, since, in a crystal of a metallic element, all the atoms must obviously be the same. The atoms may be regarded for our purpose as a large number of spheres packed together in contact in a regular way.

To describe these arrays, we use the concept of the *unit cell*, which is the simplest building block out of which the whole crystal may be assembled. The three basic ways in which a cubic unit cell may be constructed from a number of identical spheres are shown in Figure 6.1.

For the sake of clarity in the illustrations, we show the spheres with spaces between them, but in fact they are packed together so that adjacent spheres are in contact. It looks as if the unit cells contain 8, 9, or 14 atoms respectively, but the number is actually much less than this, because nearly all the atoms are shared with neighboring unit cells. Each unit

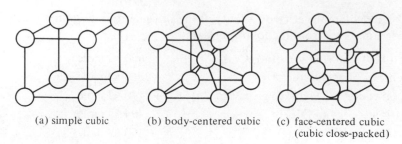

(a) simple cubic   (b) body-centered cubic   (c) face-centered cubic
                                                      (cubic close-packed)

Figure 6.1   Cubic unit cells of metal atoms.

cell is a perfect cube whose sides cut through the atoms according to their positions. Specifically:

1. In the simple cubic structure, an atom is present at each corner. Only 1/8 of the atom is actually *within* the cube, the other 7/8 being distributed among the other 7 unit cells meeting at the corner. Since there are 8 corners, the actual number of atoms *within* each unit cell cube is

$$\frac{1}{8} \times 8 = 1 \text{ atom}$$

2. In the body-centered cubic (bcc) structure, the same 8 atoms as before are present at the corners. In addition, however, there is another atom at the center of the cube (hence the name "body-centered"). This atom is entirely within this unit cell; it is not shared. So the total number of atoms within the bcc structure is

$$\frac{1}{8} \times 8 \text{ corner atoms} = 1$$

$$1 \text{ complete atom at center} = 1$$

$$\text{total} = 2 \text{ atoms}$$

3. In the face-centered cubic (fcc, also called "cubic close-packed") structure, there are again atoms in two different positions. There are eight atoms at the corners, as before, and in addition there is an atom at the center of each of the six faces of the cube (hence the name "face-centered"). The latter atoms will be equally shared between two unit cells which meet at each face, so 1/2 of each will be within the unit cell. The total within each unit cell will therefore be

$$\frac{1}{8} \times 8 \text{ corner atoms} = 1$$

$$\frac{1}{2} \times 6 \text{ face-centered atoms} = 3$$

$$\text{total} \quad 4 \text{ atoms}$$

It's interesting to work out how efficient these three structures are in "filling space." How much of the total available volume is taken up by the atoms, and how much is left for the "gaps" between them? We can work this out with a little geometry, but first we should define a couple of useful terms related to a cube:

> The *face* diagonal of a cube goes across one face between opposite corners. Since the face is a square, the length of the face diagonal is $\sqrt{2}$ × the length of the edge of the cube (by the Pythagorean theorem).

> The *body* diagonal of a cube goes through the center of the cube, running between opposite corners. Its length will always be $\sqrt{3}$ × the length of the edge of the cube.

Looking at the simple cubic structure and remembering that the spheres are actually in contact, we see that the length of the edge of the cube (l) will be equal to twice the radius of each atom (r). The corner of the unit cell, of course, is at the exact center of the atom. So we can write $l = 2r$; and the volume of the unit cell $= l^3 = 8r^3$. Now this cell actually contains only the equivalent of *one* complete atom. Knowing this to be spherical, we have

$$\text{volume of sphere} = \frac{4\pi r^3}{3}$$

This is the usefully occupied volume in the unit cell, the remainder is empty space. So our ratio is

$$\frac{\text{usefully occupied}}{\text{total volume}} = \frac{\text{volume of sphere}}{\text{volume of unit cell}} = \frac{4\pi r^3/3}{8r^3} = \frac{\pi}{6}$$

As would be expected, r^3 cancels out, giving us the result that, for any size of atom, the fraction of space usefully occupied in a simple cubic structure is $\pi/6$ or 0.524 (52.4%). This is not very efficient. Can we do better with one of the other structures?

With the bcc structure, if we imagine the spherical atoms to be in contact, it is not the adjacent atoms on the corner that touch each other. Instead, all eight of the corner atoms come into contact with the atom at the body center. This means that the length of the body diagonal is equal to $4r$, where r is the radius of the atom. Since the body diagonal is of length $\sqrt{3}\,l$, where l is the length of the edge, we have

$$\sqrt{3}\,l = 4r \quad \text{or} \quad l = 4r/\sqrt{3}$$

The volume of the cube is $l^3 = (4r/\sqrt{3})^3$. Now this contains the equivalent of two complete atoms, whose volume is $2 \times 4\pi r^3/3$. So our ratio is

$$\frac{\text{usefully occupied}}{\text{total volume}} = \frac{2 \times 4\pi r^3/3}{(4r/\sqrt{3})^3} = \frac{\sqrt{3}\,\pi}{8} = 0.680$$

So in the bcc structure we have improved our efficiency of space filling to 68.0%. A further slight improvement in efficiency is found in the face-centered cubic structure.

In the fcc unit cell, the atoms coming into contact will lie along the *face* diagonal of the cube. The length of the face diagonal will therefore be $4r$, where r is the radius of an atom, as shown in Figure 6.2.

Knowing the face diagonal is of length $\sqrt{2}\,l$, we can write

$$\sqrt{2}\,l = 4r \quad \text{or} \quad l = 4r/\sqrt{2}$$

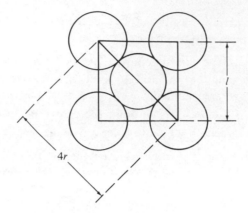

Figure 6.2 Touching spheres on one face, fcc structure.

and the volume of the unit cell $= l^3 = (4r/\sqrt{2})^3$. In this case, the unit cell contains the equivalent of four atoms, whose volume is $4 \times 4\pi r^3/3$. So our ratio is

$$\frac{\text{usefully occupied}}{\text{total volume}} = \frac{4 \times 4\pi r^3/3}{(4r/\sqrt{2})^3} = \frac{\pi}{3\sqrt{2}} = 0.741$$

The 74.1 % efficiency of this structure and the similar hexagonal close-packed structure is the best that can be achieved in packing spheres together to fill space. Both the fcc and the bcc structures are frequently found for metals, but the less efficient simple cubic structure is not found. Knowing the structure present, we have only to measure the density to find the dimensions of the unit cell and the size of the atom.

Since the dimensions of atoms, ions, and unit cells are very small, we may conveniently express them in angstrom units, symbol Å.

$$1 \text{ Å} = 10^{-8} \text{ cm} \quad (10^{-10} \text{ m})$$
$$1 \text{ Å}^3 = 10^{-24} \text{ cm}^3 \quad (10^{-30} \text{ m}^3)$$

(*Note*: The angstrom unit is not an SI unit, but is retained because it is still in wide use. The SI unit for dimensions of this type is the nanometer, nm, which is 10^{-7} cm (10^{-9} m), so 1 nm = 10 Å.)

Example 6.3 Gold crystallizes in the fcc system. The radius of the atom is 1.44 Å. Calculate:

(a) the length of the side of the unit cell
(b) the volume of the unit cell
(c) the mass of the unit cell
(d) the density of gold

SOLUTION   Figure 6.2 shows the arrangement of the atoms on one face of the unit cell.

(a) As we have previously worked out, the length of the face diagonal is four times the radius of the atom.

$$\text{face diagonal} = 4 \times 1.44 = 5.76 \text{ Å}$$

The side length, l, is related to this by a factor $\sqrt{2}$

$$l = \frac{\text{face diagonal}}{\sqrt{2}} = \frac{5.76}{\sqrt{2}} = 4.07 \text{ Å}$$

(b) In any cube, volume = (side length)^3; therefore, volume of the unit cell = $(4.07)^3$ = 67.6 Å^3.

(c) To calculate the mass of the unit cell, we need to know that, in the fcc structure, one unit cell contains four atoms.

$$\text{mass of 1 atom} = \frac{\text{molar mass}}{\text{Avogadro's number}} = \frac{197}{6.02 \times 10^{23}} = 3.27 \times 10^{-22} \text{ g}$$

So the unit cell, containing four atoms, weighs $4 \times 3.27 \times 10^{-22} = 1.31 \times 10^{-21}$ g.

(d) Density is usually expressed in g cm^{-3}, so we should put the volume of the unit cell in cm^3

$$67.6 \text{ Å}^3 = 67.6 \times 10^{-24} \text{ cm}^3 = 6.76 \times 10^{-23} \text{ cm}^3$$

$$\text{density} = \frac{\text{mass}}{\text{volume}} = \frac{1.31 \times 10^{-21} \text{ g}}{6.76 \times 10^{-23} \text{ cm}^3} = 19.4 \text{ g cm}^{-3}$$

We have, of course, calculated the density of the unit cell in this operation, but this will be the same as the density of bulk gold. *Density is a quantity that is independent of the amount of material present.*

Example 6.4 Rubidium crystallizes in the bcc system, the radius of the atom being 2.47 Å. What is the density of the metal?

SOLUTION Looking at Figure 6.1(ii), we see that, in the bcc arrangement, three atoms are in contact along the body diagonal of the cube. This has length $4r$, where r is the atomic radius, and it may also be put equal to $\sqrt{3}\,l$ (see Figure 6.3).

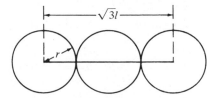

Figure 6.3 Touching spheres along body diagonal, bcc structure.

For rubidium, the body diagonal $= 4 \times 2.47 = 9.88$ Å, and since the length of the side is $1/\sqrt{3}$ times the body diagonal, we know

$$\text{side of unit cell} = \frac{9.88}{\sqrt{3}} = 5.70 \text{ Å}$$

$$\text{volume of unit cell} = (5.70)^3 = 186 \text{ Å}^3$$

$$= \frac{186}{10^{24}} = 1.86 \times 10^{-22} \text{ cm}^3$$

The unit cell in a bcc structure contains two atoms. The atomic weight of Rb is 85.5, so the mass of the unit cell (two atoms) is

$$\frac{2 \times 85.5}{6.02 \times 10^{23}} = 2.84 \times 10^{-22} \text{ g,}$$

$$\text{density of Rb} = \frac{2.84 \times 10^{-22} \text{ g}}{1.86 \times 10^{-22} \text{ cm}^3} = 1.53 \text{ g cm}^{-3}$$

Example 6.5 Sodium crystallizes in the bcc structure with a density of 0.971 g cm^{-3}. Calculate the size of the unit cell and the radius of the sodium atom.

SOLUTION   The only other information we need here is the atomic weight of sodium (23.0) and the value of Avogadro's number (6.02×10^{23}).
A mole of sodium weighs 23.0 g. This contains 6.02×10^{23} atoms, so

$$\text{mass of one atom Na} = \frac{23.0}{6.02 \times 10^{23}} \text{ g}$$

In the bcc structure, the unit cell contains the equivalent of two atoms, so

$$\text{mass of each unit cell of Na} = \frac{2 \times 23.0}{6.02 \times 10^{23}} \text{ g}$$

Since the density is 0.971 g cm^{-3}, the volume of the unit cell is

$$\frac{2 \times 23.0}{0.971 \times 6.02 \times 10^{23}} = 7.87 \times 10^{-23} \text{ cm}^3 = 7.87 \times 10^{-23} \times 10^{24} \text{ Å}^3 = 78.7 \text{ Å}^3$$

The side length, l, is the cube root of the volume

$$l = \sqrt[3]{78.7} = 4.29 \text{ Å}$$

As we showed before (Figure 6.3), in a bcc structure the atomic radius is related to the side length of the unit cell by

$$4r = \sqrt{3}\,l \quad \text{or} \quad r = \sqrt{3}\,l/4$$

for sodium, $r = (\sqrt{3} \times 4.29)/4 = 1.86$ Å.

Example 6.6 Copper crystallizes with a face-centered cubic structure having a density of 8.93 g cm^{-3}. What is the radius of a copper atom?

SOLUTION   The calculation follows a similar route to the previous example, except that we now have a fcc structure.

A mole of Cu weighs 63.5 g, so the mass of one atom is $63.5/(6.02 \times 10^{23})$g.

Since the fcc structure contains the equivalent of four atoms in the unit cell,

$$\text{mass of the unit cell} = \frac{4 \times 63.5}{6.02 \times 10^{23}} \text{ g}$$

Since the density of Cu is 8.93 g cm^{-3},

$$\text{volume of unit cell} = \frac{4 \times 63.5}{8.93 \times 6.02 \times 10^{23}} = 4.73 \times 10^{-23} \text{ cm}^3 = 47.3 \text{ Å}^3$$

$$\text{length of side cube} = \sqrt[3]{47.3} = 3.62 \text{ Å } (l)$$

In the fcc structure, three atoms are in contact along the face diagonal of the cube, which has length $\sqrt{2}\,l = 4r$:

$$r = \frac{\sqrt{2}\,l}{4} = \frac{\sqrt{2} \times 3.62}{4} = 1.28 \text{ Å}$$

Example 6.7   Solid methane, CH_4, crystallizes in a bcc lattice with a density of 0.415 g cm^{-3}. If the CH_4 molecule is regarded as essentially spherical, what is its effective radius?

SOLUTION   This is a similar calculation, except that we are now working with a compound rather than an element.

A mole of CH_4 weighs 16.0 g. The bcc unit cell contains two molecules, which weigh $(2 \times 16.0)/(6.02 \times 10^{23})$g. With a density of 0.415, the volume of the unit cell is, therefore,

$$\frac{2 \times 16.0}{0.415 \times 6.02 \times 10^{23}} = 1.28 \times 10^{-22} \text{ cm}^3 = 1.28 \times 10^{-22} \times 10^{24} \text{ Å}^3 = 128 \text{ Å}^3$$

The length of the side cube $= \sqrt[3]{128} = 5.04$ Å $(=l)$. And, since in the bcc structure we have three molecules in contact along the body diagonal,

$$4r = \sqrt{3}\,l = \sqrt{3} \times 5.04 \qquad r = 2.18 \text{ Å}$$

6.3 The Structures of Ionic Compounds

We can do very similar calculations on ionic compounds, but we must first look at some of the simplest cubic structures found in ionic systems. Of the many different possibilities, we will only consider two here; they are generally known as the cesium chloride structure and the sodium chloride structure: both are shown in Figure 6.4.

There is an obvious resemblance between the CsCl structure and the bcc lattice, and indeed CsCl is often said to have a body-centered cubic lattice. Similarly, the NaCl structure is called a face-centered cubic lattice, because it consists of a fcc array of one type of ion with a second fcc array of the other ion exactly fitted into the gaps in the first lattice.

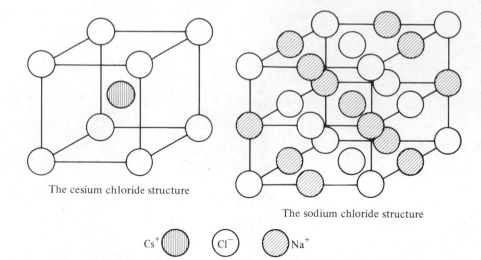

The cesium chloride structure

The sodium chloride structure

Cs^+ ⬭ Cl^- ○   ⬭ Na^+

Figure 6.4 The unit cells of CsCl and NaCl.

How many ions of each type are actually *within* each unit cell? We would expect that the number of ions of each type should be equal, to make the unit cell electrically neutral, and this turns out to be the case.

(i) In the CsCl structure, there is:

$$1 \text{ complete } Cs^+ \text{ ion at center of cube} = 1Cs^+$$
$$1/8 \text{ of } 8Cl^- \text{ ions at corners of cube} = 1Cl^-$$

$$\text{total ions} = 1Cs^+ + 1Cl^-$$

(*Note*: obviously the same result would be found if we put the Cl^- ion at the center and the Cs^+ ions at the corners of the unit cell.)

(ii) In the unit cell of the NaCl structure there are the following ions:

$$\text{the whole of one } Na^+ \text{ at cube center} = 1Na^+$$
$$1/4 \text{ of each of } 12Na^+ \text{ on cube edges} = 3Na^+$$
$$1/2 \text{ of each of } 6Cl^- \text{ at centers of faces} = 3Cl^-$$
$$1/8 \text{ of each of } 8Cl^- \text{ at cube corners} = 1Cl^-$$

$$\text{total ions} = 4Na^+ + 4Cl^-$$

(*Note*: Again, the same result would be obtained if the two ions were reversed.

All we have to do is to measure the density of an ionic compound and we can then work out the dimensions of its unit cell, as we did for the metals. However, we are not concerned with working out the radius of a single sphere in an ionic compound, but rather the distance between two ions.

Interionic distance = distance between centers of two adjacent (touching) ions of opposite charge.

The interionic distance is therefore equal to the sum of the radii of the two ions (their ionic radii). By using other information, the total interionic distance may be divided up to give effective ionic radii of the two touching ions, but we will not consider here how these are calculated.

Example 6.8 Cesium chloride has a density of 3.97 g cm^{-3}. Calculate the size of the unit cell and the interionic distance.

SOLUTION   This follows much the same lines as the previous problems on bcc structures, remembering that the unit cell contains one Cs^+ ion and one Cl^- ion.

A mole of CsCl weighs 168.4 g, so one ion pair of CsCl (i.e., one ion of each) weighs $168.4/(6.02 \times 10^{-23})$g. This is the contents of one unit cell, so the volume of the unit cell (density 3.97 g cm^{-3}) is

$$\frac{168.4}{3.97 \times 6.02 \times 10^{23}} = 7.05 \times 10^{-23} \text{ cm}^3 = 70.5 \text{ Å}^3$$

The length of the side of the unit cell cube is $\sqrt[3]{70.5} = 4.13$ Å $(= l)$.

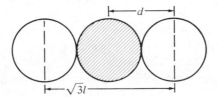

Figure 6.5 Interionic distance (d) in the CsCl structure.

Looking at the structure of the unit cell, we see that the length of the body diagonal is twice the interionic distance, since the central ion is in contact with two ions of opposite charge whose centers are at each corner of the cube as shown in Figure 6.5. If the interionic distance is d, then

$$2d = \sqrt{3}\,l \qquad d = \frac{\sqrt{3}\,l}{2} = \frac{\sqrt{3} \times 4.13}{2} = 3.57 \text{ Å}$$

Example 6.9 Potassium fluoride crystallizes with the sodium chloride structure and has a density of 2.48 g cm^{-3}. Calculate the K–F interionic distance.

SOLUTION   One mole of KF weighs 58.1 g.

In a unit cell with the NaCl structure, we have $4K^+$ ions and $4F^-$ ions. These will weigh $(4 \times 58.1)/(6.02 \times 10^{23})$g. Therefore, with a density of 2.48 g cm^{-3}, the volume of the unit cell will be

$$\frac{4 \times 58.1}{2.48 \times 6.02 \times 10^{23}} = 1.56 \times 10^{-22} \text{ cm}^3 = 1.56 \times 10^{-22} \times 10^{24} \text{ Å}^3 = 156 \text{ Å}^3$$

The length of the side of the unit cell cube is $\sqrt[3]{156} = 5.38$ Å.

In a unit cell of the sodium chloride type, the ions come into direct contact along the edge of the cube, and the edge of the cube is of length $2d$, where d is the interionic distance. So in the case of KF, $d = 5.38/2 = 2.69$ Å.

6.4   X-Ray Diffraction and Bragg's Law

The use of x-ray diffraction provides a different method for investigating interatomic distances in crystalline solids, and constitutes one of the most powerful tools for discovering the structures of molecules. The principle of the method is very similar to the phenomenon of diffraction of light waves by a diffraction grating, but uses x-rays because their wavelength is of the same order as the distances between layers of atoms in a crystal. The basic formula governing the diffraction phenomenon may be worked out very easily by considering the geometry of the situation.

Figure 6.6 shows a beam of x-rays striking a crystal at an angle of θ and being reflected. The adjacent layers of atoms (or ions) in the crystal are represented by the two lines at a separation of d. The incoming x-ray beam starts as a plane wave-front AD and strikes the crystal lattice at an angle θ. Reflection occurs at angle θ and the wavefront of the leaving beam is shown at CH. However, the x-ray beam following the path ABC has obviously followed a shorter distance than the beam that followed the path DFH.

For any intensity of reflection under these conditions, it is essential for the x-ray beams following these two paths to be in phase. If they are not in phase, they will cancel out and give zero intensity. The condition for the beams to be in phase is that the *difference* in path length between the two shall be an integral (whole number) multiple of the wavelength of the radiation.

Looking at the diagram, we can easily work out the extra distance that the lower beam had to follow. Obviously $AB = DE$ and $BC = GH$, so the additional distance is the portion $EF + FG$. From the geometry of the arrangement, we know that BEF is a right-angled triangle with $BF = d$ and $\angle EBF = \theta$. Similarly BGF is a right-angled triangle with $\angle FBG = \theta$, so we can write

$$EF = FG = d \sin \theta$$

The total extra distance $EF + FG = 2d \sin \theta$.

(In any right-angled triangle, the sine of the angle θ is defined as

$$\frac{\text{length of side of triangle opposite } \theta}{\text{length of hypotenuse of triangle}}$$

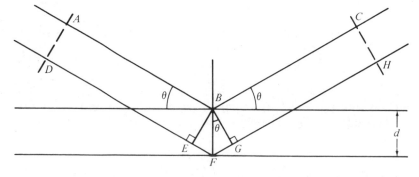

Figure 6.6   Reflection of x rays from a crystal.

(The hypotenuse is the side opposite the right angle.) A table of sines of angles, to the nearest degree, is given in Table 6.1.)

Now, our condition for the beams to be in phase was that this extra distance should be an integral number of wavelengths, which we can denote $n\lambda$, where n is an integer (whole number) and λ is the wavelength. Putting the two equal to each other gives

$$n\lambda = 2d \sin \theta$$

This important relationship is known as *Bragg's law*. It gives us the means to calculate the separation between layers of atoms in terms of the angles at which x-ray radiation of a certain wavelength will be reflected. The value of n is called the "order" of the reflection, and putting $n = 1, 2, 3$, etc. will give a group of reflection angles.

Example 6.10 We found in Example 6.6 that copper atoms had a radius of 1.28 Å. If a copper crystal contains layers of these atoms in contact, at what angles will a beam of x-rays of wavelength 1.80 Å be reflected?

SOLUTION We found previously that the distance between the centers of two adjacent copper atoms in a crystal (*twice* the radius) is 2.56 Å. This is the quantity d in Bragg's law, so we can write

$$\sin \theta = \frac{n\lambda}{2d} = \frac{n \times 1.80}{2 \times 2.56} = 0.352n$$

There are several solutions to this, depending on the value of n.

$n = 1$: $\sin \theta = 0.352$ $\theta = 21°$

$n = 2$: $\sin \theta = 0.704$ $\theta = 45°$

$n = 3$: $\sin \theta = 1.056$ $\theta = ?$

For $n = 3$, we have an impossible situation, since $\sin \theta$ cannot be more than 1.00 (Table 6.1). So only two reflections are observed from these layers of atoms.

In practice, the situation is a little more complex, since there are many other "planes" within the crystal, composed of atoms other than those directly in contact, and each set of planes will produce its own diffraction pattern. We will consider only the planes formed by layers of atoms in direct contact.

Example 6.11 Knowing that the interionic distance in CsCl is 3.57 Å, predict the angle at which a first-order reflection of x-rays of wavelength 1.75 Å will occur.

SOLUTION In this case, we can put d equal to the interionic separation, which will be the distance between adjacent layers of Cs^+ and Cl^- ions

$$\sin \theta = \frac{n\lambda}{2d} = \frac{1.75}{2 \times 3.57} \qquad n = 1 \text{ for first order}$$

$$\sin \theta = 0.245$$

$$\theta = 14°$$

Example 6.12   First-order reflection of x-rays of wavelength 1.90 Å from a metal crystal occurs at an angle of 28°.

(a) What is the spacing between layers in the crystal?
(b) Where will the second-order reflection appear?

SOLUTION

(a) In this case, we know $n = 1$ and $\lambda = 1.90$ Å. Bragg's law gives us

$$d = \frac{n\lambda}{2 \sin \theta} = \frac{1.90}{2 \sin \theta}$$

and since $\sin 28° = 0.469$,

$$d = \frac{1.90}{2 \times 0.469} = 2.02 \text{ Å}$$

(b) For the second-order reflection, we can recalculate $\sin \theta$ using the known value of d and putting $n = 2$:

$$\sin \theta = \frac{n\lambda}{2d} = \frac{2 \times 1.90}{2 \times 2.02} = 0.94$$

$$\theta = 70°$$

Information obtained by the use of Bragg's law may be combined with calculations based on density, and this approach gives us a means of estimating Avogadro's Number.

Example 6.13   X-ray diffraction measurements show that potassium crystallizes in a body-centered cubic structure with the edge length of the unit cell at 5.33 Å. If the density of potassium is 0.86 g cm^{-3}, calculate Avogadro's number.

SOLUTION   The volume of the unit cell is the cube of the edge length:

$$5.33^3 = 151 \text{ Å}^3 = 1.51 \times 10^{-22} \text{ cm}^3$$

Since a mole of potassium weighs 39.1 g, the weight of the unit cell (containing two atoms) will be $(2 \times 39.1)/N$ g (N = Avogadro's number). So the density of the unit cell will be

$$\frac{\text{mass}}{\text{volume}} = \frac{2 \times 39.1}{N \times 1.51 \times 10^{-22}} \text{ g cm}^{-3}$$

Knowing the density is 0.86, we can solve for N:

$$\frac{2 \times 39.1}{N \times 1.51 \times 10^{-22}} = 0.86 \qquad N = \frac{2 \times 39.1}{0.86 \times 1.51 \times 10^{-22}} = 6.0 \times 10^{23}$$

The above calculations using Bragg's law are simple examples of the technique of x-ray crystallography, which is one of the most powerful techniques we have for investigation of molecular structure. Provided that we can obtain a sample in crystalline form, we can determine the arrangement of atoms in the molecule of almost any substance by the way in which the crystal interacts with a beam of x-rays. For large molecules, the task is only possible with the aid of a large computer.

TABLE 6.1   Angles and Their Sines

Angle (degrees)	Sine	Angle (degrees)	Sine	Angle (degrees)	Sine
1	0.017	31	0.515	61	0.875
2	0.035	32	0.530	62	0.883
3	0.052	33	0.545	63	0.891
4	0.070	34	0.559	64	0.899
5	0.087	35	0.574	65	0.906
6	0.105	36	0.588	66	0.914
7	0.122	37	0.602	67	0.921
8	0.139	38	0.616	68	0.927
9	0.156	39	0.629	69	0.934
10	0.174	40	0.643	70	0.940
11	0.191	41	0.656	71	0.946
12	0.208	42	0.669	72	0.951
13	0.225	43	0.682	73	0.956
14	0.242	44	0.695	74	0.961
15	0.259	45	0.707	75	0.966
16	0.276	46	0.719	76	0.970
17	0.292	47	0.731	77	0.974
18	0.309	48	0.743	78	0.978
19	0.326	49	0.755	79	0.982
20	0.342	50	0.766	80	0.985
21	0.358	51	0.777	81	0.988
22	0.375	52	0.788	82	0.990
23	0.391	53	0.799	83	0.993
24	0.407	54	0.809	84	0.995
25	0.423	55	0.819	85	0.996
26	0.438	56	0.829	86	0.998
27	0.454	57	0.839	87	0.999
28	0.469	58	0.848	88	0.999
29	0.485	59	0.857	89	1.000
30	0.500	60	0.866	90	1.000

KEY WORDS

lattice energy
Born–Haber cycle
ionization energy
electron affinity
unit cell
simple cubic structure

body-centered cubic structure
face-centered cubic structure
cesium chloride structure
sodium chloride structure
interionic distance
Bragg's law

STUDY QUESTIONS

1. How are the various quantities in the Born–Haber cycle interconnected? KBr and NaCl have the same structure, but the lattice energy of NaCl is greater. Why?

2. Why do ionic compounds have high boiling points?

3. How do the various cubic structures compare in their efficiency of filling space? Why is the simple cubic structure not favored in practice?

4. What information do we need to calculate the radius of an atom in a solid metal?

5. How can we find Avogadro's number from this type of observation?

6. How does the CsCl structure (a) resemble (b) differ from the body-centered cubic lattice?

7. How does the NaCl structure (a) resemble (b) differ from the face-centered cubic lattice?

8. How is the interionic distance related to the size of the unit cell in (a) CsCl structure (b) NaCl structure? How many ions are present *inside* one unit cell of each?

9. Why are x-rays used in diffraction experiments with crystals?

10. What do we mean by "order" of a reflection? Why is there a limit to the number of different "orders" of reflections observed?

PROBLEMS

1. Calculate the lattice energy of rubidium fluoride, RbF, from the following data:

standard heat of formation of $RbF(s)$	-548 kJ mole^{-1}
heat of sublimation of $Rb(s)$	$+86$ kJ mole^{-1}
dissociation energy of $F_2(g)$	$+155$ kJ mole^{-1}
ionization energy of $Rb(g)$	$+402$ kJ mole^{-1}
electron affinity of $F(g)$	-346 kJ mole^{-1}

2. Calculate the lattice energy of calcium chloride, $CaCl_2$, from the following data:

standard heat of formation of $CaCl_2(s)$	-794 kJ mole^{-1}
heat of sublimation of $Ca(s)$	$+193$ kJ mole^{-1}

$$
\begin{array}{ll}
\text{dissociation energy of } Cl_2(g) & +242 \text{ kJ mole}^{-1} \\
\text{ionization energy of } Ca(g) \text{ to } Ca^{2+}(g) & +1725 \text{ kJ mole}^{-1} \\
\text{electron affinity of } Cl(g) & -347 \text{ kJ mole}^{-1}
\end{array}
$$

3. Calculate the electron affinity of $I(g)$ from the following data:

$$
\begin{array}{ll}
\text{standard heat of formation of } KI(s) & -327 \text{ kJ mole}^{-1} \\
\text{heat of sublimation of } K(s) & +90 \text{ kJ mole}^{-1} \\
\text{heat of sublimation of } I_2(s) & +62 \text{ kJ mole}^{-1} \\
\text{dissociation energy of } I_2(g) & +149 \text{ kJ mole}^{-1} \\
\text{ionization energy of } K(g) & +418 \text{ kJ mole}^{-1} \\
\text{lattice energy of } KI(s) & -633 \text{ kJ mole}^{-1}
\end{array}
$$

4. Calculate the electron affinity of the gaseous hydrogen atom from the following data (*Note*: The ions present in LiH are Li^+ and H^-.)

$$
\begin{array}{ll}
\text{standard heat of formation of } LiH(s) & -90 \text{ kJ mole}^{-1} \\
\text{heat of sublimation of } Li(s) & +155 \text{ kJ mole}^{-1} \\
\text{dissociation energy of } H_2(g) & +432 \text{ kJ mole}^{-1} \\
\text{ionization energy of } Li(g) & +518 \text{ kJ mole}^{-1} \\
\text{lattice energy of } LiH(s) & -906 \text{ kJ mole}^{-1}
\end{array}
$$

5. Calculate the radius of the following atoms:
   (a) cesium: crystallizes in a bcc lattice, density 1.873 g cm^{-3}
   (b) chromium: crystallizes in a bcc lattice, density 7.15 g cm^{-3}
   (c) silver: crystallizes in a fcc lattice, density 10.5 g cm^{-3}

6. Calculate the interionic distance in the following ionic crystals:
   (a) CsI: crystallizes in CsCl structure, density 4.51 g cm^{-3}
   (b) MgO: crystallizes in NaCl structure, density 3.58 g cm^{-3}
   (c) AgBr: crystallizes in NaCl structure, density 6.47 g cm^{-3}

7. (a) Ammonium bromide crystallizes in the CsCl structure with a density of 2.43 g cm^{-3}. Calculate the interionic distance.
   (b) If the radius of a bromide ion is 1.95 Å, what is the radius of an ammonium ion?

8. Nickel sulfide, NiS, crystallizes in the NaCl structure with an interionic distance of 2.40 Å. What will be the density of the crystals?

9. Lead crystallizes in the fcc structure with a density of 11.34 g cm^{-3}. The length of the unit cell is 4.95 Å. Use these figures to calculate a value of Avogadro's number.

10. Calculate the spacing between layers in a crystal that reflect x-rays in the following ways:
    (a) wavelength 1.98 Å undergo a first-order reflection at 23°
    (b) wavelength 2.06 Å undergo a second-order reflection at 35°
    (c) wavelength 2.11 Å undergo a first-order reflection at 16°

11. Calculate the angles at which first-order and second-order reflection of x-rays will occur when:
    (a) wavelength 2.05 Å hit a crystal with spacings at 3.00 Å
    (b) wavelength 1.88 Å hit a crystal with spacings at 2.50 Å
    (c) wavelength 1.93 Å hit a crystal with spacings at 2.90 Å

12. (a) What is the wavelength of an x-ray beam which gives a first-order reflection at an angle of 18° from a crystal with spacings at 2.20 Å?
    (b) At what angle will the second-order reflection be found?

SOLUTIONS TO PROBLEMS

1. Setting up a Born–Haber cycle,

$$Rb(s) + \frac{1}{2} F_2(g) \xrightarrow{-548} RbF(s)$$

$$\downarrow +86 \qquad \downarrow \frac{+155}{2} \qquad \nwarrow x \text{ (lattice energy)}$$

$$Rb(g) + F(g) \xrightarrow{+402 - 346} Rb^+(g) + F^-(g)$$

$$-548 = +86 + \frac{155}{2} + 402 - 346 + x$$

$$x = -768 \text{ kJ mole}^{-1}$$

2. Setting up a Born–Haber cycle,

$$Ca(s) + Cl_2(g) \xrightarrow{-794} CaCl_2(s)$$

$$\downarrow +193 \qquad \downarrow +242 \qquad \nwarrow x \text{ (lattice energy)}$$

$$Ca(g) + 2\,Cl(g) \xrightarrow[2 \times (-347)]{+1725} Ca^{2+}(g) + 2Cl^-(g)$$

Note that we dissociate Cl_2 to give 2Cl (atomic) and hence $2Cl^-$ for one mole $CaCl_2$.

$$-794 = +193 + 242 + 1725 - 694 + x$$

$$x = -2260 \text{ kJ mole}^{-1}$$

3. Setting up a Born–Haber cycle,

$$K(s) + \frac{1}{2} I_2(s) \xrightarrow{-327} KI(s)$$

$$\left| \frac{+62}{2} \right. \downarrow$$

$$\left| {+90} \right. \qquad \frac{1}{2} I_2(g) \qquad \nearrow^{-633}$$

$$\left| \frac{+149}{2} \right. \downarrow$$

$$K(g) + I(g) \xrightarrow{+418 + x} K^+(g) + I^-(g)$$

(x = electron affinity of I.)

$$-327 = 90 + \frac{62}{2} + \frac{149}{2} + 418 + x - 633$$

$$x = -308 \text{ kJ mole}^{-1}$$

4. Setting up a Born–Haber cycle,

$$Li(s) + \frac{1}{2} H_2(g) \xrightarrow{-90} LiH(s)$$

$$\left| {+155} \right. \downarrow \qquad \left| \frac{+432}{2} \right. \downarrow \qquad \nearrow^{-906}$$

$$Li(g) + H(g) \xrightarrow{+518 + x} Li^+(g) + H^-(g)$$

(x = electron affinity of hydrogen.)

$$-90 = 155 + \frac{432}{2} + 518 + x - 906$$

$$x = -73 \text{ kJ mole}^{-1}$$

5. (a) mass of unit cell $= \dfrac{2 \times 133}{6.02 \times 10^{23}} = 4.42 \times 10^{-22}$ g

volume $= \dfrac{4.42 \times 10^{-22}}{1.873} = 2.36 \times 10^{-22}$ cm^3 = 236 Å^3

length of side $= \sqrt[3]{236} = 6.18$ Å
body diagonal $= \sqrt{3} \times 6.18 = 10.7$ Å
atomic radius $10.7/4 = 2.68$ Å

(b)   mass of unit cell $= \dfrac{2 \times 52.0}{6.02 \times 10^{23}} = 1.73 \times 10^{-22}$ g

volume $= \dfrac{1.73 \times 10^{-22}}{7.15} = 2.42 \times 10^{-23}$ cm$^3 = 24.2$ Å^3

length of side $= \sqrt[3]{24.2} = 2.89$ Å
body diagonal $= \sqrt{3} \times 2.89 = 5.01$ Å
atomic radius $= 5.01/4 = 1.25$ Å

(c)   mass of unit cell $= \dfrac{4 \times 108}{6.02 \times 10^{23}} = 7.18 \times 10^{-22}$ g

volume $= \dfrac{7.18 \times 10^{-22}}{10.5} = 6.83 \times 10^{-23}$ cm$^3 = 68.3$ Å^3

length of side $= \sqrt[3]{68.3} = 4.09$ Å
side diagonal $= \sqrt{2} \times 4.09 = 5.78$ Å
atomic radius $= 5.78/4 = 1.45$ Å

6. (a)   mass of unit cell $= \dfrac{133 + 127}{6.02 \times 10^{23}} = 4.32 \times 10^{-22}$ g

volume $= \dfrac{4.32 \times 10^{-22}}{4.51} = 9.58 \times 10^{-23}$ cm$^3 = 95.8$ Å^3

length of side $= \sqrt[3]{95.8} = 4.58$ Å
body diagonal $= \sqrt{3} \times 4.58 = 7.92$ Å
interionic distance $= 7.92/2 = 3.96$ Å

(b)   mass of unit cell $= \dfrac{(24.3 + 16.0) \times 4}{6.02 \times 10^{23}} = 2.68 \times 10^{-22}$ g

volume $= \dfrac{2.68 \times 10^{-22}}{3.58} = 7.48 \times 10^{-23}$ cm$^3 = 74.8$ Å^3

length of side $= \sqrt[3]{74.8} = 4.21$ Å
interionic distance $= 4.21/2 = 2.11$ Å

(c)   mass of unit cell $= \dfrac{(108 + 80) \times 4}{6.02 \times 10^{23}} = 1.25 \times 10^{-21}$ g

volume $= \dfrac{1.25 \times 10^{-21}}{6.47} = 1.93 \times 10^{-22}$ cm$^3 = 193$ Å^3

length of side $= \sqrt[3]{193} = 5.78$ Å
interionic distance $= 5.78/2 = 2.89$ Å

7.   mass of unit cell $= \dfrac{18.0 + 79.9}{6.02 \times 10^{23}} = 1.63 \times 10^{-22}$ g

volume $= \dfrac{1.63 \times 10^{-22}}{2.43} = 6.69 \times 10^{-23}$ cm$^3 = 66.9$ Å^3

length of side = $\sqrt[3]{66.9}$ = 4.06 Å
body diagonal = $\sqrt{3}$ × 4.06 = 7.03 Å
interionic distance = 7.03/2 = 3.52 Å
radius of NH_4^+ ion = 3.52 − 1.95 = 1.57 Å

8. length of side of unit cell = 2 × 2.40 = 4.80 Å
volume of unit cell = $(4.80)^3$ = 111 $Å^3$ = 1.11 × 10^{-22} cm^3

$$\text{mass of unit cell} = \frac{(58.7 + 32.1) \times 4}{6.02 \times 10^{23}} = 6.03 \times 10^{-22} \text{ g}$$

$$\text{density} = \frac{6.03 \times 10^{-22} \text{ g}}{1.11 \times 10^{-22} \text{ cm}^3} = 5.45 \text{ g cm}^{-3}$$

9. volume of unit cell = $(4.95)^3$ = 121 $Å^3$ = 1.21 × 10^{-22} cm^3

$$\text{mass of unit cell} = \frac{4 \times 207}{N} = \frac{828}{N} \text{ g}$$

$$\text{density} = \frac{828}{N \times 1.21 \times 10^{-22}} = 11.34 \qquad \text{(given value)}$$

$$N = \frac{828}{11.34 \times 1.21 \times 10^{-22}} = 6.03 \times 10^{23}$$

10. From $n\lambda = 2d \sin \theta$, $d = \dfrac{n\lambda}{2 \sin \theta}$.

(a) $n = 1$, $\lambda = 1.98$ Å, $\theta = 23°$

$$d = \frac{1.98}{2 \times \sin 23°} = \frac{1.98}{2 \times 0.391} = 2.53 \text{ Å}$$

(b) $n = 2$, $\lambda = 2.06$ Å, $\theta = 35°$

$$d = \frac{2 \times 2.06}{2 \times \sin 35°} = \frac{2.06}{0.574} = 3.59 \text{ Å}$$

(c) $n = 1$, $\lambda = 2.11$ Å, $\theta = 16°$

$$d = \frac{2.11}{2 \times \sin 16°} = \frac{2.11}{2 \times 0.276} = 3.83 \text{ Å}$$

11. From $n\lambda = 2d \sin \theta$, $\sin \theta = n\lambda/2d$.
(a) $d = 3.00$ Å, $\lambda = 2.05$ Å

$$n = 1: \quad \sin \theta = \frac{2.05}{2 \times 3.00} = 0.342 \qquad \theta = 20°$$

$$n = 2: \quad \sin \theta = \frac{2 \times 2.05}{2 \times 3.00} = 0.683 \qquad \theta = 43°$$

(b) $d = 2.50$ Å, $\lambda = 1.88$ Å

$$n = 1: \quad \sin \theta = \frac{1.88}{2 \times 2.50} = 0.376 \qquad \theta = 22°$$

$$n = 2: \quad \sin \theta = \frac{2 \times 1.88}{2 \times 2.50} = 0.752 \qquad \theta = 49°$$

(c) $d = 2.90$ Å, $\lambda = 1.93$ Å

$$n = 1: \quad \sin \theta = \frac{1.93}{2 \times 2.90} = 0.333 \qquad \theta = 19°$$

$$n = 2: \quad \sin \theta = \frac{2 \times 1.93}{2 \times 2.90} = 0.666 \qquad \theta = 42°$$

12. $d = 2.20$ Å, $\theta = 18°$, and $n = 1$

$$\lambda = \frac{2d \sin \theta}{n} = \frac{2 \times 2.20 \sin 18°}{1} = 2 \times 2.20 \times 0.309 = 1.36 \text{ Å}$$

If $n = 2$, $\sin \theta = \dfrac{2 \times 1.36}{2 \times 2.20} = 0.618 \qquad \theta = 38°.$

Chapter 7

The Shapes of Covalent Molecules

From a drop of water, a logician could infer the possibility of an Atlantic or a Niagara without having seen or heard of one or the other. So all life is a great chain, the nature of which is known whenever we are shown a single link of it.
A STUDY IN SCARLET

When discussing the ionic bond, we regarded each ion as a sphere, which attracted ions of opposite charge from all directions equally. The arrangement of ions around each other depends largely on their relative sizes, and the way they can be fitted together to give the most favorable lattice energy.

With covalent molecules, the picture is quite different. If a single atom forms two or more covalent bonds to other atoms, these bonds will be at definite angles to each other. The factors determining the arrangement and angle of the bonds at the central atom will be largely a result of the number and arrangement of the electrons on the central atom, and the nature of the other atoms attached to it has only a secondary effect on the resulting structure.

Putting it simply, covalent molecules always have shape. By "shape," we mean the arrangement of the atoms and the angles between the bonds. In this chapter, we will see how considerable insight into these shapes may be found by simply counting electrons. We are not here discussing any theory of bonding, we are simply counting electrons and arranging atoms. If you can use the periodic table and count up to 12 (sometimes a little higher) then you can work out the shape of a molecule.

We should note one important qualification before going any further; we will restrict our discussions to compounds of the main-group elements only. The structures of compounds of transition elements are complicated by the presence of partly filled d orbitals, and we will not consider them in this book.

7.1 Electron Pairs

Let's consider first a typical covalent molecule AB_x, in which x atoms of some general element B (not boron!) are joined to the central atom A by x single covalent bonds. We might think at first that the structure and shape of the molecule would be fixed by the value of x. For $x = 3$, perhaps, the three atoms of B would arrange themselves in a triangle with atom A at its center. Experiment shows, however, that this is not necessarily the case. Although a flat triangle arrangement for an AB_3 molecule is sometimes found, several other structures are possible. Examination of many molecules leads to the following important result:

In a molecule AB_x, the factor determining the shape of the molecule is not the value of x, but the number of electron pairs in the valence shell of atom A in the compound.

The valence shell is the shell in atom A, containing electrons, which has the *highest* value of the principal quantum number (see Chapter 5). It is the electrons in this shell that pair up to form covalent bonds. Generally speaking, all the electrons in a compound of a main-group element will be paired (this is by no means the case with compounds of transition elements, but we are not considering them). There are very few common compounds of main-group elements where unpaired electrons are present: O_2, NO, NO_2, ClO_2, are some examples.

So we have a picture of a central atom A surrounded in its valence shell by a number of *pairs* of electrons. How will these arrange themselves? This question can easily be answered if we remember that electrons are all charged negatively, so they repel each other strongly. They will arrange themselves around the central atom in such a way as to get as far apart as possible.

Let's consider how this may be achieved for various numbers of electron pairs:

1. One electron pair. Obviously there is no problem here; the one pair can move around freely without encountering any repulsive forces from other pairs.

2. Two electron pairs. This again is a simple case. The pairs will arrange themselves on diametrically opposite sides of the atom to give an angle of 180° between them. In this way the repulsive force between them is minimized.

$$\text{-} \colon A \colon \text{-}$$

3. Three electron pairs. The arrangement is a planar triangle, with angles of 120° between the electron pairs.

4. Four electron pairs. There are two symmetrical ways of arranging four objects around another object.

| square planar | tetrahedral |

The square-planar arrangement, as the name implies, is two dimensional, with the electron pairs at the corner of a square. The tetrahedral arrangement is three dimensional, with the electrons at the corners of a tetrahedron. (*Note*: A tetrahedron is a triangular pyramid, a four-sided solid with each side an equilateral triangle. The corners of the tetrahedron are the alternate corners of a cube.)

As the diagrams show, the angles between electron pairs in the square-planar structure are 90°, whereas in the tetrahedral arrangement the angles are 109.5°, the so-called "tetrahedral angle." Obviously, the repulsions will be less in the latter arrangement, so, in practice, this is always found in compounds of main-group elements.

5. Five electron pairs. There is only one perfectly symmetrical way to arrange five objects around another object, and that is in a planar pentagon. In this arrangement, all angles between adjacent electron pairs are 72°, which means they are going to repel each other strongly. Obviously a three-dimensional arrangement would be better than a flat structure, and the most common one is based on the trigonal bipyramid (tbp).

planar pentagon   trigonal bipyramid (tbp)

The trigonal bipyramid is simply two triangular pyramids stuck together base-to-base. In other words, there are three pairs of electrons in an equilateral triangle around atom A, with one other pair above and below. The former three pairs are called "equatorial," while the latter two are called "axial" pairs. The angles between the equatorial pairs are 120°, while the axial pairs make an angle of 90° with the equatorial pairs.

Obviously, the angles between electron pairs are all much greater in the tbp arrangement than in the planar pentagon, so the repulsions are less in the tbp structure and the planar pentagon is never found.

6. Six electron pairs. Here, a very attractive, symmetrical arrangement is always found in practice—octahedral. The octahedron is an eight-sided solid, each face being an equilateral triangle. If we put one electron pair at each of the six corners of the octahedron we will minimize repulsions between them. All angles between adjacent pairs will be 90°.

7. Seven electron pairs and up. There are comparatively few compounds known with more than six electron pairs around the central atom, and the structure of each one has to be determined individually.

Table 7.1 summarizes the conclusions so far.

Now we can apply these ideas to some real compounds. In choosing examples in this chapter, we shall have to find some suitable atoms of type B to join onto our central atom A. The requirement for a type-B atom is that it shall form a single covalent bond to various other atoms, and this is easily fulfilled by choosing B atoms from hydrogen or the halogens (F, Cl, Br, I).

Looking at compounds of some first-row elements (Li—Ne), we find that Be forms the fluoride BeF_2. Since Be started with two electrons in its valence shell ($2s^2$) and has paired both up with electrons from F atoms, it now has two electron pairs in its valence shell.

TABLE 7.1   **Electron-Pair Arrangements**

Electron pairs	Arrangement	Bond angles
2	linear	180°
3	planar triangle	120°
4	tetrahedral	109.5°
5	trigonal bipyramid	120° and 90°
6	octahedral	90°

We would expect these to be at 180°, and this is indeed found. The bond angle in BeF_2 is 180° and the three atoms lie in a straight line.

$$F—Be—F$$

Boron, the next element, starts with three electrons $(2s^2 2p^1)$ and pairs them all up with fluorine atoms in forming BF_3. The three electron pairs around the B atom in BF_3 arrange themselves at angles of 120° and the molecule has a planar triangular structure.

$$
\begin{array}{c}
F \\
| \\
B \\
F \quad\quad F
\end{array}
$$

Remember that "planar" here includes the B atom as well as the three F atoms. All four are in the same plane, and the whole molecule is flat.

Carbon starts with four electrons $(2s^2 2p^2)$ and pairs them all up with, e.g., hydrogen atoms to form methane, CH_4. The four electron pairs arrange themselves tetrahedrally and the bond angles are all 109.5° (see Figure 7.1).

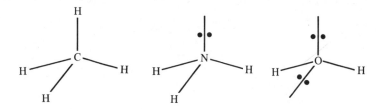

Figure 7.1   The arrangement of four electron pairs (tetrahedral).

The previous three compounds have been formed by pairing up *all* of the electrons originally present in the valence shells of the central atoms. With the next first-row element, nitrogen, we find a difference. Although nitrogen starts out with five electrons $(2s^2 2p^3)$, only three of them are paired up to form compounds such as NH_3, NF_3, etc. There is, consequently, one pair in the original valence shell of the N atom which is still present in the NH_3 molecule. This is called a "nonbonding" pair (sometimes a "lone" pair). So the N atom in NH_3 has *four* electron pairs in its valence shell, three bonding pairs, which attach the three H atoms, and one nonbonding pair. As we have carefully specified that we must count the *total* number of electron pairs around the central atom, rather than the number of bonds formed, we should expect this to affect the structure, and this is indeed the case. The molecule of NH_3 (and NF_3, etc.) is pyramidal, rather than planar, with the N atom at the top of a triangular pyramid. The structure is derived from the tetrahedral CH_4 by knocking off one corner. The nonbonding electron pair shown in the diagram cannot be directly observed, but all properties of the molecule are consistent with its presence in that position. The H—N—H bond angle is 107.3°, very similar to the tetrahedral angle.

Oxygen, the next element in the first row, starts with six electrons in its valence shell $(2s^2 2p^4)$. Only two of these pair up to form H_2O, F_2O, etc., so compounds of this type have two nonbonding pairs remaining around the oxygen atom, in addition to the two bonding pairs, for a total of four pairs. The structures reflect this arrangement. The water molecule, for example, is not linear, but is bent with an H—O—H angle of 104.5°. The structure in the diagram resembles a tetrahedron with two corners knocked off, but again the position of the nonbonding pairs is inferred, rather than directly observed.

Fluorine, the element following oxygen, forms only a single bond with hydrogen in the molecule HF. Since fluorine started with seven electrons $(2s^2 2p^5)$ in its valence shell, we can assume that six of them are left undisturbed when the seventh bonds with an atom of H, so in HF the F atom has in its valence shell one bonding pair and three nonbonding pairs. Although we may suggest a tetrahedral arrangement for these four pairs, there is no experimental means of confirming that this is the case.

We see from the summary in Table 7.2 that the geometrical arrangement of the electron pairs around the central atom will be the same as the shape of the molecule *only if* the electron pairs are all bonding. As soon as we introduce nonbonding pairs, the electron-pair arrangement must differ from the shape of the molecule.

For first-row elements, we are limited to a total of four electron pairs in the valence shell of the central atom, but with second- or third-row elements, we can exceed this value. Thus, phosphorus forms the fluorides PF_3 (similar to NF_3) and also PF_5, in which five bonding electron pairs are formed by pairing with fluorine atoms all five of the electrons originally present on P $(3s^2 3p^3)$. The five pairs arrange themselves in a tbp structure, and so also do the five F atoms.

Sulfur forms many divalent compounds, e.g., H_2S, and also forms SF_4. Since the S atom started with six valence-shell electrons $(3s^2 3p^4)$, it must have one nonbonding pair left after pairing up four with F atoms, for a total of five pairs. Here we have a tbp arrangement

TABLE 7.2   Up to Four Electron Pairs

Compound	Electrons in original atom (valence shell)	Bonding pairs formed	Nonbonding pairs	Total electron pairs	Arrangement of electrons	Shape of molecule
BeF_2	2	2	0	2	linear	linear
BF_3	3	3	0	3	planar triangle	planar triangle
CH_4	4	4	0	4	tetrahedral	tetrahedral
NH_3	5	3	1	4	tetrahedral	pyramidal
H_2O	6	2	2	4	tetrahedral	bent
HF	7	1	3	4	tetrahedral	linear

of electron pairs, the nonbonding pair being in an equatorial position. The shape of this molecule is usually described as "distorted tetrahedral" (see Figure 7.2).

Chlorine combines with fluorine to form ClF_3, in which three of the original seven electrons on the Cl atom $(3s^2 3p^5)$ have paired up, leaving two nonbonding pairs. So we have again a total of five electron pairs in a tbp arrangement, with a T-shaped molecule (both nonbonding pairs are found in equatorial positions).

Finally in this series, we have the compound XeF_2, in which two of the original eight electrons on $Xe(5s^2 5p^6)$ have paired up, leaving no less than three nonbonding pairs. The resulting molecule is linear, but the electron pairs are still in a trigonal bipyramid.

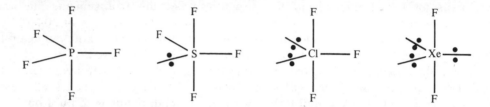

Figure 7.2   The arrangement of five electron pairs (trigonal bipyramidal).

Table 7.3 summarizes the data for five electron pairs, and the structures are shown in Figure 7.2. Note, as before, that the electron-pair arrangement is the same as the molecular geometry only when no nonbonding electron pair is present.

TABLE 7.3   Five Electron Pairs

Compound	Electrons in original atom (valence shell)	Bonding pairs formed	Nonbonding pairs	Total electron pairs	Arrangement of electrons	Shape of molecule
PF_5	5	5	0	5	tbp	tbp
SF_4	6	4	1	5	tbp	distorted tetrahedral
ClF_3	7	3	2	5	tbp	T-shaped
XeF_2	8	2	3	5	tbp	linear

Turning now to compounds in which the central atom has six electron pairs in its valence shell, we find that all six of sulfur's valence electrons can be paired up to form SF_6. Both electron-pair arrangement and molecular shape are octahedral.

When bromine forms BrF_5, five of the original seven electrons $(4s^2 4p^5)$ are paired up, leaving one nonbonding pair and a total of six pairs. The electron pairs are octahedral, but the molecule is best described as a square pyramid (see Figure 7.3).

Finally in this group, we have the xenon compound XeF_4. Of the xenon atom's original eight electrons, four are paired up, while four remain as two nonbonding pairs. In the molecule, these two pairs are at an angle of 180°, so the shape of the molecule itself is a planar square. Data on compounds with six electron pairs are summarized in Table 7.4 and the structures shown in Figure 7.3.

Seven electron pairs on a central atom is a rare occurrence, but this arrangement is found in the molecule IF_7, where all seven electrons on the iodine atom $(5s^2 5p^5)$ are

TABLE 7.4   Six Electron Pairs

Compound	Electrons in original atom (valence shell)	Bonding pairs formed	Nonbonding pairs	Total electron pairs	Arrangement of electrons	Shape of molecule
SF_6	6	6	0	6	octahedral	octahedral
BrF_5	7	5	1	6	octahedral	square pyramid
XeF_4	8	4	2	6	octahedral	square planar

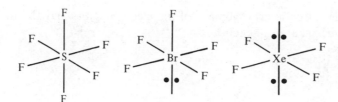

Figure 7.3   The arrangement of six electron pairs (octahedral).

paired up to give seven bonding pairs. The structure of this molecule is a pentagonal bipyramid, that is, five fluorine atoms are around the iodine atom at the corners of a pentagon, with one fluorine above and one below.

Xenon forms the fluoride XeF_6, in which six of the xenon atom's original eight electrons have been paired up, leaving one nonbonding pair for a total of seven pairs. The structure of this molecule is not yet known with certainty, but appears to be consistent with the rules we have discussed.

The preceding discussion may be compressed into a set of rules for finding the shape of a covalent molecule of the type AB_x, where there are single bonds between atoms A and B.

1. Find the number of electrons originally present in the valence shell of atom A (before combination). This information comes from the position of A in the periodic table (remember that the number of electrons in the valence shell is always equal to the number of the main group in which the element is found). Suppose we have found that A had originally y electrons in its valence shell, where y must be in the range 1–8 inclusive.

2. The number of bonds formed on atom A is x. Each of these requires *one* of the electrons from atom A, which pairs with one from a B atom to give a total of x pairs of bonding electrons.

3. After x electrons have been paired up, there are $y - x$ electrons remaining in the valence shell of atom A that are not used in bonding. These will be arranged in pairs, so we have $(y - x)/2$ pairs of nonbonding electrons. (*Note:* The value of $y - x$ will almost invariably be divisible by two.)

4. The *total* number of electron pairs in the valence shell of atom A will therefore be

$$x + \frac{y - x}{2} = \frac{x + y}{2}$$

This must give a whole number in the range $1 - 6$ (or, rarely, 7), and these electron pairs will arrange themselves as shown in Table 7.1.

5. All that remains to be done is to arrange the x atoms of B around atom A to give the molecular shape. The position of the B atoms is dictated by the location of the electron pairs that constitute the bonds holding them in position. Where all the electron pairs are bonding pairs, this is very straightforward. In cases where some pairs are to be bonding and some nonbonding, we work on the assumption that the *nonbonding* pairs take up more room. Thus in SF_4, ClF_3 or XeF_2, the nonbonding pairs all go into equatorial positions in the tbp structure, because bond angles here are $120°$, which means less repulsive forces than in the axial positions, where bond angles are $90°$. In XeF_4, the two nonbonding pairs are at opposite corners of the octahedron ($180°$ angle), where the repulsions are less than they would be if they were at adjacent corners ($90°$ angle).

Example 7.1   Work out the arrangement of the electron pairs and the molecular shape of the following. Give a sketch of each.

(a) SiF_4 (b) PBr_3
(c) $SeCl_4$ (d) OF_2

SOLUTION

(a) Silicon is in group IV; it starts with four valence-shell electrons ($3s^2 3p^2$), so $y = 4$. Four bonds are formed, so $x = 4$. The total electron pairs on Si is, therefore $(x + y)/2 = 4$. The arrangement of four electron pairs is tetrahedral; and all are bonding pairs, attaching four F atoms, so the molecular shape is also tetrahedral.

$$F$$
$$|$$
$$Si$$
$$F \diagup \; \diagup \; F$$
$$F$$

(b) Phosphorus is in group V; it starts with five valence-shell electrons ($3s^2 3p^3$), so $y = 5$. Three bonds are formed, so $x = 3$. Therefore, the total electron pairs on P is $(x + y)/2 = 4$. The arrangement of four electron pairs is tetrahedral. Only three of these are bonding pairs, attaching the three Br atoms, so the molecular shape is pyramidal.

$$\cdot\dagger$$
$$P$$
$$Br \diagup \; \diagup \; Br$$
$$Br$$

(c) Selenium is in group VI; it starts with six valence-shell electrons ($4s^2 4p^4$), so $y = 6$. Four bonds are formed, so $x = 4$. Therefore, the total electron pairs on Se is $(x + y)/2 = 5$. The arrangement of five electron pairs is trigonal bipyramid. Only four of these are bonding pairs; so the molecular shape is "distorted tetrahedral" (i.e., tbp with one corner knocked off). The nonbonding electron pair is in the equatorial position.

$$\overset{\displaystyle Cl}{\underset{\displaystyle Cl}{\underset{\diagdown}{Cl}\overset{|}{\underset{|}{Se}}-Cl}}$$

(d) Oxygen is in group VI; it starts with six valence shell electrons $(2s^2 2p^4)$ so $y = 6$. Two bonds are formed, so $x = 2$. Therefore the total electron pairs on O is $(x + y)/2 = 4$. The arrangement of four electron pairs is tetrahedral. Only two of these are bonding pairs, so the molecular shape is bent (tetrahedral with two corners knocked off).

$$\overset{\displaystyle \dagger}{\underset{\displaystyle F \diagup \hspace{-0.6em}\diagup \quad F}{\overset{O}{\diagup \quad \diagdown}}}$$

Example 7.2 Give examples of molecules having the following characteristics:

(a) Total of four electron pairs around central atom, of which three are bonding pairs.
(b) Total of six electron pairs around central atom, of which five are bonding pairs.
(c) Octahedral molecule with octahedral arrangement of electron pairs.
(d) Bent molecule with tetrahedral arrangement of electron pairs.

SOLUTION   This is the same problem in reverse. We can work backwards from the structure of the compound to find out how many electrons were present in the valence shell of the original atom, then choose a suitable example with the aid of the periodic table.

(a) The original central atom must have contained *one* electron from each of the three bonding pairs plus *two* electrons of the nonbonding pair, for a total of five electrons. It must be a Group V element, so nitrogen would be an obvious choice. Taking hydrogen for the B-type element gives NH_3 as an example of this structure.

(b) Similarly, we count *one* electron from each of five bonding pairs, plus two from the nonbonding pair for a total of seven electrons originally present. This must be a group VII element, of which Br would be an example. Taking fluorine for the other atom gives BrF_5 as an example with this structure.

(c) An octahedral molecule in which six electron pairs are arranged octahedrally contains no nonbonding pair. The only electrons present in the original atom's valence shell were the six that have now formed the six pairs by bonding. So we need a group VI element, of which sulfur is an obvious example, and our molecule would be, for example, SF_6.

(d) In a molecule with four electron pairs arranged tetrahedrally, but in a bent structure, there must be two bonding pairs holding on two atoms at an angle close to the tetrahedral angle, while the two remaining pairs are nonbonding. Hence the original central atom contained in its valence shell the four electrons that now make up the two nonbonding pairs, plus an additional two electrons used in forming the two bonds, for a total of six. Again this must be a group VI element; and, taking sulfur again, this structure would be found in, e.g., H_2S.

In selecting appropriate examples of compounds to illustrate structures, as in Example 7.2, we should always be chemically realistic. Thus, it would be wrong to select a first-row element for our central atom if it is required to accommodate more than four electron pairs in its valence shell in the molecule, for this is not possible. Similarly, when we need to form a large number of bonds to our central atom, we should choose our B-type atoms from fluorine or chlorine, because these elements form high-energy bonds to other atoms which stabilize such structures.

Thus, in Example 7.2(c), it would be optimistic to suggest SH_6 as an example, as such a compound is not known and is unlikely to be capable of stable existence. However, a student suggesting this example would be given credit for it in this context, since, if it *did* exist, it would have the structure required.

The compounds that must definitely be avoided are those involving the more electropositive metals, particularly groups I and II (except Be), since they form almost exclusively ionic, rather than covalent, compounds.

7.2 Shapes of Ions

Ions are said to be either *simple*—which means that only one atom is present (K^+, Br^-, etc.) or *complex*—when more than one atom is present (NH_4^+, BF_4^-, etc.). In the case of complex ions, the atoms within the ion are held together with covalent bonds, and the rules governing their structures are exactly the same as those for neutral molecules. A complex ion is simply a molecule with a charge on it. There is only one difference in the calculation; we have to add or subtract electrons in accordance with the charge on the ion. This modifies our rule for calculating the total number of electron pairs in the valence shell of the central atom of the molecule. Thus, the number of electron pairs in the valence shell of the central atom in the ion $[AB_x]^n$ is $(x + y - n)/2$, where x = number of bonds formed, y = number of electrons originally present in the valence shell of A, and n = the charge on the ion, with attention to sign. (*Note*: in this expression, we *subtract* n because each added electron has a negative charge.)

Example 7.3   What are the electron arrangements and shapes of the following ions:

(a) NH_4^+ (b) H_3O^+ (c) BH_4^- (d) SiF_6^{2-} (e) $HgCl_3^-$

SOLUTION
(a) Nitrogen started with five valence electrons ($y = 5$) and has formed four bonds to hydrogen ($x = 4$) in an ion with a charge of $+1$ (n). So

$$\text{electron pairs} = \frac{4 + 5 - 1}{2} = 4 \text{ pairs} \quad \text{(tetrahedral)}$$

To hold on four H atoms, all must be bonding pairs, so the ion is tetrahedral.
(b) Here we have initially six electrons on oxygen ($y = 6$) and it forms three bonds to hydrogen ($x = 3$) in an ion with a $+1$ charge (n).

$$\text{electron pairs} = \frac{6 + 3 - 1}{2} = 4 \text{ pairs} \quad \text{(tetrahedral)}$$

Since there are only three H atoms attached by bonding pairs, we have one nonbonding pair and a pyramidal ion.

(c) Boron starts with three electrons ($y = 3$) and forms four bonds ($x = 4$) in an ion of charge -1 (n).

$$\text{electron pairs} = \frac{4 + 3 - (-1)}{2} = 4 \text{ pairs} \qquad \text{(tetrahedral)}$$

All four are bonded to hydrogen, so the ion is tetrahedral.

ammonium ion   hydronium ion   borohydride ion

(d) The silicon atom (Group IV) starts with four electrons ($y = 4$) and forms six bonds ($x = 6$) in an ion with a charge of -2 (n).

$$\text{electron pairs} = \frac{6 + 4 - (-2)}{2} = 6 \text{ pairs} \qquad \text{(octahedral)}$$

All six are bonded to fluorine, so the ion is octahedral.

(e) Mercury, which, although metallic, forms many covalently bonded compounds, starts out with two electrons in its valence shell ($6s^2$; the $5d$ shell is full and Hg is not usually thought of as a transition element) so $y = 2$. In the $HgCl_3^-$ it is forming three bonds ($x = 3$) and the charge is -1 (n).

$$\text{electron pairs} = \frac{3 + 2 - (-1)}{2} = 3 \text{ pairs} \qquad \text{(planar triangle)}$$

All are bonding pairs, holding on the three Cl atoms, so the ion is planar triangular in shape.

The approach taken to molecular structure in this chapter is often called the VSEPR (pronounced "vesper") approach, meaning Valence Shell Electron Pair Repulsion. It is not intended to provide a sophisticated picture of bonding in terms of molecular orbitals, but simply to give a method for accounting for the observed structures of a large number of compounds in terms of a few simple rules.

7.3 Hybridization of Orbitals

We are mainly concerned in this book with the numerical aspects of chemistry, so we will not attempt here a discussion of the nature of the covalent bond. However, one additional feature can be added to our description of the structure of a covalent molecule or ion: namely, a note of the orbitals on the central atom which contain the valence-shell electrons. (We are not concerned with the filled inner orbitals which take little part in the bonding.) We make the following assumptions:

1. The electrons in the valence shell of the central atom are arranged in pairs (the number of compounds in which unpaired electrons are present in the valence shell is so small that we may neglect this possibility).

2. Each pair of electrons in the valence shell, whether bonding or nonbonding, will occupy an orbital. The separate orbitals are identified by sets of quantum numbers, and occupy different regions in space.

3. The orbitals in the valence shell of an atom that is bonded to other atoms will be related to the orbitals originally present in the atom before bonding occurred, but some changes in their character will be associated with bond formation.

4. The characteristics of the orbitals used by the central atom in bonding must be consistent with the observed structure of the compound. For example, if an atom forms bonds to two other atoms at a 180° angle, the orbitals used to accommodate the electron pairs forming these bonds must also be at a 180° angle.

Let's apply these ideas to a simple example, the methane molecule, CH_4. Carbon starts out with four electrons in the valence shell of the atom, $2s^2 2p^2$.

$$2s \qquad 2p$$

C ground-state atom $\boxed{\uparrow\downarrow}$ $\boxed{\uparrow\,|\,\uparrow\,|\,}$

In CH_4, all these are paired up to give four pairs (all bonding). If we try placing four pairs of electrons in the above orbitals we get

$$2s \qquad 2p$$

C in CH_4 molecule $\boxed{\uparrow\downarrow}$ $\boxed{\uparrow\downarrow\,|\,\uparrow\downarrow\,|\,\uparrow\downarrow}$

This does *not* give an adequate picture of the arrangement of the electrons in the compound, because it implies that one pair in a $2s$ orbital is in some way different from the other three pairs in $2p$ orbitals. Examination of methane shows that this is not the case; all four C—H bonds are identical in every respect.

We therefore put in an additional step in compound formation called *hybridization*, in which we mix together the characteristics of the original atomic orbitals to produce a set of equivalent *hybrid orbitals* which are used in the compound. The hybrid orbitals combine all the characteristics of the atomic orbitals from which they were formed, but there are no longer any differences between them, just as we might combine one can of black paint with three cans of white paint to produce four "equivalent" cans of grey paint.

We name the hybrid orbitals according to the atomic orbitals from which they were derived, using the letters s, p, or d and superscripts to denote the *number* of orbitals of that type which we used (the superscript being omitted if we only use one orbital of that type). Thus, in CH_4, we used one s-type orbital and three p-type orbitals, so the hybrid type is denoted sp^3. We don't bother to indicate the principal quantum number of the atomic orbitals (2 in this case) in the hybrid type, so the other group IV compounds SiH_4, GeH_4, etc. would all be called sp^3 hybrids.

Obviously the total number of atomic orbitals used to make the set of hybrid orbitals must be the same as the total number of electron pairs for which we have to find space, including both bonding and nonbonding pairs. For two, three, or four electron pairs, we can use atomic s and p orbitals to make our hybrid orbitals.

Two pairs: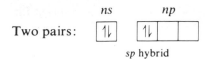

A pair of *sp* hybrid orbitals are directed at a 180° angle, so we have a linear arrangement of electron pairs.

Three pairs: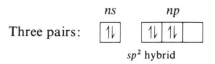

Three *sp*² hybrid orbitals are in a plane at 120° angles, so we have a planar triangular arrangement of electron pairs.

Four pairs, discussed above, give an *sp*³ hybrid set and a tetrahedral arrangement of electron pairs.

For anything in excess of four electron pairs, we must go beyond the *s* and *p* atomic orbitals and use one or more of the *d* orbitals to accommodate electrons. Note that, for main-group-element compounds, we use the *d* orbital with the same principal quantum number as the *s* and *p* orbitals that we are using. Thus, we would hybridize together the 3*s*, 3*p*, and 3*d* or the 4*s*, 4*p*, and 4*d*, etc.

It follows that, since there is no 2*d* orbital, this process is not possible when we are dealing with first-row elements (Li—Ne) where the valence shell on the central atom is formed from the 2*s* and 2*p* atomic orbitals. Hence the elements of the first row never exceed four electron pairs (i.e., four single bonds) in their compounds. (*Note*: This is the so-called "octet rule".)

Five pairs: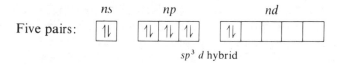

The *sp*³*d* hybrid orbitals are arranged in a trigonal bipyramidal shape.

Six pairs: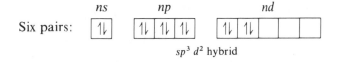

The *sp*³*d*² hybrid orbitals are arranged in an octahedral shape.

It becomes apparent that the arrangements of electron pairs predicted from the hybridization schemes discussed above are the same as those we forecast on a basis of electron-pair repulsions. Table 7.5 summarizes these conclusions.

The decision on the type of hybridization present in a molecule or ion is easily made once we have worked out the arrangement of electron pairs in the valence shell of the

TABLE 7.5   Hybridization and Arrangements of Electron Pairs

Total pairs of electrons	Atomic orbitals used in hybridization	Hybrid type	Arrangements of electrons
2	One s, one p	sp	linear
3	One s, two p	sp^2	planar triangle
4	One s, three p	sp^3	tetrahedral
5	One s, three p, one d	sp^3d	trigonal bipyramid
6	One s, three p, two d	sp^3d^2	octahedral

central atom. All we have to do is to fit them into the "boxes," one pair to each box, in the familiar framework, starting from the left. Then we count how many orbitals (boxes)

of each type have been used, and that gives us our hybrid type. The five basic geometries of electron-pair arrangement are related directly to these hybrids (as in Table 7.5) and to the number of electron pairs present.

We can end this section by noting again the essential requirement for the VSEPR treatment; that we count *all* electron pairs in the valence shell of the central atom, both bonding and nonbonding.

7.4   Polar Bonds

In discussing the covalent bond, we have talked of electron pairs being shared between two atoms. It often happens, however, that the electrons are unequally shared, that is, they are drawn to one atom more than to the other. A covalent bond in which this occurs is said to be *polar*, meaning that it has an unequal distribution of electrical charge.

Gaseous hydrogen chloride contains a polar covalent bond, because the chlorine in the molecule attracts the electrons more strongly than the hydrogen. As a result, the Cl atom carries a slight negative charge and the H atom an equal and opposite positive charge. The overall molecule, of course, remains electrically uncharged. We can indicate the polarization of the bond by writing $\delta+$ and $\delta-$ for these partial charges:

$$\text{H--Cl} \qquad \text{a polar bond}$$
$$\delta+ \quad \delta-$$

Note that this is only a *partial* separation of charge. If we moved a complete unit of charge from H to Cl, we would have two ions (H^+ and Cl^-) and an ionic bond. The properties of HCl show that this has not occurred.

Obviously we would not have a polar bond in a molecule where the atoms joined together were identical, such as H_2 or Cl_2. The electrons in such molecules are attracted equally to both atoms, and the bond is said to be *nonpolar*.

$$\text{H--H} \qquad \text{Cl--Cl} \qquad \text{O=O} \qquad \text{N≡N} \qquad \text{nonpolar bonds}$$

Generally speaking, a bond between two identical atoms will be nonpolar, while a bond between two different atoms will be polar. An obvious question is "How do we determine the direction of partial charge transfer between unlike atoms? In HCl, for example, how do we know that the partial negative charge will be on the Cl atom?"

This can only be answered by experiment. As a result of a large number of observations, the elements have been classified according to a quality called *electronegativity*, which is defined as the tendency of an atom in a molecule to attract electrons to itself. The greater the electronegativity, the more successful the atom will be in attracting electrons to itself, so if two different atoms are joined together, the bond will be polarized with the $\delta-$ charge on the more electronegative atom.

As you might expect, the electronegativity of an element depends very much on its position in the periodic table. Nonmetals are electronegative, metals are not. Of the nonmetals, fluorine is the most electronegative, followed by oxygen, then come chlorine, nitrogen, and bromine. Electronegativity is not an easy quantity to define or measure exactly, so different workers have obtained slightly varying orders for the various elements (where differences are often very small), so we will not attempt to work out the direction of polarity of every possible bond. For our purposes, we will simply assume that any bond between unlike elements is a polar bond; if one of the elements is in the highly electronegative group mentioned above, then we can be sure of the direction of polarization, otherwise we will not worry about it.

Don't confuse electronegativity with electron affinity (Chapter 6). The difference between them is that electron affinity is a property of an isolated atom, whereas electronegativity is a property of an atom covalently bonded to others in a molecule. The two do go together, however, in the sense that elements of high electron affinity values (F, Cl) also have high electronegativities.

7.5 Polar Molecules

Clearly we cannot measure the properties of an isolated bond, we can only work with whole molecules. The charge distribution in a molecule may be demonstrated by studying its interaction with an electric field (that is, by putting it between two charged plates). If the distribution of charge on the molecule is unsymmetrical, it will try to turn and align itself in response to the electrostatic forces, whereas, if it is symmetrically distributed, there will be no such interaction. (See Figure 7.4.) The unsymmetrical molecule (at left) is called a *polar* molecule, while the symmetrical molecule (at right) is a *nonpolar* molecule. Quantitatively, we express the unsymmetrical nature of the charge distribution by measuring the

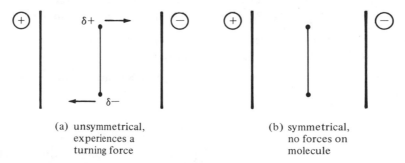

(a) unsymmetrical,
    experiences a
    turning force

(b) symmetrical,
    no forces on
    molecule

Figure 7.4

dipole moment of the molecule, defined as the product of the magnitude of each partial charge ($\delta+$ and $\delta-$) and the distance by which the charges are separated. Obviously a nonpolar molecule will have a dipole moment of zero.

From the chemist's point of view, the dipole moment of a molecule is a useful fact in working out its possible structure. We should therefore consider the connection between the polarity of a molecule and the polarity and arrangement of the bonds within it.

It should be obvious that, if all the bonds in a molecule are nonpolar, the molecule itself must be nonpolar, since there is no possibility of an unsymmetrical charge arrangement. Examples:

$$H-H \qquad \begin{array}{c} S-S-S \\ S \qquad\qquad S \\ S-S-S \end{array} \qquad \begin{array}{c} P \\ P \\ P\!-\!\!-\!\!-P \end{array}$$

$$Cl-Cl$$

nonpolar molecules with nonpolar bonds

However, the converse is *not* true: if there are polar bonds in a molecule, it does *not* always follow that the molecule will be polar. The reason for this is that, in a molecule with several polar bonds, the structure of the molecule may be such that the effect of the bonds is to cancel each other out, giving the molecule as a whole a symmetrical charge distribution. Let's illustrate this by looking at CO and CO_2. Both of these molecules are linear, and, since oxygen is more electronegative than carbon, all the carbon–oxygen bonds will be polar. Since CO has only one such bond, the molecule must be polar.

$$\overset{\delta+}{C}\!\equiv\!\overset{\delta-}{O}$$

A polar molecule with a polar bond

In CO_2, however, the polarity of the two bonds is equal and opposite, so the effects cancel out.

$$\overset{\delta-}{O}\!=\!\overset{2\delta+}{C}\!=\!\overset{\delta-}{O}$$

A nonpolar molecule with polar bonds

Putting this another way, we can say that each polar bond in a molecule has a dipole moment associated with it. The dipole moment is a vector quantity, that is, it has both magnitude and direction. The overall dipole moment of the molecule is the vector sum of these individual moments, and if their distribution is symmetrical, the vector sum will be zero.

By "symmetrical," we include the following common molecular shapes:

2 bonds at 180° in a linear molecule (e.g., $BeCl_2$)
3 bonds in a planar triangular molecule (e.g., BCl_3)
4 bonds in a square planar molecule (e.g., XeF_4) or tetrahedral molecule (e.g., CF_4)
5 bonds in a trigonal bipyramidal molecule (e.g., PF_5)
6 bonds in an octahedral molecule (e.g., SF_6).

In all these cases, the molecule will be nonpolar because the dipole moments of the individual bonds cancel out (provided, of course, that the atoms surrounding the central atom are all of the same element).

If we have two or more polar bonds that do *not* cancel each other, then the resulting molecule will be polar. Water is a good example here. Since it is a bent molecule, the polar O—H bonds do *not* cancel out.

A polar molecule with polar bonds

Experiments confirm that water molecules do tend to align themselves in an electric field, which constitutes the simplest direct proof that they have a bent structure. Similarly, in NH_3 and other pyramidal molecules, the bond dipole moments add together, rather than cancelling, and the molecules are polar.

A polar molecule with polar bonds

If you are not sure about deciding when bond dipoles will cancel, a simple mechanical analogy may help you. Imagine that the central atom is a weight, to which ropes are attached in the direction of the bonds. We pull with an equal force on all the ropes simultaneously. If the geometry of the arrangement is such that the pulling forces all cancel out, then the central weight will not move. This corresponds to a nonpolar molecule. If, on the other hand, the combined pull would move the central weight, then the molecule must be polar. Thus, in the above examples, it is clear that a simultaneous pull on the H atoms would move the O atom in H_2O or the N atom in NH_3 in a downward direction.

Example 7.4   Classify the following as polar or nonpolar:

(a) CH_4 (b) OF_2 (c) SF_4 (d) XeF_2 (e) XeF_4 (f) ClF_3
(g) BrF_5

SOLUTION

(a) CH_4 is tetrahedral. Bond dipole moments cancel out and the molecule is nonpolar.
(b) OF_2 is bent, like H_2O, and therefore polar.
(c) SF_4 is an irregular shape (usually described as distorted tetrahedral) and therefore polar.
(d) XeF_2 is linear, symmetrical, and nonpolar.
(e) XeF_4 is square planar, symmetrical, and nonpolar.
(f) ClF_3 has a T shape in which bond dipoles do not cancel. It is a polar molecule.
(g) BrF_5 is square pyramidal and polar.

(*Note*: If the above conclusions are not obvious to you at once, try drawing the various structures and thinking of which are symmetrical and which are not.)

We have to consider one final aspect of molecular dipoles, and that is the question of what happens when the atoms joined to the central atom are of more than one kind. This will usually increase the chance of getting a polar molecule, since it will reduce the symmetry of the molecule. An example will help to clarify this.

Example 7.5 Exchange of H for Cl produces the series of compounds CH_4, CH_3Cl, CH_2Cl_2, $CHCl_3$ and CCl_4. Which of these are polar and which are nonpolar?

SOLUTION   The basic bonding arrangement and geometry is the same in all these compounds—C is forming four bonds in a roughly tetrahedral arrangement.

Obviously, in CH_4 and CCl_4, the molecule is symmetrical and nonpolar. In the other three, neither the C—Cl bonds nor the C—H bonds can be arranged symmetrically, so they do not cancel out. Since Cl and H have different electronegativities, the dipole moment of a C—Cl bond cannot be canceled out with that of a C—H bond, so these three molecules are polar.

The mechanical analogy here is to think of pulling with ropes with different forces along bonds of different types. In CH_2Cl_2, for example, if the C—Cl ropes are pulled with more force than the C—H ropes, the C atom will move in the direction of the C—Cl bonds.

Example 7.6 Ethylene, C_2H_4, is a planar molecule with the structure

The H atoms may be replaced by Cl atoms to give the series C_2H_4, C_2H_3Cl, $C_2H_2Cl_2$, C_2HCl_3, and C_2Cl_4. Which of these will be polar and which nonpolar? (Note that this molecule is flat, and one end cannot rotate relative to the other end.)

SOLUTION   This looks at first similar to Example 7.5, but it contains a slight catch.

To start with, C_2H_4 is obviously symmetrical and nonpolar, while C_2H_3Cl must be polar, because it contains only one C—Cl bond and the symmetry is lost.

But what will happen when we put a second Cl atom in? Because of the structure of the molecule, there are three different ways of arranging the two Cl atoms.

$$
\begin{array}{ccc}
\underset{Cl}{\overset{Cl}{\diagdown}}C=C\underset{H}{\overset{H}{\diagup}} & \underset{H}{\overset{Cl}{\diagdown}}C=C\underset{H}{\overset{Cl}{\diagup}} & \underset{H}{\overset{Cl}{\diagdown}}C=C\underset{Cl}{\overset{H}{\diagup}} \\
(a) & (b) & (c)
\end{array}
$$

These are distinct chemical compounds, which we call *isomers* because they have the same molecular formula. As you may see, (a) and (b) are polar, because the C—Cl bonds cannot be arranged to oppose each other, so their bond dipoles do not cancel. Structure (c), on the other hand, is nonpolar, because the C—Cl bonds are directly opposite to each other, as are the C—H bonds.

The molecule C_2HCl_3 is polar, whereas C_2Cl_4 is nonpolar:

$$
\begin{array}{cc}
\underset{Cl}{\overset{Cl}{\diagdown}}C=C\underset{Cl}{\overset{H}{\diagup}} & \underset{Cl}{\overset{Cl}{\diagdown}}C=C\underset{Cl}{\overset{Cl}{\diagup}} \\
\text{polar} & \text{symmetrical, nonpolar}
\end{array}
$$

KEY WORDS

covalent bond	bonding electron pair
linear	nonbonding electron pair
planar triangle	tetrahedral
trigonal bipyramidal	octahedral
simple ion	complex ion
hybridization	polar bond
polar molecule	nonpolar bond
nonpolar molecule	electronegativity

STUDY QUESTIONS

1. Why don't all molecules with the atomic combining ratio AB_3 have the same shape?

2. What *does* determine the shape of a covalent molecule?

3. How does the covalent bond differ from the ionic bond in its influence on structures?

4. How many different combinations of (bonding pairs) + (nonbonding pairs) are found for 2, 3, 4, 5, or 6 total electron pairs on the central atom?

5. Why is the combination 3 bonding + 3 nonbonding pairs not found?

6. What effect does hybridization have on a set of orbitals?

7. Why do first-row elements never form more than four single covalent bonds?

8. What constitutes a polar bond? Give examples.

9. Give examples of
   (a) nonpolar molecules with nonpolar bonds
   (b) nonpolar molecules with polar bonds
   (c) polar molecules with polar bonds

10. Explain why the molecules in (b) above are classified as nonpolar.

11. Give a sketch of the forces acting on (b) and (c) type molecules (question 9) in an electrostatic field.

PROBLEMS

1. Complete the following table, following the example of the first entry, to show the electron-pair arrangement, shape, and hybridization of various covalent molecules and ions. (All references to "electrons" include the valence shell of the central atom only.)

	Compound	Bonding	Nonbonding	Total	Arrangement of electrons	Shape of molecule	Hybrid. at center atom
			Electron pairs				
1.	NH_3	3	1	4	tetrahedral	pyramidal	sp^3
2.	SeH_2						
3.	PCl_5						
4.	$GeCl_4$						
5.	$HgBr_2$						
6.	AsF_3						
7.	TeF_6						
8.	KrF_2						
9.	$AlBr_3$						
10.	SeF_4						
11.	IF_5						
12.	BrF_3						
13.	PH_4^+						
14.	PCl_6^-						
15.	$AlCl_4^-$						
16.	BrF_4^-						
17.	SiF_5^-						

2. The following table is laid out in the same way as in the previous problem, but this time data on the electronic arrangements have been provided. Complete the table and give a suitable example for each structure. Try to avoid using compounds given in the first problem.

| | | Electron pairs | | | | |
	Compound	Bonding	Nonbonding	Total	Arrangement of electrons	Shape of molecule	Hybrid. at center atom
1.		5	0				
2.		2	2				
3.		3		3			
4.			2	5			
5.		6		6			
6.		5		6			
7.					tetrahedral	pyramidal	
8.					trigonal bipyramid	distorted tetrahedral	
9.					trigonal bipyramid	linear	
10.					octahedral	square planar	
11.					trigonal bipyramid	T-shaped	
12.						bent	sp^3
13.						square pyramid	$sp^3 d^2$
14.			0				$sp^3 d$
15.		2					sp^2
16.			1				$sp^3 d$
17.		4					$sp^3 d^2$

3. Return to problem 1 and classify the neutral molecules in the table as polar or nonpolar.

4. Sulfur dioxide, SO_2, and nitrogen dioxide, NO_2, are both polar, whereas CO_2 is nonpolar. What does this suggest about the structures of these oxides?

5. Nitrous oxide, N_2O, is known to be a linear molecule, but is polar. Suggest an explanation in terms of its structure.

6. Mixing $SnCl_4$ and $SnBr_4$ produces the mixed halides $SnCl_3Br$, $SnCl_2Br_2$, and $SnClBr_3$. Will these be polar or nonpolar?

7. Cyclobutadiene, C_4H_4, contains a planar square arrangement of carbon atoms.

This is a nonpolar molecule. What will happen to polarity as the H atoms are successively replaced by Cl atoms to give C_4H_3Cl, $C_4H_2Cl_2$, C_4HCl_3, and C_4Cl_4?

The remaining questions require a little thought and imagination.

8. (a) What polarity would you expect for the molecules SF_4Cl_2, SF_3Cl_3, and SF_2Cl_4?
   (b) Are isomers possible for these compounds?

9. Benzene, C_6H_6, is a planar, regular hexagonal molecule. Obviously this structure is

nonpolar. What will be the polarity of the chlorobenzenes C_6H_5Cl, $C_6H_4Cl_2$, $C_6H_3Cl_3$, and $C_6H_2Cl_4$?

10. Platinum(II) chloride, $PtCl_2$, forms an adduct with ammonia of formula $PtCl_2(NH_3)_2$, in which there is a coordinate covalent bond from each N atom to the Pt. It is found that this compound can exist in two isomeric forms, one of which is polar and the other nonpolar. What does this tell us about the arrangement of the four bonds around Pt? Sketch the structures of the two isomers.

SOLUTIONS TO PROBLEMS

1.

	Compound	Bonding	Nonbonding	Total	Arrangement of electrons	Shape of molecule	Hybrid. at center atom
			Electron pairs				
1.	NH_3	3	1	4	tetrahedral	pyramidal	sp^3
2.	SeH_2	2	2	4	tetrahedral	bent	sp^3
3.	PCl_5	5	0	5	trigonal bipyramid	trigonal bipyramid	sp^3d
4.	$GeCl_4$	4	0	4	tetrahedral	tetrahedral	sp^3
5.	$HgBr_2$	2	0	2	linear	linear	sp
6.	AsF_3	3	1	4	tetrahedral	pyramidal	sp^3
7.	TeF_6	6	0	6	octahedral	octahedral	sp^3d^2
8.	KrF_2	2	3	5	trigonal bipyramid	linear	sp^3d
9.	$AlBr_3$	3	0	3	planar triangle	planar triangle	sp^2
10.	SeF_4	4	1	5	trigonal bipyramid	distorted tetrahedral	sp^3d
11.	IF_5	5	1	6	octahedral	square pyramid	sp^3d^2
12.	BrF_3	3	2	5	trigonal bipyramid	T-shaped	sp^3d
13.	PH_4^+	4	0	4	tetrahedral	tetrahedral	sp^3
14.	PCl_6^-	6	0	6	octahedral	octahedral	sp^3d^2
15.	$AlCl_4^-$	4	0	4	tetrahedral	tetrahedral	sp^3
16.	BrF_4^-	4	2	6	octahedral	square planar	sp^3d^2
17.	SiF_5^-	5	0	5	trigonal bipyramid	trigonal bipyramid	sp^3d

2.

	Compound	Electron pairs			Arrangement of electrons	Shape of molecule	Hybrid. at center atom
		Bonding	Nonbonding	Total			
1.	AsF_5	5	0	5	trigonal bipyramid	trigonal bipyramid	sp^3d
2.	$TeCl_2$	2	2	4	tetrahedral	bent	sp^3
3.	$AlCl_3$	3	0	3	planar triangle	planar triangle	sp^2
4.	BrF_3	3	2	5	trigonal bipyramid	T-shaped	sp^3d
5.	SeF_6	6	0	6	octahedral	octahedral	sp^3d^2
6.	IF_5	5	1	6	octahedral	square pyramid	sp^3d^2
7.	PCl_3	3	1	4	tetrahedral	pyramidal	sp^3
8.	$TeCl_4$	4	1	5	trigonal bipyramid	distorted tetrahedral	sp^3d
9.	$XeCl_2$	2	3	5	trigonal bipyramid	linear	sp^3d
10.	KrF_4	4	2	6	octahedral	square planar	sp^3d^2
11.	ICl_3	3	2	5	trigonal bipyramid	T-shaped	sp^3d
12.	OCl_2	2	2	4	tetrahedral	bent	sp^3
13.	AtF_5	5	1	6	octahedral	square pyramid	sp^3d^2
14.	$BiCl_5$	5	0	5	trigonal bipyramid	trigonal bipyramid	sp^3d
15.	$SnCl_2(g)^*$	2	1	3	planar triangle	bent	sp^2
16.	TeF_4	4	1	5	trigonal bipyramid	distorted tetrahedral	sp^3d
17.	XeF_4	4	2	6	octahedral	square planar	sp^3d^2

* This compound is ionic in the solid state.

3. Polar: NH_3, SeH_2, AsF_3, SeF_4, IF_5, BrF_3
   Nonpolar: PCl_5, $GeCl_4$, $HgBr_2$, TeF_6, KrF_2, $AlBr_3$

4. SO_2 and NO_2 are bent, while CO_2 is linear

5. Most AB_2 molecules have the A atom in the center and a B atom on each side, but N_2O is different. The central atom is nitrogen and the structure $N=N=O$. This is the only way a linear arrangement could give a polar molecule.

6. $SnCl_4$ and $SnBr_4$ are nonpolar, while all three mixed halides are polar. The arrangement of bonds at Sn is, of course, approximately tetrahedral for all five compounds.

7. The monochloro- and trichlorocompounds will always be polar (no isomers).

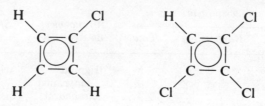

The dichlorocompounds, $C_4H_2Cl_2$, would exist in two isomeric forms.

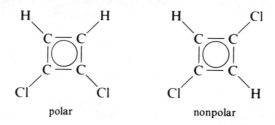

8. (a) In these compounds, the bonding at sulfur will be octahedral in all cases.
   (b) In each compound, there are two (and only two) isomers.

It is not possible, in practice, to isolate these isomers, but the separation may readily be made for a series of metal complexes of similar geometry. In a classic piece of chemical research, A. Werner deduced in 1893 that octahedral geometry must be present because of the number of isomers he could prepare.

9. This is very similar to problem 7. The various isomers that can (and do) exist are:

C_6H_5Cl

polar

$C_6H_4Cl_2$

polar   polar   nonpolar

$C_6H_3Cl_3$

polar   polar   nonpolar

$C_6H_2Cl_4$

polar   polar   nonpolar

10. There are two reasonable arrangements of the four bonds around the central Pt atom; tetrahedral and square planar. If the structure of $PtCl_2(NH_3)_2$ were tetrahedral, it would look like this:

polar

There is only one form possible for such a species (like CH_2Cl_2 and $SnCl_2Br_2$), and it must be polar.

With a square planar arrangement, however, two isomers may exist. The isolation of two isomers, one polar and the other nonpolar, proves that this is the correct arrangement of bonds around the Pt atom.

polar   nonpolar

Chapter 8

The Properties of Gases

"It is entirely a question of barometric pressure."
"I do not quite follow."
"How is the glass? Twenty-nine, I see."
THE BOSCOMBE VALLEY MYSTERY

The behavior of gases is subject to a number of relatively simple laws, which enable us to obtain important information about elements and compounds. In this chapter, we will consider application of these various laws, and the deduction of useful information from them.

8.1 The Gas Laws

There are two laws that govern the behavior of gases under changing conditions:

(a) Boyle's law. *At constant temperature, the pressure of a gas is inversely proportional to its volume.*

In other words, the product PV (P = pressure; V = volume) is a constant.

(b) Charles' law. *At constant pressure, the volume of a gas is proportional to the absolute temperature.*

At constant volume, the pressure of a gas is proportional to the absolute temperature. (*Note*: Absolute temperature is measured above absolute zero, $-273.15°$C. Temperatures on this scale are measured in kelvins, abbreviated K. Thus $0°$C $= 273.15$ K, $100°$C $= 373.15$ K, etc.)

These two laws may be combined into the relationship

$$PV = CT$$

where C is a constant proportional to the amount of gas present.

162

This relationship may be broadened by combining it with Avogadro's hypothesis, which states that, under equal conditions, equal volumes of all gases contain the same number of molecules. There will therefore be a certain volume that will contain one mole of any gas (Avogadro's number of molecules) and, for that quantity of gas, the proportionality constant C has the value R, known as the *gas constant*. Thus for one mole of gas, $PV = RT$. If we have n moles of gas present, rather than just one mole, the equation becomes:

$$PV = nRT$$

and this is known as the *ideal gas equation of state*. The name "ideal gas" refers to an imaginary gas that behaves in a manner exactly corresponding to this equation, but, in practice, real gases show some degree of departure from ideal behavior. The departures are usually small, however, and in this chapter we will assume ideal behavior at all times.

The value of the constant R is of great importance in many chemical calculations, and there are various ways of measuring it. In connection with gases, we may evaluate it by noting the value of the molar volume (volume of one mole) of an ideal gas under standard conditions of temperature and pressure (STP), which is 1 atmosphere pressure and 273.15 K. If one mole of ideal gas has a volume of 22.413 liter at STP, then

$$R = \frac{PV}{T} = \frac{1.000 \times 22.413}{273.15} = 0.0821 \text{ liter atm K}^{-1} \text{ mole}^{-1}$$

This value of R is used so often in calculations involving gases that it is worth memorizing, but it is always available to you in tables of reference. Although these are the units of R we will use in most of this chapter, it is worth noting that the atmosphere is not an SI unit of pressure. (In SI units, the standard atmosphere has the value 1.01325×10^5 newton meter^{-2} and volume is measured in meter^3. In these units, $R = 8.314 \text{ J K}^{-1} \text{ mole}^{-1}$, since 1 newton meter = 1 joule.)

We can illustrate the use of the ideal gas equation with some calculations.

Example 8.1 Calculate the volume of 2.50 mole of gas at 100°C and 4.00 atm pressure.

SOLUTION We have $n = 2.50$ mole, $T = 100°C = 373$ K, $P = 4.00$ atm. Using the ideal gas equation of state,

$$V = \frac{nRT}{P} = \frac{(2.50 \text{ mole}) \times (0.0821 \text{ liter atm K}^{-1} \text{ mole}^{-1}) \times 373 \text{ K}}{4.00 \text{ atm}}$$

$$= 19.1 \text{ liter}$$

Example 8.2 A gas has a volume of 200 ml at 1.20 atm and 27°C. How many moles are present?

SOLUTION Here we have $V = 200$ ml $= 0.200$ liter, $T = 27°C = 300$ K, and $P = 1.20$ atm. So

$$n = \frac{PV}{RT} = \frac{1.20 \times 0.200}{0.0821 \times 300} = 9.74 \times 10^{-3} \text{ mole} = 9.74 \text{ mmole}$$

As with any other molar quantity, small amounts are conveniently measured in millimoles. If V is in ml, then n will come out in millimole with the standard values of P, R, and T.

In other words, the value of R, the gas constant, would still be 0.0821 in the units ml atm mmole^{-1} K^{-1}.

In the laboratory, gas pressures are very often measured with the aid of a mercury manometer. With this instrument, the standard atmosphere is defined as the pressure exerted by a column of mercury 760 mm high, under prescribed conditions. Pressures measured with a mercury manometer are expressed in cm Hg or mm Hg (called "Torr") and, before using the ideal gas equation, they must be converted into atmospheres

$$\text{pressure in atmospheres} = \frac{\text{pressure in cm Hg}}{76.0}$$

$$= \frac{\text{pressure in mm Hg (Torr)}}{760}$$

(*Note*: Since mercury is universally used as a barometer fluid, we very often omit the Hg symbol and talk of a pressure of "51.0 cm" or "510 mm." Such figures are always understood to refer to a pressure equivalent to a column of mercury of that height.)

Special note: One of the commonest sources of student error in gas calculations is to use the wrong units. Make sure every time you use the equation that you have

(a) temperature in kelvins, K, (absolute degrees) *not* °C
(b) volume and number of moles in corresponding units, either

volume in *liters*, n in *moles*

or

volume in *milliliters*, n in *millimoles*

(c) pressure in atmospheres, obtained by either

dividing pressure in cm Hg by 76.0

or

dividing pressure in mm Hg (Torr) by 760

It is very easy to make a mistake in one of the above quantities, with disastrous consequences for the calculation.

Example 8.3   What pressure, in mm Hg, will 14.6 mmole of gas exert in a volume of 750 ml at 20°C?

SOLUTION   20°C is 293 K, while the other quantities are in corresponding units (ml and mmole).

$$P = \frac{nRT}{V} = \frac{(14.6 \text{ mmole}) \times (0.0821 \text{ ml atm K}^{-1} \text{ mmole}^{-1}) \times 293 \text{ K}}{750 \text{ ml}} = 0.469 \text{ atm}$$

To convert this to mm Hg, we multiply by 760, the number of mm Hg in one atmosphere:

$$P = 0.469 \times 760 = 356 \text{ mm Hg}$$

Example 8.4 A certain quantity of gas is under a pressure of 65.0 cm Hg in a volume of 500 ml at 0°C. What pressure will the same quantity of gas exert in a volume of 700 ml at 100°C?

SOLUTION We could solve this by finding n, the number of moles of gas, from the first set of data, then using this in the changed conditions (n remains constant) to find the pressure. However, the calculation may be shortened somewhat.

For a given quantity of gas, the number of moles is constant, and R, of course, is always constant. So we can always write $nR = PV/T$. In other words, if we have the same quantity of gas occupying a volume of V_1 under pressure P_1 and temperature T_1 and, after changing conditions, occupying volume V_2 under pressure P_2 at temperature T_2, then we can write

$$\frac{P_1 V_1}{T_1} = \frac{P_2 V_2}{T_2}$$

since both of these are equal to the product nR. This simple, easily remembered, relationship is very useful for converting gas volumes or pressures from one set of conditions to another. In the present problem, we have

$$P_1 = 65.0 \text{ cm Hg} \qquad V_1 = 500 \text{ ml} \qquad T_1 = 273 \text{ K}$$

$$P_2 = \text{unknown} \qquad V_2 = 700 \text{ ml} \qquad T_2 = 373 \text{ K}$$

So we can write

$$\frac{65.0 \times 500}{273} = \frac{P_2 \times 700}{373} \qquad P_2 = \frac{65.0 \times 500 \times 373}{273 \times 700} = 63.4 \text{ cm Hg}$$

The units of P_2 must be the same as the units of P_1.

Example 8.5 A quantity of gas has a volume of 175 ml at 150 mm Hg pressure and 23°C. (a) What volume will it have at STP? (b) How many moles of gas are present?

SOLUTION
(a) Reduction to standard conditions is a very common type of gas calculation. As before, we can use

$$\frac{P_1 V_1}{T_1} = \frac{P_2 V_2}{T_2}$$

In this case,

$$P_1 = 150 \text{ mm Hg} \qquad T_1 = 23°C = 296 \text{ K} \qquad V_1 = 175 \text{ ml}$$

$$P_2 = 760 \text{ mm Hg} \qquad T_2 = 273 \text{ K} \qquad V_2 = \text{unknown}$$

So we can write

$$\frac{150 \times 175}{296} = \frac{760 \times V_2}{273}$$

(*Note*: P_1 and P_2 may be left in the units of mm Hg in this calculation, since we are only interested in the *ratio* in which the pressure changes.)

$$V_2 = \frac{150 \times 175 \times 273}{296 \times 760} = 31.9 \text{ ml}$$

(b) To calculate n, we convert P_1 to atmospheres: $P_1 = 150/760$ atm. Then, using V_1 and T_1 we get n from

$$n = \frac{P_1 V_1}{RT_1} = \frac{150}{760} \times \frac{175}{0.0821 \times 296} = 1.42 \text{ mmole}$$

Obviously, we could get the same value of n from P_2, T_2 and V_2, since n does not change with the conditions.

8.2   Gas Density and Molecular Weight

So far, we have expressed the ideal gas equation solely in terms of n, the number of moles of gas present. Can we relate this to the mass of gas? We have previously seen that mass and number of moles are always related by the equation

$$\text{number of moles} = \frac{\text{mass of compound}}{\text{molar mass}}$$

If we are dealing with a gas, "number of moles" is simply n in our ideal gas equation, so we can write

$$PV = nRT = \frac{m}{M}(RT)$$

where m is the mass of gas present and M is its molar mass. We can arrange this equation to calculate either of these quantities from the other:

$$m = \frac{MPV}{RT} \quad \text{or} \quad M = \frac{mRT}{PV}$$

Example 8.6   What is the mass of 5.60 liter of gaseous O_2 at 100°C and 0.500 atm?

SOLUTION O_2 has a molar mass of 32.0 g, so

$$m = \frac{32.0 \times 0.500 \times 5.60}{0.0821 \times 373} = 2.93 \text{ g}$$

Example 8.7   500 ml of a gas weighs 0.838 g, measured at 27°C and 650 mm pressure. What is its molecular weight?

SOLUTION   Be careful with the units!

$$P = \frac{650}{760} \text{ atm} \qquad V = 0.500 \text{ liter} \qquad T = 300 \text{ K}$$

$$M = \frac{mRT}{PV} = \frac{0.838 \times 0.0821 \times 300}{(650/760) \times 0.500} = 48.3 \text{ g}$$

$M.W. = 48.3$

Simple measurements on gaseous samples give us a quick, easy way of determining the $M.W.$ of an unknown compound.

We can combine the mass and volume of a sample to find its density, which in turn will be related to its molar mass. Using d for density, we can write

$$d = \frac{m}{V} = \frac{MP}{RT} \quad \text{or} \quad M = \frac{RTd}{P}$$

With the usual units of grams for m and liters for V, d will be in g liter^{-1}.

Example 8.8   What is the density of gaseous SO_2 at 47°C and 62.4 cm pressure?

SOLUTION   Since 62.4 cm is 62.4/76.0 atm, and the molar mass of SO_2 is 64.1,

$$d = \frac{MP}{RT} = \frac{64.1 \times (62.4/76.0)}{0.0821 \times 320} = 2.00 \text{ g liter}^{-1}$$

(*Note*: Volume was neither specified nor required in this problem; in other words, density is independent of volume.)

Example 8.9   An unknown gas has a density of 4.80 g liter^{-1} at 50.0 cm pressure and 27°C. What is its $M.W.$?

SOLUTION

$$P = \frac{50.0}{76.0} \text{ atm}$$

$$M = \frac{RTd}{P} = \frac{0.0821 \times 300 \times 4.80}{50.0/76.0} = 180 \text{ g}$$

$M.W. = 180$

Example 8.10 At what temperature will the density of gaseous CO_2 be 2.00 g liter^{-1} at a pressure of 1.00 atm?

SOLUTION Rearranging our equation gives

$$T = \frac{MP}{Rd}$$

in this case, $M = 44.0$ g and $P = 1.00$ atm, so

$$T = \frac{44.0 \times 1.00}{0.0821 \times 2.00} = 268 \text{ K} \quad (-5°\text{C})$$

Example 8.11 A hot-air balloon has a volume of 10.0 m^3 and contains air at 100°C. The balloon fabric (excluding the air inside) weighs 1.0 kg. If the outside air is at 27°C, what is the maximum load the balloon can lift? Assume a pressure of 1.00 atm throughout, and take the *M.W.* of air as 29.

SOLUTION The upthrust on the balloon is the difference between its weight and the weight of the air it displaces. We can calculate both from

$$m = \frac{MPV}{RT} \quad \text{where} \quad V = 10.0 \text{ m}^3 = 1.00 \times 10^4 \text{ liter}$$

For the hot air inside the balloon

$$m = \frac{29 \times 1.00 \times 1.00 \times 10^4}{0.0821 \times 373} = 9.5 \times 10^3 \text{ g} = 9.5 \text{ kg}$$

Adding 1.0 kg for the balloon fabric gives a total weight of 10.5 kg for the balloon. And for the displaced cold air at 300 K (27°C),

$$m = \frac{29 \times 1.00 \times 1.00 \times 10^4}{0.0821 \times 300} = 1.18 \times 10^4 \text{ g} = 11.8 \text{ kg}$$

Since this is greater than the mass of the ballon, the upthrust will be $11.8 - 10.5 = 1.3$ kg, which is the maximum additional load the balloon could lift.

It is not necessary to memorize the various versions of the ideal gas law used in calculations of this type. All we have to remember is the one equation, $PV = nRT$, then we can rewrite the equation in terms of mass, density, or molar mass by remembering the simple ways in which these quantities are related.

8.3   Dalton's Law of Partial Pressures

The final law governing the behavior of gases is concerned with a mixture of gases. *Dalton's law* states:

The total pressure of a gaseous mixture is the sum of the partial pressures of the gases present. The partial pressure of each gas is the pressure it would exert if it were the only gas present in that volume.

This may be expressed as

$$P = p_a + p_b + p_c + p_d \cdots$$

where P, the total pressure, is the sum of the *partial pressures* p_a, p_b, etc. of the various components. For each component, the partial pressure may be related to temperature and volume by the equation

$$p_a = \frac{n_a RT}{V}$$

where n_a = number of moles of component a etc. So

$$P = \frac{RT}{V}(n_a + n_b + n_c + n_d \cdots)$$

This relationship may be used to calculate the total pressure in a gaseous mixture. Obviously R, T, and V will be the same for all the gases present together. We assume, of course, that no chemical reaction occurs between the gases in the mixture.

Example 8.12   A mixture of 5.00 g O_2, 15.00 g N_2, and 12.00 g CO_2 is contained in a volume of 1.00 liter at 27°C. What is the total pressure?

SOLUTION   First we convert to mole quantities:

$$O_2: \quad \frac{5.00}{32.0} = 0.156 \text{ mole}$$

$$N_2: \quad \frac{15.00}{28.0} = 0.536$$

$$CO_2: \quad \frac{12.00}{44.0} = 0.273$$

$$\text{total} = \overline{0.965} \text{ mole}$$

We have $V = 1.00$ liter and $T = 300$ K:

$$P = \frac{0.0821 \times 300 \times 0.965}{1.00} = 23.8 \text{ atm}$$

Example 8.13   A container of volume 2.48 liter contains gas at 200 mm Hg pressure at 300 K. An additional 0.048 mole of another gas is added. What will the pressure become?

SOLUTION   The identities of the gases do not matter here. The partial pressure that the additional gas will exert will be the same as that it would exert if it were alone in the container:

$$P = \frac{nRT}{V} = \frac{0.048 \times 0.0821 \times 300}{2.48} = 0.477 \text{ atm}$$

This pressure is equivalent to $0.477 \times 760 = 362$ mm Hg. The total pressure therefore becomes

$$200 + 362 = 562 \text{ mm Hg}$$

In this type of calculation, we always assume that no chemical reaction occurs between the gases that are mixed together.

Example 8.14   A mixture of gases at 20°C has a partial pressure of 81 mm Hg of O_2, 104 mm CO, and 250 mm CO_2. What is the density of the mixture?

SOLUTION   We want the density in g liter^{-1}, so the easiest way to find this is to calculate the number of moles of each gas present in a volume of 1 liter, and hence the mass of each. So in our calculation, V will be 1.00 liter for each gas and $T = 293$ K.

for O_2: $P = \dfrac{81}{760}$ atm, $n(O_2) = \dfrac{PV}{RT} = \dfrac{81 \times 1.00}{760 \times 0.0821 \times 293} = 4.43 \times 10^{-3}$ mole

for CO: $P = \dfrac{104}{760}$ atm, $n(CO) = \dfrac{104 \times 1.00}{760 \times 0.0821 \times 293} = 5.69 \times 10^{-3}$ mole

for CO_2: $P = \dfrac{250}{760}$ atm, $n(CO_2) = \dfrac{250 \times 1.00}{760 \times 0.0821 \times 293} = 1.37 \times 10^{-2}$ mole

The masses of each gas are found by multiplying the numbers of moles present by the respective molar masses:

$$\text{mass } O_2 = 4.43 \times 10^{-3} \times 32.0 = 0.142 \text{ g}$$
$$\text{mass CO} = 5.69 \times 10^{-3} \times 28.0 = 0.159$$
$$\text{mass } CO_2 = 1.37 \times 10^{-2} \times 44.0 = 0.603$$
$$\overline{}$$
$$\text{total mass} = 0.904 \text{ g}$$

This is the total mass of gas in a volume of 1 liter, which is, of course, the density in g liter^{-1}. Note that it would not do any good to add up the three partial pressures to find the total pressure on the gaseous mixture, since we could not use this figure in a gas calculation to find the density. Each component must be considered separately.

Example 8.15   A gaseous mixture is known to contain only N_2 and H_2. At STP, it has a density of 0.785 g liter^{-1}. What is the partial pressure of each gas present? Express the composition on a molar percentage basis *and* on a mass percentage basis.

SOLUTION   We know the total pressure is 1.00 atm. Suppose the partial pressure of H_2 is y, then the partial pressure of nitrogen is $(1.00 - y)$. So in 1 liter,

mass of H_2 present: $\dfrac{PV}{RT} \times M.W. = \dfrac{y \times 1.00 \times 2.02}{0.0821 \times 273} = 0.0901y$ g

mass of N_2 present: $\dfrac{(1.00 - y) \times 1.00 \times 28.0}{0.0821 \times 273} = 1.249(1.00 - y)$ g

The density of the mixture will be the sum of these two masses, in g liter^{-1}, and we are told this comes to 0.785 g liter^{-1}, we can then solve for y:

$$0.0901y + 1.249(1.00 - y) = 0.785$$

$$1.249 - 0.785 = 1.249y - 0.0901y$$

$$0.464 = 1.159y$$

$$y = 0.400$$

So the partial pressure of H_2 is 0.400 atm, and the partial pressure of N_2 is $1.00 - y = 0.600$ atm.

Since the numbers of moles of each gas present are directly proportional to the respective partial pressures, the percent composition of the mixture on a *mole* basis follows directly from these figures; it is 40% H_2 and 60% N_2.

On a *mass* basis, we must multiply each of these by the respective molecular weights:

H_2: $0.400 \times 2.02 = 0.808$ g

N_2: $0.600 \times 28.0 = 16.80$ g

This is the relative proportion by weight. Converting to a percent basis, we divide each by the total mass (17.6 g), giving

H_2: $\dfrac{0.808}{17.6} \times 100 = 4.6\%$

N_2: $\dfrac{16.80}{17.6} \times 100 = 95.4\%$

8.4   Corrections for the Presence of Water Vapor

An important example of Dalton's law occurs in connection with gases collected over water. When the pressure of gas in such a system is measured, it will contain two components: the pressure due to dry gas (or a gaseous mixture) *plus* the pressure due to water vapor. So to calculate the actual amount of dry gas present, we must first *subtract* from the observed pressure the vapor pressure of water. This quantity has been accurately measured at various temperatures, and some values are given in Table 8.1. Application of these figures in gas law calculations can be shown in a few examples.

TABLE 8.1   The Vapor Pressure of Water

Temp. °C	Pressure mm Hg	Temp. °C	Pressure mm Hg	Temp. °C	Pressure mm Hg	Temp. °C	Pressure mm Hg
0	4.6	13	11.2	26	25.2	55	118
1	4.9	14	12.0	27	26.7	60	149
2	5.3	15	12.8	28	28.3	65	188
3	5.7	16	13.6	29	30.0	70	233
4	6.1	17	14.5	30	31.8	75	289
5	6.5	18	15.5	31	33.7	80	355
6	7.0	19	16.5	32	35.7	85	434
7	7.5	20	17.5	33	37.7	90	526
8	8.0	21	18.7	34	39.9	95	634
9	8.6	22	19.8	35	42.2	100	760
10	9.2	23	21.1	40	55.3	105	906
11	9.8	24	22.4	45	71.9		
12	10.5	25	23.8	50	92.5		

Example 8.16   Oxygen is collected over water at 24°C. The volume is 880 ml and the total pressure 758 mm Hg. If water has a vapor pressure of 22.4 mm at this temperature,

(a) what is the volume of dry oxygen collected, measured at STP?
(b) how many moles of oxygen are present?
(c) how many moles of water vapor?

SOLUTION   Dalton's law tells us the total pressure is the sum of the pressure due to O_2 and that due to water vapor. So,

(a) pressure of dry O_2 = 758 − 22.4 = 736 mm Hg. Correcting to STP conditions, we have

$$\frac{736 \times 880}{297} = \frac{760 \times V_2}{273}$$

$$V_2 = \frac{736 \times 880 \times 273}{297 \times 760} = 783 \text{ ml dry } O_2 \text{ at STP}$$

(b) number of moles of oxygen: $n = \dfrac{PV}{RT} = \dfrac{1.00 \times 0.783}{0.0821 \times 273} = 0.0350$ mole

(*Note*: Remember to put P in atm and V in liters.)

(c) The original conditions had

$$P(\text{H}_2\text{O}) = 22.4 \text{ mm} = \frac{22.4}{760} \text{ atm}$$

so

$$n(\text{H}_2\text{O}) = \frac{22.4 \times 0.880}{760 \times 0.0821 \times 297} = 1.06 \times 10^{-3} \text{ mole} = 1.06 \text{ mmole}$$

(*Note*: We use the original value of V and T because the water was present in that state.)

Example 8.17 What volume would 1.00 g of nitrogen occupy if collected over water at a total pressure of 750 mm Hg at 22°C? The vapor pressure of water is 20 mm at that temperature.

SOLUTION   The partial pressure of the N_2 will be $750 - 20 = 730$ mm Hg. Using this value in the usual formula will give us the volume of the gas. In this case $n = 1.00/28.0$ mole and $P = 730/760$ atm so

$$V = \frac{nRT}{P} = \frac{1.00 \times 760 \times 0.0821 \times 293}{28.0 \times 730} = 0.894 \text{ liter}$$

An obvious question would be: "Shouldn't we add on the volume of the water vapor?" The answer is no, however, because the water vapor is contained in the *same* volume as the nitrogen. Addition of water to dry nitrogen increases the total pressure (from 730 to 750 mm in this case) but does *not* change the volume.

Example 8.18 Compare the densities of dry CO_2 and water-saturated CO_2 at 1.00 atm *total* pressure and 22°C.

SOLUTION   For dry CO_2, we know

$$d = \frac{P \times M.W.}{RT} = \frac{1.00 \times 44.0}{0.0821 \times 295} = 1.82 \text{ g liter}^{-1}$$

For water-saturated CO_2, we have two components. First, the CO_2, for which P is now $760 - 20 = 740$ mm or $740/760$ atm, and

$$d = \frac{740 \times 44.0}{760 \times 0.0821 \times 295} = 1.77 \text{ g liter}^{-1}$$

But the water also will contribute slightly, with $P = 20/760$ atm, $M.W. = 18.0$, and

$$d = \frac{20 \times 18.0}{760 \times 0.0821 \times 295} = 0.020 \text{ g liter}^{-1}$$

So the total mass of a liter of "wet" CO_2 is $1.77 + 0.020 = 1.79$ g liter^{-1}.

8.5 Molecular Velocity

If we make a few simple assumptions about the behavior of molecules in a gas, we can construct a mathematical model that is quite successful in predicting the properties of a gas. The most important assumption is that the gaseous molecules are in continuous motion, making elastic collisions with each other and the walls of the container, and that each molecule possesses kinetic energy. The theory is therefore known as the *kinetic theory of gases*.

Without developing this theory in detail, we will make use of an important result. It may be shown that, for one mole of gas, the molecular velocity is related to the other properties of the gas by the equation

$$PV = \frac{1}{3} Nmu^2$$

where N = Avogadro's number, m = mass of one molecule, and u is the average molecular velocity. (This average velocity is calculated by squaring all the individual molecular velocities, averaging them, and taking the square root, so its full title is the "root mean square" molecular velocity.)

From the ideal gas equation of state, we know that, for one mole of gas, $PV = RT$. Combining this with the previous equation gives

$$PV = \frac{1}{3} Nmu^2 = RT$$

If m is the molecular mass, the product Nm is the molar mass, M, so we can write

$$\frac{1}{3} Mu^2 = RT \qquad u = \sqrt{\frac{3RT}{M}}$$

The last equation shows that average molecular velocity is proportional to the square root of the temperature and *inversely* proportional to the square root of the molar mass. However, it does *not* depend on the pressure of the gas.

Remember also that this equation only tells us the *average* molecular velocity. Within any gas, there will always be a wide range of molecular velocities.

It is interesting to work out some specific examples using this equation.

Example 8.19   Calculate the average velocity of molecules of (a) oxygen, (b) hydrogen, at both 25°C and 500°C.

SOLUTION   The arithmetic is fairly simple here, as we can use $u = \sqrt{3RT/M}$. However, we have to be careful about units. The easiest approach is to use strict SI units and put R in joule mole^{-1} K^{-1} and M, the molar mass, in kg rather than g. The definition of the joule is 1 newton × 1 m, where the newton (unit of force) is mass × acceleration = kg m s^{-2}. The joule has units (kg m s^{-2}) × m = kg m^2 s^{-2}. Therefore

$$R = 8.314 \text{ J mole}^{-1} \text{ K}^{-1} = 8.314 \text{ kg m}^2 \text{ s}^{-2} \text{ mole}^{-1} \text{ K}^{-1}$$

(*Note*: Yes, this *is* the same R that we were using before as 0.0821 liter atm mole^{-1} K^{-1}. All we have done is changed its units.) Having worked this out, we can substitute the data for O_2 at 25°C (298 K), where M = 32.0 g mole^{-1} = 0.0320 kg mole^{-1}:

$$u = \sqrt{\frac{3 \times (8.314 \text{ kg m}^2 \text{ s}^{-2} \text{ mole}^{-1} \text{ K}^{-1}) \times (298 \text{ K})}{0.0320 \text{ kg mole}^{-1}}}$$

$$= \sqrt{2.32 \times 10^5 \text{ m}^2 \text{ s}^{-2}} = 482 \text{ m s}^{-1}$$

This is a considerable speed, greater than the velocity of sound (about 350 m s^{-1} in air). At 500°C, the average velocity of O_2 molecules is

$$u = \sqrt{\frac{3 \times 8.314 \times 773}{0.0320}} = \sqrt{6.03 \times 10^5} = 776 \text{ m s}^{-1}$$

For H_2, the molar mass is 2.02×10^{-3} kg, so

at 25°C: $u = \sqrt{\dfrac{3 \times 8.314 \times 298}{2.02 \times 10^{-3}}} = \sqrt{3.68 \times 10^6} = 1.92 \times 10^3 \text{ m s}^{-1}$

at 500°C: $u = \sqrt{\dfrac{3 \times 8.314 \times 773}{2.02 \times 10^{-3}}} = \sqrt{9.54 \times 10^6} = 3.09 \times 10^3 \text{ m s}^{-1}$

As we would expect, the average velocity of H_2 molecules is much greater than that of O_2, while both molecular velocities are increased by raising the temperature.

8.6   Gaseous Diffusion: Graham's Law

Although it is interesting to be able to calculate molecular velocities, these numbers are not of much practical value. However, a simple application of the above principle leads to a result of direct practical significance.

Suppose we have two gases, A and B, whose molar masses are M_a and M_b. Their molecular velocities will be

$$u_a = \sqrt{\frac{3RT}{M_a}} \qquad u_b = \sqrt{\frac{3RT}{M_b}}$$

If both are at the same temperature, the ratio of their molecular velocities will be

$$\frac{u_a}{u_b} = \frac{\sqrt{3RT/M_a}}{\sqrt{3RT/M_b}} = \sqrt{\frac{M_b}{M_a}}$$

Knowing that molar mass is directly proportional to molecular weight, we can rewrite this as

$$\frac{u_a}{u_b} = \sqrt{\frac{(M.W.)_b}{(M.W.)_a}}$$

So any property of the gases that depends on the actual molecular velocity will be governed by a relationship of this form. One such property is that of *diffusion* (or effusion), in which the gas passes through a small hole. The rate of diffusion is directly proportional to the average molecular velocity, so its dependence on $M.W.$ will be according to *Graham's law* of gaseous diffusion:

The relative rate of diffusion or effusion of two gases under the same conditions is *inversely proportional* to the square root of the ratio of their molecular weights.

We can make use of this in finding molecular weights.

Example 8.20   Calculate the relative rate of effusion of hydrogen and oxygen under the same conditions.

SOLUTION   The molecular weights are: H_2, 2.0; O_2, 32.0. Using Graham's law, we have

$$\frac{\text{rate}(H_2)}{\text{rate}(O_2)} = \sqrt{\frac{M(O_2)}{M(H_2)}} = \sqrt{\frac{32.0}{2.0}} = \sqrt{16.0} = 4.0$$

The ratio of average molecular velocity $H_2:O_2$ is 4.0, and this is the ratio also of the rates of effusion of the two gases.

Example 8.21   Gaseous NH_3 and HBr start to diffuse toward each other along a narrow tube. Whereabouts in the tube will they meet to form NH_4Br?

SOLUTION   First we calculate the relative rates of diffusion of the two gases, whose *M.W.*'s are: NH_3, 17; HBr, 81.

$$\frac{\text{rate }(NH_3)}{\text{rate }(HBr)} = \sqrt{\frac{M(HBr)}{M(NH_3)}} = \sqrt{\frac{81}{17}} = \sqrt{4.76} = 2.18$$

So in the time it takes them to meet, the NH_3 will diffuse 2.18 times the distance that the HBr diffuses. If the total length of the tube is 3.18 units, the gases will meet at a point 2.18 units along the tube from the end where the NH_3 started, as shown in Figure 8.1. The NH_4Br will form at a point 2.18/3.18 = 0.69 or 69 % of the length of the tube from the end at which the NH_3 started.

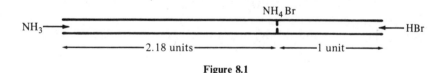

Figure 8.1

Example 8.22   A piston falls in a cylinder of gas which has a small exit hole through which gas effuses into a vacuum. With oxygen in the cylinder, the time taken for the piston to fall a certain distance is 38.3 seconds. With a second gas under the same conditions, the piston falls the same distance in 82.0 s. What is the molecular weight of the second gas?

SOLUTION   The data here refer to time for effusion to occur. Under the same conditions, time for effusion will be *inversely* proportional to rate of effusion (the faster the gas molecules move, the shorter the time they will take to get out). Calling the unknown gas X, we have

$$\frac{\text{time }(X)}{\text{time }(O_2)} = \frac{\text{rate }(O_2)}{\text{rate }(X)} = \sqrt{\frac{M(X)}{M(O_2)}}$$

$$\frac{82.0}{38.3} = \sqrt{\frac{M(X)}{32.0}} \qquad \frac{M(X)}{32.0} = \left(\frac{82.0}{38.3}\right)^2 = 4.58$$

$$M(X) = 32.0 \times 4.58 = 147$$

KEY WORDS

Boyle's law
ideal gas equation of state
standard conditions (STP)
Graham's law
partial pressure

Charles' law
gas constant
gas density
Dalton's law

STUDY QUESTIONS

1. How do we combine Boyle's law, Charles' law, and Avogadro's hypothesis into one equation?

2. How does the ideal gas equation change when we go from 1 mole of gas to n moles?

3. In the ideal gas equation, what units do we usually employ for (a) pressure, (b) volume, and (c) temperature?

4. How do we adapt the ideal gas equation to find (a) mass of gas present, (b) gas density, and (c) $M.W.$ from gas density?

5. What happens to total pressure when gases are mixed?

6. How do we determine the composition of a mixture of two gases from measurement of its density? Could this be applied to a mixture containing three or more gases?

7. How do we find the amount of gas present when its volume and pressure have been measured over water?

8. How does average gaseous molecular velocity vary with (a) temperature and (b) molecular weight?

9. How are average molecular velocities related for two gases of different $M.W.$?

PROBLEMS

Take $R = 0.0821$ liter atm K^{-1} mole^{-1} = 8.314 J K^{-1} mole^{-1}.
1. Calculate the volume of the following:
    (a) 1.28 mole of gas at 100°C and 560 mm Hg pressure
    (b) 0.0443 mole of gas at 30°C and 745 mm pressure
    (c) 0.668 mole of gas at 200K and 0.511 atm pressure

2. Calculate the pressure in atmospheres and mm Hg of the following gases:
   (a) 0.029 mole in 5.0 liter at 0°C
   (b) 1.44×10^{-3} mole in 0.200 liter at 28°C
   (c) 0.697 mole in 10.0 liter at -50°C

3. How many moles of gas are present in the following?
   (a) 400 ml of gas at 105°C and 0.600 atm pressure
   (b) 2.81 liter of gas at 24°C and 748 mm pressure
   (c) 0.28 liter of gas at -78°C and 50 mm pressure

4. Calculate the volume that each of the gases in problem 3 would occupy at STP.

5. What is the mass of the following?
   (a) 450 ml of methane, CH_4, at 25°C and 0.250 atm
   (b) 6.68 liter of sulfur hexafluoride, SF_6, at 100°C and 2.00 atm

6. At what temperature would 5.00 g of hydrogen gas occupy a volume of 100 liters at 2.00 atm pressure?

7. Calculate the densities of the following:
   (a) NH_3 gas at 1.20 atm pressure and 50°C
   (b) He gas at -200°C and 0.800 atm pressure
   (c) H_2S gas at STP

8. (a) Calculate the pressure at which air (*M.W.* 29) would have a density equal to that of water (1.00 g cm^{-3}) at 0°C.
   (b) Use (a) to work out the depth in the ocean below which air bubbles should *sink*. Assume the ocean to be an incompressible liquid, density 1.00 g cm^{-3}, temperature 0°C, in which air will not dissolve. The density of mercury is 13.7 g cm^{-3}.

9. Calculate the molecular weights of the following gases:
   (a) density 1.14 g liter^{-1} at 22°C and 750 mm Hg pressure
   (b) 250 ml weighs 0.670 g at STP
   (c) 1.18 liter weigh 3.87 g at 100°C and 765 mm Hg pressure

10. What pressure is produced when 0.400 g H_2, 2.00 g N_2, and 10.5 g CO_2 are put into a volume of 10.0 liters at 273 K?

11. What is the density of a mixture at 25°C in which the partial pressures are HCl, 200 mm; HBr, 100 mm; HI, 50 mm?

12. A mixture of oxygen and nitrogen has a density of 1.38 g liter^{-1} at STP. What is the partial pressure of each gas present?

13. Deuterium, the isotope of hydrogen with atomic weight 2.00, may be progressively isolated from naturally occurring hydrogen by various means. If a sample of hydrogen has a density at STP of 0.161 g liter^{-1} after enrichment, what percentage of deuterium is present?

14. A balloon has a volume of 200 m^3 and is filled with hydrogen. The fabric of the balloon (excluding the gas) weighs 2.0 kg. Take the pressure as 1.00 atm throughout, the temperature as 25°C, and the *M.W.* of air as 29.
    (a) What is the maximum load the balloon could lift?
    (b) What would be the effect of filling the balloon with helium instead of hydrogen?

15. A sample of gas is collected over water at 22°C. It has volume 228 ml at a total pressure of 748 mm.
    (a) What would the volume of dry gas be at STP?
    (b) How many moles of dry gas are present?

16. Oxygen is collected over water at 26°C and a total pressure of 755 mm.
    (a) If the volume of gas is 1.68 liter, what mass of O_2 is present?
    (b) What mass of water is present in the gas phase?

17. Calculate average molecular velocities for the following gases:
    (a) hydrogen at -50°C
    (b) nitrogen at 1000°C
    (c) methane, CH_4, at 298 K

18. A helium atom challenges a molecule of uranium hexafluoride, UF_6, to a race. To try and give a fair handicap, they each agree to travel at the temperature of their normal boiling points, which are 4.18 K for He and 56°C for UF_6. Which would you back to win?

19. At what temperature will the average molecular velocity in a gas be double its value at 0°C? Does this vary with the nature of the gas?

20. Calculate the average molecular velocities of the noble gases Ar, Ne, and Xe relative to He under the same conditions.

21. A sample of hydrogen takes 150 seconds to effuse out of a small hole. How long would the same volume of gaseous SO_2 take to effuse through the same hole under the same conditions?

22. An unknown gas diffuses through a certain apparatus at a rate of 14.0 ml min^{-1}. Under the same conditions, nitrogen diffuses at a rate of 29.9 ml min^{-1}. What is the *M.W.* of the unknown?

23. Neon contains the isotopes ^{20}Ne and ^{22}Ne (atomic weights 20.0 and 22.0 respectively). What will be the ratio of their rates of diffusion under the same conditions?

24. The uranium isotopes ^{235}U and ^{238}U are separated by utilizing the difference in rate of effusion of their gaseous hexafluorides. Calculate the relative rate of effusion of $^{235}UF_6$ to $^{238}UF_6$ (take the atomic weights as 235 and 238 for the two isotopes).

25. NH_3 and BF_3 gases react to form a white solid. If the two gases start to diffuse toward each other along a narrow tube, whereabouts in the tube will the solid appear?

SOLUTIONS TO PROBLEMS

1. Use $V = nRT/P$:

   (a) $V = \dfrac{1.28 \times 0.0821 \times 373}{560/760} = 53.2$ liter

   (b) $V = \dfrac{0.0443 \times 0.0821 \times 303}{745/760} = 1.12$ liter

   (c) $V = \dfrac{0.668 \times 0.0821 \times 200}{0.511} = 21.5$ liter

2. Use $P = nRT/V$:

   (a) $P = \dfrac{0.029 \times 0.0821 \times 273}{5.0} = 0.13$ atm $0.13 \times 760 = 99$ mm Hg

   (b) $P = \dfrac{1.44 \times 10^{-3} \times 0.0821 \times 301}{0.200} = 0.178$ atm $0.178 \times 760 = 135$ mm Hg

   (c) $P = \dfrac{0.697 \times 0.0821 \times 223}{10.0} = 1.28$ atm $1.28 \times 760 = 970$ mm Hg

3. Use $n = PV/RT$:

   (a) $n = \dfrac{0.600 \times 0.400}{0.0821 \times 378} = 7.73 \times 10^{-3}$ mole   (7.73 mmole)

   (b) $n = \dfrac{(748/760) \times 2.81}{0.0821 \times 297} = 0.113$ mole

   (c) $n = \dfrac{(50/760) \times 0.280}{0.0821 \times 195} = 1.15 \times 10^{-3}$ mole   (1.15 mmole)

4. At STP, $P = 1.00$ atm, $T = 273$ K, and

$$V = \frac{nRT}{P} = (0.0821 \times 273)n = 22.4n$$

   (a) $22.4 \times 7.73 \times 10^{-3} = 0.173$ liter
   (b) $22.4 \times 0.113 = 2.53$ liter
   (c) $22.4 \times 1.15 \times 10^{-3} = 2.58 \times 10^{-2}$ liter $= 25.8$ ml

5. Use $m = MPV/RT$:

   (a) $m = \dfrac{16.0 \times 0.250 \times 0.450}{0.0821 \times 298} = 7.36 \times 10^{-2}$ g

   (b) $m = \dfrac{146 \times 2.00 \times 6.68}{0.0821 \times 373} = 63.7$ g

6. Rearranging gives $T = MPV/Rm$:

$$T = \frac{2.02 \times 2.00 \times 100}{0.0821 \times 5.00} = 984 \text{ K} \quad (711 \text{ }^\circ\text{C})$$

7. Using $m/V = MP/RT$:

(a) $\dfrac{m}{V} = \dfrac{17.0 \times 1.20}{0.0821 \times 323} = 0.769 \text{ g liter}^{-1}$

(b) $\dfrac{m}{V} = \dfrac{4.00 \times 0.800}{0.0821 \times 73} = 0.534 \text{ g liter}^{-1}$

(c) $\dfrac{m}{V} = \dfrac{34.1 \times 1.00}{0.0821 \times 273} = 1.52 \text{ g liter}^{-1}$

8. (a) Rearranging gives $P = mRT/VM$, so, if $m/V = 1.00 \text{ g cm}^{-3} = 1.00 \times 10^3 \text{ g liter}^{-1}$,

$$P = 1.00 \times 10^3 \times \frac{0.0821 \times 273}{29} = 7.7 \times 10^2 \text{ atm}$$

(b) Above a pressure of 770 atm, air should be denser than water. One atmosphere corresponds to 76.0 cm Hg (density 13.7 g cm^{-3}). Since water has a density of only 1.00 g cm^{-3}, one atmosphere pressure corresponds to a column of water of height

$$76.0 \times 13.7 = 1.04 \times 10^3 \text{ cm} = 10.4 \text{ m}$$

So our 770 atm pressure will be found at an ocean depth of

$$10.4 \times 770 = 8.0 \times 10^3 \text{ m}$$

Below this depth, at higher pressures, air bubbles would sink, according to this approach (in practice, of course, the bubbles would dissolve).

9. (a) $M = \dfrac{RTd}{P}$ $M = \dfrac{0.0821 \times 295 \times 1.14}{750/760} = 28.0 \text{ g}$ M.W. = 28.0

(b) $M = \dfrac{RTm}{PV}$ $M = \dfrac{0.0821 \times 273 \times 0.670}{1.00 \times 0.250} = 60.0 \text{ g}$ M.W. = 60.0

(c) $M = \dfrac{0.0821 \times 373 \times 3.87}{(765/760) \times 1.18} = 100 \text{ g}$ M.W. = 100

10. 0.400 g H$_2$: $\dfrac{0.400}{2.02} = 0.198 \text{ mole}$

2.00 g N$_2$: $\dfrac{2.00}{28.0} = 0.0714$

10.5 g CO$_2$: $\dfrac{10.5}{44.0} = 0.239$

total $= 0.508$ mole

$$P = \frac{nRT}{V} = \frac{0.508 \times 0.0821 \times 273}{10.0} = 1.14 \text{ atm}$$

11. Use $m = PVM/RT$; then, in a volume of 1.00 liter of the mixture,

for HCl: $m = \dfrac{(200/760) \times 1.00 \times 36.5}{0.0821 \times 298} = 0.393$ g

for HBr: $m = \dfrac{(100/760) \times 1.00 \times 80.9}{0.0821 \times 298} = 0.435$

for HI: $m = \dfrac{(50/760) \times 1.00 \times 128}{0.0821 \times 298} = 0.34$

$$\text{total mass} = 1.17 \text{ g}$$

density: 1.17 g liter^{-1}

12. Let the partial pressure of O_2 be y atm, then that of N_2 is $(1.00 - y)$ atm, since we know the total pressure is 1.00 atm.

use $m = \dfrac{PVM}{RT}$ and put $V = 1.00$ liter

for O_2: $m = \dfrac{y \times 1.00 \times 32.0}{0.0821 \times 273} = 1.43y$

for N_2: $m = \dfrac{(1.00 - y) \times 1.00 \times 28.0}{0.0821 \times 273} = 1.25(1.00 - y)$

total mass: $1.43y + 1.25(1.00 - y) = 1.25 + 0.18y$
density: $(1.25 + 0.18y)$ g liter^{-1}
putting this equal to the given density gives

$$1.25 + 0.18y = 1.38$$

$$0.18y = 1.38 - 1.25 = 0.13$$

partial pressure O_2: $y = \dfrac{0.13}{0.18} = 0.72$ atm

partial pressure N_2: $(1.00 - y) = 0.28$ atm

13. Let the fraction of H_2 be y and the fraction of deuterium (D_2) be $(1 - y)$; use $V = 1$ liter.

mass of H_2: $\dfrac{y \times 1.00 \times 2.02}{0.0821 \times 273} = (9.01 \times 10^{-2})y$

mass of D_2: $\dfrac{(1 - y) \times 1.00 \times 4.00}{0.0821 \times 273} = 0.179(1 - y)$

density: $(9.01 \times 10^{-2})y + 0.179(1 - y) = 0.179 - (8.9 \times 10^{-2})y$

putting this equal to the given density gives

for H_2: $0.179 - (8.9 \times 10^{-2})y = 0.161$

$$(8.9 \times 10^{-2})y = 0.179 - 0.161 = 0.018$$

$$y = \dfrac{0.018}{8.9 \times 10^{-2}} = 0.20$$

for D_2: $1 - y = 0.80$
     The mixture is 20% H_2, 80% D_2.

14. (a) For the weight of H_2 in the balloon (volume 10.0 m³ = 1.00×10^4 liter),

$$m = \frac{MPV}{RT} = \frac{2.02 \times 1.00 \times 1.00 \times 10^4}{0.0821 \times 298} = 826 \text{ g}$$

Adding the mass of the balloon fabric (2.00 kg) gives the total mass of balloon $+ H_2$ gas:

$$2.00 + 0.826 = 2.83 \text{ kg}$$

The displaced air mass is

$$m = \frac{MPV}{RT} = \frac{29 \times 1.00 \times 1.00 \times 10^4}{0.0821 \times 298} = 1.19 \times 10^4 \text{ g} = 11.9 \text{ kg}$$

This is equal to the total upthrust on the balloon, so the net lifting power is the difference between the upthrust and the weight of the balloon:

$$\text{maximum load} = 11.9 - 2.8 = 9.1 \text{ kg}$$

(b) With He instead of H_2, the weight of gas in the balloon would be

$$m = \frac{4.00 \times 1.00 \times 1.00 \times 10^4}{0.0821 \times 298} = 1.64 \times 10^3 \text{ g}$$

total weight $2.00 + 1.64 = 3.64$ kg
    net upthrust (maximum load): $11.9 - 3.64 = 8.3$ kg
(Although the He-filled balloon has less lifting power, it is preferable to hydrogen in practice because it will not burn.)

15. (a) vapor pressure of water at 22°C = 20 mm
    partial pressure of O_2 = 748 − 20 = 728 mm

Use $\dfrac{P_1 V_1}{T_1} = \dfrac{P_2 V_2}{T_2}$, where

$P_1 = 760$ mm	$V_1 =$ unknown	$T_1 = 273$ K
$P_2 = 728$ mm	$V_2 = 228$ ml	$T_2 = 295$ K

$$\frac{760 \times V_1}{273} = \frac{728 \times 228}{295}$$

$$V_1 = \frac{728 \times 228 \times 273}{760 \times 295} = 202 \text{ ml}$$

(b) moles of $O_2 = \dfrac{P_1 V_1}{RT_1} = \dfrac{1.00 \times 0.202}{0.0821 \times 273} = 9.02 \times 10^{-3}$ mole

16. vapor pressure of water at 26°C = 25.2 mm
    partial pressure of O_2 = 755 − 25 = 730 mm
    use $m = MPV/RT$

(a) mass of O_2: $\dfrac{32.0 \times (730/760) \times 1.68}{0.0821 \times 299} = 2.10$ g

(b) mass of H_2O: $\dfrac{18.0 \times (25.2/760) \times 1.68}{0.0821 \times 299} = 0.041$ g

17. Use $u = \sqrt{3RT/M}$ (remember, $R = 8.314$ J, M in kg):

   (a) H_2 at $-50°C$: $u = \sqrt{\dfrac{3 \times 8.314 \times 223}{2.02 \times 10^{-3}}} = \sqrt{2.75 \times 10^6} = 1.66 \times 10^3 \text{ m s}^{-1}$

   (b) N_2 at $1000°C$: $u = \sqrt{\dfrac{3 \times 8.314 \times 1273}{0.0280}} = \sqrt{1.13 \times 10^6} = 1.07 \times 10^3 \text{ m s}^{-1}$

   (c) CH_4 at 298 K: $u = \sqrt{\dfrac{3 \times 8.314 \times 298}{0.0160}} = \sqrt{4.65 \times 10^5} = 682 \text{ m s}^{-1}$

18. For He (molar mass 4.00 g) at 4.18 K:

$$u = \sqrt{\dfrac{3 \times 8.314 \times 4.18}{4.00 \times 10^{-3}}} = \sqrt{2.61 \times 10^4} = 161 \text{ m s}^{-1}$$

   For UF_6 (molar mass 352 g) at $56°C$ (329 K):

$$u = \sqrt{\dfrac{3 \times 8.314 \times 329}{0.352}} = \sqrt{2.33 \times 10^4} = 153 \text{ m s}^{-1}$$

   Helium wins by a short head.

19. Suppose the gas has molar mass M. At 273 K,

$$u_1 = \sqrt{\dfrac{3 \times 8.314 \times 273}{M}} = \sqrt{\dfrac{6.81 \times 10^3}{M}}$$

   At some other temperature T K,

$$u_2 = \sqrt{\dfrac{3 \times 8.314 \times T}{M}} = \sqrt{\dfrac{24.9T}{M}}$$

   If the average velocity at temperature T is twice that at 273 K, we know that

$$\dfrac{u_2}{u_1} = 2 = \dfrac{\sqrt{24.9T/M}}{\sqrt{6.81 \times 10^3/M}} = \sqrt{\dfrac{24.9T}{6.81 \times 10^3}} = \sqrt{(3.66 \times 10^{-3})T}$$

   Squaring both sides gives $2^2 = 4 = (3.66 \times 10^{-3})T$ and

$$T = \dfrac{4}{3.66 \times 10^{-3}} = 1092 \text{ K} \quad (819°C)$$

   Note that M canceled out in the calculation, so this temperature will be the same for any gas.

20. Use Graham's law ($u = $ molecular velocity):

$$\dfrac{u(\text{Ar})}{u(\text{He})} = \sqrt{\dfrac{4.00}{39.9}} = \sqrt{0.100} = 0.317$$

$$\dfrac{u(\text{Ne})}{u(\text{He})} = \sqrt{\dfrac{4.00}{20.2}} = \sqrt{0.198} = 0.445$$

$$\dfrac{u(\text{Xe})}{u(\text{He})} = \sqrt{\dfrac{4.00}{131}} = \sqrt{3.05 \times 10^{-2}} = 0.175$$

21. $\dfrac{u(H_2)}{u(SO_2)} = \sqrt{\dfrac{64}{2.02}} = \sqrt{31.7} = 5.63$

    Since the H_2 is moving *faster*, the SO_2 will take much *longer* to effuse through the hole.
    time for SO_2: $150 \times 5.63 = 844$ s

22. Let the unknown gas have molar mass M_x and average molecular velocity u_x. Then

    $$\frac{u_x}{u(N_2)} = \sqrt{\frac{28.0}{M_x}}$$

    The rate of diffusion is directly proportional to the molecular velocity, so

    $$\frac{u_x}{u(N_2)} = \sqrt{\frac{14.0}{29.9}} = \sqrt{\frac{28.0}{M_x}} \qquad \frac{28.0}{M_x} = \left(\frac{14.0}{29.9}\right)^2 \qquad M_x = 28.0 \times \left(\frac{29.9}{14.0}\right)^2 = 128$$

23. $\dfrac{u(^{20}Ne)}{u(^{22}Ne)} = \sqrt{\dfrac{22.0}{20.0}} = \sqrt{1.10} = 1.05$

    The ratio of the rates of diffusion is 1.05.

24. *M.W.*'s are $^{235}UF_6$, 349; $^{238}UF_6$, 352, so

    $$\frac{u(^{235}UF_6)}{u(^{238}UF_6)} = \sqrt{\frac{352}{349}} = \sqrt{1.009} = 1.004$$

    (The two isotopes are successfully separated by repeating this diffusion many times in a tedious and expensive process.)

25. $$\frac{u(NH_3)}{u(BF_3)} = \sqrt{\frac{67.8}{17.0}} = \sqrt{3.99} = 2.00$$

    NH_3 moves 2.00 units while BF_3 moves 1.00 unit, so if they move together along a tube 3.00 units long, they meet at a point 2.00 units (66.7%) along from the end where the NH_3 started.

Chapter 9

The Properties of Solutions

You will remember, Watson, how the dreadful business of the Abernetty family was first brought to my notice by the depth which the parsley had sunk into the butter upon a hot day.   THE ADVENTURE OF THE SIX NAPOLEONS

9.1   Mole Fraction and Molality

In Chapter 3, we considered some methods of expressing the concentration of a solution: in molarity and in grams per liter. We start this chapter by defining two additional ways of expressing concentration.

The first of these is as a *mole fraction.* The mole fraction of a substance in a solution is defined as

$$\frac{\text{moles of that substance present}}{\text{total moles present, all substances}}$$

This is a very general definition, applicable to liquid- or gas-phase solutions. It has one obvious advantage in that it does not require any distinction to be made between solute and solvent. This is very useful in dealing with liquid mixtures, when two or more liquids may form a mixed solution in various amounts, where no liquid is in a great excess, and it would be artificial to designate one or another as solvent.

If we work out the mole fractions of all substances present, they must add up to exactly unity. Note that mole fraction is a dimensionless ratio.

Example 9.1   5.00 mole of methanol, 2.00 mole of ethanol, and 6.00 mole of water are mixed. What is the mole fraction of each substance present?

SOLUTION   The total moles present is $5.00 + 2.00 + 6.00 = 13.00$. The mole fraction of each will be found by dividing the number of moles of each by the total moles:

methanol: $5.00/13.0 = 0.385$

ethanol: $2.00/13.0 = 0.154$

water: $6.00/13.0 = 0.461$

(*Check*: The total of the three mole fractions is 1.000.)

Example 9.2   10.0 g of chloroform, $CHCl_3$, are mixed with 15.0 g of benzene, C_6H_6. What is the mole fraction of each present?

SOLUTION   First, we convert to mole quantities:

$$CHCl_3 \; (M.W. \; 119.5): \quad \frac{10.0}{119.5} = 0.0837 \text{ mole}$$

$$C_6H_6 \; (M.W. \; 78.0): \quad \frac{15.0}{78.0} = 0.192 \text{ mole}$$
$$\text{total moles present} = \overline{0.276} \text{ mole}$$

$$\text{mole fraction of } CHCl_3 \text{ is } \frac{0.0837}{0.276} = 0.303$$

$$\text{mole fraction of } C_6H_6 \text{ is } \frac{0.192}{0.276} = 0.696$$

Example 9.3   What is the composition by weight of a mixture of chloroform and benzene in which the mole fraction of each is 0.500?

SOLUTION   The mole fractions are converted to weights by multiplying by the respective *M.W.*'s.

$$CHCl_3 \text{ has } M.W. \; 119.5, \text{ so } 0.500 \text{ mole is } 0.500 \times 119.5 = 59.8 \text{ g}$$
$$C_6H_6 \text{ has } M.W. \; 78.0, \text{ so } 0.500 \text{ mole is } 0.500 \times 78.0 = 39.0 \text{ g}$$
$$\text{total weight of solution} = \overline{98.8} \text{ g}$$

Of this total,

$$CHCl_3 \text{ amounts to } \frac{59.8}{98.8} \times 100 = 60.5 \% \text{ by weight}$$

$$C_6H_6 \text{ amounts to } \frac{39.0}{98.8} \times 100 = 39.5 \% \text{ by weight}$$

The second method of expressing concentrations is called *molality*, defined as the number of moles of solute dissolved in *one kilogram* of solvent. Note that we use the mass of the solvent, *not* the mass of the solution. Obviously molality is going to be a useful method of expressing concentrations when we make up a solution by weight, whereas molarity is useful when a solution is made up by volume. The units of *molality* are mole kg^{-1}, and the symbol for it is *m*.

Example 9.4   0.448 g naphthalene ($C_{10}H_8$) is dissolved in 11.4 g of benzene. What is the molality of the solution?

SOLUTION

$C_{10}H_8$ has *M.W.* 128

$$0.448 \text{ g} = \frac{0.448}{128} = 3.50 \times 10^{-3} \text{ mole}$$

3.50×10^{-3} mole $C_{10}H_8$ are dissolved in 11.4 g of benzene

$$\frac{3.50 \times 10^{-3} \times 1000}{11.4} = 0.307 \text{ mole naphthalene in 1 kg benzene}$$

The solution is 0.307*m*.

Example 9.5   0.288 g of an unknown solute is dissolved in 15.2 g of hexane and the solution is found to be 0.221*m*. What is the *M.W.* of the unknown?

SOLUTION   If 0.288 g is dissolved in 15.2 g of solvent, then in 1 kg of solvent there is dissolved

$$\frac{0.288 \times 1000}{15.2} = 18.9 \text{ g of solute}$$

The solution is known to be 0.221*m*, so 0.221 mole is equal to 18.9 g and

$$1 \text{ mole is equal to } \frac{18.9}{0.221} = 85.7 \text{ g}$$

M.W. = 85.7.

Although the molality of a solution may be defined without reference to the nature of the solvent, we need to know the molar mass of the solvent if we want to relate mole fraction and molality in a solution.

Suppose a solute is present in a solution in mole fraction *f* while the solvent (the only other component of the solution) has molar mass *M* g.

Since the two mole fractions must add up to 1, the number of moles of solvent containing f mole of solute is $(1 - f)$ mole.

$$(1 - f) \text{ mole solvent has mass } M(1 - f) \text{ g}$$

If f mole solute is in $M(1 - f)$ g solvent, then

$$\frac{f}{M(1 - f)} \text{ mole solute is in 1 g solvent}$$

$$\frac{1000f}{M(1 - f)} \text{ mole solute is in 1000 g solvent}$$

Since the last quantity is, by definition, the molality of the solution, we can write

$$m = \frac{1000f}{M(1 - f)}$$

(Note: Remember, M is the molar mass of the *solvent*.)

Example 9.6   A solution of naphthalene, $C_{10}H_8$, in benzene, C_6H_6, has a mole fraction of naphthalene of 0.100. What is the molality of the solution?

SOLUTION   Since the naphthalene has a mole fraction of 0.100, the benzene has mole fraction $(1 - 0.100) = 0.900$.

0.900 mole benzene ($M.W.$ 78.0) is $0.900 \times 78.0 = 70.2$ g
0.100 mole naphthalene is dissolved in 70.2 g benzene

$$\frac{0.100 \times 1000}{70.2} \text{ mole is dissolved in 1000 g benzene}$$

this is 1.42 mole, so molality is $1.42m$

It is not often necessary to convert from molality to molarity, but this can be easily done provided that the density of the solution is known.

Example 9.7   A solution of nitric acid in water is $1.50M$ and has a density of 1.049 g cm^{-3}. What would the concentration of nitric acid be in molality?

SOLUTION   A liter of the solution weighs $1.049 \times 1000 = 1049$ g. This contains 1.50 mole of HNO_3, which is $1.50 \times 63.0 = 94.5$ g HNO_3. The amount of water present, by difference, is $1049 - 94.5 = 954.5$ g. So we have 1.50 mole HNO_3 dissolved in 954.5 g of water, hence the molality (number of moles dissolved in 1000 g water) is

$$1.50 \times \frac{1000}{954.5} = 1.57m$$

Example 9.8 Distillation of hydrochloric acid gives a mixture known as "constant-boiling HCl," which contains 20.22% by weight of HCl (the remainder being water), and has a density of 1.096 g cm^{-3}.

(a) Express the concentration of this solution in terms of mole fraction, molality, and molarity.
(b) What weight of the constant boiling acid would be needed to make one liter of 1.000M solution?

SOLUTION   The simplest approach to this problem is to work out what is present in 1.000 liter of the constant-boiling acid.

mass of 1.000 liter: $1000 \times 1.096 = 1096$ g

HCl present: $\dfrac{1096 \times 20.22}{100} = 221.6$ g HCl

moles HCl present: $\dfrac{221.6}{36.46} = 6.078$ mole

water present: $\dfrac{1096 \times (100 - 20.22)}{100} = 874.4$ g H$_2$O

moles H$_2$O present: $\dfrac{874.4}{18.02} = 48.52$ mole

(a) From the above data, we can express the concentration in any way we desire:

total moles present: $6.078 + 48.52 = 54.60$ mole

mole fraction of HCl: $\dfrac{6.078}{54.60} = 0.1113$

mole fraction of H$_2$O: $\dfrac{48.52}{54.60} = 0.8887$

6.078 mole HCl are dissolved in 874.4 g H$_2$O

molality of HCl: $\dfrac{6.078 \times 1000}{874.4} = 6.952m$

6.078 mole HCl are dissolved in 1.000 liter solution

molarity of HCl solution: 6.078M

(b) To make 1.000 liter of 1.000M HCl we need 1.00 mole HCl. Since we have 6.078 mole HCl in 1096 g of the constant-boiling acid, there is 1.000 mole HCl in

$$\frac{1096}{6.078} = 180.3 \text{ g of acid}$$

(This is in fact a useful way of preparing solutions of HCl of accurately known concentration.)

To be successful in calculations of the above type, all one needs is a clear grasp of the definitions of the various ways in which solution concentrations are expressed.

We should note one final point about mole fraction and molality: they are independent of temperature. The only measurement is that of mass, which does not change with temperature, whereas molarity depends on volume and is therefore temperature dependent.

9.2 Raoult's Law and Vapor Pressure

We will start our treatment of the properties of solutions by considering the vapor pressure of a mixture of two liquids. (In all cases when talking of mixtures, we assume that the liquids in question mix freely to give a homogeneous solution.) The vapor pressure of such a mixture is given by *Raoult's law*:

The vapor pressure of a liquid mixture is equal to the sum of the partial vapor pressures of the component liquids. The partial vapor pressure of each component is equal to its vapor pressure at that temperature, multiplied by the mole fraction of that liquid present in the solution.

There's an obvious resemblance here to Dalton's law of partial pressures for gaseous mixtures. Note, however, that in Raoult's law we are concerned with the mole fraction of each component present in the *liquid* phase, not the gas phase.

Example 9.9 At 25°C, cyclohexane, C_6H_{12}, has a vapor pressure of 100 mm, while octane, C_8H_{18} has a vapor pressure of 20 mm. What will be the vapor pressure of a mixture of 120 g cyclohexane and 80 g octane at that temperature?

SOLUTION We first find the mole fractions of each:

$$\text{cyclohexane, } C_6H_{12}, \text{ } M.W. \text{ 84:} \quad 120 \text{ g} = \frac{120}{84} = 1.43 \text{ mole}$$

$$\text{octane, } C_8H_{18}, \text{ } M.W. \text{ 114:} \quad 80 \text{ g} = \frac{80}{114} = 0.70 \text{ mole}$$

$$\text{total} = \overline{2.13} \text{ mole}$$

mole fraction C_6H_{12}: 1.43/2.13 = 0.67
mole fraction C_8H_{18}: 0.70/2.13 = 0.33

Partial vapor pressures are the product of mole fraction and vapor pressure:

$$\begin{array}{ll} \text{partial vp cyclohexane:} & 0.67 \times 100 = 67 \text{ mm} \\ \text{partial vp octane:} & 0.33 \times 20 = \underline{6.6 \text{ mm}} \\ & \text{total vp of the solution} = \overline{74 \text{ mm}} \end{array}$$

9.3 Dilute Solutions and Depression of Vapor Pressure

Raoult's law applies only to solutions that behave in an ideal manner. In practice, real liquids show departures from ideal behavior (just as do real gases) so vapor pressures found experimentally will often differ from those calculated. However, in dilute solution, the behavior of the solution approximates better to ideal behavior, so many measurements are made on dilute solutions.

Let's consider the case of a nonvolatile solute in dilute solution in a volatile solvent. By "nonvolatile," we mean that the solute is making no appreciable contribution to the vapor pressure of the solution, so the measured vp will be entirely contributed by the solvent. Since we have added solute to the pure solvent, the mole fraction of solvent present in the solution will be less than unity. Hence the vapor pressure of a solution containing a non-volatile solute will always be *less* than that of the pure solvent. The difference is called the *vapor pressure depression*.

Suppose we have n_a moles of A, a nonvolatile solute, dissolved in n_b moles of B, the volatile solvent.

The mole fraction of B will be: $\dfrac{n_b}{n_a + n_b}$

The vapor pressure of B will be: $\dfrac{n_b p_b}{n_a + n_b}$ (p_b = vp of pure B)

If the other component is nonvolatile, this will be the total vp of the solution. So the *depression* of the vp (difference between pure solvent and solution) is

$$p_b - \frac{n_b p_b}{n_a + n_b} = \frac{p_b(n_a + n_b) - n_b p_b}{n_a + n_b} = \frac{p_b n_a}{n_a + n_b}$$

Knowing that $n_a/(n_a + n_b)$ is the mole fraction of component A (the nonvolatile part), we see that the depression of vp is given by:

(vp of pure solvent) × (mole fraction of nonvolatile solute)

We had previously related molality and mole fraction by the equation

$$m = \frac{1000f}{M(1 - f)}$$

and we can simplify this for a dilute solution by assuming that, if f is small

$$(1 - f) \approx 1 \quad \text{and} \quad m \approx \frac{1000f}{M}$$

This rearranges to give $f = Mm/1000$, so we can relate vp depression to the molality of the solution by the equation

$$\text{vp depression} = \frac{p_b M_b m}{1000}$$

where m is the molality of the nonvolatile solute. Note that this equation will only be true under two conditions; firstly, that the solution is dilute, and secondly, that the solute does not dissociate (e.g., by ionization) in solution. We shall return to the second point in more detail later.

Example 9.10   Water has a vp of 22.4 mm at 24°C. What is the vp of an 0.1 molal solution of a nonvolatile solute?

SOLUTION   We should first consider whether a solution of 0.1 molal concentration qualifies as "dilute." For water, the *M.W.* is 18.0, so 1000 g amounts to 55.5 mole. If we have 0.1 mole of solute, this is obviously negligible beside 55.5 mole of solvent. The assumption we are making is that $55.5 + 0.1 \approx 55.5$, which is quite acceptable within the accuracy of the given data.

Using the relationship derived above, we have

$$\text{vp depression} = \frac{p_b M_b}{1000} \times \text{molality} = \frac{22.4 \times 18.0}{1000} \times 0.1 = 0.04 \text{ mm}$$

This is a very small difference; within the accuracy of the given data, the vp is not altered, but is still 22.4 mm.

9.4 Colligative Properties

Notice in the above calculation that the identity of the solute does not matter. Any 0.1 *m* solution in water would have shown the same effect. Properties of this type, which depend only on the concentration of the solute, rather than on its nature, are called *colligative properties*. Examples are depression of the freezing point, elevation of the boiling point, and osmotic pressure, properties in which the same effect will be seen for equal concentrations of different solutes in a given solvent. Some properties that are *not* colligative are color, odor, or hydrogen-ion concentration, which all depend very much on the nature of the solute we put in.

One of the principal applications of colligative properties is in the measurement of molecular weight. Although, in theory, vp depression could be used to measure the molality of a solution and hence the *M.W.* of an unknown solute, this calculation shows that the effect is too small to be useful. We cannot measure vp of a solution to small fractions of a mm Hg.

9.5 Boiling-Point Elevation and Freezing-Point Depression

The normal boiling point (bp) of a liquid is the temperature at which its vp equals one atmosphere. For a solution, where the vp at a given temperature is lower, we are obviously going to need a *higher* temperature to get the vp up to one atmosphere.

The difference between the boiling point of the solvent and the boiling point of the solution is called the *boiling-point elevation*. Since this is a colligative property, the ratio (bp elevation)/(molality of solution), for any solution in a given solvent, will be the same. This value is called the *molal boiling-point constant* (or molal ebullioscopic constant), K_b for that solvent. It varies very much from one solvent to another; selected values are shown in Table 9.1.

A related phenomenon is found at the low-temperature end of the liquid range. Since the liquid range of a solution is extended further in both directions, compared with that of the solvent, the freezing point (fp) of the solution is always *lower* than that of the solvent. The difference is called the *freezing-point depression*, and the ratio fp depression/molality for any solution in a given solvent is called the *molal freezing-point depression constant* (or molal cryoscopic constant), K_f, for that solvent.

Since small differences in temperature may be measured quite accurately, measurements of changes in fp and bp will enable us to find the molality of solutions, and hence the

TABLE 9.1   Molal bp elevation (K_b) and fp depression (K_f) constants in K mole^{-1} kg

Solvent	K_f	K_b
Water	1.86	0.512
Benzene	5.12	2.53
Acetic acid	3.90	3.07
Chloroform	4.68	3.63
Ethyl alcohol	1.99	1.22
Naphthalene	6.8	—
Camphor	39.7	—

M.W. of unknown solutes. For solutions other than 1 molal, the fp depression, ΔT_f, will be related to K_f by simple proportionality:

$$\Delta T_f = mK_f$$

where m is molality of solution. Similarly, the bp elevation, ΔT_b, will be given by

$$\Delta T_b = mK_b$$

Example 9.11   Chloroform freezes at $-63.5°C$ and boils at $61.2°C$. What will be the fp and bp of a $1.6m$ solution in chloroform?

SOLUTION   The fp will be *lower* by ΔT_f, where $\Delta T_f = mK_f = 1.6 \times 4.68 = 7.5°C$. This will depress the fp from -63.5 to $-63.5 - 7.5 = -71.0°C$. In the same solution, the bp will be *higher* by ΔT_b, where $\Delta T_b = mK_b = 1.6 \times 3.63 = 5.8°C$. This will raise the bp from 61.2 to $61.2 + 5.8 = 67.0°C$. (*Note*: The nature of the solute does not enter into this calculation.)

Example 9.12   What will be the bp of a benzene solution that freezes at $4.0°C$? Pure benzene has fp $5.5°C$ and bp $80.1°C$.

SOLUTION   A benzene solution freezing at $4.0°C$ has $\Delta T_f = 5.5 - 4.0 = 1.5°C$. The molality is therefore

$$\frac{\Delta T_f}{K_f} = \frac{1.5}{5.12} = 0.29m$$

Such a solution will have $\Delta T_b = mK_b = 0.29 \times 2.53 = 0.74°C$, which will raise the bp from $80.1°C$ to $80.8°C$.

In an experiment conducted to determine the *M.W.* of an unknown from fp or bp changes, we always measure the fp or bp of the pure solvent at the same time, using the same thermometer. It would be a mistake to rely on published values of the fp or bp of a solvent, since our experimental conditions may be different from those under which published values were measured.

Example 9.13 A solution of 1.04 g of unknown Q in 25.3 g of benzene had a bp of 80.78°C. Under the same conditions, pure benzene boiled at 80.06°C. Calculate the *M.W.* of Q.

SOLUTION The bp elevation is the difference between the two temperatures:

$$\Delta T_b = 80.78 - 80.06 = 0.72°C$$

For benzene, $K_b = 2.53$ (Table 9.1), so in this solution

$$\text{molality} = \frac{\Delta T_b}{K_b} = \frac{0.72}{2.53} = 0.285m$$

The solution was made up with 1.04 g in 25.3 g solvent, that is

$$\frac{1.04 \times 1000}{25.3} = 41.1 \text{ g per kg of solute}$$

$$41.1 \text{ g of solute Q} = 0.285 \text{ mole}$$

$$\frac{41.1}{0.285} = 144 \text{ g of Q} = 1 \text{ mole}$$

The molecular weight of Q is 1.4×10^2 (only 2 significant figures are justified).

Example 9.14 4.00 g of an unknown solute dissolved in 60.0 g of water freezes at −0.688°C. Under the same conditions, pure water freezes at −0.015°C. What is the *M.W.* of the unknown?

SOLUTION (Remember, we *subtract* the fp of the solution *from* the fp of the solvent to find ΔT_f.)

$$\Delta T_f = -0.015 - (-0.688) = 0.673°C$$

K_f for water is 1.86

$$\text{molality} = \frac{\Delta T_f}{K_f} = \frac{0.673}{1.86} = 0.361m$$

$$\text{concentration is } \frac{(4.00 \text{ g solute}) \times 1000}{60.0 \text{ g water}} = 66.7 \text{ g solute per kg solvent}$$

66.7 g solute is 0.361 mole

$$M.W. \text{ of unknown is } 66.7/0.361 = 184$$

Example 9.15   The fp of a pure camphor sample is found to be 177.88°C. Under the same conditions, a solution of 1.08 mg of substance Y in 0.206 g of camphor had a fp of 175.34°C. What is the *M.W.* of substance Y?

SOLUTION

$$\Delta T_f = 177.88 - 175.34 = 2.54°C$$

$$\text{molality} = \frac{\Delta T_f}{K_f} = \frac{2.54}{39.7} = 0.0640m$$

1.08 mg (i.e., 1.08×10^{-3} g) is dissolved in 0.206 g of solvent

$$\frac{1.08 \times 10^{-3} \times 1000}{0.206} = 5.24 \text{ g is dissolved in 1 kg of solvent}$$

5.24 g is 0.0640 mole

  M.W. of unknown is $5.24/0.0640 = 81.9$

(*Note*: Camphor may seem an odd choice for a solvent, but it has the advantage of a very large value of K_f.)

9.6   Degree of Dissociation

If we measure the molecular weight of a solution of sodium chloride in water, the result will be close to one-half of the value we would expect from the apparent molecular weight of NaCl (58.5). The reason is apparent if we consider the nature of such a solution. Instead of containing NaCl "molecules," it contains the hydrated ions $Na^+(aq)$ and $Cl^-(aq)$. Since a mole of NaCl produces, on dissociation, one mole *each* of Na^+ and Cl^- ions, the total moles of solute will be twice the value we would expect from the formula of the salt, giving *twice* the expected value of ΔT_f. If we try to calculate the *M.W.* from such data, our answer will be half the expected value.

This leads us to a slightly broader definition of the term "solute" when dealing with a colligative property:

A colligative property of a solution is one which is proportional to the total number of moles of particles of solute present, and not on their nature. "Particles" may be molecules, positive ions, or negative ions.

Ionic salts such as NaCl are completely dissociated in solution, although, in high concentrations, interionic attractions make the solutions behave as if the degree of dissociation were less than 100%, and we speak of the "apparent" degree of dissociation. For many substances, however, a covalent molecule with a partial degree of dissociation exists in solution.

Let's consider a general substance AB, which dissociates:

$$AB \rightleftharpoons A^+ + B^-$$

Say we start with C mole kg^{-1} of solute AB, out of which x mole kg^{-1} dissociates. At equilibrium, the concentrations will be

$$\underset{(C-x)}{AB} \rightleftharpoons \underset{x}{A^+} + \underset{x}{B^-}$$

So the total number of mole kg^{-1}, for all three species present in solution, will be the sum of these three

$$(C - x) + x + x = C + x \text{ mole kg}^{-1}$$

Students are often puzzled to see that the concentration of the solution is *greater* after dissociation has occurred than it was originally. The reason for this is, of course, that each AB molecule gives *two* ions on dissociating, so the total concentration of all particles (molecules *plus* ions) goes *up* as dissociation occurs.

When we measure the molality of the solution, using a colligative property, the result we obtain will be the value of the total particle concentration, $C + x$. Knowing the value of C, the number of moles of undissociated AB with which we started, we can easily calculate x and hence the degree of dissociation, defined as:

$$\frac{\text{amount of substance which has dissociated}}{\text{amount originally present}}$$

Example 9.16 What molal concentration of sodium chloride, NaCl, is needed to lower the fp of water to $-5.00°C$? (Assume complete dissociation.) What weight of solid sodium chloride would have to be spread on a road to melt 1000 kg of ice at this temperature?

SOLUTION   The fp depression is 5.00°C, so the molality, m, is

$$\frac{\Delta T}{K_f} = \frac{5.00}{1.86} = 2.69m$$

Since the dissociation of sodium chloride produces *two* moles of ions for each mole of solute, we only need half this number of moles NaCl for each kg of solvent, so the NaCl should be made up in $2.69/2 = 1.34$ molal concentration.

We need 1.34 mole of NaCl for each kg of solvent, so to melt 1000 kg of ice we need 1.34×1000 mole NaCl $= 1.34 \times 58.5$ kg NaCl $= 78.6$ kg.

Example 9.17 5.0 g of HX, a weak acid with *M.W.* of 150, is dissolved in 100 g of water and fp measurements show the solution to be 0.54 molal. What is the degree of dissociation of HX in this solution?

SOLUTION   We started out with 5.0 g of HX in 100 g water, which is 50 g in 1000 g water. If the *M.W.* of HX is 150, this solution is originally (before dissociation) $50/150 = 0.33m$.

Suppose x mole kg^{-1} of HX dissociate, then $(0.33 - x)$ will remain.

$$\underset{(0.33-x)}{HX} \rightleftharpoons \underset{x}{H^+} + \underset{x}{X^-}$$

Total mole kg^{-1} is $(0.33 - x) + x + x = (0.33 + x)m$. But this is found by experiment to be $0.54m$.

$$0.33 + x = 0.54 \qquad x = 0.21$$

The degree of dissociation is $0.21/0.33 = 0.64$, or 64%. (*Note*: We divide by the amount of HX originally present in calculating degree of dissociation. We do *not* divide by the amount of HX remaining at equilibrium.)

9.7   Osmotic Pressure

If a solution and the corresponding pure solvent are separated by a semipermeable barrier, such as a membrane, through which the solvent but not the solute may pass, a flow of solvent into the solution will be observed. The effect of this is that the solution tends to become more dilute—this phenomenon is called *osmosis*. If a pressure is applied to the solution, the flow can be prevented, and the pressure needed to prevent flow of solvent into the solution through the barrier is called the *osmotic pressure* of that solution.

Osmotic pressure is a colligative property, and its dependence on concentration is given by the simple equation

$$\pi = MRT$$

where π is the osmotic pressure; M is the molarity of the solution (*Note*: *Molarity*, not molality); R is the gas constant; and T is the temperature in kelvins. If we put in the usual value for R of 0.0821 liter atm K^{-1} $mole^{-1}$, then units of π will be

$$(\text{mole liter}^{-1}) \times (\text{liter atm K}^{-1} \text{ mole}^{-1}) \times K = atm$$

We can therefore readily calculate the osmotic pressure to be expected in a solution.

Example 9.18   Calculate the osmotic pressure of an $0.100M$ solution of sucrose in water at 25°C.

SOLUTION   Using $\pi = MRT$: $\pi = 0.100 \times 0.0821 \times 298 = 2.45$ atm.
(*Note*: As usual with a colligative property, the nature of the solute does not matter.)

We see that the osmotic pressure is quite large, even though this is a fairly dilute solution. This has a useful application in the determination of *M.W.* by using solutions that are very dilute, since it is often possible to measure an osmotic pressure quite accurately when a solution is too dilute to give observable fp depression values.

Example 9.19   A complex organic substance has a *M.W.* of 1.0×10^4. If 0.60 g of it is dissolved in water and made up to 100 ml volume, what will be the osmotic pressure of the solution at 25°C? Give your answer in cm of water. Would ΔT_f for this solution be observable?

SOLUTION

(a) The molarity of the solution is

$$\frac{0.60 \text{ g}}{(1.0 \times 10^4 \text{ g mole}^{-1}) \times (0.100 \text{ liter})} = 6.0 \times 10^{-4}M$$

so

$$\pi = MRT = 6.0 \times 10^{-4} \times 0.0821 \times 298 = 1.5 \times 10^{-2} \text{ atm}$$

Osmotic pressures are often measured in cm of water because, in practice, the pressure necessary to stop osmosis occurring is conveniently supplied by a column of water.

Since mercury has a density of 13.6 g cm^{-3}, whereas water has a density of 1.00 g cm^{-3}, the height of a column of water equivalent to 1 atm pressure (the "water barometer") will be greater in the inverse ratio of the densities.

$$1 \text{ atm} = 76.0 \text{ cm Hg} = 76.0 \times 13.6 \text{ cm H}_2\text{O} = 1.03 \times 10^3 \text{ cm H}_2\text{O}$$

So our osmotic pressure of 1.5×10^{-2} atm would be equal to a column of water of height

$$(1.5 \times 10^{-2}) \times (1.03 \times 10^3) = 15 \text{ cm}$$

(b) To calculate the fp depression, we need to know the molality of the solution. In this very dilute solution, we can, without appreciable error, put molality equal to molarity, so

$$\Delta T_f = 6.0 \times 10^{-4} \times 1.86 = 1.1 \times 10^{-3} \text{°C}$$

The fp depression is indetectably small.

This last calculation shows that fp depression and osmotic pressure measurements are complementary techniques for measuring $M.W.$, that is, one may be used for this purpose under conditions where the other may not be.

Example 9.20 1.25 g of an organic substance is dissolved in 300 ml of water and the osmotic pressure of the solution is 28 cm H_2O at 20°C. What is the $M.W.$ of the solute?

SOLUTION

$$28 \text{ cm H}_2\text{O} = \frac{28}{1.03 \times 10^3} = 0.027 \text{ atm}$$

$$\pi = MRT$$

$$M = \frac{\pi}{RT} = \frac{0.027}{0.0821 \times 293} = 1.13 \times 10^{-3} M$$

$$1.25 \text{ g in 300 ml} = \frac{1.25}{0.300} = 4.17 \text{ g liter}^{-1}$$

$$\text{molar mass:} \quad \frac{4.17 \text{ g liter}^{-1}}{1.13 \times 10^{-3} M} = 3.7 \times 10^3 \text{ g}$$

$M.W.$:   about 3,700.

The emphasis in this chapter has been on the determination of molecular weight, which is by far the most important application of colligative properties.

KEY WORDS

mole fraction	molality
colligative property	Raoult's law
partial vapor pressure	vapor-pressure depression
freezing-point depression	boiling-point elevation
bp elevation constant	fp depression constant
degree of dissociation	osmotic pressure

STUDY QUESTIONS

1. Why is "mole fraction" convenient for expressing the concentrations of liquid mixtures?

2. How is molality related to mole fraction?

3. Can a concentration in molarity be converted into molality, and vice versa? What additional information is needed?

4. For practical purposes, what advantages does molality have over molarity? What disadvantages?

5. How would you set about doing a titration of solutions whose concentrations were given in molalities?

6. Vapor pressure depression is of very little use for finding molecular weights. Why is this?

7. What are the requirements of a solvent suitable for determining $M.W.$'s by fp depression? Is it necessary to have the solvent absolutely pure for this measurement?

8. Why is the fp depression always increased when a solute dissociates?

9. Could any colligative property be used to determine degree of dissociation of a solute? Suppose we used fp depression and bp elevation on the same solution, would we expect to get the same result from each? Explain.

10. How is osmotic pressure related to concentration? What units are used here?

11. What class of substances may conveniently have $M.W.$'s determined by osmotic pressure measurements? Explain.

PROBLEMS

1. Calculate the mole fraction of each component present in the following mixtures:
   (a) 5.00 g hexane, C_6H_{14}; 15.0 g carbon tetrachloride, CCl_4; 17.5 g dichloromethane, CH_2Cl_2.
   (b) Equal weights of water, ethyl alcohol (C_2H_5OH), and acetic acid (CH_3COOH).
   (c) Concentrated nitric acid, which is 70% HNO_3 by weight, the rest being water.

2. What weight of formic acid, HCOOH, should be dissolved in 10.0 g of water to make the mole fraction of the acid 0.400?

3. Express the concentration of the following solutions in terms of molality:
   (a) 10.0 g of sulfuric acid, H_2SO_4, in 100 g of water
   (b) 5.0 g of white phosphorus, P_4, in 60 g of carbon disulfide
   (c) a solution of oxalic acid, $(COOH)_2$, in water, the mole fraction of oxalic acid being 0.0500.

4. Calculate the molecular weights of the solutes in the following solutions:
   (a) 4.81 g is dissolved in 88.7 g solvent and the solution is found to be 0.173m
   (b) 0.112 g is dissolved in 8.51 g solvent and the solution is found to be 0.0161m
   (c) 28.3 g of solute is dissolved in 59.0 g of ethyl alcohol and the mole fraction of solute is 0.180.

5. A solution of phosphoric acid, H_3PO_4, is 1.074M and has a density of 1.053 g ml^{-1}. What is the molality of H_3PO_4 in this solution?

6. Sodium hydroxide solution, 3.410m, has a density of 1.131 g ml^{-1}. What is the molarity of this solution?

7. A solution contains 14.0% by weight of a certain solute. The concentration of solute has the same numerical value when expressed as molality or molarity. What is the density of the solution? (*Hint*: Use symbols to represent any unknowns and follow the same logic as before.)

8. Sodium nitrate, $NaNO_3$, is soluble in water to the extent of making a solution which is 45.0% by weight of $NaNO_3$ and has a density of 1.368 g cm^{-3}. Express the composition of this solution in terms of the mole fraction, molality, and molarity of $NaNO_3$.

9. A solution of potassium bromide, KBr, in water contains 30.0% by weight of KBr. A 10.0 ml portion is withdrawn and treated with an excess of silver nitrate, when the precipitate of silver bromide, AgBr, weighs 5.97 g. What is the density of the KBr solution?

10. At 27°C, ethyl acetate $(CH_3COOC_2H_5)$ has a vp of 100 mm, while ethyl propionate $(C_2H_5COOC_2H_5)$ has a vp of 40 mm. What is the vp of a mixture of equal weights of the two compounds?

11. At 35°C, $SiCl_4$ has a vp of 370 mm, while $SnCl_4$ has a vp of 39 mm. What is the vp of a mixture of 10.0 g $SiCl_4$ with 20.0 g $SnCl_4$?

12. Chloroform, $CHCl_3$, has a vp of 120 mm at a certain temperature. What is the vp of an $0.200m$ solution of a nonvolatile solute in chloroform at the same temperature?

Refer to Table 9.1 for values of constants needed in the following problems.

13. 5.00 g of glucose, $C_6H_{12}O_6$, is dissolved in 72.8 g of water. What will be the fp depression and bp elevation in this solution?

14. A solution of 0.228 g of unknown in 14.8 g of benzene freezes at 5.117°C, while pure benzene in the same apparatus freezes at 5.449°C. What is the *M.W.* of the unknown?

15. The bp elevation in a solution of 0.919 g of unknown in 12.1 g of acetic acid is found to be 1.08 K. Calculate the *M.W.* of the unknown.

16. When 8.81×10^{-3} g of a solid is dissolved in 0.118 g of naphthalene, the fp is lowered from 80.128°C to 78.987°C. What is the *M.W.* of the solid?

17. 16.5 mg of an unknown compound is dissolved in 208 mg of camphor and the mp is 172.81°C. Under the same conditions, pure camphor freezes at 175.74°C. What is the molecular weight of the unknown?

18. A solution of iodic acid, HIO_3, is made by dissolving 6.51 g of the acid in 100 g water. The freezing-point depression of this solution is found to be 1.17 K. What is the degree of dissociation of HIO_3 in this solution?

19. A compound AB_2 ionizes into $A^+ + 2B^-$ in acetic acid solution. If a solution is initially made up to be 0.25 molal and undergoes 40% dissociation, what will the fp depression be?

20. What is the osmotic pressure at 25°C of a solution in water whose concentration is $2.00 \times 10^{-3}M$? Give your answer in atm and in cm H_2O (density of Hg is 13.6 g cm^{-3}).

21. An aqueous solution freezes at −0.010°C. What would its osmotic pressure be at this temperature? (Assume the density is 1.00 g ml^{-1}.)

22. 0.51 g of a compound with a *M.W.* of 2.4×10^4 is dissolved in water and made up to 125 ml volume.
    (a) What will the osmotic pressure be at 25°C.?
    (b) Calculate a rough value for the fp depression to be expected.

23. A solution of an unknown compound contains 0.460 g in 250 ml water. The osmotic pressure is 51 cm H_2O at 25°C. What is the *M.W.* of the unknown?

SOLUTIONS TO PROBLEMS

1. (a)   Convert to molar quantities:

$$\text{hexane:} \quad \frac{5.00}{86.1} = 0.0581 \qquad mf = \frac{0.0581}{0.361} = 0.161$$

$$\text{carbon tetrachloride:} \quad \frac{15.0}{154} = 0.0974 \qquad mf = \frac{0.0974}{0.361} = 0.270$$

$$\text{dichloromethane:} \quad \frac{17.5}{85.0} = 0.206 \qquad mf = \frac{0.206}{0.361} = 0.570$$

$$\text{total} \qquad\qquad 0.361 \text{ mole} \qquad\qquad\qquad 1.00$$

(b)   Take 100 g of each and convert to moles:

$$\text{water:} \quad \frac{100}{18.0} = 5.56 \qquad mf = \frac{5.56}{9.40} = 0.592$$

$$\text{ethyl alcohol:} \quad \frac{100}{46.0} = 2.17 \qquad mf = \frac{2.17}{9.40} = 0.231$$

$$\text{acetic acid:} \quad \frac{100}{60.0} = 1.67 \qquad mf = \frac{1.67}{9.40} = 0.178$$

$$\text{total} \qquad\qquad 9.40 \text{ mole} \qquad\qquad\qquad 1.00$$

(*Note*: Since we are only interested in the *ratio* when calculating mole fraction, the calculation would give the same result for any weight taken.)

(c)   100 g of concentrated acid will contain 70 g of HNO_3 and 30 g of H_2O.

$$\text{nitric acid:} \quad \frac{70}{63.0} = 1.11 \qquad mf = \frac{1.11}{2.78} = 0.40$$

$$\text{water:} \quad \frac{30}{18.0} = 1.67 \qquad mf = \frac{1.67}{2.78} = 0.60$$

$$\text{total} \qquad\qquad 2.78 \text{ mole} \qquad\qquad\qquad 1.00$$

2.   10.0 g of water is 0.556 mole. Since the *water* is to amount to 0.600 mf, the *total* moles present must be x, where

$$\frac{0.556}{x} = 0.600 \qquad x = \frac{0.556}{0.600} = 0.926 \text{ mole}$$

Taking out the water (0.556 mole) from this total leaves

$$0.926 - 0.556 = 0.370 \text{ mole HCOOH}$$

This is 46.0 × 0.370 = 17.0 g HCOOH.

3. (a) 10.0 g H_2SO_4 in 100 g H_2O is 100 g H_2SO_4 in 1000 g H_2O.

$$\text{molality} = \frac{100 \text{ g kg}^{-1}}{98.0 \text{ g mole}^{-1}} = 1.02m$$

(b) 5.0 g P_4 in 60 g CS_2 is $(5.0 \times 1000)/60 = 83$ g kg^{-1}.
P_4 has molar mass $4 \times 31.0 = 124$ g.

$$\text{molality:} \quad \frac{83}{124} = 0.67m$$

(c) If the mf of oxalic acid is 0.0500, the mf of water is $1 - 0.0500 = 0.950$.
0.0500 mole of acid is dissolved in 0.950 mole of water, which is $0.950 \times 18.0 = 17.1$ g H_2O.
In 1 kg H_2O, there will be $(0.0500 \times 1000)/17.1 = 2.92$ mole acid.

$$\text{molality } (COOH)_2: \quad 2.92m$$

4. (a) concentration: $\dfrac{4.81 \times 1000}{88.7} = 54.2$ g kg^{-1}

molar mass: $\dfrac{54.2 \text{ g kg}^{-1}}{0.173 \text{ mole kg}^{-1}} = 313$ g mole^{-1}

(b) concentration: $\dfrac{0.112 \times 1000}{8.51} = 13.2$ g kg^{-1}

molar mass: $\dfrac{13.2 \text{ g kg}^{-1}}{0.0161 \text{ mole kg}^{-1}} = 817$ g mole^{-1}

(c) 59.0 g of ethyl alcohol, C_2H_5OH, is $59.0/46.0 = 1.28$ mole.
The mf of ethyl alcohol (solvent) is $1 - 0.180 = 0.820$.
Total moles present (solute + solvent) is x, where

$$\frac{1.28}{x} = 0.820 \qquad x = \frac{1.28}{0.820} = 1.56$$

Subtracting the moles of ethyl alcohol gives $1.56 - 1.28 = 0.28$ mole of solute present.
28.3 g of solute is 0.28 mole.

$$M.W. = \frac{28.3}{0.28} = 101$$

5. One liter of the solution weighs $1000 \times 1.053 = 1053$ g.
This contains 1.074 mole of H_3PO_4 ($M.W.$ 98.0).
The mass of H_3PO_4 is $1.074 \times 98.0 = 105.3$ g.
By difference, mass of water present is $1053 - 105.3 = 948$ g $= 0.948$ kg.

$$\text{molality of } H_3PO_4: \quad \frac{1.074 \text{ mole}}{0.948 \text{ kg}} = 1.133m$$

6. 3.410 mole of NaOH (*M.W*. 40.00) weigh 3.410 × 40.00 = 136.4 g.
    Dissolved in 1 kg H_2O, this makes the total mass of the solution 1000 + 136.4 = 1136 g.
    The density of the solution is 1.131 g ml^{-1}, so the mass of 1000 ml is 1131 g.
    Since 3.410 mole are dissolved in 1136 g, the amount of NaOH in 1131 g is

$$3.410 \times \frac{1131}{1136} = 3.395 \text{ mole}$$

solution is 3.395*M*

7. Let the solution density be *d* g ml^{-1} and the molar mass of the solute be *M* g. One liter of solution has mass 1000*d* g, so it contains

$$\frac{1000d \times 14.0}{100} = 140d \text{ g solute} = \frac{140d}{M} \text{ mole solute}$$

The mass of solvent in 1 liter of solution is

$$1000d - 140d = 860d \text{ g} = 0.860d \text{ kg}$$

The molality is

$$\frac{140d/M \text{ mole}}{0.860d \text{ kg}} = \frac{140}{0.860M} \, m$$

Knowing the molarity is 140*d*/*M*, we equate molarity and molality to give

$$\frac{140d}{M} = \frac{140}{0.860M} \qquad d = \frac{1}{0.860} = 1.163 \text{ g ml}^{-1}$$

(*Note*: M cancels out; the nature of the solute does not affect the result.)

8. In one liter of solution (1368 g):

$NaNO_3$: $\dfrac{1368 \times 45.0}{100} = 616 \text{ g} = \dfrac{616}{85.0} \text{ mole} = 7.24 \text{ mole}$

H_2O: $\dfrac{1368 \times 55.0}{100} = 752 \text{ g} = \dfrac{752}{18.0} \text{ mole} = 41.8 \text{ mole}$

total: 7.24 + 41.8 = 49.0 mole

mole fraction $NaNO_3$: $\dfrac{7.24}{49.0} = 0.148$

mole fraction H_2O: $\dfrac{41.8}{49.0} = 0.852$

molality of $NaNO_3$: $\dfrac{7.24 \text{ mole}}{0.752 \text{ kg}} = 9.63m$

molarity of $NaNO_3$: $\dfrac{7.24 \text{ mole}}{1.00 \text{ liter}} = 7.24M$

9. 5.97 g of AgBr is $5.97/188 = 3.18 \times 10^{-2}$ mole. This comes from 3.18×10^{-2} mole KBr, mass $3.18 \times 10^{-2} \times 119 = 3.78$ g. Our 10.0 ml sample contained 3.78 g of KBr. Since the composition was 30.0% by weight of KBr, the total mass of the sample was $(3.78 \times 100)/30.0 = 12.6$ g.

$$\text{density:} \quad \frac{12.6 \text{ g}}{10.0 \text{ ml}} = 1.26 \text{ g ml}^{-1}$$

10. Take 100 g of each:

$$\text{ethyl acetate:} \quad \frac{100}{88.0} = 1.14 \text{ mole} \qquad mf = \frac{1.14}{2.12} = 0.54$$

$$\text{ethyl propionate:} \quad \frac{100}{102} = 0.98 \text{ mole} \qquad mf = \frac{0.98}{2.12} = 0.46$$

$$\text{total} = \overline{2.12} \text{ mole}$$

partial vp ethyl acetate: $100 \times 0.54 = 54$ mm
partial vp ethyl propionate: $40 \times 0.46 = \underline{18 \text{ mm}}$
total vapor pressure $= \overline{72 \text{ mm}}$

11. 10.0 g SiCl$_4$: $\quad \dfrac{10.0}{170} = 0.0588$ mole $\qquad mf = \dfrac{0.0588}{0.135} = 0.434$

20.0 g SnCl$_4$: $\quad \dfrac{20.0}{261} = 0.0766$ mole $\qquad mf = \dfrac{0.0766}{0.135} = 0.566$

$$\text{total} = \overline{0.135} \text{ mole}$$

partial vp SiCl$_4$: $0.434 \times 370 = 161$ mm
partial vp SnCl$_4$: $0.566 \times 39 \;= \underline{22 \text{ mm}}$
total vapor pressure $= \overline{183 \text{ mm}}$

12. The solution contains 0.200 mole solute in 1000 g of CHCl$_3$.

$$\text{moles of solvent present:} \quad \frac{1000}{119.5} = 8.37$$

total moles present: $8.37 + 0.200 = 8.57$

$$mf \text{ of CHCl}_3: \quad \frac{8.37}{8.57} = 0.977$$

partial vp of CHCl$_3$: $0.977 \times 120 = 117$ mm

13. Glucose has a molar mass of 180 g.

$$\text{solution molality:} \quad \frac{5.00 \text{ g}}{(180 \text{ g mole}^{-1}) \times (0.0728 \text{ kg})} = 0.382 m$$

bp elevation: $0.512 \times 0.382 = 0.195$ K

fp depression: $1.86 \times 0.382 = 0.711$ K

14. fp depression: $5.449 - 5.117 = 0.332$ K

   molality: $\dfrac{0.332}{2.53} = 131m$

   concentration: $\dfrac{0.228 \text{ g}}{0.0148 \text{ kg}} = 15.4 \text{ g kg}^{-1}$

   molar mass: $\dfrac{15.4 \text{ g kg}^{-1}}{0.131 \text{ mole kg}^{-1}} = 118 \text{ g mole}^{-1}$

15. bp elevation of acetic acid ($K_b = 3.07$) is 1.08 K

   molality: $\dfrac{1.08}{3.07} = 0.352m$

   concentration: $\dfrac{0.919 \text{ g}}{0.0121 \text{ kg}} = 76.0 \text{ g kg}^{-1}$

   molar mass: $\dfrac{76.0 \text{ g kg}^{-1}}{0.352 \text{ mole kg}^{-1}} = 216 \text{ g mole}^{-1}$

16. fp depression: $80.128 - 78.987 = 1.141$ K

   molality: $\dfrac{1.141}{6.8} = 0.168m$

   concentration: $\dfrac{8.81 \times 10^{-3} \text{ g}}{0.118 \times 10^{-3} \text{ kg}} = 74.7 \text{ g kg}^{-1}$

   molar mass: $\dfrac{74.7 \text{ g kg}^{-1}}{0.168 \text{ mole kg}^{-1}} = 4.4 \times 10^2 \text{ g mole}^{-1}$

17. fp depression: $175.74 - 172.81 = 2.93$ K

   molality: $\dfrac{2.93}{39.7} = 0.0738m$

   concentration: $\dfrac{16.5 \times 10^{-3} \text{ g}}{208 \times 10^{-6} \text{ kg}} = 79.3 \text{ g kg}^{-1}$

   molar mass: $\dfrac{79.3 \text{ g kg}^{-1}}{7.38 \times 10^{-2} \text{ mole kg}^{-1}} = 1.07 \times 10^3 \text{ g mole}^{-1}$

18. Before dissociation, the molality of HIO_3 is $6.51/(176 \times 0.100) = 0.370m$. Suppose x mole kg^{-1} dissociate:

$$HIO_3 \rightleftharpoons H^+ + IO_3^-$$
$$(0.370 - x) \qquad\qquad x \quad\ x$$

   total moles (all particles):

$$0.370 - x + 2x = 0.370 + x$$

From fp depression, molality at equilibrium is $1.17/1.86 = 0.629m$. Therefore,

$$0.370 + x = 0.629 \qquad x = 0.259$$

degree of dissociation: $\dfrac{0.259}{0.370} = 0.700 \quad (70.0\%)$

19. moles of AB_2 dissociated (per kg solvent): $\dfrac{0.25 \times 40}{100} = 0.10$ mole kg^{-1}

undissociated AB_2 remaining: $0.25 - 0.10 = 0.15$ mole kg^{-1}
dissociation of 0.10 mole AB_2 gives 0.10 mole A^{2+} and 0.20 mole B^- (from equation)
total moles kg^{-1}($AB_2 + A^{2+} + B^-$) will be $0.15 + 0.10 + 0.20 = 0.45$ mole kg^{-1}
fp depression: $0.45 \times 3.90 = 1.8$ K

20. Use $\pi = MRT$:

$$\pi = (2.00 \times 10^{-3}) \times 0.0821 \times 298 = 4.89 \times 10^{-2} \text{ atm}$$

$$1 \text{ atm} = 76.0 \text{ cm Hg} = 76.0 \times 13.6 \text{ cm H}_2\text{O} = 1.03 \times 10^3 \text{ cm H}_2\text{O}$$

$$\pi = (4.89 \times 10^{-2}) \times (1.03 \times 10^3) = 50.6 \text{ cm H}_2\text{O}$$

21. The molality is $0.010/1.86 = 5.4 \times 10^{-3}m$. At this dilution, molality and molarity are very nearly equal (i.e., 1 kg of solvent is very nearly the same as 1 liter of solution).

$$\pi = (5.4 \times 10^{-3}) \times 0.0821 \times 273 = 0.12 \text{ atm}$$

22. molarity: $\dfrac{0.51}{(2.4 \times 10^4) \times 0.125} = 1.7 \times 10^{-4}M$

(a) $\pi = (1.7 \times 10^{-4}) \times 0.0821 \times 298 = 4.2 \times 10^{-3} \text{ atm}$

$$= (4.2 \times 10^{-3}) \times (1.03 \times 10^3) \text{ cm H}_2\text{O} = 4.3 \text{ cm H}_2\text{O}$$

(b) Assuming molality = molarity at this dilution,

$$\Delta T_f = 1.86 \times (\text{molality}) \approx 1.86 \times 1.7 \times 10^{-4} = 3.2 \times 10^{-4} \text{ K}$$

Such a small depression is not detectable.

23. $\pi = MRT \qquad M = \dfrac{\pi}{RT} \qquad (\pi \text{ in atm})$

$$\pi = 51 \text{ cm H}_2\text{O} = \dfrac{51}{1.03 \times 10^3} \text{ atm} = 5.0 \times 10^{-2} \text{ atm}$$

molarity: $\dfrac{5.0 \times 10^{-2}}{0.0821 \times 298} = 2.0 \times 10^{-3}M$

molar mass: $\dfrac{0.460 \text{ g}}{(0.250 \text{ liter}) \times (2.0 \times 10^{-3}M)} = 9.1 \times 10^2 \text{ g mole}^{-1}$

Chapter 10

Chemical Kinetics: Rate of Reaction

"Just look up the trains in Bradshaw," said he, and turned back to his chemical studies.   THE ADVENTURE OF THE COPPER BEECHES

Note: Familiarity with exponential and logarithmic relationships is essential for understanding the material of this chapter. You are advised to study Appendix A.3 before tackling these problems.

10.1   Concentration Effects and Rate Laws

Different chemical reactions proceed at different rates. Some are so fast as to be instantaneous, such as the neutralization of an acid with a base, while some are so slow that no sign of reaction is apparent in months or years. Can you see diamonds burning in air at room temperature? Thermodynamics tells us that such a reaction is energetically very favorable, but its *rate* is indetectably slow.

Apart from the nature of the reactants, there are three factors that influence the rate of a reaction: concentration of reactants, temperature, and the presence of a catalyst. We shall consider these effects in turn in this chapter.

We will first write a perfectly general equation for a reaction between several substances:

$$A + B + C + D + \ldots \longrightarrow \text{products}$$

For the moment, we are considering only the rate of the forward reaction, so we are not concerned with the nature of the products. The rate of the forward reaction will be given by a *rate law* of the form

$$\text{rate} = k[A]^a[B]^b[C]^c[D]^d \ldots$$

where k is a constant called the *specific rate constant* and the terms in square brackets, $[A]$, $[B]$, etc., represent the molar concentrations of reactants A, B, etc. The letters a, b, c, etc., represent the *exponents* of [A], [B], etc., in the rate law, that is, the power to which each concentration must be raised to give the correct dependence of rate on concentration. These exponents must be determined by actual experiment; they cannot be found by theory. In particular, *there is no connection between these exponents and the coefficients of A, B, etc. in the balanced equation for the overall reaction.*

The exponents are usually simple whole numbers such as 1 or 2, but they can be fractional. In chemical kinetics, we call each of these exponents the *order of reaction* in that particular reactant, for example

If exponent $a = 1$, the reaction is first order in reactant A.

If exponent $a = 2$, the reaction is second order in reactant A, etc.

It is possible for exponent a to be zero, which means that the rate of the reaction is independent of the concentration of component A (zero order in A).

When we have determined the exponent of each reactant in the rate law, we can add them together to give the *overall order of reaction.*

The method of deducing orders of reaction and rate laws from experimental data may be illustrated by some examples.

Example 10.1   An experiment is conducted on the rate of decomposition of N_2O_5, that is, the reaction

$$N_2O_5(g) \longrightarrow N_2O_4(g) + \frac{1}{2}O_2(g)$$

The following data are found:

	Concentration of N_2O_5 (mole liter^{-1})	Rate of reaction (mole liter^{-1} s^{-1})
(i)	2.40×10^{-3}	4.02×10^{-5}
(ii)	7.20×10^{-3}	1.21×10^{-4}
(iii)	1.44×10^{-2}	2.41×10^{-4}

Calculate the complete rate law for the reaction.

SOLUTION   We have three sets of data. Comparing the value of $[N_2O_5]$ in (i) and (ii), we see that it has increased, in the ratio

$$\frac{7.20 \times 10^{-3}}{2.40 \times 10^{-3}} = 3.00$$

At the same time, the rate of reaction has increased by a factor of

$$\frac{1.21 \times 10^{-4}}{4.02 \times 10^{-5}} = 3.01$$

Within experimental error, the increase in rate is in direct proportion to the increase in $[N_2O_5]$, so the exponent of $[N_2O_5]$ in the rate law is 1.00 (first order).

As a check on this, we can compare the changes occurring between experiments (ii) and (iii):

$$\frac{[N_2O_5](iii)}{[N_2O_5](ii)} = \frac{1.44 \times 10^{-2}}{7.20 \times 10^{-3}} = 2.00 \qquad \frac{rate(iii)}{rate(ii)} = \frac{2.41 \times 10^{-4}}{1.21 \times 10^{-4}} = 1.99$$

Again, we see a direct proportionality between $[N_2O_5]$ and the rate of reaction.

(*Note*: An excellent alternative to the above discussion would be to plot a graph of rate versus $[N_2O_5]$. A straight line would be obtained, showing that the rate is directly proportional to $[N_2O_5]$. See discussion of linear relationships, Appendix A.5.)

The rate law is

$$rate = k[N_2O_5]$$

(an exponent of 1 is not usually written down in an equation).

To find the value of k, the specific rate constant, we put in a corresponding set of data for $[N_2O_5]$ and rate. It doesn't matter whether we use the data set (i) or (ii), they should both give the same value for k (within experimental error) if we have the correct value for the exponent.

Using the data from (i)

$$rate = k[N_2O_5]$$

$$k = \frac{rate}{[N_2O_5]} = \frac{4.02 \times 10^{-5} \text{ mole liter}^{-1} \text{ s}^{-1}}{2.40 \times 10^{-3} \text{ mole liter}^{-1}} = 0.0168 \text{ s}^{-1}$$

We can use the third data set as a check on this result. If $[N_2O_5] = 1.44 \times 10^{-2}$, our rate law would forecast

$$rate = 0.0168[N_2O_5] = 0.0168 \times (1.44 \times 10^{-2}) = 2.42 \times 10^{-4} \text{ mole liter}^{-1} \text{ s}^{-1}$$

This agrees well with the reported experimental result.

Example 10.2   The following data are found for the reaction:

$$2H_2 + NO \longrightarrow H_2O + \frac{1}{2}N_2$$

	$[H_2]$ (mole liter^{-1})	$[NO]$ (mole liter^{-1})	Rate (mole liter^{-1} s^{-1})
(i)	1.8×10^{-3}	2.1×10^{-2}	5.4×10^{-5}
(ii)	3.6×10^{-3}	2.1×10^{-2}	10.8×10^{-5}
(iii)	1.8×10^{-3}	6.3×10^{-2}	4.9×10^{-4}

Calculate the complete rate law for this reaction. What would the rate be for $[H_2] = 4.0 \times 10^{-3}$ and $[NO] = 5.5 \times 10^{-2}$ mole liter^{-1}?

SOLUTION   Here we have two reactants to consider and the rate law will be of the form

$$rate = k[H_2]^a[NO]^b$$

Looking at the data provided, we see that, on going from (i) to (ii), [NO] does not change, so any change in rate will be solely due to changes in [H$_2$]. The ratio of [H$_2$] in (ii) to that in (i) is

$$\frac{3.6 \times 10^{-3}}{1.8 \times 10^{-3}} = 2.0$$

At the same time, the ratio of the rate in (ii) to the rate in (i) is

$$\frac{10.8 \times 10^{-5}}{5.4 \times 10^{-5}} = 2.0$$

Since the rate has changed in the same ratio as the concentration of [H$_2$], the exponent of [H$_2$] in the rate law must be 1.00 (first order in [H$_2$]).

Comparing experiments (i) and (iii), we see that this time it is [H$_2$] which has been held constant, while [NO] changes in the ratio

$$\frac{[NO](iii)}{[NO](i)} = \frac{6.3 \times 10^{-2}}{2.1 \times 10^{-2}} = 3.0$$

At the same time, the rate changes, and

$$\frac{rate(iii)}{rate(i)} = \frac{4.9 \times 10^{-4}}{5.4 \times 10^{-5}} = 9.1$$

Within experimental error, $9.1 = (3.0)^2$, in other words,

$$\frac{rate(iii)}{rate(i)} = \left(\frac{[NO](iii)}{[NO](i)}\right)^2$$

This tells us that b, the exponent of [NO] in the rate law, must be 2.0 (second order in [NO]). So our rate law is

$$rate = k[H_2][NO]^2$$

As before, we find k by putting in a corresponding set of data

$$k = \frac{rate}{[H_2][NO]^2} = \frac{5.4 \times 10^{-5} \text{ mole liter}^{-1} \text{ s}^{-1}}{(1.8 \times 10^{-3} \text{ mole liter}^{-1}) \times (2.1 \times 10^{-2} \text{ mole liter}^{-1})^2}$$

$$= 68 \text{ mole}^{-2} \text{ liter}^2 \text{ s}^{-1}$$

The complete rate law is

$$rate = 68[H_2][NO]^2$$

From this, we can find the rate for any other concentration values. Thus, if

$$[H_2] = 4.0 \times 10^{-3} \quad \text{and} \quad [NO] = 5.5 \times 10^{-2}$$

then

$$rate = 68 \times (4.0 \times 10^{-3})(5.5 \times 10^{-2})^2 = 8.2 \times 10^{-4} \text{ mole liter}^{-1} \text{ s}^{-1}$$

In practice, we measure the rate of a reaction by observing the rate of disappearance of one of the starting materials or the rate of production of one of the reaction products.

The above example is made up using obviously artificial numbers that are arranged to lead easily to a numerical result. The rate changes by a factor 9, while [NO] changes by a factor of 3, and we all know that $9 = 3^2$. However, the problem illustrates the most important principle of calculations of this type; that we should vary *one* concentration at a time in order to evaluate the exponents *separately*, then combine the exponents to give a complete rate law.

When the numbers don't work out quite so neatly, we can work out the exponent quite easily by using logarithms. Suppose we have measured rate (i) and rate (ii) at different concentrations [A](i) and [A](ii). We work out the ratios

$$\frac{\text{rate(ii)}}{\text{rate(i)}} = P \qquad \frac{[A](ii)}{[A](i)} = Q$$

This gives us two numbers P and Q, and we know they are related by an equation

$$P = (Q)^a$$

If P and Q are simple numbers, as in the example above where $P = 9$ and $Q = 3$, then we can solve for a by inspection (that is, by guesswork). But suppose we have $P = 1.90$ and $Q = 1.38$. How would we solve the equation $1.90 = (1.38)^a$?

This is solved by taking logarithms of both sides of the equation. So, if we have $P = (Q)^a$, taking logs of each side gives

$$\log P = \log[(Q)^a] = a \log Q$$

(*Note*: The logarithm of a number raised to an exponent is the logarithm of the number, *multiplied by* the exponent.) Then, putting $Q = 1.38$ and $P = 1.90$ gives

$$\log 1.90 = a(\log 1.38)$$

and

$$a = \frac{\log 1.90}{\log 1.38} = \frac{0.279}{0.140} = 1.99 \approx 2.0$$

It is most important to realize that we are *dividing* one logarithm by the other, *not* subtracting one log from the other. (Further discussion of exponentials and logarithms will be found in Appendixes A.2 and A.3.)

Example 10.3   The following data are found for the reaction A + B → products.

	[A] (mole liter^{-1})	[B] (mole liter^{-1})	Rate (mole liter^{-1} s^{-1})
(i)	1.78×10^{-2}	2.81×10^{-3}	6.00×10^{-5}
(ii)	2.85×10^{-2}	2.81×10^{-3}	7.59×10^{-5}
(iii)	1.78×10^{-2}	4.07×10^{-3}	1.26×10^{-4}

Calculate the complete rate law, including the value of the specific rate constant. What will be the rate for $[A] = 2.13 \times 10^{-2}$ mole liter^{-1} and $[B] = 5.16 \times 10^{-3}$ mole liter^{-1}?

SOLUTION   This problem is tackled in the same way as the previous example. From (i) to (ii), [B] is kept constant, so any change in rate will be solely due to changes in [A].

$$\frac{\text{rate(ii)}}{\text{rate(i)}} = \frac{7.59 \times 10^{-5}}{6.00 \times 10^{-5}} = 1.26 \qquad \frac{[A](ii)}{[A](i)} = \frac{2.85 \times 10^{-2}}{1.78 \times 10^{-2}} = 1.60$$

If the exponent of [A] is a, then

$$1.26 = (1.60)^a \quad \text{or} \quad \log 1.26 = a(\log 1.60)$$

and

$$a = \frac{\log 1.26}{\log 1.60} = \frac{0.102}{0.204} = 0.500$$

The rate depends on $[A]^{0.5}$, that is, the rate is proportional to the *square root* of [A]. From (i) to (iii), [B] changes but [A] does not.

$$\frac{\text{rate(iii)}}{\text{rate(i)}} = \frac{1.26 \times 10^{-4}}{6.00 \times 10^{-5}} = 2.10 \qquad \frac{[B](iii)}{[B](i)} = \frac{4.07 \times 10^{-3}}{2.81 \times 10^{-3}} = 1.45$$

$$2.10 = (1.45)^b \quad \text{or} \quad \log 2.10 = b(\log 1.45)$$

and

$$b = \frac{\log 2.10}{\log 1.45} = \frac{0.322}{0.161} = 2.00$$

The rate law is

$$\text{rate} = k[A]^{0.5}[B]^2$$

To find k, we write

$$k = \frac{\text{rate}}{[A]^{0.5} \times [B]^2}$$

and putting in the numbers from data set (i) gives

$$k = \frac{6.00 \times 10^{-5} \text{ mole liter}^{-1} \text{ s}^{-1}}{(1.78 \times 10^{-2} \text{ mole liter}^{-1})^{0.5} \times (2.81 \times 10^{-3} \text{ mole liter}^{-1})^2}$$

$$= 56.9 \text{ mole}^{-1.5} \text{ liter}^{1.5} \text{ s}^{-1}$$

We use the complete rate law to find the rate under other conditions. If $[A] = 2.13 \times 10^{-2}$ mole liter^{-1} and $[B] = 5.16 \times 10^{-3}$ mole liter^{-1}, then

$$\text{rate} = 56.9[A]^{0.5}[B]^2$$

$$= 56.9(2.13 \times 10^{-2})^{0.5} \times (5.16 \times 10^{-3})^2$$

$$= 2.21 \times 10^{-4} \text{ mole liter}^{-1} \text{ s}^{-1}$$

10.2 Effect of Temperature: The Arrhenius Equation

Having considered the effect of concentration on the rate of a chemical reaction, we now look at the influence of temperature. Almost every reaction goes faster when the temperature is raised. To make a quantitative estimate of the effect that temperature has on rate, we must examine the changes in potential energy that occur when two molecules meet and react. This is shown in Figure 10.1 for the process

$$A + B \longrightarrow C + D$$

This diagram is a plot of the potential energy possessed by the molecules as a function of the "reaction coordinate," which can be regarded as the distance between the two molecules. At the left of the diagram, reactants A and B are a long way apart. As they come together, the mutual repulsion between their electron clouds has to be overcome by converting kinetic energy into potential energy, so the potential energy of the system rises.

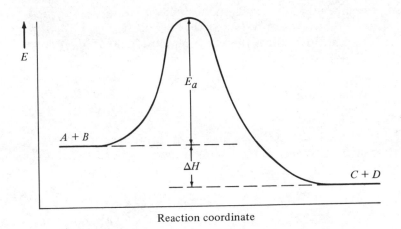

Figure 10.1 Activation energy.

Potential energy of the system reaches a maximum in the *activated complex*, which is a short-lived compound of A and B. This complex may decompose either by returning to A + B, or by going to the products, C + D. In the example in the diagram, the energy level (enthalpy) of C + D is below that of A + B, so the reaction is exothermic.

The difference in energy between the level of A + B and that of the activated complex represents an energy barrier which the reactant molecules must have in order for reaction to occur, and this is called the *activation energy*, E_a, for this particular reaction. As we raise the temperature of the reaction mixture, we increase the fraction of the reactant molecules that possess this minimum energy, so the rate of reaction increases. This is expressed quantitatively in the *Arrhenius equation*:

$$k = Ae^{-E_a/RT}$$

where k is the specific rate constant, A is a constant for the particular reaction, e is the base of natural logarithms (2.71828...), E_a is the activation energy, R is the gas constant, and T is the absolute temperature.

If we have a value for k at a particular temperature, we cannot find E_a from this equation, because we do not know the value of A. However, if we measure k at two different temperatures, we can solve the resulting two equations, assuming that the value of A does not change with temperature.

Suppose we find a value of the rate constant of k_1 at temperature T_1 and k_2 at temperature T_2. We can write two equations:

$$k_1 = Ae^{-E_a/RT_1} \qquad k_2 = Ae^{-E_a/RT_2}$$

It's easier to handle these equations if we take logarithms of both sides. Taking logarithms to base e (natural logs)

$$\ln k_1 = \ln A - \frac{E_a}{RT_1} \qquad \ln k_2 = \ln A - \frac{E_a}{RT_2}$$

Remember, $\ln x$ means *natural* logarithm, to base e; $\log x$ means *common* logarithm, to base 10. Also remember that the right side in each equation was the *product* of two terms, A and $e^{-E_a/RT}$, so the log of this product is the *sum* of the logs of the two terms. For the exponential, $\ln e^x = x$. Now, converting to common logs (base 10) for ease in calculation gives

$$\log k_1 = \log A - \frac{E_a}{2.303RT_1} \qquad \log k_2 = \log A - \frac{E_a}{2.303RT_2}$$

(The factor 2.303 is $\ln 10$, and converts logarithms from base e to base 10.) If we subtract the first equation from the second, the common term $\log A$ cancels out, leaving

$$\log k_2 - \log k_1 = \frac{E_a}{2.303RT_1} - \frac{E_a}{2.303RT_2}$$

We can tidy this equation up a bit. Remembering that the difference between two logs is the log of the *ratio* of the two numbers, we know that

$$\log k_2 - \log k_1 = \log \frac{k_2}{k_1}$$

On the right side, we can take out a common factor $E_a/2.303R$, so the equation becomes

$$\log \frac{k_2}{k_1} = \frac{E_a}{2.303R} \left(\frac{1}{T_1} - \frac{1}{T_2} \right)$$

Hence, if we find the corresponding rate constants and temperatures, we can solve for the activation energy of a reaction. However, this may be simplified by remembering that, if we keep all the concentrations the same, the changes in rate of reaction will be due solely to changes in the rate constant. Since we are interested only in the *relative* rate constants, we can find the ratio from

$$\frac{k_2}{k_1} = \frac{\text{rate}(2)}{\text{rate}(1)}$$

Example 10.4   The following data are found for a reaction as the temperature is changed, concentrations being kept constant:

temperature: 20°C   30°C

rate: 1.5   2.4 mole liter^{-1} s^{-1}

Calculate the activation energy. What will the rate be at 50°C, other things being kept the same?

SOLUTION   We have $T_1 = 20°C = 293$ K and $T_2 = 30°C = 303$ K, while

$$\frac{k_2}{k_1} = \frac{\text{rate}(2)}{\text{rate}(1)} = \frac{2.4}{1.5}$$

Substituting in the Arrhenius equation gives

$$\log \frac{k_2}{k_1} = \log \frac{2.4}{1.5} = \frac{E_a}{2.303R} \left(\frac{1}{293} - \frac{1}{303} \right)$$

$$\log 1.6 = \frac{E_a}{2.303R} [(3.413 \times 10^{-3}) - (3.300 \times 10^{-3})]$$

$$0.204 = \frac{E_a}{2.303R} (1.13 \times 10^{-4})$$

$$E_a = \frac{0.204 \times 2.303R}{1.13 \times 10^{-4}} = 4.2 \times 10^3 R$$

Having got so far, we must stop to think about units. The log term is dimensionless, so is 2.303, so the dimensions of the 4.2×10^{-3} part of our answer are derived only from the temperature part, which has the units of kelvins (K). Since we are measuring an energy, our answer should come out in the units we usually employ to express energy, namely joule mole^{-1}, so we should put in a value of R in joule mole^{-1} K^{-1}.

In these units, $R = 8.314$ joule mole^{-1} K^{-1}, so we can calculate this particular activation energy as

$$E_a = 4.2 \times 10^3 R = 4.2 \times 10^3 \times 8.314 = 3.5 \times 10^4 \text{ J mole}^{-1}$$

The activation energy is, therefore, 35 kJ mole^{-1}.

Recall from Chapter 8 that R, the gas constant, can appear in different sets of units as needed. Be careful to remember that the value 8.314 is in *joules*, not kilojoules, so any equations using this value of R must always have E_a or other energy quantities in joules also. A very common mistake is to use R in joules and forget that E_a is in kJ, with disastrous results for the calculation.

If we further increase the temperature to 50°C (323 K) the rate will increase to some value rate(3). Substituting the known value of E_a, we can work out the ratio of rate(3) to rate(1):

$$\log \frac{\text{rate}(3)}{\text{rate}(1)} = \frac{E_a}{2.303R} \left(\frac{1}{T_1} - \frac{1}{T_3} \right)$$

$$= \frac{34,600}{2.303 \times 8.314} \left(\frac{1}{293} - \frac{1}{323} \right)$$

$$= 1807(3.413 \times 10^{-3} - 3.096 \times 10^{-3})$$

$$= 1807 \times 3.17 \times 10^{-4} = 0.573$$

Taking antilogs of both sides gives

$$\frac{\text{rate}(3)}{\text{rate}(1)} = 3.74$$

Knowing that rate(1) = 1.5 mole liter^{-1} s^{-1} gives

$$\text{rate}(3) = 3.74 \times 1.5 = 5.6 \text{ mole liter}^{-1} \text{ s}^{-1}$$

Example 10.5   A reaction has $E_a = 84$ kJ mole^{-1}. What is the effect on the rate (other things equal) of raising the temperature from (a) 20°C to 30°C and (b) 100°C to 110°C?

SOLUTION

(a) $T_1 = 293$ K and $T_2 = 303$ K

$$\frac{1}{T_1} - \frac{1}{T_2} = \frac{1}{293} - \frac{1}{303} = 1.13 \times 10^{-4}$$

$$\log \frac{k_2}{k_1} = \frac{E_a}{2.303R} \left(\frac{1}{T_1} - \frac{1}{T_2} \right)$$

$$= \frac{84 \times 10^3 \times 1.13 \times 10^{-4}}{2.303 \times 8.314} = 0.496 \qquad \frac{k_2}{k_1} = 3.13$$

The rate will increase by a factor of 3.13.

(b) In this case, $T_1 = 373$ K and $T_2 = 383$ K.

$$\frac{1}{T_1} - \frac{1}{T_2} = \frac{1}{373} - \frac{1}{383} = 7.00 \times 10^{-5}$$

$$\log \frac{k_2}{k_1} = \frac{84 \times 10^3 \times 7.00 \times 10^{-5}}{2.303 \times 8.314} = 0.307 \qquad \frac{k_2}{k_1} = 2.03$$

The rate of reaction increases by a factor 2.03.

We see that the effect on the rate of a ten-degree rise in temperature from 100°C is less than the effect of the same temperature difference, starting at 20°C.

Example 10.6

(a) What is E_a for a reaction whose rate is doubled in going from 20°C to 30°C?
(b) At what temperature will the rate be three times that found at 20°C, other things being equal?

SOLUTION

(a) If the rate is doubled, then

$$\frac{k_2}{k_1} = \frac{\text{rate(2)}}{\text{rate(1)}} = 2.0 \qquad \log \frac{k_2}{k_1} = 0.301$$

$T_1 = 293$ K and $T_2 = 303$ K

$$\frac{1}{T_1} - \frac{1}{T_2} = \frac{1}{293} - \frac{1}{303} = 1.13 \times 10^{-4}$$

$$0.301 = \frac{E_a}{2.303R} \times 1.13 \times 10^{-4}$$

$$E_a = \frac{0.301 \times 2.303 \times 8.314}{1.13 \times 10^{-4}} = 5.12 \times 10^4 \text{ J mole}^{-1} = 51.2 \text{ kJ mole}^{-1}$$

(b) Suppose the rate at temperature T_3 is three times that found at 20°C.

$$\log \frac{\text{rate(3)}}{\text{rate(1)}} = \log 3 = \frac{E_a}{2.303R} \left(\frac{1}{T_1} - \frac{1}{T_3} \right)$$

$$0.477 = \frac{5.12 \times 10^4}{2.303 \times 8.314} \left(\frac{1}{293} - \frac{1}{T_3} \right)$$

$$(3.413 \times 10^{-3}) - \frac{1}{T_3} = \frac{0.477 \times 2.303 \times 8.314}{5.12 \times 10^4} = 1.78 \times 10^{-4}$$

$$\frac{1}{T_3} = (3.413 \times 10^{-3}) - (1.78 \times 10^{-4}) = 3.23 \times 10^{-3}$$

$$T_3 = \frac{1}{3.23 \times 10^{-3}} = 309 \text{ K} = 36°C$$

At 36°C, the rate of reaction will be three times as fast as it was at 20°C.

10.3   Effect of a Catalyst

The effect that a *catalyst* has on the rate of a reaction may be simply described in terms of the concepts used previously in this chapter. A catalyst lowers the activation energy, so the rate will be speeded up according to the Arrhenius equation.

Example 10.7   A catalyst lowers the activation energy for a certain reaction from 75 to 20 kJ mole^{-1}. What will be the effect on the rate of the reaction at 20°C, other things being equal?

SOLUTION Suppose the rate constants are k_1 and k_2 for activation energies $E_a(1)$ and $E_a(2)$, respectively. We can write two corresponding versions of the Arrhenius equation:

$$\log k_1 = \log A - \frac{E_a(1)}{2.303RT} \qquad \log k_2 = \log A - \frac{E_a(2)}{2.303RT}$$

Subtracting the first equation from the second gives

$$\log k_2 - \log k_1 = \frac{E_a(1)}{2.303RT} - \frac{E_a(2)}{2.303RT}$$

As before, we can tidy up an equation of this type, to get

$$\log \frac{k_2}{k_1} = \frac{E_a(1) - E_a(2)}{2.303RT}$$

Other things being equal, rate of reaction will be proportional to the value of the specific rate constant, so we can write

$$\log \frac{\text{rate}(2)}{\text{rate}(1)} = \log \frac{k_2}{k_1} = \frac{E_a(1) - E_a(2)}{2.303RT}$$

In this case, $E_a(1) = 75$ and $E_a(2) = 20$ kJ mole^{-1}:

$$\log \frac{\text{rate}(2)}{\text{rate}(1)} = \frac{(75 - 20) \times 10^3}{2.303 \times 8.314 \times 293} = 9.80 \qquad \frac{\text{rate}(2)}{\text{rate}(1)} = 6.4 \times 10^9$$

The reaction speeds up by a very large factor when the catalyst lowers the activation energy.

Example 10.8 A reaction is found to have an activation energy of 50 kJ mole^{-1} at 25°C. On addition of a catalyst, the rate speeds up by a factor of 10^6. What is the activation energy with the catalyst present?

SOLUTION We can use the same equation as in the previous example, with the ratio of the rates equal to 10^6 and $E_a(1) = 5.0 \times 10^4$ J:

$$\log(10^6) = 6.0 = \frac{(5.0 \times 10^4) - E_a(2)}{2.303 \times 8.314 \times 298}$$

$$(5.0 \times 10^4) - E_a(2) = 6.0 \times 2.303 \times 8.314 \times 298 = 3.4 \times 10^4$$

$$E_a(2) = (5.0 \times 10^4) - (3.4 \times 10^4) = 1.6 \times 10^4 \text{ J mole}^{-1}$$

The activation energy has been lowered from 50 kJ mole^{-1} to 16 kJ mole^{-1}.

10.4   Exponential Decay and the Concept of Half-Life

Up to now, we have been considering the rate at which a reaction proceeds for given concentrations of the various reactants taking part. In this section, we change our point of view to look at how much of a reactant is left after the reaction has been proceeding for a certain length of time.

We can't find this out by simply multiplying the initial rate of reaction by the time elapsed since the reaction was started, because the rate of reaction changes as time goes by and the concentration of reactant falls. To sum a continuously varying quantity over a period of time, we have to use integral calculus. We also have to deal with a specific type of rate law, and we are going to consider here a first-order rate law only (a similar treatment may be applied to other types of rate law).

Consider a substance A undergoing reaction

$$A \longrightarrow products$$

according to a first-order rate law. The rate law is

$$rate = k[A]$$

We may express "rate" in rather more precise terms by calling it "the rate of decrease of [A] with respect to time." In the symbolism of calculus,

$$rate = \frac{-d[A]}{dt} = k[A] \qquad (t = time)$$

(the minus sign indicates a rate of *de*crease). Rearranging gives

$$\frac{d[A]}{[A]} = -k \, dt$$

Integrating the equation gives

$$\ln[A] = -kt + C \qquad (natural\ logarithm)$$

where C is a constant of integration. Putting in the conditions at the start of reaction

$$[A] = [A]_0 \quad at \quad t = 0$$

where $[A]_0$ is the initial concentration of A, we obtain

$$\ln[A]_0 = C$$

Substituting in the integrated equation

$$\ln[A] = -kt + \ln[A]_0 \quad or \quad \ln \frac{[A]}{[A]_0} = -kt$$

Raise e to power of each side

$$\frac{[A]}{[A]_0} = e^{-kt}$$

This equation, known as the *integrated rate law* for a first-order reaction, tells us what we want to know; how does the concentration of A change as time goes by? The particular form of dependence is called *exponential decay*, because it involves an exponential relationship between t and [A].

If $t = 0$, $e^{-kt} = 1$ and $[A] = [A_0]$ as we would expect. When does [A] become zero, in other words, when is all this reactant consumed? If we put $[A] = 0$ and attempt to find t, using the above equation, there is no finite answer. Only as t tends to infinity does [A] tend to zero.

This gives us the important conclusion that there is always *some* quantity of the reactant left after the reaction has been proceeding for any finite time. Because we cannot give an answer to the question of when it is all gone, we compromise by quoting the time when *half* of the initial amount is gone. This is called the *half-life* of the particular process, and we can readily calculate its relationship to k using the above equation.

Call the half-life $t_{1/2}$, at which time

$$[A] = \frac{[A]_0}{2} \quad \text{so that} \quad \frac{[A]}{[A]_0} = \frac{1}{2}$$

then

$$e^{-kt_{1/2}} = 1/2$$

taking logarithms (base e) of both sides,

$$-kt_{1/2} = \ln 1/2 = -\ln 2$$

giving the very simple result:

$$t_{1/2} = \frac{\ln 2}{k} = \frac{0.693}{k}$$

So for any first-order reaction we may calculate k from the half-life, or vice versa. Knowing k, we can calculate the amount of substance remaining at different times from the exponential decay equation. However, if reaction proceeds for a time that is a number of half-lives, there is a simpler connection between time and amount remaining:

After 1 half-life, amount remaining = 1/2 original

After 2 half-lives, amount remaining = 1/4 original.

After 3 half-lives, amount remaining = 1/8 original.

After 4 half-lives, amount remaining = 1/16 original.

After n half-lives, amount remaining = $(1/2)^n$ original.

Example 10.9 The reaction

$$A \longrightarrow B + C$$

is first-order in [A] and has a half-life of 30 min. Calculate the specific rate constant. If [A] is initially 0.10 mole liter^{-1}, what will be its value after (a) 1.0 hour and (b) 24.0 hour?

SOLUTION   The rate law is of the form

$$\text{rate} = k[A]$$

Putting [A] in mole liter^{-1} and rate in units of mole liter^{-1} min^{-1}, we have

$$k = \frac{\text{rate}}{[A]} \text{ min}^{-1}$$

Knowing that $t_{1/2} = 30$ min, we can write

$$k = \frac{0.693}{t_{1/2}} = \frac{0.693}{30} = 2.31 \times 10^{-2} \text{ min}^{-1}$$

(a) 1.0 hour is 2.0 half-lives, so the amount of A present decreases by a factor

$$(1/2)^2 = 1/4$$

Initially, [A] = 0.10, so after 2.0 half-lives

$$[A] = 0.10 \times 1/4 = 0.025 \text{ mole liter}^{-1}$$

(b) After 24.0 hour, 48 half-lives have elapsed, so [A] has decreased by a factor

$$(1/2)^{48} = 3.55 \times 10^{-15}$$

If [A] was originally 0.10 mole liter^{-1}, it is now

$$0.10 \times 3.55 \times 10^{-15} = 3.55 \times 10^{-16} \text{ mole liter}^{-1}$$

(*Note*: How does one evaluate $(1/2)^{48}$? You could spend an hour dividing by two 48 times, but it's quicker to use logs. Remember log $1/2 = -$log 2, so we look up log 2 (0.3010), change the sign (-0.3010), multiply by 48 (-14.45 or $\overline{15}.55$) and look up the antilog (3.55×10^{-15}). More examples in Appendix A.3.)

Although the concept of half-life is completely general, applicable to any reaction, it is particularly useful in connection with radioactive-decay processes. Although a nuclear transformation is not usually regarded as a chemical reaction, the kinetic laws governing its progress are precisely similar to those applied to any first-order chemical reaction.

Example 10.10   Potassium has a natural radioactive isotope $^{40}_{19}K$ with a half-life of 1.3×10^9 years. Out of a sample of 1 mole of ^{40}K, how much will remain after (a) 10^{10} years and (b) 10^{11} years?

SOLUTION   Obviously this is very similar to the previous example, except that we are dealing with time intervals that are not exact numbers of half-lives.

(a) 10^{10} years is $10^{10}/(1.3 \times 10^9) = 7.69$ half-lives. The generalization that after n half-lives, the amount remaining is

$$(1/2)^n \times \text{original amount}$$

is true for any value of n, not just for integral values, so we can evaluate

$$(1/2)^n = (1/2)^{7.69} = 4.83 \times 10^{-3}$$

So, out of our original 1 mole, we have 4.83×10^{-3} mole remaining after 10^{10} years. (*Note*: How do you raise a number to the power 7.69? Look up the log, multiply by 7.69, look up the antilog.)

An alternative calculation route is to first evaluate the specific rate constant

$$k = \frac{0.693}{t_{1/2}} = \frac{0.693}{1.3 \times 10^9} = 5.33 \times 10^{-10} \text{ year}^{-1}$$

Now we can use the equation

$$\frac{[A]}{[A]_0} = e^{-kt}$$

In this case $[A]_0 = 1$ mole, and $t = 10^{10}$ year:

$$kt = 5.33 \times 10^{-10} \times 10^{10} = 5.33$$

so

$$[A] = e^{-5.33} = 4.84 \times 10^{-3} \text{ mole}$$

(*Note*: A value of e^x or e^{-x} may be obtained from tables of these functions in reference books, or, failing this, by using ordinary logarithms: That is, if $y = e^x$, then $2.303 \log_{10} y = x$.)

(b) After 10^{11} years, the calculation is the same, except that we have now waited $10^{11}/(1.3 \times 10^9) = 76.9$ half-lives. So the amount of ^{40}K present has changed by a factor of

$$(1/2)^{76.9} = 7.0 \times 10^{-24}$$

Therefore, of our original 1 mole, we have remaining

$$7.0 \times 10^{-24} \text{ mole} = 7.0 \times 10^{-24} \times 6.02 \times 10^{23} \text{ atoms} = 4.2 \text{ atoms}$$

Approximately four atoms would be left, according to theory. (*Note*: In practice, the concepts of half-life and rate law can only be applied to fairly large numbers of atoms, where a statistical averaging effect occurs; we cannot accurately deal with small numbers of atoms on this basis.)

Example 10.11   The above isotope ^{40}K occurs to the extent of 0.0118 % in natural potassium. How many atoms decay in one second in a mole of natural potassium?

SOLUTION   It's more convenient to work throughout in seconds:

$$t_{1/2} = 1.3 \times 10^9 \text{ yr}$$

$$= 1.3 \times 10^9 \times 365 \times 24 \times 3600 \text{ s}$$

$$= 4.1 \times 10^{16} \text{ s}$$

$$k = \frac{0.693}{t_{1/2}} = \frac{0.693}{4.1 \times 10^{16}} = 1.7 \times 10^{-17} \text{ s}^{-1}$$

Our rate law for the first-order reaction is

$$\text{rate} = 1.7 \times 10^{-17} [^{40}\text{K}]$$

where rate is in mole s^{-1}, [^{40}K], the concentration of the radioactive isotope, is 0.0118 % (0.0118/100 = 1.18×10^{-4} mole of ^{40}K in each mole of the natural potassium). Since a mole of ^{40}K contains 6.02×10^{23} atoms, we have

$$[^{40}\text{K}] = 1.18 \times 10^{-4} \times 6.02 \times 10^{23} = 7.1 \times 10^{19} \text{ atoms of } ^{40}\text{K per } \textit{mole} \text{ of natural potassium}$$

If we use this concentration of ^{40}K in our rate equation, we obtain the desired result in terms of the number of *atoms* of the radioactive isotope decaying in each second in a mole of natural potassium.

$$\text{rate} = (1.7 \times 10^{-17})(7.1 \times 10^{19})$$

$$= 1.2 \times 10^3 \text{ atoms s}^{-1} \text{ mole}^{-1}$$

There are about 1200 atoms decaying in each second, but it takes 1300 million *years* for half the initial amount to disappear! This concept reminds us again of the very large value of Avogadro's number.

10.5   Radiocarbon Dating

An interesting and valuable application of the above principles is the use of methods of *dating* old materials by studying the extent to which radioactive isotopes have decayed. For example, the age of the earth itself cannot be as much as 10^{11} years, or there would be no detectable amount of ^{40}K left, according to our calculation above (in fact, current estimates put the age of the earth around 4.5×10^9 years).

Radiocarbon is the rare isotope ^{14}C, which has a half-life of 5760 years. Obviously any of this isotope produced at the time the earth was formed would have disappeared long ago, but a very small concentration of ^{14}C is produced continuously by the action of cosmic rays on the upper atmosphere. Having chemical properties identical with the other carbon isotopes ^{12}C and ^{13}C (nonradioactive), ^{14}C is incorporated into the normal carbon cycle and finds its way into atmospheric CO_2, plants, animals, etc. But once a living organism dies, its carbon content no longer undergoes exchange with the carbon in nature, so any ^{14}C will slowly decay without being renewed.

We can, within limits, determine the age of a dead object by comparing its ^{14}C content with that of a live object. Obviously, we have to have an object sufficiently old that its ^{14}C content has decreased significantly, but not so old that there is hardly any detectable amount left.

Example 10.12   A sample of "new" wood shows a count due to ^{14}C of 15.3 disintegrations per minute for each gram of natural carbon present, whereas a sample of "old" wood shows a count of 5 disintegrations per minute per g of natural C. If ^{14}C has a half-life of 5760 years, calculate:

(a) the fraction of ^{14}C in natural carbon
(b) the age of the old sample
(c) Could radiocarbon dating be used to find the age of petroleum samples in the earth?

SOLUTION

(a) This time we can work in units of minutes:

$$t_{1/2} = 5760 \text{ yr} = 3.03 \times 10^9 \text{ min} \quad \text{and} \quad k = \frac{0.693}{t_{1/2}} = \frac{0.693}{3.03 \times 10^9} = 2.29 \times 10^{-10} \text{ min}^{-1}$$

Our rate law is rate = $2.29 \times 10^{-10}[^{14}C]$.
  Given that rate = 15.3 disintegrations $\text{min}^{-1} \text{ g}^{-1}$,

$$[^{14}C] = \frac{15.3}{2.29 \times 10^{-10}} = 6.68 \times 10^{10} \text{ atoms } ^{14}C \text{ per gram natural carbon}$$

But 1 g natural carbon (atomic weight 12.0) contains

$$\frac{6.02 \times 10^{23}}{12.0} = 5.02 \times 10^{22} \text{ atoms}$$

and the fraction that is ^{14}C is

$$\frac{6.68 \times 10^{10}}{5.02 \times 10^{22}} = 1.33 \times 10^{-12}$$

Only about 1 atom in 10^{12} of natural carbon is the radioactive ^{14}C.

(b) In the old wood, the rate had decreased to 5 disintegrations $\text{min}^{-1} \text{ g}^{-1}$. This is due to the decrease in concentration of ^{14}C, so

$$\frac{\text{amount } ^{14}C \text{ present}}{\text{amount } ^{14}C \text{ present initially}} = \frac{5}{15.3} = 0.33$$

This decay will occur in n half-lives, where

$$(0.5)^n = 0.33$$

We solve this equation by taking logs:

$$n(\log 0.5) = \log 0.33 \qquad n = \frac{\log 0.33}{\log 0.5} = \frac{-0.481}{-0.301} = 1.6$$

1.6 half-lives is $1.6 \times 5760 = 9200$ years (note that the accuracy of this result is not very good, perhaps $\pm$ 1000 years).

(c) Radiocarbon dating is useless for petroleum deposits, which have been in the earth for millions of years (i.e., hundreds of half-lives). No detectable amounts of ^{14}C will be found after this length of time.

10.6   Rate Laws and the Reaction Mechanism

One of the most important and interesting questions the chemist tries to answer is the determination of the actual mechanism by which a reaction occurs. In other words, what is actually happening on the molecular scale? What species must collide to start the reaction? What bonds are broken first? What is the nature of the activated complex? Of course, we can't actually watch these processes going on, but a study of the rate law for a particular reaction will often enable us to deduce the probable mechanism.

The following assumptions form the basis of the logic we use in deducing *reaction mechanisms*:

1. The balanced equation representing the overall stoichiometry of a reaction does *not* necessarily indicate what molecules are actually interacting with each other.

2. Although some simple reactions may occur in one step, many reactions occur in a *series* of steps. Frequently, this reaction sequence will involve an *intermediate* which is formed in one step and consumed in another, and hence does not appear in the initial reactants, nor in the final products of the reaction.

3. If a reaction occurs in a series of steps, the sum of the equations representing these steps must give the overall equation for the reaction.

4. Out of a series of steps, one will be the *slow*, or *rate-determining*, step. *The overall rate of a reaction occurring in a series of steps will be that of the slow step.*

5. The rate law for each separate step in a reaction sequence will be proportional to the concentrations of the reactants in that step, raised to exponents which are equal to their coefficients *in that step.*

6. Combining the previous two statements, we see that: *The rate law for the overall reaction will be determined by the stoichiometry of the slow step* (not by the stoichiometry of the overall reaction).

Let's first consider some possible "simple steps" that might form part of a reaction sequence, and decide what form of rate law we would expect for each.

The simplest type is the process

$$A \longrightarrow B + C$$

for which all we need is a molecule of A. No collision with other species occurs; the molecule simply falls apart. The rate will be directly proportional to [A], in other words we have a reaction that is first order in A:

$$\text{rate} = k[A]$$

This is known as a *unimolecular* process, because only one molecule is involved.

The next simplest case involves the collision of A with B

$$A + B \longrightarrow C + D$$

Note that, when we talk of a step in an actual reaction sequence, we interpret an equation in a slightly different way. The above equation does not say "one mole of A reacts with one mole of B" (although this is, of course, still true); it tells us that, as reaction occurs, one *molecule* (or atom) of A must actually make a physical collision with one *molecule* (or atom) of B. Under these conditions, the rate law is of the form

$$\text{rate} = k[\text{A}][\text{B}]$$

This is first order in A, first order in B, for an overall order of reaction of two. Such a reaction is said to be *bimolecular*, because it involves the collision of *two* molecules.

A simple variation of the previous case occurs when A = B, in other words, two molecules of A must collide to bring about reaction

$$\text{A} + \text{A} \longrightarrow \text{A}_2$$

or

$$2\text{A} \longrightarrow \text{A}_2$$

The rate law here will be second order in A:

$$\text{rate} = k[\text{A}]^2$$

and again the process is said to be *bi*molecular.

The previous three examples represent the only important types of step that may form part of a reaction sequence. More complicated steps, such as

$$\text{A} + \text{B} + \text{C} \longrightarrow \text{D} + \text{E}$$

are very unlikely to occur, because they involve a three-body collision. Molecules A, B, and C would have to arrive at the same point at the same time, and such an event is not common in a mixture of molecules all moving in random directions.

So we may simply conclude that an overall reaction, however complex it appears, may always be broken into a series of steps, each of which involves, at most, the collision of two molecules.

Before proceeding to some actual examples, we should mention one further process that may make up part of a reaction sequence; a rapidly established *equilibrium* involving a reactant and an intermediate taking part in a subsequent step. Very often, this type of equilibrium is a dissociation. Consider the following sequence for the overall reaction:

$$\text{A}_2 + \text{B} \longrightarrow \text{C} + \text{D}$$

(i) $\qquad\qquad\qquad \text{A}_2 \rightleftharpoons 2\text{A} \qquad$ (fast equilibrium)

(ii) $\qquad\qquad\qquad \text{A} + \text{B} \longrightarrow \text{C} + \text{D} \qquad$ (slow)

The rate of the overall process will be that of the slow step, which will be first order in A and in B, so we can write

$$\text{rate} = k[\text{A}][\text{B}]$$

This rate law is written in terms of A, the intermediate, rather than A_2, the original reactant. But we always write our rate laws in terms of the original reactants, rather than intermediates which cannot be isolated, so we have to relate [A] to [A_2]. We do this by using an equilibrium constant for the dissociation of A_2 to 2A (The subject of equilibrium constants is taken up in detail in chapter 11.) Thus, for

$$A_2 \rightleftharpoons 2A$$

we have

$$K = \frac{[A]^2}{[A_2]}$$

where K is the equilibrium constant. So we can write

$$[A] = K^{0.5}[A_2]^{0.5}$$

The concentration of A is proportional to the *square root* of the concentration of A_2.
   Substituting in the rate law gives

$$\text{rate} = kK^{0.5}[A_2]^{0.5}[B] = k'[A_2]^{0.5}[B]$$

where k' combines k and $K^{0.5}$. This gives a rate law with overall order 1.5.
   It may also happen that a reaction sequence includes an *associative* equilibrium as one step, for example,

(i) $\qquad\qquad\qquad A + A \rightleftharpoons A_2 \qquad$ (fast equilibrium)

(ii) $\qquad\qquad B + A_2 \longrightarrow C + D \qquad$ (slow)

Here, we may write, from step (ii)

$$\text{rate} = k[B][A_2]$$

and from the equilibrium of step (i)

$$K = \frac{[A_2]}{[A]^2} \quad \text{or} \quad [A_2] = K[A]^2$$

and substituting in the rate law gives

$$\text{rate} = kK[B][A]^2 = k'[B][A]^2$$

This reaction is first order in B, second order in A, overall order three.
   Let's try to apply these principles to some actual reactions.

Example 10.13   Deduce the rate law to be expected for the overall gaseous reaction

$$2NO + Cl_2 \longrightarrow 2NOCl$$

for each of the possible mechanisms below:

(a) (i) $\qquad\qquad\qquad 2NO \rightleftharpoons N_2O_2 \qquad$ (fast equilibrium)

(ii) $\qquad\qquad N_2O_2 + Cl_2 \longrightarrow 2NOCl \qquad$ (slow)

(b) (i) $\qquad\qquad\qquad Cl_2 \rightleftharpoons 2Cl \qquad$ (fast equilibrium)

(ii) $\qquad\qquad Cl + NO \longrightarrow NOCl \qquad$ (slow)

SOLUTION

(a) Write the rate law for the *slow* step.

$$rate = k[N_2O_2][Cl_2]$$

Now use the equilibrium to relate $[N_2O_2]$ to $[NO]$:

$$K = \frac{[N_2O_2]}{[NO]^2} \qquad [N_2O_2] = K[NO]^2$$

and substitute in the rate law:

$$rate = kK[NO]^2[Cl_2] = k'[NO]^2[Cl_2]$$

(b) For the alternative mechanism, we follow the same sequence.

slow step: rate $= k[Cl][NO]$

equilibrium: $K = \dfrac{[Cl]^2}{[Cl_2]} \qquad [Cl] = K^{0.5}[Cl_2]^{0.5}$

substitution: rate $= kK^{0.5}[Cl_2]^{0.5}[NO] = k'[Cl_2]^{0.5}[NO]$

We see that the different mechanisms give quite different rate laws. By doing an experiment on the system, we may determine the actual rate law of the reaction, which may enable us to choose between these alternate hypothetical mechanisms (or, of course, they may *both* be shown to be incorrect!)

Example 10.14   The rate law for the gaseous reaction

$$CO + NO_2 \longrightarrow CO_2 + NO$$

is found to be rate $= k[NO_2]^2$. Which of the following mechanisms would be consistent with this rate law?

(a) (i) $2NO_2 \rightleftharpoons N_2O_4$ (fast equilibrium)

(ii) $N_2O_4 + 2CO \longrightarrow 2CO_2 + 2NO$ (slow)

(b) (i) $2NO_2 \longrightarrow NO_3 + NO$ (slow)

(ii) $NO_3 + CO \longrightarrow NO_2 + CO_2$ (fast)

(c) (i) $2NO_2 \longrightarrow N_2 + 2O_2$ (slow)

(ii) $2CO + O_2 \longrightarrow 2CO_2$ (fast)

(iii) $N_2 + O_2 \longrightarrow 2NO$ (fast)

SOLUTION

(a) slow step: rate $= k[N_2O_4][CO]^2$
  equilibrium: $[N_2O_4] = K[NO_2]^2$
  substitution: rate $= kK[NO_2]^2[CO]^2$
  Not consistent with the observed rate law.

(b) slow step: rate $= k[NO_2]^2$
  This *is* consistent with the observed rate law, since it does not depend on [CO].

(c) slow step: rate $= k[NO_2]^2$
  This also is consistent with the observed rate law. In other words, this experiment alone cannot distinguish between mechanisms (b) and (c).

Example 10.15   For the overall reaction

$$A_2 + B_2 \longrightarrow 2AB$$

deduce mechanisms that are consistent with the following rate laws:

(a) rate $= k[A_2][B_2]$
(b) rate $= k[A_2]$
(c) rate $= k[A_2]^{0.5}[B_2]^{0.5}$

SOLUTION

(a) The reaction's slow step must involve the collision of one molecule of A_2 with one of B_2. The overall equation written above is consistent with this type of reaction, but a slightly more complex mechanism involving an intermediate complex A_2B_2 is likely.

(i) $A_2 + B_2 \longrightarrow A_2B_2$ (slow)

(ii) $A_2B_2 \longrightarrow 2AB$ (fast)

(*Question*: What difference would it make if (i) were fast and (ii) were slow? Or if (i) were a fast equilibrium and (ii) were slow?)
  (b) Since the rate is independent of $[B_2]$, the slow step must not involve B_2, so we think of

(i) $A_2 \longrightarrow 2A$ (slow)

(ii) $A + B_2 \longrightarrow AB + B$ (fast)

(c) When we see a dependence on the square root of a concentration, we think of a dissociative equilibrium. In this case, both A_2 and B_2 must be dissociating

(i) $A_2 \rightleftharpoons 2A$ (fast equilibrium)

(ii) $B_2 \rightleftharpoons 2B$ (fast equilibrium)

(iii) $A + B \longrightarrow AB$ (slow)

$$\text{rate} = k[A][B] = k'[A_2]^{0.5}[B_2]^{0.5}$$

As you can see from these examples, the deduction of a possible mechanism from a rate law requires a reasonable guess as to what might be occurring in the reaction sequence. Although the experimental rate law can sometimes show that a postulated mechanism is wrong, it can never show unambiguously that it is correct, because there may be more than one mechanism that would give the same rate law. All we can say is that the postulated mechanism is *consistent with* the observed rate law.

KEY WORDS

rate law	specific rate constant
order of reaction	Arrhenius equation
activated complex	activation energy
catalyst	exponential decay
half-life	radiocarbon dating
reaction mechanism	rate-determining step
unimolecular	bimolecular
associative equilibrium	dissociative equilibrium

STUDY QUESTIONS

1. What do we mean by the "rate law" of a reaction?

2. Can we deduce the form of a rate law from the balanced equation for a reaction? (No!)

3. How do we determine a rate law?

4. How does an increase in temperature affect the rate of most reactions? What particular quantity associated with the reaction is important here?

5. Once an activated complex is formed, is the reaction between two molecules certain to go to completion?

6. Why do we have to measure the rate of reaction at more than one temperature in order to find the activation energy?

7. What does a catalyst do?

8. What does "exponential decay" mean? How long must exponential decay continue before a reactant is completely consumed?

9. How is half-life defined? What is the relationship between half-life and the specific rate constant k?

10. Why does radioactive decay always follow a first-order rate law?

11. Does measurement of half-lives tell us anything about the age of the earth?

12. What rate laws do we find for reactions that are (a) unimolecular and (b) bimolecular?

13. If a reaction occurs in a series of steps, which one is important in determining the rate law?

14. How can a reaction's rate be proportional to the square root of the concentration of one reactant?

15. Can we be certain that we have deduced the mechanism by which a reaction proceeds from experimental observations of rate laws?

PROBLEMS

1. The following data are found for the reaction

$$A + B \longrightarrow products$$

	[A] (mole liter^{-1})	[B] (mole liter^{-1})	Rate (mole liter^{-1} s^{-1})
(i)	2.88×10^{-2}	7.11×10^{-3}	8.83×10^{-5}
(ii)	2.88×10^{-2}	1.04×10^{-2}	1.89×10^{-4}
(iii)	7.11×10^{-2}	1.04×10^{-2}	4.66×10^{-4}

(a) Calculate the complete rate law.
(b) What will the rate of reaction be for $[A] = 5.61 \times 10^{-2}$ and $[B] = 3.01 \times 10^{-3}$ mole liter^{-1}?

2. In the gas phase decomposition of ozone to oxygen,

$$2O_3 \longrightarrow 3O_2$$

the following data were obtained:

	Pressure of ozone (mm Hg)	Rate of reaction (mole liter^{-1} s^{-1})
(i)	21	2.60×10^{-6}
(ii)	15	1.33×10^{-6}
(iii)	11	7.13×10^{-7}

What is the order of reaction in $[O_3]$?

3. Acetaldehyde, CH_3CHO, decomposes on heating by the reaction

$$CH_3CHO \longrightarrow CH_4 + CO$$

The following data were recorded for this process:

	$[CH_3CHO]$ (mole liter^{-1})	Rate (mole liter^{-1} s^{-1})
(i)	1.2×10^{-3}	6.70×10^{-5}
(ii)	2.7×10^{-3}	2.26×10^{-4}
(iii)	4.1×10^{-3}	4.23×10^{-4}

(a) Calculate the rate law for this reaction, including the value of the specific rate constant.
(b) What would the rate be for $[CH_3CHO] = 1.0 \times 10^{-2}$ mole liter^{-1}?

4. The following data are found for the reaction

$$P + Q + R \longrightarrow products$$

	$[P]$ (mole liter^{-1})	$[Q]$ (mole liter^{-1})	$[R]$ (mole liter^{-1})	Rate (mole liter^{-1} s^{-1})
(i)	1.2×10^{-3}	2.6×10^{-4}	4.0×10^{-3}	6.2×10^{-6}
(ii)	2.9×10^{-3}	2.6×10^{-4}	4.0×10^{-3}	15.0×10^{-6}
(iii)	1.2×10^{-3}	2.6×10^{-4}	6.8×10^{-3}	6.2×10^{-6}
(iv)	1.2×10^{-3}	5.9×10^{-4}	4.0×10^{-3}	9.3×10^{-6}

Calculate the complete rate law. What will the rate be for $[P] = 3.3 \times 10^{-3}$, $[Q] = 3.8 \times 10^{-4}$, $[R] = 5.5 \times 10^{-3}$ mole liter^{-1}?

5. The following data are found for a reaction:

Temperature	Rate
0°C	1.5×10^{-3} mole liter^{-1} s^{-1}
20°C	7.5×10^{-3} mole liter^{-1} s^{-1}

(a) Calculate the activation energy.
(b) What would the rate be at 50°C?

6. A reaction increases in rate by a factor of 50 when the temperature is raised from 20°C to 70°C.
(a) What is the activation energy?
(b) At what temperature would the rate be 100 times its value at 20°C, other things being equal?

7. A reaction has an activation energy of 50 kJ mole^{-1}.
(a) If the rate at 30°C is 1.2 mole liter^{-1} s^{-1}, what is the rate at 60°C, other things being equal?

(b) What would the rate at 30°C become if a catalyst were added which halved the activation energy?

8. A reaction has an activation energy of 65 kJ mole^{-1}. The rate at 100°C is found to be 7.8 × 10^{-2} mole liter^{-1} s^{-1}.
   (a) At what temperature would the rate be one tenth of this value?
   (b) What would the rate be at 20°C, other things being equal?

9. A reaction has an activation energy of 48 kJ mole^{-1} at 25°C. A catalyst increases the rate of reaction by a factor of 1000. What is the activation energy in the presence of the catalyst?

10. What would be the effect on the rate of a reaction at 0°C of adding a catalyst which lowered the activation energy from 60 to 35 kJ mole^{-1}?

11. A reactant is decomposing in a first-order reaction. How much remains after (a) 0.10 half-life, (b) 0.50 half-life, (c) 1.50 half-lives?

12. The thorium isotope ^{232}Th is radioactive with a half-life of 1.4 × 10^{10} years. If the age of the earth is 4.5 × 10^9 years, what fraction of the ^{232}Th originally present has by now decayed?

13. Repeat the above calculation for ^{235}U, half-life 7.1 × 10^8 years.

14. The longest-lived isotope of radon, heaviest of the noble gases, is ^{222}Rn, with a half-life of 3.8 days. Suppose I have a sample of the pure gas and wish to determine its density, and for this purpose it must be 99% pure. Within what time must I complete my measurements?

15. How many atomic disintegrations occur each second in a 1.0 g sample of the radon referred to in the previous example? (Take the atomic weight as 222.)

16. A first-order reaction has a specific rate constant of 4.81 × 10^{-5} s^{-1}. What fraction of a sample of reactant will remain after (a) 30 minutes, (b) 4.0 hours, (c) 12 hours?

17. The half-life of ^{14}C is 5,760 years and a sample of new wood has a count of 15 disintegrations s^{-1} g^{-1} due to ^{14}C.
   (a) If a sample of wood is 12,000 years old, how many disintegrations s^{-1} will be observed for each gram of carbon?
   (b) At what age will the count rate fall to 2 per second?

18. The reaction $H_2 + I_2 \rightarrow 2HI$ is found experimentally to be first order in H_2 and first order in I_2. For each of the reaction schemes below, work out the expected rate law. Which scheme (or schemes) is consistent with experiment?

(a) $H_2 \rightleftharpoons 2H$ (fast equilibrium)
    $I_2 \rightleftharpoons 2I$ (fast equilibrium)
    $H + I \longrightarrow HI$ (slow)

(b) $H_2 \longrightarrow 2H$ (slow)
    $H + I_2 \longrightarrow HI + I$ (fast)
    $I + H \longrightarrow HI$ (fast)

(c) $H_2 + I_2 \longrightarrow H_2I_2$ (slow)
    $H_2I_2 \longrightarrow 2HI$ (fast)

19. The reaction $2NO + Cl_2 \rightarrow 2NOCl$ is found experimentally to be second order in NO and first order in Cl_2. Which of the following schemes is likely for the reaction?

(a) $NO + NO + Cl_2 \longrightarrow 2NOCl$

(b) $Cl_2 \longrightarrow 2Cl$ (slow)
    $NO + Cl \longrightarrow NOCl$ (fast)

(c) $Cl_2 \rightleftharpoons 2Cl$ (fast equilibrium)
    $NO + Cl \longrightarrow NOCl$ (slow)

(d) $2NO \rightleftharpoons N_2O_2$ (fast equilibrium)
    $Cl_2 + N_2O_2 \longrightarrow 2NOCl$ (slow)

20. The reverse reaction from the previous example:

$$2NOCl \longrightarrow 2NO + Cl_2$$

is second order in NOCl. Which of the following is a possible mechanism?

(a) $NOCl \longrightarrow NO + Cl$ (slow)
    $2Cl \longrightarrow Cl_2$ (fast)

(b) $2NOCl \longrightarrow N_2O_2Cl_2$ (slow)
    $N_2O_2Cl_2 \longrightarrow N_2O_2 + Cl_2$ (fast)
    $N_2O_2 \rightleftharpoons 2NO$ (fast equilibrium)

(c) $NOCl \longrightarrow NO + Cl$ (slow)
    $Cl + NOCl \longrightarrow NO + Cl_2$ (fast)

21. The following mechanism has been suggested for the decomposition of ozone, O_3, to oxygen:

$$O_3 \rightleftharpoons O_2 + O \quad \text{(fast equilibrium)}$$
$$O + O_3 \longrightarrow 2O_2 \quad \text{(slow)}$$

What rate law would be consistent with the above scheme, including a dependence on $[O_2]$ as well as $[O_3]$?

22. The following reaction occurs in alkaline solution:

$$I^- + OCl^- \longrightarrow OI^- + Cl^-$$

and is first order in $[I^-]$ and $[OCl^-]$. A determination of dependence of rate on $[OH^-]$ gives the following result:

$[OH^-]$ (mole liter^{-1})	Rate (mole liter^{-1} sec^{-1})
1.00	2.44×10^{-4}
0.75	3.25×10^{-4}
0.50	4.88×10^{-4}

(a) What is the order of reaction in $[OH^-]$? Write the complete rate law.
(b) The following mechanism is suggested for this reaction:

$$H^+ + OCl^- \rightleftharpoons HOCl \qquad \text{(fast equilibrium)}$$
$$I^- + HOCl \longrightarrow OI^- + Cl^- + H^+ \qquad \text{(slow)}$$

Is this consistent with the above rate law? Explain your reasoning.

SOLUTIONS TO PROBLEMS

1. (a) From (i) to (ii), $[A]$ is constant:

$$\frac{[B](ii)}{[B](i)} = \frac{1.04 \times 10^{-2}}{7.11 \times 10^{-3}} = 1.46 \qquad \frac{\text{rate}(ii)}{\text{rate}(i)} = \frac{1.89 \times 10^{-4}}{8.83 \times 10^{-5}} = 2.14$$

If exponent of $[B]$ is b,

$$(1.46)^b = 2.14 \qquad b = \frac{\log 2.14}{\log 1.46} = 2.0$$

second order in B.

From (ii) to (iii), $[B]$ is constant:

$$\frac{[A](iii)}{[A](ii)} = \frac{7.11 \times 10^{-2}}{2.88 \times 10^{-2}} = 2.47 \qquad \frac{\text{rate}(iii)}{\text{rate}(ii)} = \frac{4.66 \times 10^{-4}}{1.89 \times 10^{-4}} = 2.47$$

first order in A.
rate law: rate $= k[A][B]^2$

(b) rate constant: $k = \dfrac{\text{rate}}{[A][B]^2} = \dfrac{8.83 \times 10^{-5} \text{ mole liter}^{-1} \text{ s}^{-1}}{(2.88 \times 10^{-2} \text{ mole liter}^{-1}) \times (7.11 \times 10^{-3} \text{ mole liter}^{-1})^2}$

$= 60.7 \text{ mole}^{-2} \text{ liter}^2 \text{ s}^{-1}$

If $[A] = 5.61 \times 10^{-2}$ and $[B] = 3.01 \times 10^{-3}$,
rate: $60.7(5.61 \times 10^{-2}) \times (3.01 \times 10^{-3})^2 = 3.08 \times 10^{-5} \text{ mole liter}^{-1} \text{ s}^{-1}$

2. The concentration of ozone in mole liter^{-1} is directly proportional to its pressure, so we use the pressure data directly. From (i) to (ii)

$$\frac{\text{pressure(i)}}{\text{pressure(ii)}} = \frac{21}{15} = 1.40 \qquad \frac{\text{rate(i)}}{\text{rate(ii)}} = \frac{2.60 \times 10^{-6}}{1.33 \times 10^{-6}} = 1.95$$

If exponent of $[O_3]$ is a,

$$(1.40)^a = 1.95 \qquad a = \frac{\log 1.95}{\log 1.40} = 2.0$$

Reaction is second order in $[O_3]$ (the third data set may be used to confirm this).

3. (a) $$\frac{[CH_3CHO](\text{ii})}{[CH_3CHO](\text{i})} = \frac{2.7 \times 10^{-3}}{1.2 \times 10^{-3}} = 2.25 \qquad \frac{\text{rate(ii)}}{\text{rate(i)}} = \frac{2.26 \times 10^{-4}}{6.70 \times 10^{-5}} = 3.37$$

If exponent of $[CH_3CHO]$ is a,

$$(2.25)^a = 3.37 \qquad a = \frac{\log 3.37}{\log 2.25} = 1.50$$

Reaction is order 1.5 in $[CH_3CHO]$.
$$\text{rate} = k[CH_3CHO]^{1.5}$$

$$k = \frac{\text{rate}}{[CH_3CHO]^{1.5}} = \frac{6.7 \times 10^{-5} \text{ mole liter}^{-1} \text{ s}^{-1}}{(1.2 \times 10^{-3} \text{ mole liter}^{-1})^{1.5}}$$

$$k = 1.6 \text{ mole}^{-0.5} \text{ liter}^{0.5} \text{ s}^{-1}$$

Check using third data set:
$$\text{rate} = 1.6[CH_3CHO]^{1.5} \text{ and } [CH_3CHO] = 4.1 \times 10^{-3}$$

$$\text{rate} = 1.6(4.1 \times 10^{-3})^{1.5} = 1.6 \times 2.6 \times 10^{-4}$$

$$= 4.2 \times 10^{-4} \text{ mole liter}^{-1} \text{ s}^{-1}$$

Good agreement with given figure.

(b) $\text{rate} = 1.6(1.0 \times 10^{-2})^{1.5} = 1.6 \times 10^{-3} \text{ mole liter}^{-1} \text{ s}^{-1}$

4. From (i) to (ii), $[Q]$ and $[R]$ are constant:

$$\frac{[P](\text{ii})}{[P](\text{i})} = \frac{2.9 \times 10^{-3}}{1.2 \times 10^{-3}} = 2.4 \qquad \frac{\text{rate(ii)}}{\text{rate(i)}} = \frac{15.0 \times 10^{-6}}{6.2 \times 10^{-6}} = 2.4$$

first order in $[P]$
From (i) to (iii), $[P]$ and $[Q]$ are constant:
$[R]$ increases, but rate does not change.
Rate is independent of $[R]$ (zero order in $[R]$).
From (i) to (iv), $[P]$ and $[R]$ are constant:

$$\frac{[Q](\text{iv})}{[Q](\text{i})} = \frac{5.9 \times 10^{-4}}{2.6 \times 10^{-4}} = 2.3 \qquad \frac{\text{rate(iv)}}{\text{rate(i)}} = \frac{9.3 \times 10^{-6}}{6.2 \times 10^{-6}} = 1.5$$

If exponent of [Q] is a,

$$(2.3)^a = 1.5 \qquad a = \frac{\log 1.5}{\log 2.3} = 0.5$$

order one-half in [Q]
rate $= k[\text{P}][\text{Q}]^{0.5}$ (R is not included)

$$k = \frac{\text{rate}}{[\text{P}][\text{Q}]^{0.5}} = \frac{6.2 \times 10^{-6} \text{ mole liter}^{-1} \text{ s}^{-1}}{(1.2 \times 10^{-3} \text{ mole liter}^{-1}) \times (2.6 \times 10^{-4} \text{ mole liter}^{-1})^{0.5}}$$

$$= 0.32 \text{ mole}^{-0.5} \text{ liter}^{0.5} \text{ s}^{-1}$$

When $[\text{P}] = 3.3 \times 10^{-3}$, $[\text{Q}] = 3.8 \times 10^{-4}$ mole liter^{-1},

rate $= 0.32(3.3 \times 10^{-3})(3.8 \times 10^{-4})^{0.5} = 2.1 \times 10^{-5}$ mole liter^{-1} s^{-1}

5. (a) $$\log \frac{7.5 \times 10^{-3}}{1.5 \times 10^{-3}} = \frac{E_a}{2.303R} \left(\frac{1}{273} - \frac{1}{293} \right)$$

$$\log 5.0 = \frac{E_a}{2.303 \times 8.314} (3.66 \times 10^{-3} - 3.41 \times 10^{-3})$$

$$E_a = \frac{0.699 \times 2.303 \times 8.314}{2.5 \times 10^{-4}} = 5.4 \times 10^4 \text{ J mole}^{-1} = 54 \text{ kJ mole}^{-1}$$

(b) $$\log \left(\frac{\text{rate}(50°\text{C})}{\text{rate}(0°\text{C})} \right) = \frac{5.4 \times 10^4}{2.303 \times 8.314} \left(\frac{1}{273} - \frac{1}{323} \right)$$

$$= 2.82 \times 10^3 \times (5.67 \times 10^{-4}) = 1.60$$

$$\frac{\text{rate}(50°\text{C})}{\text{rate}(0°\text{C})} = \text{antilog } 1.60 = 40$$

rate at $50°\text{C} = 40 \times 1.5 \times 10^{-3} = 6.0 \times 10^{-2}$ mole liter^{-1} s^{-1}

6. (a) $$\frac{\text{rate(ii)}}{\text{rate(i)}} = 50$$

$$\log 50 = \frac{E_a}{2.303R} \left(\frac{1}{293} - \frac{1}{343} \right)$$

$$1.70 = \frac{E_a}{2.303 \times 8.314} (3.41 \times 10^{-3} - 2.92 \times 10^{-3})$$

$$E_a = \frac{1.70 \times 2.303 \times 8.314}{4.9 \times 10^{-4}} = 6.6 \times 10^4 \text{ J mole}^{-1} = 66 \text{ kJ mole}^{-1}$$

(b) Suppose at temperature T K the rate is 100 times its value at 293 K (20°C):

$$\log \left(\frac{\text{rate}(T)}{\text{rate}(293)} \right) = \log 100 = \frac{6.6 \times 10^4}{2.303 \times 8.314} \left(\frac{1}{293} - \frac{1}{T} \right)$$

$$\frac{1}{293} - \frac{1}{T} = \frac{2.00 \times 2.303 \times 8.314}{6.6 \times 10^4} = 5.8 \times 10^{-4}$$

$$\frac{1}{T} = \frac{1}{293} - 5.8 \times 10^{-4} = (3.41 \times 10^{-3}) - (5.8 \times 10^{-4}) = 2.83 \times 10^{-3}$$

$$T = 353 \text{ K } (80°\text{C})$$

7. (a) $\log\left(\dfrac{\text{rate}(60°)}{\text{rate}(30°)}\right) = \dfrac{5.0 \times 10^4}{2.303 \times 8.314}\left(\dfrac{1}{303} - \dfrac{1}{333}\right)$

$= (2.6 \times 10^3) \times (2.97 \times 10^{-4}) = 0.78$

$\dfrac{\text{rate}(60°)}{\text{rate}(30°)} = \text{antilog } 0.78 = 6.0$

If rate at 30° is 1.2 mole liter^{-1} s^{-1},
rate at 60° = 6.0 × 1.2 = 7.2 mole liter^{-1} s^{-1}

(b) Using $\log\left(\dfrac{\text{rate(ii)}}{\text{rate(i)}}\right) = \dfrac{E_a(\text{i}) - E_a(\text{ii})}{2.303RT}$,

where rate(ii) is rate *with* a catalyst,
$E_a(\text{i}) = 50$ kJ   and $E_a(\text{ii}) = 25$ kJ

$\log\left(\dfrac{\text{rate(ii)}}{\text{rate(i)}}\right) = \dfrac{(50 - 25) \times 10^3}{2.303 \times 8.314 \times 303} = 4.31$

$\dfrac{\text{rate(ii)}}{\text{rate(i)}} = \text{antilog } 4.31 = 2.0 \times 10^4$

If rate(i) = 1.2 mole liter^{-1} s^{-1},
   rate(ii) = 1.2 × 2.0 × 10^4 = 2.4 × 10^4 mole liter^{-1} s^{-1}

8. (a) Suppose the lower temperature is T K:

$\log\left(\dfrac{\text{rate}(373)}{\text{rate}(T)}\right) = \dfrac{6.5 \times 10^4}{2.303 \times 8.314}\left(\dfrac{1}{T} - \dfrac{1}{373}\right)$

We are told $\dfrac{\text{rate}(373)}{\text{rate}(T)} = 10$ and $\log 10 = 1.0$.

$\dfrac{6.5 \times 10^4}{2.303 \times 8.314}\left(\dfrac{1}{T} - \dfrac{1}{373}\right) = 1.0$

$\dfrac{1}{T} - \dfrac{1}{373} = \dfrac{2.303 \times 8.314}{6.5 \times 10^4} = 3.0 \times 10^{-4}$

$\dfrac{1}{T} = \dfrac{1}{373} + 3.0 \times 10^{-4} = (2.68 \times 10^{-3}) + (3.0 \times 10^{-4}) = 2.98 \times 10^{-3}$

$T = 336$ K   (63°C)

(b) At 20°C   (293 K):

$\log\left(\dfrac{\text{rate}(373)}{\text{rate}(293)}\right) = \dfrac{6.5 \times 10^4}{2.303 \times 8.314}\left(\dfrac{1}{293} - \dfrac{1}{373}\right)$

$= (3.4 \times 10^3)(7.32 \times 10^{-4}) = 2.5$

$\dfrac{\text{rate}(373)}{\text{rate}(293)} = \text{antilog } 2.5 = 3.1 \times 10^2$

If rate(373) = 7.8 × 10^{-2} mole liter^{-1} s^{-1},

$\text{rate}(293) = \dfrac{7.8 \times 10^{-2}}{3.1 \times 10^2} = 2.6 \times 10^{-4}$ mole liter^{-1} s^{-1}

9. If the activation energy with catalyst is E_a kJ,

$$\log\left(\frac{\text{rate(catalyzed)}}{\text{rate(uncatalyzed)}}\right) = \log 1000 = \frac{(48 - E_a) \times 10^3}{2.303 \times 8.314 \times 298}$$

$$\frac{(48 - E_a) \times 10^3}{5.70 \times 10^3} = 3.00 \qquad 48 - E_a = \frac{5.70 \times 3.00 \times 10^3}{10^3} = 17$$

$$E_a = 48 - 17 = 31 \text{ kJ mole}^{-1}$$

10. $$\log\left(\frac{\text{rate(catalyzed)}}{\text{rate(uncatalyzed)}}\right) = \frac{(60 - 35) \times 10^3}{2.303 \times 8.314 \times 273} = 4.78$$

$$\frac{\text{rate(catalyzed)}}{\text{rate(uncatalyzed)}} = \text{antilog } 4.78 = 6.1 \times 10^4$$

11. We have to evaluate $(0.5)^n$ to find the fraction remaining after n half-lives. This is done by looking up the log of 0.5 (-0.301) multiplying this by n, and looking up the antilog.

 (a) $n = 0.10 \qquad -0.301n = -0.030$
   $\text{antilog}(-0.030) = \text{antilog}(\bar{1}.97) = 0.93$
     93% remains after 0.10 half-life

 (b) $n = 0.50 \qquad -0.301n = -0.151$
   $\text{antilog}(-0.151) = \text{antilog}(\bar{1}.849) = 0.71$
     71% remains after 0.50 half-life

 (c) $n = 1.50 \qquad -0.301n = -0.452$
   $\text{antilog}(-0.452) = \text{antilog}(\bar{1}.548) = 0.35$
     35% remains after 1.50 half-life

12. elapsed half-lives: $\dfrac{4.5 \times 10^9}{1.4 \times 10^{10}} = 0.32 = n$ and $(0.5)^{0.32} = 0.80$

 80% of original ^{232}Th remains, so 20% has so far decayed.

13. elapsed half-lives: $\dfrac{4.5 \times 10^9}{7.1 \times 10^8} = 6.3$ and $(0.5)^{6.3} = 1.3 \times 10^{-2}$

 1.3% of ^{235}U remains, so 99% has so far decayed.

14. The amount remaining will be 99% (fraction 0.99) when n half-lives have elapsed and $(0.5)^n = 0.99$
 Take logs: $n \log(0.5) = \log(0.99)$

 $$n \times (-0.301) = -4.36 \times 10^{-3}$$

 $$n = \frac{-4.36 \times 10^{-3}}{-0.301} = 1.45 \times 10^{-2} \text{ half-lives}$$

 Since half-life is 3.8 days, this is $1.45 \times 10^{-2} \times 3.8 = 0.055 \text{ d} = 1.3 \text{ h}$.
 Alternatively, we may evaluate the specific rate constant for the process, k, from

 $$k = \frac{0.693}{t_{1/2}} = \frac{0.693}{3.8} = 0.182 \text{ d}^{-1}$$

If $\dfrac{[A]}{[A]_0} = 0.99$ $e^{-kt} = 0.99$

taking natural logarithms:

$-kt = \ln(0.99) = 2.303 \log(0.99)$

$\qquad = 2.303(-4.36 \times 10^{-3}) = -1.01 \times 10^{-2}$

$t = \dfrac{-1.01 \times 10^{-2}}{-k} = \dfrac{-1.01 \times 10^{-2}}{-0.182} = 5.5 \times 10^{-2}\,\text{d} = 1.3\,\text{h}$

Conclusion: Radon is difficult stuff to work with!

15. Using the value of k obtained above,

rate $= k[\text{Rn}] = 0.182[\text{Rn}]$ (k in d^{-1})

If we have 1.0 g of Rn (at.wt 222), we have $1.0/222 = 4.5 \times 10^{-3}$ mole.
This is $4.5 \times 10^{-3} \times 6.0 \times 10^{23}$ atoms $= 2.7 \times 10^{21}$ atoms.
Using this for [Rn],

rate $= (0.182\,\text{d}^{-1}) \times (2.7 \times 10^{21}\,\text{atoms}) = 4.9 \times 10^{20}$ atoms d^{-1}

Every day, in a 1.0 g sample, this number of atoms disintegrate.
In each second, the number of disintegrations is

$$\frac{4.9 \times 10^{20}}{24 \times 3600} = 5.7 \times 10^{15} \text{ disintegrations s}^{-1}$$

16. Use $\dfrac{[A]}{[A]_0} = e^{-kt}$

the logarithmic form is

$$-kt = \ln \frac{[A]}{[A]_0} = 2.303 \log \frac{[A]}{[A]_0}$$

so, if $k = 4.81 \times 10^{-5}\,\text{s}^{-1}$,

$$\log \frac{[A]}{[A]_0} = \frac{-kt}{2.303} = \frac{-4.81 \times 10^{-5}\,t}{2.303}$$

where t is in seconds
(a) $t = 30$ min $= 1800$ s

$$\log \frac{[A]}{[A]_0} = \frac{-4.81 \times 10^{-5} \times 1800}{2.303} = -3.76 \times 10^{-2} \qquad (\bar{1}.962)$$

$$\frac{[A]}{[A]_0} = 0.92 \qquad 92\% \text{ remains}$$

(b) $t = 4.0$ h $= 1.44 \times 10^{4}$ s

$$\log \frac{[A]}{[A]_0} = \frac{-4.81 \times 10^{-5} \times 1.44 \times 10^{4}}{2.303} = -0.301 \qquad (\bar{1}.699)$$

$$\frac{[A]}{[A]_0} = 0.50 \qquad 50\% \text{ remains (half-life 4 h)}$$

(c) $t = 12 \text{ h} = 4.32 \times 10^4 \text{ s}$

$$\log \frac{[A]}{[A]_0} = \frac{-4.81 \times 10^5 \times 4.32 \times 10^4}{2.303} = -0.902 \quad (\bar{1}.098)$$

$$\frac{[A]}{[A]_0} = 0.13 \qquad 13\% \text{ remains}$$

17. (a) 12,000 years is $\dfrac{12,000}{5,760} = 2.08$ half-lives.

fraction remaining is $(0.5)^{2.08}$ (see problem 11) $= 0.237$
disintegration rate will have dropped from original count to
$\quad 15 \times 0.237 = 3.5 \text{ disintegrations s}^{-1} \text{ g}^{-1}$

(b) If count rate has fallen to 2 per second, the fraction remaining is $2/15 = 0.13$.
This occurs after n half-life, where

$$0.5^n = 0.13 \qquad n = \frac{\log 0.13}{\log 0.5} = 2.9$$

$\quad$ 2.9 half-lives are $2.9 \times 5,760 = 1.7 \times 10^4 \text{ yr}$

18. (a) rate $= k[H_2]^{0.5}[I_2]^{0.5}$; not consistent
(b) rate $= k[H_2]$; not consistent
(c) rate $= k[H_2][I_2]$; consistent

19. (a) This has the correct rate law, but is most unlikely because it involves a three-body collision.
(b) rate $= k[Cl_2]$; not consistent
(c) rate $= k[NO][Cl_2]^{0.5}$; not consistent
(d) rate $= k[NO]^2[Cl_2]$; consistent and most likely

20. (a) and (c) rate $= k[NOCl]$; not consistent
(b) rate $= k[NOCl]^2$; consistent, but the existence of the intermediate $N_2O_2Cl_2$ is speculative.

21. rate $= k[O_3]^2[O_2]^{-1}$
As oxygen concentration increases, the rate of reaction slows down because of the effect $[O_2]$ has on the equilibrium step.

22. (a) The rate is *inversely* proportional to $[OH^-]$, so the complete rate law is

$$\text{rate} = k[I^-][OCl^-][OH^-]^{-1}$$

(b) Yes, this mechanism is consistent.
slow step: rate $= k[I^-][HOCl]$

$$\text{equilibrium:} \quad K = \frac{[HOCl]}{[H^+][OCl^-]} \quad \text{or} \quad [HOCl] = K[H^+][OCl^-]$$

substituting: rate $= kK[I^-][H^+][OCl^-]$
But $[H^+]$ is *inversely* proportional to $[OH^-]$, so our rate law is
rate $= k'[I^-][OCl^-][OH^-]^{-1}$

Chapter 11

Equilibrium in Gas Phase Processes

"I generally have chemicals about, and occasionally do experiments. Would that annoy you?"
"By no means."   A STUDY IN SCARLET

Chemical equilibrium may be established in several different ways, but two of the most common and important are in the gas phase and in solution. In this chapter, we consider gas phase equilibrium, while subsequent chapters consider various aspects of solution reactions.

11.1   The Equilibrium Constant

A reaction mixture at equilibrium is in a state where forward and reverse reactions are proceeding at the same rate, so the concentrations of all species present remain constant. It is a dynamic condition. For a given reaction mixture at a given temperature, there is only *one* possible equilibrium position, and this may be approached from either direction.

In other words, if we consider the process

$$a\text{A} + b\text{B} \rightleftharpoons p\text{P} + q\text{Q}$$

(where a, b, p, q are the coefficients of the reactants A, B, P, Q in the balanced equation) then we can start out with pure A + B, or pure P + Q, or some mixture of all four, and we will reach the same equilibrium position. When the system has reached equilibrium, we always find that

$$K_c = \frac{[\text{P}]^p[\text{Q}]^q}{[\text{A}]^a[\text{B}]^b}$$

where K_c is a constant called the *equilibrium constant* for the reaction in question. Its value is constant only if the temperature remains constant (the manner in which it changes with temperature is considered in Chapter 16). The subscript c reminds us that we have put in the values of [A], [B], etc. as *concentrations* in the units mole liter^{-1}.

Note that the products of the reaction as written go on the top of the fraction (remember to*P*-Products) while the reactants go on the bottom. Each reactant and product concentration is raised to a power equal to its coefficient in the balanced equation. The custom of putting the fraction that way up is simply a convention agreed upon by chemists. If the reaction were written the other way round:

$$pP + qQ \rightleftharpoons aA + bB$$

then the terms in the equilibrium constant expression would be inverted and its numerical value would, of course, be the reciprocal of that previously found.

It follows that a *small* value for K_c will correspond to an equilibrium that favors the reactants (left side of the equation) while a *large* value for K_c favors the products (right side).

Example 11.1 For the gas-phase reaction $H_2 + I_2 \rightleftharpoons 2HI$, the concentrations found at 490°C in a certain experiment are, in mole liter^{-1},

$$[H_2] = 8.62 \times 10^{-4} \qquad [I_2] = 2.63 \times 10^{-3} \qquad [HI] = 1.02 \times 10^{-2}$$

(a) Calculate K_c for the equilibrium as written.
(b) What would be the value of K_c for the reaction $2HI \rightleftharpoons H_2 + I_2$?

SOLUTION For the reaction as written, $H_2 + I_2 \rightleftharpoons 2HI$,

(a)
$$K_c = \frac{[HI]^2}{[H_2][I_2]} \qquad (\text{*P*roducts on to*P*})$$

$$K_c = \frac{(1.02 \times 10^{-2})^2}{(8.62 \times 10^{-4})(2.63 \times 10^{-3})} = \frac{45.9 \ (\text{mole liter}^{-1})^2}{(\text{mole liter}^{-1})^2} = 45.9$$

(*Note*: The units cancel out and K_c is dimensionless.)
(b) For the reaction written in the reverse direction, $2HI \rightleftharpoons H_2 + I_2$,

$$K_c' = \frac{[H_2][I_2]}{[HI]^2} = \frac{1}{K_c} = \frac{1}{45.9} = 2.18 \times 10^{-2}$$

where K_c' is used to distinguish it from K_c, the previous equilibrium constant.

Example 11.2 One mole H_2 and one mole I_2 are heated to 490°C in a 1.00 liter volume. What will be the equilibrium concentrations of H_2, I_2, and HI? (*Note*: K_c for the reaction $H_2 + I_2 \rightleftharpoons 2HI$ is 45.9 at this temperature.)

SOLUTION Initially, we have 1 mole liter^{-1} of each reactant, H_2 and I_2. Suppose x mole liter^{-1} of each react to give HI. The equilibrium concentrations may then be worked out in terms of x by reference to the balanced equation. For the two reactants, the equilibrium concentration is the initial concentration *less* the amount that has reacted, so we can write

$$[H_2] = (1 - x) \text{ mole liter}^{-1}$$

$$[I_2] = (1 - x) \text{ mole liter}^{-1}$$

From the equation, we know that one mole H_2 reacts with one mole I_2 to give two mole HI, so obviously when x mole of each react we will have $2x$ mole of HI.

Summarizing:

$$H_2 \;+\; I_2 \;\;\rightleftharpoons\;\; 2HI$$

concentrations: $(1-x)$ $(1-x)$ $2x$ mole liter^{-1}

If this system is at equilibrium, we can write

$$K_c = \frac{[\mathrm{HI}]^2}{[\mathrm{H_2}][\mathrm{I_2}]} = \frac{(2x)^2}{(1-x)(1-x)} = 45.9$$

This is a quadratic equation in x, but it can easily be solved by noting that it is in the form of a perfect square. Taking the square root of each side gives

$$\frac{2x}{(1-x)} = \sqrt{45.9} = 6.77$$

$$2x = 6.77 - 6.77x \qquad 8.77x = 6.77 \qquad x = 0.772$$

Substituting for x in the concentration expressions gives

$$[\mathrm{H_2}] = [\mathrm{I_2}] = (1-x) = 0.228 \text{ mole liter}^{-1}$$

$$[\mathrm{HI}] = 2x = 1.544 \text{ mole liter}^{-1}$$

Note that the total mole of gas present is

$$1.544 + 0.228 + 0.228 = 2.000 \text{ mole liter}^{-1}$$

This is the same as it was initially, and the form of the reaction equation tells us that this must be so, since there are two moles of gas on each side. Whatever the position of equilibrium, the *total* amount of gas remains the same.

We also note that the same equilibrium position would have been reached if we had started with two mole of HI under the same conditions. (Try the calculation for yourself and check this!)

Example 11.3   0.500 mole of HI is heated to 490°C in a container of volume 4.5 liter. What will be the degree of dissociation at equilibrium? K_c for the reaction $2HI \rightleftharpoons H_2 + I_2$ is 2.18×10^{-2} at this temperature.

SOLUTION   Since the data are given for a volume other than one liter, we must first convert to concentrations in mole liter^{-1}. We can then carry out the rest of the calculation in mole liter^{-1}:

$$\frac{0.500 \text{ mole}}{4.50 \text{ liter}} = 0.111 \text{ mole liter}^{-1}$$

This is our initial concentration. Suppose that, to reach equilibrium, $2x$ mole liter^{-1} of HI undergo reaction. This will produce x mole of H_2 and x mole of I_2, so equilibrium concentrations will be

$$2HI \rightleftharpoons H_2 + I_2$$
$$(0.111 - 2x) \qquad x \quad x$$

$$K_c = \frac{[H_2][I_2]}{[HI]^2} = \frac{x^2}{(0.111 - 2x)^2} = 2.18 \times 10^{-2}$$

Again, we have a quadratic in x which is a perfect square.

$$\frac{x}{0.111 - 2x} = \sqrt{2.18 \times 10^{-2}} = 0.148$$

$$x = 0.148(0.111 - 2x) = 0.0164 - 0.296x$$

$$1.296x = 0.0164 \qquad x = 0.0127$$

The amount of dissociation was $2x = 0.0254$ mole liter^{-1}. So the degree of dissociation of HI is

$$\frac{\text{amount dissociated}}{\text{amount originally present}} = \frac{0.0254}{0.111} = 0.229 \quad \text{or} \quad 22.9\%$$

Note in this calculation that we chose $2x$ mole as the amount of HI that dissociated, because we know from the equation that this would produce x mole of H_2 and I_2. If we had started with x mole dissociating, we would have produced $x/2$ mole of H_2 and I_2, making the calculation a little clumsier (though the answer would be the same, of course).

A common type of problem involves the displacement of a gaseous equilibrium by the addition of a further quantity of one of the reactants. The best way to solve this situation is to treat it in two stages: first, to calculate the concentrations of all substances present immediately after the addition of the additional reactant (when the system is *not* at equilibrium), and second, to calculate how much reaction must occur for equilibrium to be restored.

Example 11.4   The mixture of Example 11.2 has an additional 0.200 mole of H_2 added to it, keeping volume and temperature constant. What are the concentrations when equilibrium is restored?

SOLUTION   We first calculate the concentrations of the three substances immediately after the addition of the H_2, before any reaction has occurred:

$$[HI] = 1.544 \text{ mole liter}^{-1} \qquad \text{(unchanged)}$$

$$[I_2] = 0.228 \text{ mole liter}^{-1} \qquad \text{(unchanged)}$$

$$[H_2] = 0.228 + 0.200 = 0.428 \text{ mole liter}^{-1}$$

To restore equilibrium, H_2 will react with I_2 to give HI. Suppose x mole liter^{-1} react, then the equilibrium concentrations will be:

$$[H_2] = 0.428 - x \qquad [I_2] = 0.228 - x \qquad [HI] = 1.544 + 2x$$

(Remember, x mole H_2 and I_2 give $2x$ mole HI.)

The value of K will be unchanged, so we can write

$$K_c = \frac{[HI]^2}{[H_2][I_2]} = \frac{(1.544 + 2x)^2}{(0.428 - x)(0.228 - x)} = 45.9$$

This is a quadratic equation in x, which is a little more difficult than that in Example 11.2, because it is not a perfect square. Rearranging and collecting terms gives

$$41.9x^2 - 36.29x + 2.095 = 0$$

from which x is found by the usual formula (see appendix) to be 0.0622 or 0.804. The second value may be rejected as physically impossible, since it would lead to negative concentrations of H_2 and I_2, so $x = 0.0622$ and the equilibrium concentrations are, in mole liter^{-1},

$$[H_2] = 0.428 - x = 0.366 \qquad [I_2] = 0.228 - x = 0.166; \qquad [HI] = 1.544 + 2x = 1.668$$

Note that the total moles of gas is still 2.200 mole, equal to the sum of the original 2.00 mole plus the 0.200 mole added.

An alternative approach to this problem would have been to say that we were in effect starting with $1.00 + 0.200 = 1.20$ mole of pure H_2 and 1.00 mole of pure I_2 and letting them equilibrate, as in Example 11.2. Of course, we would have ended up at the same equilibrium position. (Try it for yourself!)

11.2   Effect of Changes in Pressure or Volume on Equilibrium

In the above calculations, we expressed all the concentrations in mole liter^{-1}. Since the volume must obviously be the same for every component in a gaseous mixture, we can express these concentrations in terms of the actual number of moles of each gas present.

Suppose the volume of the mixture of H_2, I_2, and HI is V liter and contains

$$n_1 \text{ mole } H_2 \qquad n_2 \text{ mole } I_2 \qquad n_3 \text{ mole HI}$$

Then our concentrations are

$$[H_2] = \frac{n_1}{V} \qquad [I_2] = \frac{n_2}{V} \qquad [HI] = \frac{n_3}{V}$$

$$K_c = \frac{[HI]^2}{[H_2][I_2]} = \frac{(n_3/V)^2}{(n_1/V)(n_2/V)} = \frac{(n_3)^2}{n_1 n_2}$$

We see that V cancels out in the equilibrium expression, showing that, *for this reaction*, the position of equilibrium is independent of the volume. If we halve the size of our reaction vessel, the *amounts* of the three reactants will be unchanged (of course, each of their *concentrations* will be doubled as V is halved). So we cannot favor the production of H_2 or HI by changing the volume of the container.

Is this always going to be the case? Let's consider the gaseous equilibrium

$$N_2 + 3H_2 \rightleftharpoons 2NH_3$$

As before, we put the volume of the containing vessel at V liter and call the concentrations (in mole liter^{-1})

$$[N_2] = \frac{n_1}{V} \qquad [H_2] = \frac{n_2}{V} \qquad [NH_3] = \frac{n_3}{V}$$

For the reaction as written above,

$$K_c = \frac{[NH_3]^2}{[N_2][H_2]^3} = \frac{(n_3/V)^2}{(n_1/V)(n_2/V)^3} = \frac{n_3^2 V^2}{n_1 n_2^3}$$

where K_c has the units mole^{-2} liter^2. We see that V does *not* cancel out in this equilibrium. The position of equilibrium will be very much affected by the volume occupied by the equilibrium mixture.

Suppose we halve the volume, i.e., we go from volume V to volume $V/2$ liter. Since the term V^2 appears in the equilibrium expression, halving V will reduce this to $V^2/4$. If the system is to remain at equilibrium, the other part of the equilibrium expression, $n_3^2/n_1 n_2^3$, must *increase* by a factor of 4 to compensate for the change in V. This is accomplished by displacement of the equilibrium position, that is, by reaction occurring preferentially in one direction.

If we want to *increase* the quantity $n_3^2/n_1 n_2^3$, the reaction must go in such a direction as to *increase* n_3 and *decrease* n_1 and n_2. In other words, N_2 and H_2 react together to produce more NH_3, to restore equilibrium. If we had originally *increased* V, the equilibrium would have been displaced in the opposite direction.

If we look at the equations representing the two equilibria we are comparing (HI and NH_3), we see an important difference between them. The HI equilibrium was a constant-volume reaction; we started and ended with two moles of gas. By contrast, the NH_3 equilibrium shows a change in gas volumes from four moles on the left to two moles on the right. From the method we use to define the equilibrium constant, we can deduce the following very important general result:

In a *constant volume* gas reaction (by which we mean one in which there are equal numbers of moles of gaseous compounds on each side of the equation) the equilibrium constant has no units and the position of equilibrium is not influenced by changes in the volume (pressure) of the system.

In a gas reaction where the volume of gas is *not* constant, the equilibrium constant will have units, and the position of the equilibrium will be displaced by changes in volume (pressure). An *increase* in volume (decrease in pressure) will displace the equilibrium in the direction of *larger* gas volume; a *decrease* in volume (increase in pressure) will displace the equilibrium in the direction of *smaller* gas volume.

The last sentence is often quoted as an example of *Le Chatelier's principle*, which states that a system reacts to an applied stress in such a way as to minimize the effect of that stress. We saw earlier, however, that the direction in which the equilibrium position moves is entirely determined by the need to return to equilibrium at a constant value of K_c. The system does not react in this way merely because it is obeying laws laid down by Le Chatelier!

The effects described above may be illustrated by a numerical example.

Example 11.5   Gaseous N_2O_4 at 30°C is in equilibrium with NO_2 according to the equation

$$N_2O_4(g) \rightleftharpoons 2NO_2(g)$$

An initial concentration of 1.00×10^{-2} mole liter^{-1} of N_2O_4 is 33.5% dissociated into NO_2 at equilibrium. Calculate the value of K_c for this reaction. What would be the degree of dissociation for an initial concentration of N_2O_4 of 2.00×10^{-2} mole liter^{-1}?

SOLUTION   Of the original 1.00×10^{-2} mole liter^{-1} of N_2O_4, 33.5% represents 3.35×10^{-3} mole liter^{-1}. Equilibrium concentrations will be

$$[N_2O_4] = (1.00 \times 10^{-2}) - (3.35 \times 10^{-3}) = 6.65 \times 10^{-3}$$

$$[NO_2] = 2 \times (3.35 \times 10^{-3}) = 6.70 \times 10^{-3}$$

(*Note*: Remember that each mole of N_2O_4 gives *two* mole NO_2 on dissociation.)

$$K_c = \frac{[NO_2]^2}{[N_2O_4]} = \frac{(6.70 \times 10^{-3})^2}{6.65 \times 10^{-3}} = 6.75 \times 10^{-3} \text{ mole liter}^{-1}$$

(*Note*: As always in an equilibrium where there is a change in volume, K_c has units).

Suppose that, for an initial concentration of N_2O_4 of 2.00×10^{-2} mole liter^{-1}, the amount of dissociation is x mole liter^{-1}. Then the equilibrium concentrations are

$$[N_2O_4] = (2.00 \times 10^{-2}) - x \qquad [NO_2] = 2x$$

$$K_c = \frac{(2x)^2}{(2.00 \times 10^{-2}) - x} = 6.75 \times 10^{-3}$$

$$4x^2 = (1.35 \times 10^{-4}) - (6.75 \times 10^{-3})x$$

$$4x^2 + (6.75 \times 10^{-3})x - (1.35 \times 10^{-4}) = 0$$

This is a quadratic equation in x, of which the roots are $x = 5.03 \times 10^{-3}$ or $x = -6.71 \times 10^{-3}$. We may reject the latter, since x cannot be negative (we cannot end up with more N_2O_4 than we started with!) so equilibrium concentrations are

$$[N_2O_4] = (2.00 \times 10^{-2}) - x = (2.00 \times 10^{-2}) - (5.03 \times 10^{-3})$$

$$= 1.50 \times 10^{-2} \text{ mole liter}^{-1}$$

$$[NO_2] = 2x = 2 \times (5.03 \times 10^{-3}) = 1.01 \times 10^{-2} \text{ mole liter}^{-1}$$

The degree of dissociation is equal to the amount of N_2O_4 that has dissociated (x) divided by the amount originally present.

$$\text{dissociation:} \quad \frac{5.03 \times 10^{-3}}{2.00 \times 10^{-2}} \times 100 = 25.2\%$$

As we would expect, the increase in the concentration (pressure) of N_2O_4 has led to a decrease in the amount of dissociation, as the equilibrium shifts in the direction of smaller volume (towards N_2O_4). Doubling the initial pressure has reduced the degree of dissociation from 33.5% to 25.2%. The equilibrium constant, K_c, has of course remained unchanged.

Example 11.6   At a temperature of 257°C, the equilibrium

$$N_2 + 3H_2 \rightleftharpoons 2NH_3$$

has a value of 100 for K_c, the equilibrium constant. Work out the equilibrium concentrations of the three substances present when 1 mole N_2 and 3 mole H_2 are put into a volume of (a) 1 liter (b) 2 liter. Show how the change of volume affects the position of the equilibrium by calculating in each case the percent conversion of nitrogen into ammonia.

SOLUTION   In each case, our basic equation will be

$$K_c = \frac{[NH_3]^2}{[N_2][H_2]^3} = 100$$

(a) Starting with 1 mole N_2 and 3 mole H_2 in a volume of 1 liter, suppose x mole N_2 react with $3x$ mole of H_2 to give $2x$ mole of NH_3 (the equation tells us that this must be the ratio in which they react). Equilibrium concentrations will be

$$N_2 \quad + \quad 3H_2 \quad \rightleftharpoons \quad 2NH_3$$
$$(1 - x) \quad (3 - 3x) \quad\quad\quad 2x \quad \text{mole liter}^{-1}$$

$$K_c = \frac{(2x)^2}{(1 - x)(3 - 3x)^3} = \frac{(2x)^2}{27(1 - x)^4} = 100$$

This is a fourth-order equation in x, but it can be simplified to a quadratic by noting that it is a perfect square:

$$\frac{(2x)^2}{(1 - x)^4} = 2700 \qquad \frac{2x}{(1 - x)^2} = \sqrt{2700} = 51.96$$

Collecting terms gives the quadratic equation

$$51.96x^2 - 105.9x + 51.96 = 0$$

Solving by the usual formula gives two roots:

$$x = 1.216 \quad \text{or} \quad x = 0.822$$

The first can be rejected, since x cannot be greater than 1 (which would give a negative N_2 concentration). Putting $x = 0.822$ gives (in mole liter^{-1})

$$[N_2] = 1 - x = 0.178 \qquad [H_2] = 3(1 - x) = 0.534 \qquad [NH_3] = 2x = 1.64$$

The conversion of N_2 into NH_3 is

$$\frac{x \text{ (amount converted)}}{1 \text{ (amount originally present)}} \times 100 = 82.2\%$$

(b) When the volume is 2 liter, the calculation is very similar, except that the initial concentrations are halved, that is,

$$[N_2] = 0.5 \text{ mole liter}^{-1} \qquad [H_2] = 1.5 \text{ mole liter}^{-1}$$

Suppose that out of this y mole N_2 react with $3y$ mole H_2 to give a concentration of $2y$ mole liter^{-1} of NH_3. Equilibrium concentrations will be

$$N_2 \quad + \quad 3H_2 \quad \rightleftharpoons \quad 2NH_3$$
$$(0.5 - y) \quad (1.5 - 3y) \qquad\qquad 2y \quad \text{mole liter}^{-1}$$

$$K_c = \frac{(2y)^2}{(0.5 - y)(1.5 - 3y)^3} = \frac{(2y)^2}{27(0.5 - y)^4} = 100$$

As before, we take the square root of each side

$$\frac{(2y)^2}{(0.5 - y)^4} = 2700 \qquad \frac{2y}{(0.5 - y)^2} = \sqrt{2700} = 51.96$$

which rearranges to

$$51.96y^2 - 53.96y + 12.99 = 0$$

from which $y = 0.659$ or $y = 0.379$. The first root is impossible (y cannot exceed 0.5), so we can use the value $y = 0.379$ to calculate concentrations.

$$[N_2] = (0.5 - y) = 0.121 \qquad [H_2] = 3(0.5 - y) = 0.363 \qquad [NH_3] = 2y = 0.758$$

in mole liter^{-1}. The conversion of N_2 into ammonia is

$$\frac{0.379}{0.50} \times 100 = 75.8\%$$

Comparing this with the value previously obtained, we see that the conversion of N_2 to NH_3 is *less* in the larger volume, in accordance with our expectation that the equilibrium would be displaced to the larger gaseous volume (i.e., to the left as written) by going to the larger volume (lower pressure).

11.3 Calculation of Equilibrium Constants from Pressure Measurements

In the case of an equilibrium where a change in gaseous volume occurs, the total pressure on the system will enable us to calculate the position of equilibrium. This gives a valuable method of calculating equilibrium constants.

Example 11.7 1.00 mole of N_2 and 3.00 mole of H_2 are put in a volume of 1.00 liter and allowed to equilibrate at 257°C. The equilibrium pressure is 102.5 atm. Calculate the equilibrium concentrations and K_c for the equilibrium

$$N_2 + 3H_2 \rightleftharpoons 2NH_3$$

SOLUTION Using the ideal gas equation and given values of P, V, and T, we can calculate n at equilibrium

$$n = \frac{PV}{RT} = \frac{102.5 \times 1.00}{0.0821 \times 530} = 2.356 \text{ mole}$$

This is the *total* amount of gas present. We started with 1 mole N_2 and 3 mole H_2, out of which suppose x mole N_2 reacted with $3x$ mole H_2 to give $2x$ mole NH_3.

equilibrium concentrations:

$$N_2 \quad + \quad 3H_2 \quad \rightleftharpoons \quad 2NH_3$$
$$(1-x) \quad (3-3x) \qquad\qquad 2x \quad \text{mole liter}^{-1}$$

The total of the three gases present is

$$(1-x) + (3-3x) + 2x = 4 - 2x \text{ mole}$$

If we know the total amount of gas is 2.356 mole, then we can write

$$4 - 2x = 2.356 \qquad x = 0.822 \text{ mole}$$

equilibrium concentrations:

$$[N_2] = 1 - x = 0.178 \qquad [H_2] = 3 - 3x = 0.534 \qquad [NH_3] = 2x = 1.644$$

The value of the equilibrium constant may now be calculated

$$K_c = \frac{[NH_3]^2}{[N_2][H_2]^3} = \frac{(1.644)^2}{0.178(0.534)^3} = 100$$

Obviously, this is an alternative approach to the same problem we examined in the previous example.

Example 11.8 The equilibrium

$$PCl_5(g) \quad \rightleftharpoons \quad PCl_3(g) + Cl_2(g)$$

is set up by heating 10.4 g of PCl_5 to 150°C in a container of volume 1.00 liter. If the equilibrium pressure is 1.91 atm, calculate the equilibrium concentrations, the value of K_c, and the percent dissociation of the PCl_5.

SOLUTION As before, we first calculate the number of moles of gas present in 1 liter at equilibrium.

$$n = \frac{PV}{RT} = \frac{1.91 \times 1.00}{0.0821 \times 423} = 5.50 \times 10^{-2} \text{ mole}$$

This is total gas, all species.

Originally, we had 10.4 g of PCl_5 (*M.W.* 208) which is $10.4/208 = 5.00 \times 10^{-2}$ mole. Suppose x mole of this dissociate, then equilibrium concentrations will be

$$PCl_5 \quad \rightleftharpoons \quad PCl_3 + Cl_2$$
$$(5.00 \times 10^{-2} - x) \qquad\qquad x \quad\; x$$

so the total mole of gas is

$$5.00 \times 10^{-2} - x + x + x = 5.00 \times 10^{-2} + x$$

Knowing from our first calculation that the total mole of gas is 5.50×10^{-2}, we write

$$5.00 \times 10^{-2} + x = 5.50 \times 10^{-2}$$
$$x = 5.0 \times 10^{-3}$$

Using this value of x, we calculate concentrations

$$[PCl_5] = 5.00 \times 10^{-2} - x = 4.50 \times 10^{-2} \text{ mole liter}^{-1}$$

$$[PCl_3] = [Cl_2] = x = 5.0 \times 10^{-3} \text{ mole liter}^{-1}$$

$$K_c = \frac{[PCl_3][Cl_2]}{[PCl_5]} = \frac{(5.0 \times 10^{-3})^2}{4.50 \times 10^{-2}} = 5.55 \times 10^{-4} \text{ mole liter}^{-1}$$

percent dissociation: $\dfrac{5.0 \times 10^{-3}}{5.00 \times 10^{-2}} \times 100 = 10\%$

Example 11.9   The above mixture, still at 150°C, is compressed into a volume of 0.500 liter. What happens? Calculate equilibrium concentrations and the percent dissociation of PCl_5. What is the total pressure at equilibrium?

SOLUTION   This type of problem, like Example 11.4, is best tackled in two stages. First we calculate the concentrations immediately after the volume change, then we work out what must occur to restore equilibrium.

Immediately after the volume is changed to 0.500 liter, the concentrations are all doubled (the system is *not* at equilibrium)

$$[PCl_5] = 2 \times 4.50 \times 10^{-2} = 9.00 \times 10^{-2} \text{ mole liter}^{-1}$$
$$[PCl_3] = [Cl_2] = 2 \times 5.0 \times 10^{-3} = 1.00 \times 10^{-2} \text{ mole liter}^{-1}$$

Suppose that y mole liter^{-1} of PCl_3 and Cl_2 combine, giving y mole liter^{-1} of PCl_5, to restore equilibrium. (*Note*: Although we can see from the form of the equation that this will be the direction in which equilibrium is displaced, it would not matter if we made the wrong assumption here. If, in fact, the equilibrium moved in the opposite direction, we should simply obtain a negative value of y.)

Then concentrations at equilibrium will be

$$[PCl_5] = (9.00 \times 10^{-2}) + y \qquad [PCl_3] = [Cl_2] = (1.00 \times 10^{-2}) - y$$

Since K_c remains at the same value, we can write

$$K_c = \frac{[PCl_3][Cl_2]}{[PCl_5]} = \frac{[(1.00 \times 10^{-2}) - y]^2}{(9.00 \times 10^{-2}) + y} = 5.55 \times 10^{-4}$$

This rearranges to $y^2 - (2.06 \times 10^{-2})y + (5.0 \times 10^{-5}) = 0$. And, as usual, we find two roots to the quadratic,

$$y = 1.78 \times 10^{-2} \quad \text{or} \quad y = 0.28 \times 10^{-2}$$

The first may be rejected, since y cannot be greater than 1.00×10^{-2}. Using the latter value gives the concentrations

$$[PCl_5] = (9.00 \times 10^{-2}) + y = 9.28 \times 10^{-2} \text{ mole liter}^{-1}$$

$$[PCl_3] = [Cl_2] = (1.00 \times 10^{-2}) - y = 7.2 \times 10^{-3} \text{ mole liter}^{-1}$$

The percent dissociation of PCl_5 is calculated by dividing the amount that has dissociated by the total amount that would be present if *none* of it had dissociated:

$$\text{percent dissociation} = \frac{7.2 \times 10^{-3}}{(9.28 \times 10^{-2}) + (7.2 \times 10^{-3})} \times 100 = 7.2\%$$

To calculate the pressure, we need the total concentration of all three gases present:

$$[PCl_5] + [PCl_3] + [Cl_2] = (9.28 \times 10^{-2}) + 2(7.2 \times 10^{-3}) = 10.72 \times 10^{-2} \text{ mole liter}^{-1}$$

This is, of course, n/V in our gas equation

$$P = \frac{n}{V} RT = 10.72 \times 10^{-2} \times 0.0821 \times 423 = 3.73 \text{ atm}$$

(*Note*: The gas concentrations are expressed in mole per liter. Although we only have a volume of 0.500 liter, we do *not* insert $V = 0.500$ liter in the equation.)

Comparing the results of Examples 11.8 and 11.9, we see that the dissociation of PCl_5 decreased from 10% to 7.2% when we halved the volume containing the reaction mixture, in accord with the displacement of equilibrium in the direction of lower gas volume.

At the same time, the pressure changed from 1.91 atm to 3.73 atm, increasing by a factor of 1.95. If we had halved the volume of an ideal gas (where no reaction was occurring) the pressure would, of course, have doubled, so we see that the reaction mixture, by moving its equilibrium position, has been successful in partially reducing the effect of the applied stress.

11.4   The Concept of K_p

So far in this chapter, we have expressed all our concentrations in mole per liter. This is a very convenient unit, and we shall be using it extensively when working with reactions in solution. However, when dealing with a gaseous equilibrium, the *partial pressure* of each gas in the mixture is a measure of the concentration of that gas, and may be used as such in the equilibrium expression.

We use the symbol K_p for an equilibrium constant when gas concentrations are expressed in terms of the partial pressures of gases in the mixture. Let's work out the relationship between K_p and K_c, our previous equilibrium constant values where concentrations were expressed in mole liter^{-1}.

Example 11.10   Return to Examples 11.1 and 11.5 and calculate K_p for these equilibria.

SOLUTION   In each case, we have the concentration of gas in mole liter^{-1}, and this must be converted to a partial pressure by using the relationship

$$P = \frac{n}{V} \times RT$$

where n/V is the concentration in mole liter^{-1}. In Example 11.1, the HI equilibrium, concentrations, in mole liter^{-1}, were

$$[H_2] = 8.62 \times 10^{-4} \qquad [I_2] = 2.63 \times 10^{-3} \qquad [HI] = 1.02 \times 10^{-2}$$

At the temperature of the equilibrium (490°C), the partial pressures are

$$[H_2] = 8.62 \times 10^{-4} \times 0.0821 \times 763 = 0.0540 \text{ atm}$$

$$[I_2] = 2.63 \times 10^{-3} \times 0.0821 \times 763 = 0.165 \text{ atm}$$

$$[HI] = 1.02 \times 10^{-2} \times 0.0821 \times 763 = 0.638 \text{ atm}$$

$$K_p = \frac{[HI]^2}{[H_2][I_2]} = \frac{(0.638)^2}{0.0540 \times 0.165} = 45.9 \text{ (no units)}$$

For the equilibrium in Example 11.5, the concentrations at 30°C are

$$[N_2O_4] = 6.65 \times 10^{-3} \text{ mole liter}^{-1} \qquad [NO_2] = 6.70 \times 10^{-3} \text{ mole liter}^{-1}$$

As before, we multiply each by the factor RT to convert to atmospheres:

$$[N_2O_4] = 6.65 \times 10^{-3} \times 0.0821 \times 303 = 0.165 \text{ atm}$$

$$[NO_2] = 6.70 \times 10^{-3} \times 0.0821 \times 303 = 0.167 \text{ atm}$$

$$K_p = \frac{[NO_2]^2}{[N_2O_4]} = \frac{(0.167)^2}{0.165} = 0.168 \text{ atm}$$

We notice at once that, for the HI equilibrium, K_c and K_p are the same, while for the N_2O_4 case they are different. A moment's thought shows why this is so. For the HI equilibrium, the number of moles of gas on each side of the equation is the same, so all our factors of RT cancel out when we evaluate K_p. Putting this another way, we know that K has no units for the HI equilibrium, or any other gas-phase equilibrium involving equal total moles of gas on both sides of the equation, so it is not surprising that it makes no difference whether we express concentrations in mole liter^{-1} or in atmospheres, since the units cancel anyway.

For the N_2O_4 equilibrium, however, the equilibrium constant *does* have units, and it will make a difference to its numerical value if we change these units. To summarize, we can say:

For an equilibrium in which the total number of moles of gas does not change, K_c and K_p are identical. For an equilibrium in which there are n_a moles of gas on the left and n_b moles of gas on the right of the equation, the relationship is

$$K_p = \frac{K_c}{(RT)^n} \quad \text{where} \quad n = n_a - n_b$$

So for the N_2O_4/NO_2 equilibrium above, where $n = 1 - 2 = -1$,

$$K_p = \frac{K_c}{(RT)^{-1}} = K_c RT = 6.75 \times 10^{-3} \times 0.0821 \times 303 = 0.168 \text{ atm}$$

Example 11.11   At temperature of 1000 K, the value of K_c for the gas-phase equilibrium

$$N_2 + 3H_2 \rightleftharpoons 2NH_3$$

is 2.00×10^{-2} mole^{-2} liter^2. What is the value of K_p?

SOLUTION For all the substances reacting, R and T have the same value. If we have n_1 mole N_2, n_2 mole H_2, and n_3 mole NH_3 in volume V, then

$$K = \frac{(n_3/V)^2}{(n_1/V)(n_2/V)^3}$$

Each concentration term is converted to a partial pressure by multiplying it by RT, so

$$K_p = \frac{(n_3 RT/V)^2}{(n_1 RT/V)(n_2 RT/V)^3} = \frac{(n_3/V)^2}{(n_1/V)(n_2/V)^3 (RT)^2} = \frac{K_c}{(RT)^2}$$

K_p differs from K_c by a factor $(RT)^2$, since we have four moles of gas on the left of the equation and two moles on the right:

$$K_p = \frac{2.00 \times 10^{-2}}{(0.0821 \times 1000)^2} = 2.96 \times 10^{-6} \text{ atm}^{-2}$$

Example 11.12 For the reaction

$$H_2O(g) \rightleftharpoons H_2(g) + 1/2 O_2(g)$$

K_p has the value 1.59×10^{-5} atm$^{1/2}$ at 1500°C. What is the value of K_c?

SOLUTION In this reaction, with partial pressures in atm,

$$K_p = \frac{[H_2][O_2]^{1/2}}{[H_2O]} = 1.59 \times 10^{-5} \text{ atm}^{1/2}$$

To convert the pressures in atm to concentrations in mole liter^{-1}, we *divide* each by RT according to the gas equation

$$\frac{n}{V} = \frac{P}{RT}$$

In this particular equilibrium, we have units

$$\frac{(\text{atm}) \times (\text{atm})^{1/2}}{(\text{atm})}$$

so our conversion factor overall is

$$\frac{1/RT \times (1/RT)^{1/2}}{1/RT} = (1/RT)^{1/2} \qquad K_c = \frac{K_p}{(RT)^{1/2}}$$

Putting in $R = 0.0821$ and $T = 1773\text{K}$ (1500°C),

$$K_c = \frac{1.59 \times 10^{-5}}{(0.0821 \times 1773)^{1/2}} = 1.32 \times 10^{-6} \text{ mole}^{1/2} \text{ liter}^{-1/2}$$

In the above calculations, we expressed partial pressures in atmospheres. We could have used other units, such as cm Hg or mm Hg, and the resulting K_p values would have differed by a constant factor depending on the units used (but only in cases where K_p itself has units). For simplicity in the gas law calculations, we will restrict ourselves to units of atmospheres for partial pressures in the following chapter problems.

KEY WORDS

equilibrium constant   equilibrium concentration
Le Chatelier's principle K_c
homogeneous equilibrium K_p

STUDY QUESTIONS

1. What is the general definition of an equilibrium constant for a gaseous reaction?

2. Which way up does the ratio for K_c go? Why?

3. How is an equilibrium constant related to the equilibrium constant for the reverse reaction?

4. What is the significance of a value of K_c that is (a) very small, (b) about 1, and (c) very large?

5. Under what conditions does K_c for a gaseous reaction have *no* units?

6. For what type of gaseous equilibrium will a compression of the mixture affect the position of equilibrium? In which direction will equilibrium be shifted? Why?

7. An additional quantity of one reactant is added to a gaseous equilibrium mixture. What happens?

8. Could we determine K_c for the reaction

$$H_2(g) + Br_2(g) \rightleftharpoons 2HBr(g)$$

by measuring the pressure in a mixture of the three gases? Why not?

9. When does K_p for a gaseous equilibrium differ from K_c, and when are they the same?

10. If K_p and K_c differ, how may we calculate one from the other? What experimental condition must be known to make the conversion?

11. If a gas dissociates on heating, will the *total* number of moles present become more, become less, or stay constant? Is this always true?

12. What happens to the equilibrium degree of dissociation if the mixture is (a) expanded, (b) compressed?

PROBLEMS

Note: Problems that require the solution of a quadratic equation by use of the quadratic formula (see Appendix A.5) are marked by an asterisk (*). In no case is it necessary to solve a higher-order equation.

1. For the reaction

$$H_2(g) + CO_2(g) \rightleftharpoons H_2O(g) + CO(g)$$

   the following concentrations (in mole liter^{-1}) are found:

$$[H_2] = 0.600 \quad [CO_2] = 0.459 \quad [H_2O] = 0.500 \quad [CO] = 0.425$$

   What is the value of K_c?

2. For the reaction above, what would the equilibrium concentrations be if initially 2.00 mole H_2 and 2.00 mole CO_2 were put into a volume of 10.0 liter?

*3. The equilibrium in problem 2 is displaced by adding an additional 1.00 mole of H_2 to the mixture, still in a volume of 10.0 liter. What will concentrations be when equilibrium is restored?

4. For the equilibrium

$$H_2(g) + I_2(g) \rightleftharpoons 2HI(g)$$

   the equilibrium constant, K_c, is 54.8 at 425°C.
   (a) What will be the equilibrium concentrations if 0.600 mole HI is heated to this temperature in a volume of 1.00 liter?
   (b) What is the percent dissociation of the HI?

*5. The experiment in problem 4 is repeated, except that the reaction vessel initially contains 0.100 mole H_2 in addition to 0.600 mole HI.
   (a) What will equilibrium concentrations be under these conditions?
   (b) How will this affect the percent dissociation of the HI?

*6. In the ammonia equilibrium

$$N_2(g) + 3H_2(g) \rightleftharpoons 2NH_3(g)$$

   the equilibrium constant K_c has the value 2.00 at 400°C. If 2.00 mole NH_3 are heated to this temperature in a volume of 10.0 liter,
   (a) what will be the equilibrium concentrations of all reactants?
   (b) what will be the degree of dissociation of NH_3?

7. For the equilibrium

$$COCl_2(g) \rightleftharpoons CO(g) + Cl_2(g)$$

   K_p has the value 6.7×10^{-9} atm at 100°C. What is the value of K_c at this temperature?

8. On heating, NO_2 dissociates according to

$$NO_2(g) \rightleftharpoons NO(g) + \frac{1}{2}O_2(g)$$

     A gaseous sample of pure NO_2 has a concentration of 0.200 mole liter^{-1}. After equilibrium is established, the NO_2 has undergone 10% dissociation. What is the value of K_c under these conditions?

9. Under the same conditions as problem 8, what is the value of K_c for the equilibrium

$$2NO_2(g) \rightleftharpoons 2NO(g) + O_2(g)$$

10. In a study of the equilibrium

$$2SO_3(g) \rightleftharpoons 2SO_2(g) + O_2(g)$$

a sample of 0.6365g of pure SO_3 is placed in a 1.00 liter container and heated to a temperature of 827°C. The equilibrium total pressure at this temperature is 1.00 atm. Calculate the percent dissociation of SO_3 under these conditions and the values of K_c and K_p.

11. Nitrosyl chloride, NOCl, dissociates on heating:

$$NOCl(g) \rightleftharpoons NO(g) + \frac{1}{2}Cl_2(g)$$

When a sample of pure NOCl weighing 1.50 g is heated to 350°C in a volume of 1.00 liter, the degree of dissociation is 57.2%. Calculate K_c and K_p and the total pressure in the system at this temperature.

12. Nitrosyl bromide, NOBr, dissociates on heating:

$$NOBr(g) \rightleftharpoons NO(g) + \frac{1}{2}Br_2(g)$$

When 1.79 g of NOBr is placed in a 1.00 liter container and heated to 100°C, the equilibrium pressure is 0.657 atm. Calculate
(a) the partial pressures of the three gases
(b) the value of K_p
(c) the degree of dissociation of NOBr.

SOLUTIONS TO PROBLEMS

1. $K_c = \dfrac{0.500 \times 0.425}{0.600 \times 0.459} = 0.772$ (no units)

2. If x mole H_2 (in 10.0 liter) react x mole CO_2 giving x mole CO and x mole H_2O, then at equilibrium

$$\frac{x^2}{(2.00-x)^2} = K_c = 0.772$$

solving as a perfect square, $x = 0.935$

$$[H_2] = [CO_2] = \frac{2.00-x}{10.0} = 0.107 \text{ mole liter}^{-1}$$

$$[H_2O] = [CO] = \frac{x}{10.0} = 9.35 \times 10^{-2} \text{ mole liter}^{-1}$$

3. Adding the 1.00 mole H_2 to the original H_2/CO_2 mixture gives 3.00 mole H_2. Let x mole (in 10 liter) react, as before, then

$$\frac{x^2}{(3.00-x)(2.00-x)} = 0.772$$

This must be solved as a quadratic. Collecting terms gives

$$0.228x^2 + 3.86x - 4.632 = 0$$

for which $x = +1.125$ or -18.0. Taking the first root gives the concentrations in mole liter^{-1}:

$$[H_2] = 0.188 \quad [CO_2] = 0.0875 \quad [CO] = [H_2O] = 0.113$$

4. (a) Initially $[HI] = 0.600$ mole liter^{-1}. If $2x$ mole liter^{-1} dissociate, then at equilibrium

$$\frac{(0.600-2x)^2}{x^2} = K_c = 54.8$$

Solving as a perfect square, $x = 0.0638$
$[H_2] = [I_2] = x = 0.0638$ mole liter^{-1}
$[HI] = 0.600 - 2x = 0.600 - 0.128 = 0.472$ mole liter^{-1}

(b) dissociation $= \frac{2 \times 0.0638}{0.600} \times 100 = 21.3\%$

5. (a) This is similar, but initially, $[H_2] = 0.100$, so at equilibrium $[H_2] = 0.100 + x$. The equation is

$$\frac{(0.600-2x)^2}{x(0.100+x)} = 54.8$$

which gives $50.8x^2 + 7.88x - 0.360 = 0$ from which $x = -0.192$ or $+0.0369$, and, since x cannot be negative, we take 0.0369
$[HI] = 0.600 - 2x = 0.526$ mole liter^{-1}
$[H_2] = 0.100 + x = 0.137$ mole liter^{-1}
$[I_2] = x = 0.0369$ mole liter^{-1}

(b) dissociation $= \frac{2 \times 0.0369}{0.600} \times 100 = 12.3\%$

(*Note*: The additional H_2 represses dissociation.)

6. (a) Suppose $2x$ out of 0.200 mole liter^{-1} of NH_3 have dissociated, giving x mole N_2 and $3x$ mole H_2. Then, at equilibrium,

$$K_c = \frac{(0.200 - 2x)^2}{(x)(3x)^3} = 2.00$$

This is a perfect square. Taking the square root,

$$\frac{0.200 - 2x}{x^2} = \sqrt{2.00 \times 27}$$

Solving as a quadratic, the positive root is $x = 7.78 \times 10^{-2}$
so $[NH_3] = 0.0444$, $[N_2] = 0.0778$, and $[H_2] = 0.233$ mole liter^{-1}

(b) dissociation: $\dfrac{2 \times 7.78 \times 10^{-2}}{0.200} \times 100 = 77.8\%$

7. We go from 1 to 2 mole of gas, $n_a - n_b = 1 - 2 = -1$

$$K_p = \frac{K_c}{(RT)^{-1}} = K_c RT$$

$$K_c = \frac{K_p}{RT} = \frac{6.7 \times 10^{-9}}{0.0821 \times 373} = 2.19 \times 10^{-10} \text{ mole liter}^{-1}$$

8. Out of 0.200 mole liter^{-1} NO_2, 10% represents 0.0200 mole liter^{-1} so 0.180 mole liter^{-1} remain,

$$[NO_2] = 0.180 \qquad [NO] = 0.0200 \qquad [O_2] = 0.0100$$

$$K_c = \frac{0.0200 \times (0.0100)^{1/2}}{0.180} = 1.11 \times 10^{-2} \text{ mole}^{1/2} \text{ liter}^{-1/2}$$

9. For the equilibrium with doubled coefficients

$$K_c' = \frac{[NO]^2[O_2]}{[NO_2]^2} = (K_c)^2 = 1.23 \times 10^{-4} \text{ mole liter}^{-1}$$

Whenever an equation's coefficients are doubled, its equilibrium constant value is *squared*.

10. 0.6365 g of SO_3 is 7.95×10^{-3} mole
1.00 atm in 1.00 liter at 827°C (1100 K), so

$$\text{number of moles} = \frac{1.00 \times 1.00}{0.0821 \times 1100} = 1.107 \times 10^{-2} \text{ mole}$$

suppose SO_3 dissociation is $2x$ mole liter^{-1}
total mole $(7.95 \times 10^{-3}) + x = 1.107 \times 10^{-2}$ and $x = 3.12 \times 10^{-3}$
$[SO_3] = (7.95 \times 10^{-3}) - 2x = 1.72 \times 10^{-3}$ mole liter^{-1}
$[SO_2] = 2x = 6.23 \times 10^{-3}$ mole liter^{-1}
$[O_2] = x = 3.12 \times 10^{-3}$

$$K_c = \frac{(6.23 \times 10^{-3})^2 \times (3.12 \times 10^{-3})}{(1.72 \times 10^{-3})^2} = 4.09 \times 10^{-2} \text{ mole liter}^{-1}$$

going from 2 mole to 3 mole, $n_a - n_b = 2 - 3 = -1$
$K_p = K_c RT = 4.09 \times 10^{-2} \times 0.0821 \times 1100 = 3.70$ atm

$$\text{dissociation of } SO_3: \frac{6.23 \times 10^{-3}}{7.95 \times 10^{-3}} \times 100 = 78.3\%$$

11. *M.W.*(NOCl) is 65.5, so 1.50 g is 2.29×10^{-2} mole NOCl
57.2% dissociation leaves 9.80×10^{-3} mole NOCl and gives
1.31×10^{-2} mole NO and 6.55×10^{-3} mole Cl_2

$$K_c = \frac{[NO][Cl_2]^{1/2}}{[NOCl]} = \frac{(1.31 \times 10^{-2}) \times (6.55 \times 10^{-3})^{1/2}}{9.80 \times 10^{-3}} = 0.108 \text{ mole}^{1/2} \text{ liter}^{-1/2}$$

going from 1 mole gas to 1.5 mole, $n_a - n_b = -0.5$
$\quad K_p = K_c(RT)^{1/2} = 0.108(0.0821 \times 623)^{1/2} = 0.772 \text{ atm}^{1/2}$
total mole of gas $2.95 \times 10^{-2} \ (= n)$

$$\text{total pressure:} \quad \frac{nRT}{V} = \frac{2.95 \times 10^{-2} \times 0.0821 \times 623}{1.00} = 1.51 \text{ atm}$$

(Alternatively, calculate partial pressures of each gas and hence K_p and total pressure.)

12. *M.W.* = 109.9, so 1.79 g is 1.63×10^{-2} mole NOBr

$$\text{at equilibrium:} \quad n = \frac{0.657 \times 1.00}{0.0821 \times 373} = 2.15 \times 10^{-2} \text{ mole}$$

If x mole NOBr dissociate, total mole is $(1.63 \times 10^{-2}) + x/2$

$$\frac{x}{2} = (2.15 - 1.63) \times 10^{-2} = 5.2 \times 10^{-3} \text{ mole}$$

$x = 2(5.2 \times 10^{-3}) = 1.04 \times 10^{-2}$ mole
$[NOBr] = (1.63 \times 10^{-2}) - x = 5.90 \times 10^{-3}$ mole liter^{-1}
$[NO] = x = 1.04 \times 10^{-2}$ mole liter^{-1}

$$[Br_2] = \frac{x}{2} = 5.2 \times 10^{-3} \text{ mole liter}^{-1}$$

To convert to partial pressures, multiply by RT:
(a) $P(NOBr) = 5.90 \times 10^{-3} \times 0.0821 \times 373 = 0.181$ atm
$P(NO) = 1.04 \times 10^{-2} \times 0.0821 \times 373 = 0.318$ atm
$P(Br_2) = 5.2 \times 10^{-3} \times 0.0821 \times 373 = 0.159$ atm

$$\text{(b)} \quad K_p = \frac{0.318 \times (0.159)^{1/2}}{0.181} = 0.701 \text{ atm}^{1/2}$$

$$\text{(c)} \quad \text{dissociation:} \quad \frac{1.04 \times 10^{-2}}{1.63 \times 10^{-2}} \times 100 = 64\%$$

Chapter 12

Equilibrium in Solution: Solubility Product

His face showed the quiet and interested composure of the chemist who sees the crystals falling into position from his oversaturated solution.
"Remarkable!" said he. "Remarkable!"   THE VALLEY OF FEAR

12.1   The Ion Product and the Solubility Product

Equilibrium systems may be divided into two types: *Homogeneous equilibrium*, where all the reactants and products are in the same phase and *Heterogeneous equilibrium*, where more than one phase (solid, liquid, gas) is present.

In this chapter, we are going to be concerned with a very common type of heterogeneous equilibrium; that involving a solid, ionic substance and its solution in water. (The concepts are, of course, applicable to any solvent system, but since water is the best and commonest solvent for ionic substances, we will keep to aqueous solutions throughout.)

When an ionic substance dissolves in water, it loses its identity. Thus, although we would commonly speak of "a solution of potassium nitrate," the species actually present are the separate potassium (K^+) and nitrate (NO_3^-) ions, each surrounded by a large number of water molecules. There is very little interaction between them (in dilute solution) and the substance "potassium nitrate" is not really present (though of course we'd recover it if we evaporated the solution to dryness).

Similarly, if we dissolved potassium nitrate and sodium chloride in the same solution, we would have the four ions (K^+, Na^+, NO_3^-, and Cl^-) present together. We could have made *exactly* the same solution by dissolving potassium chloride and sodium nitrate together.

Obviously, if we make up more complex mixtures with many different ions present together, it could become very difficult to say what ionic salts are present in a solution. How can we carry out calculations on a mixture when there are many different possible "pairings" of cation/anion combinations which we could assume?

We get around this confusion by using the concept of the *ion product* for substances present in solution. The concentration of each ion (both cations and anions) in solution is fixed by the way we made up the solution or, for an unknown solution, it may be found by analytical techniques. So, for any solution, we may list the ion concentrations for all species present, using the usual square bracket notation to indicate mole liter^{-1}. Analysis of a complicated solution might give the following result:

Cations	*Anions*
$[Na^+] = 0.100M$	$[NO_3^-] = 0.0500M$
$[K^+] = 0.0500M$	$[Cl^-] = 0.120M$
$[Ca^{2+}] = 0.0200M$	$[SO_4^{2-}] = 0.0100M$

By taking different cation/anion combinations, we could say that nine different ionic salts are present, but such a concept is not very useful. When we have to do calculations on such a solution, we use the ion product of each substance.

The ion product of a substance is the product of the concentrations of all ions present in that substrate, each raised to the power equal to its coefficient in the formula of that substance.

So the ion products of some of the combinations taken from the above mixed solution would be:

$NaNO_3$ (1:1 ion ratio), ion product $= [Na^+][NO_3^-]$
K_2SO_4 (2:1 ion ratio), ion product $= [K^+]^2[SO_4^{2-}]$
$Ca(NO_3)_2$ (1:2 ion ratio), ion product $= [Ca^{2+}][NO_3^-]^2$
$CaSO_4$ (1:1 ion ratio), ion product $= [Ca^{2+}][SO_4^{2-}]$

Note that is the coefficient of the ion in the formula that is used in determining the exponent in the ion product, *not* the charge on the ion. Although the ions in calcium sulfate have charges of $2+$ and $2-$, they are raised to the power of one in the ion product because they are present in a 1:1 ratio.

The numerical value of the ion product is not usually fixed, but can vary over a wide range. To evaluate the value of the ion product in a particular case, all we have to do is to insert the appropriate ion concentrations into the ion product expression. It is very important to realize that, in any given solution, there is only *one* possible value for the concentration of each ion. When the same ion appears in several different ion product expressions, this same value is used in each expression. Thus, in the example quoted above, $[SO_4^{2-}] = 0.0100M$, and this value will be used in all expressions for sulfates in this solution:

Na_2SO_4, ion product $= [Na^+]^2[SO_4^{2-}] = (0.100)^2(0.0100) = 1.00 \times 10^{-4}$
K_2SO_4, ion product $= [K^+]^2[SO_4^{2-}] = (0.0500)^2(0.0100) = 2.50 \times 10^{-5}$
$CaSO_4$, ion product $= [Ca^{2+}][SO_4^{2-}] = (0.0200)(0.0100) = 2.00 \times 10^{-4}$

To repeat, the numerical values obtained for these ion products are not constants. Analysis of a different solution would give different ion concentrations and a corresponding variation in the ion product values. You might therefore wonder why we bother to calculate these quantities. When are they of any value to us?

This question brings us back to the original purpose of this chapter; to consider the equilibrium between an ionic solid and its aqueous solution. Let's consider what happens if we start with pure water (in which the ion products of all solutes are zero) and add an ionic salt. As it dissolves, the ion concentrations and the ion product increase, but eventually we reach a limit of solubility where undissolved solid remains, no matter how long we stir the solution.

An equilibrium exists between the solid (AB) and its ions in solution (A^+ and B^-), and we can represent this as

$$AB(s) + nH_2O \rightleftharpoons A^+(aq) + B^-(aq)$$

(Note: We ignore the possibility that any AB can dissolve without dissociating into ions.)

Since this is an equilibrium system, we can write an equilibrium constant, leaving out $[H_2O]$ which is constant:

$$K = \frac{[A^+][B^-]}{[AB(s)]}$$

Since $[AB(s)]$ refers to the concentration of solid AB in the *solid* phase, it is a constant and may be omitted, giving the slightly different constant known as the *solubility product*, K_{sp}:

$$K_{sp} = [A^+][B^-]$$

We see at once that the form of this expression is the same as that of the ion product, but there is one important difference. Since K_{sp} is measured in an equilibrium situation, its value is constant (at a given temperature). We see that:

For any substance, the solubility product at a given temperature is the value of the ion product of the substance in a solution that is in equilibrium with the solid substance at that temperature. Such a solution is said to be *saturated* with respect to that substance.

Obviously, it is possible for the ion product to be less than the solubility product. This is a common stable situation, which obtains in the great majority of reagent solutions used in the laboratory. Such solutions are said to be *unsaturated solutions*, implying that, if more solid is added, it will dissolve in the solution.

The reverse situation, where the ion product is greater than the solubility product, is an unstable nonequilibrium situation, usually produced by carefully cooling a saturated solution; it is said to be *supersaturated*. On disturbing it by mechanical shock, addition of solid matter, etc., solid crystals will separate to lower the ion product until it reaches the value of K_{sp} and equilibrium is restored. Although supersaturation is quite a common occurrence, we will ignore it in this chapter, since we are concerned only with systems at equilibrium.

A glance at the table of K_{sp} values (Table 12.1) shows that most of the numbers are small, ranging from about 10^{-3} down to 10^{-53}. This is because we only use the solubility-product concept for substances with low solubilities ($=$ small K_{sp} values). For freely soluble substances, ion concentrations in their saturated solutions will be large, and our assumption that no interaction occurs between cations and anions will no longer be valid.

Since we are going to be talking about substances of low solubility for the remainder of this chapter, you should have some idea of what classes of substance are and are not likely to be freely soluble. The following generalizations, taken in conjunction with your laboratory experience, should familiarize you with the behavior of the common cations and anions encountered every day in chemistry.

1. All common sodium (Na^+), potassium (K^+), and ammonium (NH_4^+) salts are soluble.
2. All nitrates (NO_3^-) are soluble.

TABLE 12.1   Solubility Products (measured between 18° and 25°)

barium carbonate	$BaCO_3$	8.1×10^{-9}	lithium carbonate	Li_2CO_3	1.7×10^{-3}
barium chromate	$BaCrO_4$	1.6×10^{-10}	magnesium carbonate	$MgCO_3$	2.6×10^{-5}
barium fluoride	BaF_2	1.7×10^{-6}	magnesium fluoride	MgF_2	7.1×10^{-9}
cadmium sulfide	CdS	3.6×10^{-29}	magnesium hydroxide	$Mg(OH)_2$	1.2×10^{-11}
calcium sulfate	$CaSO_4$	2.4×10^{-4}	manganese hydroxide	$Mn(OH)_2$	4×10^{-14}
calcium oxalate	CaC_2O_4	2.6×10^{-9}	mercuric sulfide	HgS	4×10^{-53}
iron(II) hydroxide	$Fe(OH)_2$	1.6×10^{-14}	silver bromate	$AgBrO_3$	5.8×10^{-5}
iron (III) hydroxide	$Fe(OH)_3$	1.1×10^{-36}	silver bromide	$AgBr$	7.7×10^{-13}
iron (II) sulfide	FeS	3.7×10^{-19}	silver chloride	$AgCl$	1.6×10^{-10}
lead carbonate	$PbCO_3$	3.3×10^{-14}	silver iodide	AgI	1.5×10^{-16}
lead fluoride	PbF_2	3.2×10^{-8}	strontium carbonate	$SrCO_3$	1.6×10^{-9}
lead iodide	PbI_2	1.4×10^{-8}	strontium fluoride	SrF_2	2.8×10^{-9}
lead oxalate	PbC_2O_4	2.7×10^{-11}	strontium sulfate	$SrSO_4$	2.5×10^{-7}
lead sulfide	PbS	3.4×10^{-28}	zinc hydroxide	$Zn(OH)_2$	1.2×10^{-17}
lead sulfate	$PbSO_4$	1.06×10^{-8}	zinc sulfide	ZnS	1.2×10^{-23}

3. Most chlorides (Cl^-) and bromides (Br^-) are soluble (exceptions: Pb^{2+}, Cu^+, Ag^+, Hg_2^{2+})
4. Most sulfates (SO_4^{2-}) are soluble (exceptions: Ca^{2+}, Sr^{2+}, Ba^{2+}, Pb^{2+}).
5. Apart from group I metals, NH_4^+, and Ba^{2+}, metal hydroxides have *low* solubilities.
6. Apart from group I metals, and NH_4^+, most sulfides (S^{2-}) and carbonates (CO_3^{2-}) have *low* solubilities.

Although we could always indicate the solubility of a substance by quoting its K_{sp} value, it is useful to have the simple idea of its solubility available also.

The solubility of a substance in a given solution is the amount of that substance that would have to be dissolved in each liter of the solution to produce a solution saturated in that substance. "Amount" may be expressed in moles or in grams.

An alternative way of looking at solubility is to say it is the amount of the substance in question that could be recovered from a liter of saturated solution.

When we work out the relationship between solubility, ion concentrations, and solubility product, there are two possible situations to consider: (1) equilibrium is between the ionic solid and its solution in water, with no other solutes present; (2) the solution in equilibrium with the solid contains various other solutes which enter into the equilibrium.

Let's consider the first case, where we need consider only the ionic solid and water. Suppose the solubility of the substance is S mole liter^{-1}. What are the ion concentrations produced in solution? To answer this, we must know the formula of the substance, because this affects the number of ions produced on dissociation.

In the salt AB, the ions are present in a 1:1 ratio. Dissociation of one mole of AB will give one mole A^+ and one mole B^-, so S mole AB will give S mole of A^+ and S mole of B^-. Ion concentrations produced by dissolving S mole liter^{-1} will be

$$[A^+] = S \qquad [B^-] = S$$

Note that it would be *wrong* to write $[AB] = S$, because we are assuming that there is no undissociated AB left in the solution.

For this substance, therefore, the solubility product may be easily related to its solubility

$$K_{sp} = [A^+][B^-] = S^2 \qquad S = \sqrt{K_{sp}}$$

This relationship may be applied to any substance which has ions in a 1:1 ratio.

Example 12.1   The solubility of calcium sulfate is 2.09 g liter^{-1} at 30°C. Calculate the ion concentrations and the value of K_{sp}.

SOLUTION   Calcium sulfate, $CaSO_4$, has *M.W.* 136, so 2.09 g liter^{-1} is

$$\frac{2.09}{136} = 1.54 \times 10^{-2} \text{ mole liter}^{-1} \qquad \text{(this is } S\text{)}$$

Dissociation gives the corresponding amount of each ion:

$$CaSO_4(s) \rightleftharpoons Ca^{2+} + SO_4^{2-}$$

$$[Ca^{2+}] = [SO_4^{2-}] = 1.54 \times 10^{-2} M$$

$$K_{sp} = [Ca^{2+}][SO_4^{2-}] = S^2 = (1.54 \times 10^{-2})^2 = 2.4 \times 10^{-4}$$

Although K_{sp} always has units (mole^2 liter^{-2} in this case), they are usually omitted.

Example 12.2   The solubility product of silver bromide, AgBr, at 25°C is 7.7×10^{-13}. Calculate the solubility in g liter^{-1}.

SOLUTION   From the definition of K_{sp}

$$K_{sp} = [Ag^+][Br^-] = 7.7 \times 10^{-13}$$

The two ions are produced in equal amounts, so

$$[Ag^+] = [Br^-] = S = \sqrt{K_{sp}} = \sqrt{7.7 \times 10^{-13}} \qquad S = \sqrt{77 \times 10^{-14}} = 8.8 \times 10^{-7} M$$

AgBr has *M.W.* 188, so 8.8×10^{-7} mole is $8.8 \times 10^{-7} \times 188 = 1.7 \times 10^{-4}$ g. The solubility is 1.7×10^{-4} g liter^{-1}.

The above two examples show the method of interconversion between solubility and K_{sp} for salts of formula AB. For salts of formula AB_2 or A_2B, the method is similar. Suppose the solubility of AB_2 is S mole liter^{-1}.

$$AB_2(s) \rightleftharpoons A^{2+} + 2B^-$$

Since each mole of AB_2 produces one mole of A^{2+} and *two* mole of B^-, the concentrations will be

$$[A^{2+}] = S \qquad [B^-] = 2S$$

Therefore, for salt AB_2,

$$K_{sp} = [A^{2+}][B^-]^2 = S \times (2S)^2 = 4S^3$$

The A_2B salt gives the same result:

$$A_2B(s) \rightleftharpoons 2A^+ + B^{2-}$$

$$K_{sp} = [A^+]^2[B^{2-}] = (2S)^2 \times S = 4S^3$$

Example 12.3 2.0×10^{-4} mole of CaF_2 dissolves in 1.00 liter of water to make a saturated solution. What is K_{sp} for this compound?

SOLUTION   Dissociation of CaF_2 gives

$$CaF_2(s) \rightleftharpoons Ca^{2+} + 2F^-$$

Concentrations of the ions will be

$$[Ca^{2+}] = 2.0 \times 10^{-4}M \qquad [F^-] = 2 \times (2.0 \times 10^{-4}) = 4.0 \times 10^{-4}M$$

$$K_{sp} = [Ca^{2+}][F^-]^2 = (2.0 \times 10^{-4})(4.0 \times 10^{-4})^2 = 3.2 \times 10^{-11}$$

(*Note:* We could have simply said $K_{sp} = 4S^3$, where $S = 2.0 \times 10^{-4}$.)

Example 12.4   Lead chloride, $PbCl_2$, has $K_{sp} = 1.7 \times 10^{-5}$. What is its solubility in g liter^{-1}?

SOLUTION   If the solubility is S mole liter^{-1},

$$[Pb^{2+}] = S \qquad [Cl^-] = 2S \qquad K_{sp} = 4S^3$$

$$1.7 \times 10^{-5} = 4S^3 \qquad S^3 = \frac{1.7 \times 10^{-5}}{4} = 4.2 \times 10^{-6} \qquad S = 1.6 \times 10^{-2}$$

Since $PbCl_2$ has *M.W.* 278, 1.6×10^{-2} mole is $1.6 \times 10^{-2} \times 278 = 4.4$ g.
Solubility of $PbCl_2$ is 4.4 g liter^{-1}.

The above treatments of AB, AB_2, and A_2B salts cover the majority of common ionic compounds, and the same approach can easily be extended to other stoichiometries such as AB_3 or A_2B_3.

12.2   The Common Ion Effect

We must now consider the second possibility for an equilibrium between an ionic solid and its solution; that where another solute is present to enter into the equilibrium. In this case, the calculation will follow a slightly different course.

Taking a specific example, suppose we have solid silver chloride in equilibrium with its saturated solution in water and we add some solid sodium chloride. What happens? Sodium chloride is freely soluble and of course dissociates, so the concentration of chloride ions increases sharply. In order to keep the system at equilibrium, the silver ion concentration must be reduced by a corresponding amount, and this can only occur by silver chloride precipitating. The solubility of silver chloride is very much reduced. This is a general effect:

The solubility of a slightly soluble salt will be much less in a solution containing an excess of one of the ions present in the salt.

This is known as the *common ion effect*, because the slightly soluble salt and the other solute contain one ion in common (chloride in the above example). We could equally well have reduced the solubility of silver chloride by adding an excess of silver nitrate (giving Ag^+ as the common ion), whereas sodium nitrate would have had no effect, having no ion in common.

It is most important to realize that the addition of another solute does *not* change K_{sp} for the slightly soluble salt. The ion product still equals the same K_{sp} value when there is equilibrium between the solid and the solution, and we use this fact in calculating solubilities and ion concentrations.

Example 12.5   Calculate the maximum concentration of Ag^+ and the solubility of AgBr in g liter^{-1} in 0.10M sodium bromide solution. Compare your answers with Example 12.2. (K_{sp} for AgBr is 7.7×10^{-13}).

SOLUTION   Sodium bromide (NaBr) is completely dissociated, so we know $[Br^-] = 0.10M$. (Any additional Br^- coming from dissolved AgBr will be quite negligible beside this.)

In a solution saturated with AgBr, we know

$$[Ag^+][Br^-] = 7.7 \times 10^{-13} \qquad [Ag^+] = \frac{7.7 \times 10^{-13}}{[Br^-]}$$

If $[Br^-] = 0.10$, then $[Ag^+] = (7.7 \times 10^{-13})/0.10 = 7.7 \times 10^{-12}$.

We see that the two ion concentrations are very different, whereas previously, in pure water, they were equal. So the solubility, regarded as the maximum amount of silver bromide that we could obtain from a liter of this solution, will be limited by the ion in *smaller* concentration (Ag^+).

From 7.7×10^{-12} mole of AgBr come 7.7×10^{-12} mole of Ag^+, so this is the solubility (S) of AgBr in this solution. AgBr has *M.W.* 188, so

$$7.7 \times 10^{-12} \text{ mole liter}^{-1} = 7.7 \times 10^{-12} \times 188 = 1.5 \times 10^{-9} \text{ g liter}^{-1}$$

The solubility of AgBr has decreased by a factor of about 10^5 in going from pure water (Example 12.2) to 0.10M sodium bromide solution.

The common ion effect is made use of in quantitative analysis, when we want to precipitate as much as possible of one ion from a solution in a convenient form for weighing. Thus, if we were trying to estimate the concentration of Ag^+ in a solution, we could add an excess of a soluble bromide (such as NaBr) and weigh the precipitated AgBr. The above calculation tells us that the Ag^+ remaining in solution would be quite negligible. Of course, if we had to determine the amount of *bromide* ion in a solution, we would add to it an excess of a soluble silver salt (such as $AgNO_3$) to ensure complete precipitation of Br^- from the solution as AgBr.

Example 12.6   Fluoride ion is to be estimated in a solution by precipitation as calcium fluoride, CaF_2. A 500 ml sample is known to contain about 0.2 g of fluoride ion. If exactly the right amount of calcium nitrate is added to convert all this to calcium fluoride, will the error in weight of the precipitate due to incomplete precipitation be serious? What improvement would result from adding a sufficient excess of $Ca(NO_3)_2$ to make the solution 0.10M in Ca^{2+}? (The solubility of CaF_2 in water is $2.0 \times 10^{-4}M$.)

SOLUTION   In our sample of solution, we have $0.2/19 = 1.05 \times 10^{-2}$ mole of F^-. Since the volume of solution is 500 ml (0.500 liter), the original concentration is

$$[F^-] = \frac{1.05 \times 10^{-2}}{0.500} = 2.1 \times 10^{-2} \text{ mole liter}^{-1}$$

We add the exact stoichiometric amount of calcium ion, which is

$$\frac{1.05 \times 10^{-2}}{2} = 0.53 \times 10^{-2} \text{ mole}$$

and convert all of this to CaF_2, some of which precipitates.

Since we are told the solubility (S) of CaF_2 is $2.0 \times 10^{-4}M$, the equilibrium concentration of F^- will be $2S$, or $4.0 \times 10^{-4}M$. As a percentage of the original F^- concentration, this is

$$\frac{4.0 \times 10^{-4}}{2.1 \times 10^{-2}} \times 100 = 1.9\%$$

which is obviously a significant error.

We have already calculated K_{sp} from the solubility figure (Example 12.3) to be 3.2×10^{-11}. If we increase the concentration of Ca^{2+} to $0.1M$, the concentration of F^- must *decrease* to keep the solution at equilibrium:

$$[Ca^{2+}][F^-]^2 = 3.2 \times 10^{-11} \qquad [F^-]^2 = \frac{3.2 \times 10^{-11}}{[Ca^{2+}]}$$

$$[F^-]^2 = \frac{3.2 \times 10^{-11}}{0.1} = 3.2 \times 10^{-10} \qquad [F^-] = 1.8 \times 10^{-5}M$$

As a percentage of the original F^- concentration, this now represents

$$\frac{1.8 \times 10^{-5}}{2.1 \times 10^{-2}} \times 100 = 0.09\% \text{ error}$$

a considerable improvement on the previous figure.

A knowledge of K_{sp} values will often enable us to predict whether or not a precipitate will appear in a certain solution.

Example 12.7 A solution is $1.0 \times 10^{-5}M$ in lead nitrate, $Pb(NO_3)_2$. Potassium iodide is added until $[I^-] = 1.0 \times 10^{-2}M$. Will PbI_2 precipitate? (K_{sp} for PbI_2 is 1.4×10^{-8}).

SOLUTION   This type of problem should be approached by assuming that precipitation does *not* occur, then seeing what consequences follow. If it turns out that an impossible situation would be produced, then the assumption is wrong.

If no precipitation occurs

$$[Pb^{2+}] = 1.0 \times 10^{-5} \quad \text{and} \quad [I^-] = 1.0 \times 10^{-2}$$

The ion product $[Pb^{2+}][I^-]^2 = (1.0 \times 10^{-5}) \times (1.0 \times 10^{-2})^2 = 1.0 \times 10^{-9}$, which is less than K_{sp}, hence no precipitation would occur.

Suppose $[I^-]$ is increased to 0.10. Then, without precipitation occurring,

$$[Pb^{2+}][I^-]^2 = (1.0 \times 10^{-5}) \times (0.10)^2 = 1.0 \times 10^{-7}$$

which is greater than K_{sp}, hence the concentration of iodide ion could not be increased to this value without precipitation occurring.

Note: A very common student error in calculations of the above type is to double the concentration of the ion of which two are used in the salt (I^- in the above example), in other words to write

$$\text{ion product} = [Pb^{2+}][2I^-]^2$$

It should be clear from our previous discussion that this is incorrect. It is the actual concentration of each of the ions which is used in ion product expressions, and there is only *one* value possible for a given ion concentration in any solution.

Another common type of problem related to solubilities takes the form: "What will happen if solutions of AB and CD are mixed?" In solving this, all we have to do is to think of the possible results that could occur. We know the solution contains four ions; two cations (A^+ and C^+) and two anions (B^- and D^-). Their concentrations are fixed by the amounts of AB and CD which we took, and any precipitate formed will be a new compound resulting from a different combination of these ions, either A^+D^- or C^+B^-. If the ion product of one of these combinations exceeds its K_{sp}, then it will precipitate.

Example 12.8   0.050 mole of K_2CO_3 is dissolved in 1.00 liter of $1.0 \times 10^{-3}M$ strontium chloride ($SrCl_2$) solution. Does anything precipitate?

SOLUTION   If *no* precipitate forms, then we have

$$[K^+] = 2 \times 0.050 = 0.10 \qquad [CO_3^{2-}] = 0.050$$

$$[Sr^{2+}] = 1.0 \times 10^{-3} \qquad [Cl^-] = 2 \times 1.0 \times 10^{-3} = 2.0 \times 10^{-3}$$

Apart from the original substances present ($SrCl_2$ and K_2CO_3), the only combinations of ions possible are KCl and $SrCO_3$. Which of these is likely to precipitate? All those who don't know that KCl is soluble in water should be ashamed of themselves, but everybody can see that *KCl does not appear in the* K_{sp} *table* whereas $SrCO_3$ does ($K_{sp} = 1.6 \times 10^{-9}$). So if anything precipitates, it will be $SrCO_3$. In the above solution, the ion product for $SrCO_3$ would be

$$[Sr^{2+}][CO_3^{2-}] = (1.0 \times 10^{-3}) \times (0.050) = 5.0 \times 10^{-5}$$

This is greater than K_{sp}, so $SrCO_3$ *would* precipitate in the above mixture.

Example 12.9   A solution is $1.0 \times 10^{-3}M$ in both barium and strontium chlorides. Potassium chromate is added until a precipitate just starts to form. Calculate:

(a) What is the precipitate?
(b) What is the concentration of chromate ion at this point?
(c) What concentration of the other metal ion would be needed for both chromates to precipitate at the same point?
(d) Is this a reasonable composition for a solution?

($K_{sp} = 1.6 \times 10^{-10}$ for $BaCrO_4$ and 3.6×10^{-5} for $SrCrO_4$.)

SOLUTION

(a) The concentrations of Ba^{2+} and Sr^{2+} are both $1.0 \times 10^{-3}M$, so we can evaluate the necessary concentrations of chromate ion (CrO_4^{2-}) to reach the point of precipitation for each according to their K_{sp} values:

for Ba^{2+}:
$$[CrO_4^{2-}] = \frac{K_{sp}}{[Ba^{2+}]} = \frac{1.6 \times 10^{-10}}{1.0 \times 10^{-3}} = 1.6 \times 10^{-7}$$

for Sr^{2+}:
$$[CrO_4^{2-}] = \frac{K_{sp}}{[Sr^{2+}]} = \frac{3.6 \times 10^{-5}}{1.0 \times 10^{-3}} = 3.6 \times 10^{-2}$$

Obviously the first chromate to precipitate will be the one that requires the *lower* concentration of CrO_4^{2-}, that is, $BaCrO_4$.

(b) $BaCrO_4$ precipitates when $[CrO_4^{2-}] = 1.6 \times 10^{-7}M$.

(c) If the other chromate (strontium) is to precipitate at this same chromate concentration, then $[Sr^{2+}]$ must be increased until

$$[Sr^{2+}][CrO_4^{2-}] = 3.6 \times 10^{-5} \qquad [Sr^{2+}] = \frac{3.6 \times 10^{-5}}{[CrO_4^{2-}]} = \frac{3.6 \times 10^{-5}}{1.6 \times 10^{-7}} = 2.3 \times 10^{2}M$$

(d) This is obviously *not* a reasonable composition for any solution, since it would contain more strontium chloride than solvent. In other words, in practice, we could not achieve this concentration.

Example 12.10   A solution contains chloride ion, Cl^-, and chromate ion, CrO_4^{2-}, both in about $0.050M$ concentration. A solution of silver nitrate is added slowly.

(a) Which will precipitate first, $AgCl$ or Ag_2CrO_4?
(b) What will be the concentrations of Ag^+, Cl^-, and CrO_4^{2-} at the point where both $AgCl$ and Ag_2CrO_4 begin to precipitate together?

  K_{sp} for $AgCl$: 1.6×10^{-10}
  K_{sp} for Ag_2CrO_4: 2.5×10^{-12}

SOLUTION

(a) At first sight, we might think that silver chromate would precipitate first, because it has the lower value of K_{sp}, which was the case in Example 12.9. However, calculation shows that this is not true here.

For $AgCl$ to precipitate, we must have

$$[Ag^+][Cl^-] = 1.6 \times 10^{-10} \qquad [Ag^+] = \frac{1.6 \times 10^{-10}}{[Cl^-]}$$

$$[Ag^+] = \frac{1.6 \times 10^{-10}}{0.050} = 3.2 \times 10^{-9}M$$

For $AgCrO_4$ (which gives $2Ag^+$ on dissociation) we need

$$[Ag^+]^2[CrO_4^{2-}] = 2.5 \times 10^{-12} \qquad [Ag^+]^2 = \frac{2.5 \times 10^{-12}}{[CrO_4^{2-}]}$$

$$[Ag^+]^2 = \frac{2.5 \times 10^{-12}}{0.050} = 5.0 \times 10^{-11} \qquad [Ag^+] = 7.1 \times 10^{-6}M$$

Because of the different ion ratios in the two compounds, it is $AgCl$ which precipitates first, although it has the greater K_{sp} value.

(b) The Ag_2CrO_4 will not precipitate until we have added enough silver nitrate solution to increase $[Ag^+]$ to $7.1 \times 10^{-6}M$. At this point, the solution is still in equilibrium with solid $AgCl$, so we can calculate $[Cl^-]$ from the K_{sp} value for $AgCl$

$$[Cl^-] = \frac{1.6 \times 10^{-10}}{[Ag^+]} = \frac{1.6 \times 10^{-10}}{7.1 \times 10^{-6}} = 2.3 \times 10^{-5}M$$

We see that no precipitation of Ag_2CrO_4 can occur until the *chloride* ion concentration has fallen to a very low value. The amount of our original Cl^- left in solution is only

$$\frac{2.3 \times 10^{-5}}{0.050} \times 100 = 0.05\%$$

so that 99.95 % of the chloride has been removed from solution as $AgCl(s)$.

This calculation is the basis of the *Mohr titration*, a volumetric procedure used for the determination of chloride by titration with silver nitrate solution. Potassium chromate is added as an indicator, and the end point is shown by the appearance of dark red silver chromate, showing that all the chloride has been removed from solution.

As you should realize, the key to solving problems based on solubility product is to write down the expression for K_{sp} for the compound or compounds involved. Look on it as an equation relating several terms, the numerical K_{sp} value and the ion concentrations. Decide which of these are knowns and which has to be found, then solve the equation in the usual way. That's all there is to it!

Before leaving the topic of solubility product, we should note one simplifying assumption that is implicit in the above calculations. We have assumed throughout this chapter that we can calculate ion concentrations in solution directly from the amounts of various solutes that have been added, and we have ignored the possibility that these concentrations could be changed by reactions occurring between the ion and the solvent. In practice, such effects can be important, particularly for ions such as sulfide, carbonate, oxalate, etc., which are derived from weak acids. We will return to consider this problem in more detail in Chapter 14, but we must take time out first to consider acids and bases in Chapter 13.

KEY WORDS

heterogeneous equilibrium	solubility product
ion product	unsaturated solution
saturated solution	supersaturated solution
solubility by weight	common ion effect

STUDY QUESTIONS

1. What is the difference between the concept of "ion product" and that of "solubility product"?   (Yes, there *is* a difference!)

2. Under what conditions is the ion product (a) less than K_{sp}; (b) greater than K_{sp}; (c) equal to K_{sp}? Are all these stable, equilibrium situations?

3. What variable(s) influence the value of K_{sp} for a given solute in water?

4. Why do we regard the concentration of a solid ionic substance as constant in equilibrium with its solution?

5. How do we convert solubility by weight into solubility product, and vice versa?

6. What happens to the solubility of substance AB when a large excess of ion A^+ *or* B^- is supplied from another source? Is the value of K_{sp} for substance AB changed?

7. For analytical purposes, what do we do to ensure that as much as possible of a certain ion is precipitated from solution as a slightly soluble salt?

8. Give examples of several different classes of ionic salts that are (a) generally freely soluble in water and (b) generally not soluble in water.

9. Solutions of some compounds unfamiliar to you are mixed together. How would you go about finding out what, if anything, is likely to precipitate?

PROBLEMS

(Necessary K_{sp} values are given in Table 12.1.)

1. Calculate the solubility products of the following substances from the given solubilities in g liter^{-1}:
   (a) $Cd(OH)_2$, 2.6×10^{-3} \quad (b) $Ca(IO_3)_2$, 1.0 \quad (c) $PbCrO_4$, 5.8×10^{-5}

2. Using the figures in Table 12.1, calculate the solubilities of the following substances in g liter^{-1}:
   (a) $BaCO_3$ \quad (b) $Fe(OH)_2$ \quad (c) PbI_2

3. Calculate the solubilities of the following in g liter^{-1}:
   (a) $BaCrO_4$ in $0.020M$ $BaCl_2$ solution
   (b) $CaSO_4$ in $0.10M$ $Ca(NO_3)_2$ solution
   (c) PbF_2 in $0.25M$ KF solution
   (d) $Mg(OH)_2$ in $0.010M$ NaOH solution

4. Calculate the minimum concentration that the added metal ion must reach for precipitation to occur in the following solutions:
   (a) Ba^{2+} added to $0.010M$ NaF solution
   (b) Ag^+ added to $5.0 \times 10^{-3}M$ KI solution
   (c) Mn^{2+} added to $0.030M$ KOH solution
   (d) Zn^{2+} added to $8.0 \times 10^{-7}M$ Na_2S solution

5. Will precipitation occur in the following cases? If so, name the precipitate which forms.
   (a) NaF is added to $0.10M$ $Pb(NO_3)_2$ solution until $[F^-] = 5.0 \times 10^{-4}M$.
   (b) 2.0×10^{-5} mole of sodium oxalate, $Na_2C_2O_4$, is added to one liter of $0.10M$ $Pb(NO_3)_2$ solution.

(c) Strontium nitrate is dissolved in a saturated solution of $CaSO_4$ in water until $[Sr^{2+}] = 1.0 \times 10^{-2}M$.

(d) 2.5×10^{-5} mole of silver nitrate, $AgNO_3$, is dissolved in one liter of saturated PbI_2.

(e) 1.5×10^{-2} g $ZnCl_2$ and 3.4×10^{-6} g Na_2S are added to 1.0 liter of water.

(f) 0.040 g $MgCl_2$ and 0.16 g KOH are added to 500 ml of water.

6. Calculate the minimum volume of water needed to dissolve 1.0 g of the following: (a) $PbSO_4$, (b) SrF_2.

7. A solution is $1.2M$ in $MgSO_4$ and $2.0 \times 10^{-3}M$ in $Zn(NO_3)_2$. Solid KOH is added until a precipitate just starts to form. Determine:
(a) The nature of the precipitate
(b) The concentration of OH^- at the point where precipitation starts

8. A solution is $0.012M$ in $Pb(NO_3)_2$ and $0.20M$ in $Sr(NO_3)_2$. Solid Na_2SO_4 is added until a precipitate just starts to form.
(a) What is the precipitate?
(b) What is the concentration of sulfate ion at this point?

9. The solubility product of magnesium oxalate, MgC_2O_4, is 1.0×10^{-8}. A solution is known to be about $0.020M$ in Mg^{2+}, and a student attempts to precipitate the magnesium in a 1 liter sample by adding 0.020 mole of sodium oxalate, $Na_2C_2O_4$. Calculate:
(a) the concentration of Mg^{2+} remaining in solution (*Hint*: remember that the precipitate removes both Mg^{2+} *and* $C_2O_4^{2-}$ from the solution.)
(b) the percent of the original Mg^{2+} that has not precipitated
(c) the percent of the Mg^{2+} that would remain unprecipitated if the student added a further 0.020 mole $Na_2C_2O_4$

10. A solution is made by dissolving 33.1 g of lead nitrate, $Pb(NO_3)_2$, in water and making up to 1.00 liter volume. Three 100 ml portions of this solution are withdrawn and concentrated sulfuric acid, H_2SO_4, is added to each in the following amounts: (a) 5.0×10^{-3} mole, (b) 1.0×10^{-2} mole, and (c) 2.0×10^{-2} mole. Calculate for each of these mixtures $[Pb^{2+}]$, $[SO_4^{2-}]$, and the percentage of lead that has *not* precipitated. Assume that the addition of H_2SO_4 makes no significant change in volume.

11. (a) Develop a general expression relating solubility in mole liter$^{-1}(S)$ to K_{sp} for a salt of formula A_2B_3, in which the ions are A^{3+} and B^{2-}.
(b) Use your result to find the solubility in mole liter^{-1} for In_2S_3, for which $K_{sp} = 6 \times 10^{-74}$, and all ion concentrations in its saturated solution in water.

12. A swimming pool measures 50 m by 10 m by 2 m. If the water in it is saturated with mercuric sulfide, how many mercury ions will be present?

13. The concentration of bromine (as Br^-) in sea water is about 0.07 gram per liter, while the concentration of chlorine (as Cl^-) is about 19 gram per liter.
(a) What will precipitate first if silver nitrate is added to seawater?
(b) Would this represent a useful method for extracting bromine from the sea?

SOLUTIONS TO PROBLEMS

1. (a) $Cd(OH)_2$ has *M.W.* 146, so 2.6×10^{-3} g liter^{-1} is

$$\frac{2.6 \times 10^{-3}}{146} = 1.8 \times 10^{-5}M$$

$[Cd^{2+}] = 1.8 \times 10^{-5}$ and $[OH^-] = 2 \times 1.8 \times 10^{-5} = 3.6 \times 10^{-5}$
$K_{sp} = [Cd^{2+}][OH^-]^2 = (1.8 \times 10^{-5}) \times (3.6 \times 10^{-5})^2 = 2.3 \times 10^{-14}$

(b) $Ca(IO_3)_2$ has *M.W.* 390, so 1.0 g liter^{-1} is

$$\frac{1.0}{390} = 2.6 \times 10^{-3}M$$

$[Ca^{2+}] = 2.6 \times 10^{-3}$ and $[IO_3^-] = 2 \times 2.6 \times 10^{-3} = 5.2 \times 10^{-3}$
$K_{sp} = [Ca^{2+}][IO_3^-]^2 = (2.6 \times 10^{-3}) \times (5.2 \times 10^{-3})^2 = 6.7 \times 10^{-8}$

(c) $PbCrO_4$ has *M.W.* 323, so 5.8×10^{-5} g liter^{-1} is

$$\frac{5.8 \times 10^{-5}}{323} = 1.8 \times 10^{-7}M$$

$[Pb^{2+}] = [CrO_4^{2-}] = 1.8 \times 10^{-7}$
$K_{sp} = [Pb^{2+}][CrO_4^{2-}] = (1.8 \times 10^{-7})^2 = 3.2 \times 10^{-14}$

2. (a) Suppose x mole liter^{-1} of $BaCO_3$ dissolve:

$$[Ba^{2+}] = [CO_3^{2-}] = x$$

$K_{sp} = [Ba^{2+}][CO_3^{2-}] = x^2$ $K_{sp} = 8.1 \times 10^{-9} = x^2$
$x = 9.0 \times 10^{-5}$ (molarity of solution)
$BaCO_3$ has *M.W.* 197, so $9.0 \times 10^{-5}M$ is

$9.0 \times 10^{-5} \times 197 = 0.018$ g liter^{-1}

(b) Suppose x mole liter^{-1} of $Fe(OH)_2$ dissolve:

$$[Fe^{2+}] = x \qquad [OH^-] = 2x$$

$K_{sp} = [Fe^{2+}][OH^-]^2 = x(2x)^2 = 4x^3$
$K_{sp} = 1.6 \times 10^{-14} = 4x^3$

$$x^3 = \frac{1.6 \times 10^{-14}}{4} = 4.0 \times 10^{-15}$$

$x = 1.6 \times 10^{-5}$ (molarity of solution)
$Fe(OH)_2$ has *M.W.* 90, so $1.6 \times 10^{-5}M$ is

$1.6 \times 10^{-5} \times 90 = 1.4 \times 10^{-3}$ g liter^{-1}

(c) Suppose x mole liter^{-1} of PbI_2 dissolve:

$$[Pb^{2+}] = x \qquad [I^-] = 2x$$

$K_{sp} = [Pb^{2+}][I^-]^2 = x(2x)^2 = 4x^3$
$K_{sp} = 1.4 \times 10^{-8} = 4x^3$

$$x^3 = \frac{1.4 \times 10^{-8}}{4} = 3.5 \times 10^{-9}$$

$x = 1.5 \times 10^{-3}$ (molarity of solution)
PbI_2 has *M.W.* 461, so $1.5 \times 10^{-3}M$ is

$1.5 \times 10^{-3} \times 461 = 0.70$ g liter^{-1}

3. (a) $[Ba^{2+}] = 0.020$ from $BaCl_2$

$$[CrO_4^{2-}] = \frac{K_{sp}}{[Ba^{2+}]} = \frac{1.6 \times 10^{-10}}{0.020} = 8.0 \times 10^{-9}M$$

This is number of moles of $BaCrO_4$ that dissolve. $BaCrO_4$ has *M.W.* 253, so $8.0 \times 10^{-9}M$ is

$$8.0 \times 10^{-9} \times 253 = 2.0 \times 10^{-6} \text{ g liter}^{-1}$$

(b) $[Ca^{2+}] = 0.10$ from $Ca(NO_3)_2$

$$[SO_4^{2-}] = \frac{K_{sp}}{[Ca^{2+}]} = \frac{2.4 \times 10^{-4}}{0.10} = 2.4 \times 10^{-3}M$$

This is number of moles of $CaSO_4$ that dissolve. $CaSO_4$ has *M.W.* 136, so $2.4 \times 10^{-3}M$ is

$$2.4 \times 10^{-3} \times 136 = 0.33 \text{ g liter}^{-1}$$

(c) $[F^-] = 0.25$ from KF solution

$$[Pb^{2+}] = \frac{K_{sp}}{[F^-]^2} = \frac{3.2 \times 10^{-8}}{(0.25)^2} = 5.1 \times 10^{-7}M$$

(*Note:* Do *not* double $[F^-]$ here!)
This is number of moles in PbF_2 that dissolve. PbF_2 has *M.W.* 245, so $5.1 \times 10^{-7}M$ is

$$5.1 \times 10^{-7} \times 245 = 1.3 \times 10^{-4} \text{ g liter}^{-1}$$

(d) $[OH^-] = 0.010$ from NaOH

$$[Mg^{2+}] = \frac{K_{sp}}{[OH^-]^2} = \frac{1.2 \times 10^{-11}}{(0.010)^2} = 1.2 \times 10^{-7}M$$

This is number of moles of $Mg(OH)_2$ that dissolve. $Mg(OH)_2$ has *M.W.* 58, so $1.2 \times 10^{-7}M$ is

$$1.2 \times 10^{-7} \times 58 = 7.0 \times 10^{-6} \text{ g liter}^{-1}$$

4. (a) $[F^-] = 0.010$ from NaF
BaF_2 will precipitate when
$[Ba^{2+}][F^-]^2 = K_{sp} = 1.7 \times 10^{-6}$

$$[Ba^{2+}] = \frac{1.7 \times 10^{-6}}{[F^-]^2} = \frac{1.7 \times 10^{-6}}{(0.010)^2} = 1.7 \times 10^{-2}M$$

(b) $[I^-] = 5.0 \times 10^{-3}$ from KI
AgI will precipitate when
$[Ag^+][I^-] = K_{sp} = 1.5 \times 10^{-16}$

$$[Ag^+] = \frac{1.5 \times 10^{-16}}{[I^-]} = \frac{1.5 \times 10^{-16}}{5.0 \times 10^{-3}} = 3.0 \times 10^{-14}M$$

(c) $[OH^-] = 0.030$ from KOH
$Mn(OH)_2$ will precipitate when
$[Mn^{2+}][OH^-]^2 = K_{sp} = 4 \times 10^{-14}$

$$[Mn^{2+}] = \frac{4 \times 10^{-14}}{[OH^-]^2} = \frac{4 \times 10^{-14}}{(0.030)^2} = 4 \times 10^{-11}M$$

(d) $[S^{2-}] = 8.0 \times 10^{-7}$ from Na_2S
ZnS will precipitate when
$[Zn^{2+}][S^{2-}] = K_{sp} = 1.2 \times 10^{-23}$

$$[Zn^{2+}] = \frac{1.2 \times 10^{-23}}{[S^{2-}]} = \frac{1.2 \times 10^{-23}}{8.0 \times 10^{-7}} = 1.5 \times 10^{-17} M$$

5. (a) $[Pb^{2+}] = 0.10$ and $[F^-] = 5.0 \times 10^{-4}$
Ion product for PbF_2 in solution is
$[Pb^{2+}][F^-]^2 = 0.10 \times (5.0 \times 10^{-4})^2 = 2.5 \times 10^{-8}$
which is less than $K_{sp}(3.2 \times 10^{-8})$, so there is no precipitation.

(b) $[Pb^{2+}] = 0.10$ and $[C_2O_4^{2-}] = 2.0 \times 10^{-5}$ from $Na_2C_2O_4$
Ion product for PbC_2O_4 in solution is
$[Pb^{2+}][C_2O_4^{2-}] = 0.10 \times 2.0 \times 10^{-5} = 2.0 \times 10^{-6}$
which is greater than $K_{sp}(2.7 \times 10^{-11})$, so PbC_2O_4 precipitates.

(c) In saturated $CaSO_4$ solution,
$[Ca^{2+}] = [SO_4^{2-}] = \sqrt{K_{sp}} = \sqrt{2.4 \times 10^{-4}} = 1.6 \times 10^{-2}$
If $[Sr^{2+}] = 1.0 \times 10^{-2}$, the ion product for $SrSO_4$ in solution is
$[Sr^{2+}][SO_4^{2-}] = (1.0 \times 10^{-2}) \times (1.6 \times 10^{-2}) = 1.6 \times 10^{-4}$
which is greater than $K_{sp}(2.5 \times 10^{-7})$, so $SrSO_4$ precipitates.

(d) In saturated PbI_2 solution (see problem 2(c) above), the solubility is 1.5×10^{-3} mole PbI_2 per liter, so

$$[I^-] = 2 \times 1.5 \times 10^{-3} = 3.0 \times 10^{-3}$$

If $[Ag^+] = 2.5 \times 10^{-5}$, the ion product for AgI is

$$[Ag^+][I^-] = (2.5 \times 10^{-5}) \times (3.0 \times 10^{-3}) = 7.5 \times 10^{-8}$$

which is greater than $K_{sp}(1.5 \times 10^{-16})$, so AgI precipitates.

(e) 1.5×10^{-2} g $ZnCl_2$: $\dfrac{1.5 \times 10^{-2}}{136} = 1.1 \times 10^{-4}$ mole

3.4×10^{-6} g Na_2S: $\dfrac{3.4 \times 10^{-6}}{78} = 4.4 \times 10^{-8}$ mole

If these are both in 1.0 liter,

$[Zn^{2+}] = 1.1 \times 10^{-4}$ and $[S^{2-}] = 4.4 \times 10^{-8}$

The ion product for ZnS in solution is

$$[Zn^{2+}][S^{2-}] = (1.1 \times 10^{-4}) \times (4.4 \times 10^{-8}) = 4.8 \times 10^{-12}$$

which is greater than $K_{sp}(1.2 \times 10^{-23})$, so ZnS precipitates.

(f) 0.040 g $MgCl_2$: $0.040/95 = 4.2 \times 10^{-4}$ mole
0.16 g KOH: $0.16/56 = 2.9 \times 10^{-3}$ mole
If these are both in 500 ml (0.50 liter),

$$[Mg^{2+}] = \frac{4.2 \times 10^{-4}}{0.50} = 8.4 \times 10^{-4} M \qquad [OH^-] = \frac{2.9 \times 10^{-3}}{0.50} = 5.8 \times 10^{-3} M$$

The ion product for $Mg(OH)_2$ in solution is

$$[Mg^{2+}][OH^-]^2 = (8.4 \times 10^{-4}) \times (5.8 \times 10^{-3})^2 = 2.8 \times 10^{-8}$$

which is greater than $K_{sp}(1.2 \times 10^{-11})$, so $Mg(OH)_2$ precipitates.

6. (a)   In saturated $PbSO_4$ solution,

$$[Pb^{2+}] = [SO_4^{2-}] = \sqrt{K_{sp}} = \sqrt{1.06 \times 10^{-8}} = 1.03 \times 10^{-4}M$$

This is the number of moles $PbSO_4$ in each liter.
1.0 g $PbSO_4$ is $1.0/303 = 3.3 \times 10^{-3}$ mole.
Volume of $1.03 \times 10^{-4}M$ solution needed to dissolve this amount is

$$\frac{3.3 \times 10^{-3} \text{ mole}}{1.03 \times 10^{-4} \text{ mole liter}^{-1}} = 32 \text{ liter}$$

(b)   Suppose x mole liter^{-1} SrF_2 dissolve:

$$[Sr^{2+}] = x \qquad [F^-] = 2x$$

$$K_{sp} = [Sr^{2+}][F^-]^2 = x(2x)^2 = 4x^3$$

$$4x^3 = 2.8 \times 10^{-9} \qquad x^3 = \frac{2.8 \times 10^{-9}}{4} = 7.0 \times 10^{-10}$$

$x = 8.9 \times 10^{-4}$
This is number of moles SrF_2 in each liter.
1.0 g SrF_2 is $1.0/126 = 7.9 \times 10^{-3}$ mole.
Volume of $8.9 \times 10^{-4}M$ solution needed to dissolve this amount is

$$\frac{7.9 \times 10^{-3}}{8.9 \times 10^{-4}} = 8.9 \text{ liter}$$

7. (a)   The substances likely to precipitate are the hydroxides $Mg(OH)_2$ and $Zn(OH)_2$ (*not* K_2SO_4 or KNO_3).

(b)   For $Mg(OH)_2$, given $[Mg^{2+}] = 1.2M$:

$$[OH^-]^2 = \frac{K_{sp}}{[Mg^{2+}]} = \frac{1.2 \times 10^{-11}}{1.2} = 1.0 \times 10^{-11}$$

required for precipitation: $[OH^-] = 3.2 \times 10^{-6}$
For $Zn(OH)_2$, given $[Zn^{2+}] = 2.0 \times 10^{-3}M$:

$$[OH^-]^2 = \frac{K_{sp}}{[Zn^{2+}]} = \frac{1.2 \times 10^{-17}}{2.0 \times 10^{-3}} = 6.0 \times 10^{-15}$$

required for precipitation: $[OH^-] = 7.8 \times 10^{-8}$
Since $Zn(OH)_2$ requires a *lower* value of $[OH^-]$ to cause precipitation, it is this compound which precipitates first as KOH is added, and the precipitate appears when $[OH^-] = 7.8 \times 10^{-8}M$.

8. (a)   This is similar to the preceding problem. The possible precipitates are $PbSO_4$ and $SrSO_4$, each of which will precipitate when the value of $[SO_4^{2-}]$ is high enough.

(b)   For $PbSO_4$, given $[Pb^{2+}] = 0.012M$:

$$[SO_4^{2-}] = \frac{K_{sp}}{[Pb^{2+}]} = \frac{1.06 \times 10^{-8}}{0.012} = 8.8 \times 10^{-7}$$

For $SrSO_4$, given $[Sr^{2+}] = 0.20M$:

$$[SO_4^{2-}] = \frac{K_{sp}}{[Sr^{2+}]} = \frac{2.5 \times 10^{-7}}{0.20} = 1.3 \times 10^{-6}$$

$PbSO_4$ precipitates first, when $[SO_4^{2-}] = 8.8 \times 10^{-7}M$.

9. (a) Mg^{2+} and $C_2O_4^{2-}$ ions are present in equal amounts, so, after precipitation of solid MgC_2O_4, we have in solution

$$[Mg^{2+}] = [C_2O_4^{2-}] = \sqrt{K_{sp}} = \sqrt{1.0 \times 10^{-8}} = 1.0 \times 10^{-4}$$

   (b)   Unprecipitated Mg^{2+} is $1.0 \times 10^{-4} M$ out of an original concentration of $0.020 M$.

   unprecipitated: $\dfrac{1.0 \times 10^{-4}}{0.020} \times 100 = 0.50\%$

   (c)   The additional $C_2O_4^{2-}$ remains in solution:

$$[Mg^{2+}] = \frac{K_{sp}}{[C_2O_4^{2-}]} = \frac{1.0 \times 10^{-8}}{0.020} = 5.0 \times 10^{-7} M$$

   unprecipitated: $\dfrac{5.0 \times 10^{-7}}{0.020} \times 100 = 2.5 \times 10^{-3}\%$

10.   The initial solution of $Pb(NO_3)_2$ (*M.W.* 331) is

$$\frac{33.1 \text{ g liter}^{-1}}{331 \text{ g mole}^{-1}} = 0.100 M$$

Each 100 ml portion contains 1.0×10^{-2} mole Pb^{2+}, which will react in a 1:1 molar ratio with SO_4^{2-} to precipitate $PbSO_4$.

   (a)   We have 5.0×10^{-3} mole SO_4^{2-}, so SO_4^{2-} is limiting.
   5.0×10^{-3} mole $PbSO_4$ precipitates, and leaves
   $(1.0 \times 10^{-2}) - (5.0 \times 10^{-3}) = 5.0 \times 10^{-3}$ mole Pb^{2+}
   This is in 100 ml solution, so $[Pb^{2+}] = 5.0 \times 10^{-2} M$

$$[SO_4^{2-}] = \frac{K_{sp}}{[Pb^{2+}]} = \frac{1.06 \times 10^{-8}}{5.0 \times 10^{-2}} = 2.1 \times 10^{-7} M$$

   percent Pb^{2+} unprecipitated: $\dfrac{5.0 \times 10^{-2}}{0.100} \times 100 = 50\%$

   (b)   We have 1.0×10^{-2} mole of each, the stoichiometric ratio, so after precipitation

$$[Pb^{2+}] = [SO_4^{2-}] = \sqrt{K_{sp}} = \sqrt{1.06 \times 10^{-8}} = 1.0 \times 10^{-4} M$$

   percent Pb^{2+} unprecipitated: $\dfrac{1.0 \times 10^{-4}}{0.100} \times 100 = 0.10\%$

   (c)   Now the amount of SO_4^{2-} is greater than the amount of Pb^{2+}, so Pb^{2+} is limiting.
   SO_4^{2-} remaining after precipitation: $(2.0 \times 10^{-2}) - (1.0 \times 10^{-2}) = 1.0 \times 10^{-2}$ mole SO_4^{2-}
   This is in 100 ml, so $[SO_4^{2-}] = 0.10 M$

$$[Pb^{2+}] = \frac{K_{sp}}{[SO_4^{2-}]} = \frac{1.06 \times 10^{-8}}{0.10} = 1.1 \times 10^{-7} M$$

   percent Pb^{2+} unprecipitated: $\dfrac{1.1 \times 10^{-7}}{0.100} \times 100 = 1.1 \times 10^{-4}\%$

11.   (a)   If the solubility of A_2B_3 is S mole liter^{-1},
   $[A^{3+}] = 2S$ and $[B^{2-}] = 3S$ and $K_{sp} = [A^{3+}]^2[B^{2-}]^3 = (2S)^2(3S)^3 = 108S^5$

   (b)   For In_2S_3: $S^5 = \dfrac{6 \times 10^{-74}}{108} = 5.6 \times 10^{-76}$ $S = 9 \times 10^{-16}$ mole liter^{-1}

   $[In^{3+}] = 1.8 \times 10^{-15} M$ and $[S^{2-}] = 2.7 \times 10^{-15} M$

12. $S = \sqrt{4 \times 10^{-53}} = 6.3 \times 10^{-27} M = [Hg^{2+}]$
     volume: $50 \times 10 \times 2 = 10^3 \text{ m}^3 = 10^6 \text{ liter}$
     total Hg^{2+} ions: $6.3 \times 10^{-27} \times 10^6 = 6.3 \times 10^{-21} \text{ mole}$
     $= 6.3 \times 10^{-21} \times 6.02 \times 10^{23} \text{ ions} = 3.8 \times 10^3 \text{ ions}$

13. (a)   Concentrations of the ions:

$$[Br^-] = \frac{0.07}{80} = 9 \times 10^{-4} M \qquad [Cl^-] = \frac{19}{35.5} = 0.54 M$$

AgBr precipitates when $[Ag^+] = \dfrac{K_{sp}}{[Br^-]} = \dfrac{7.7 \times 10^{-13}}{9 \times 10^{-4}} = 9 \times 10^{-10} M$

AgCl precipitates when $[Ag^+] = \dfrac{K_{sp}}{[Cl^-]} = \dfrac{1.6 \times 10^{-10}}{0.54} = 3.0 \times 10^{-10} M$

Although K_{sp} is much lower for AgBr, the higher concentration of Cl^- means that AgCl precipitates first, so this would not be an efficient way of extracting bromine from sea water. (In practice, bromine is obtained from sea water by other routes.)

Chapter 13
Equilibrium in Solution: Acids and Bases

He dipped the litmus-paper into the test-tube and it flushed at once into a dull, dirty crimson. "Hum! I thought as much!" he cried. THE NAVAL TREATY

In the previous chapter, we considered the equilibrium between a solid ionic compound and the separate ions in solution, being careful to exclude the possibility of finding undissociated molecules in solution. We now must consider the case of equilibria in which all species are in solution, whether as undissociated covalent molecules or as ions. Some of the most important compounds and equilibria in solution are those affecting the degree of acidity or basicity, because virtually every type of chemical reaction, whether occurring in the laboratory or in a living organism, will be very much influenced by the acidity of the solution in which it takes place. This and the following chapter are therefore devoted to equilibria involving acids, bases, and other substances that may affect acidity.

We shall be referring frequently to the ions present in solution; and, using water as a solvent, these will be the hydrogen ion and the hydroxide ion. In the simplest case, these may be regarded as resulting from the ionization of water:

$$H_2O \rightleftharpoons H^+ + OH^-$$

But a more correct picture is to regard the reaction as a transfer of a proton from one water molecule to another:

$$H_2O + H_2O \rightleftharpoons H_3O^+ + OH^-$$

The ion H_3O^+ is called the *hydronium ion*, and is often used to indicate the species present in an acid solution. However, the actual structure of an aqueous solution is very complicated and not fully understood. It appears that, in addition to H_3O^+, there are more complex ions present in an acid solution, such as $H_9O_4^+$ (that is, $H^+ + 4H_2O$). We will therefore adopt a simple and widely used convention and use the symbol H^+ to represent *all* species in solution formed by the reaction of a proton with liquid water and refer collectively to them as "hydrogen ion concentration." It should be clearly understood that this does *not* imply that an acid solution contains bare protons that have not interacted with the solvent.

Because of the small size of the proton, it has a very high attraction for a nonbonding electron pair on a molecule of water (or any other suitable solvent) and the equilibrium

$$H^+ + H_2O \rightleftharpoons H_3O^+$$

is very much in favor of H_3O^+.

Strictly speaking, we should denote the fact that we are dealing with ions in aqueous solution by putting (*aq*) after each symbol, but we often omit this in the interests of simplicity. It would be important, of course, to include such terms if we were concerned with equilibria in different phases, e.g., between gaseous Cl^- and Cl^-(*aq*) in solution. You should realize that there is always some interaction between a dissolved ion and the solvent molecules. If no such interaction were present, the substance would not be able to go into solution.

13.1   Acids and Bases

The terms "acid" and "base" are very difficult to define rigorously, and many different suggestions have been advanced. The names of Arrhenius, Brønsted, Lowry, and Lewis, among others, will always be associated with this problem. For our purposes, however, we may adopt simple functional definitions of acids and bases in terms of their effect on the solution:

An *acid* is a substance which, when put into aqueous solution, *increases* the hydrogen ion concentration.

A *base* is a substance which, when put into aqueous solution, *decreases* the hydrogen ion concentration. (We might use the hydroxide ion concentration when referring to a base, but, since the concentrations of the two ions are inseparable, this would be effectively the same definition.)

Our definition of an acid will include the obvious case of hydrogen-containing covalent molecules such as HCl, HNO_3, etc., which dissociate into ions in solution. We must also include compounds such as CO_2 and SO_2 which, although not themselves containing hydrogen, can react with water to increase $[H^+]$. Such compounds are often called "acid anhydrides." For example,

$$SO_2 + H_2O \rightleftharpoons H_2SO_3 \rightleftharpoons H^+ + HSO_3^- \rightleftharpoons 2H^+ + SO_3^{2-}$$

We shall also see in Chapter 14 that an ion such as NH_4^+ can act as an acid through an hydrolysis reaction.

Similarly, bases include ionic metal hydroxides (for example, NaOH, KOH), neutral molecules such as NH_3, and ions such as CH_3COO^- (See chapter 14).

13.2   Strong and Weak Acids: Dissociation Constants

The typical acid, which we may call HA, is, in the pure state, a covalent molecule. On dissolving this covalent molecule in water, a proton is transferred from the acid molecule to a water molecule, giving an acidic solution:

$$HA + H_2O \rightleftharpoons H_3O^+ + A^-$$

TABLE 13.1   Acid Ionization Constants (near 25°C)

Acid	K_a	Acid	K_a
acetic, CH_3COOH	1.8×10^{-5}	hydrocyanic, HCN	4.9×10^{-10}
benzoic, C_6H_5COOH	6.4×10^{-5}	hydrofluoric, HF	6.7×10^{-4}
chloracetic, $CH_2ClCOOH$	1.4×10^{-3}	iodic, HIO_3	0.17
formic, HCOOH	1.8×10^{-4}	nitrous, HNO_2	4.5×10^{-4}
hydrazoic, HN_3	1.9×10^{-5}	propionic, C_2H_5COOH	1.3×10^{-5}

(*Note*: The donation of a proton *from* the acid molecule *to* a water molecule leaves a negative ion A^-, which we call an *anion*.)

In writing an equilibrium constant for the above reaction we may leave out $[H_2O]$, which is virtually constant. Making the further simplification of writing $[H^+]$ for $[H_3O^+]$ gives

$$K_a = \frac{[H^+][A^-]}{[HA]}$$

K_a is an important constant called the *ionization constant* of the acid HA. Clearly, it represents an equilibrium constant for the simplified acid dissociation equation or ionization equation:

$$HA \rightleftharpoons H^+ + A^-$$

(*Note*: We use the terms "dissociation" and "ionization" interchangeably. Strictly, *ionization* is a special kind of dissociation.)

The value of K_a enables us to give a more precise meaning to the terms "strong" and "weak" acid. A *strong acid* is one which is highly dissociated in solution; it will have a relatively *large* ionization constant. A *weak acid* will have a smaller tendency to dissociate, and will therefore have a *smaller* value of K_a.

For the ordinary strong acids used in the laboratory, such as HCl, HNO_3 etc., the value of K_a cannot be measured accurately. It is usual to assume that, in all but the most concentrated solutions, these acids are completely dissociated. This implies an infinitely large value for K_a, which is unreasonable; but it can be shown that, if K_a has a value around 10 or greater, the error introduced by assuming 100% dissociation will be negligible.

The great majority of acids, however, are classified as "weak," and are not fully dissociated in solution. Their K_a values range from fairly large (for example, HIO_3, $K_a = 1.7 \times 10^{-1}$) through small (for example, acetic acid, $K_a = 1.8 \times 10^{-5}$) to very small (for example, HCN, $K_a = 4.9 \times 10^{-10}$). Like any other equilibrium constant, these values will change slightly with temperature, but we will assume that all solutions considered are at room temperature and ignore this possibility.

Dissociation constants of various acids in water have been measured and tabulated, and we may use these values to calculate the degree of dissociation and the resulting hydrogen ion concentration in solutions of acids. (Some typical K_a values are given in Table 13.1.)

Example 13.1   The ionization constant of acetic acid, CH_3COOH, is 1.8×10^{-5}. Calculate the hydrogen ion concentration and percentage dissociation at concentrations (a) $1.0M$, (b) $1.0 \times 10^{-2}M$, and (c) $1.0 \times 10^{-3}M$.

SOLUTION

(a) Suppose x mole liter^{-1} dissociate. This will produce x mole liter^{-1} of CH_3COO^- and H^+ and leave $(1.0 - x)$ mole liter^{-1} of CH_3COOH.

$$CH_3COOH \rightleftharpoons CH_3COO^- + H^+$$

$$(1.0 - x) \qquad\qquad x \qquad\qquad x \text{ mole liter}^{-1}$$

$$K_a = 1.8 \times 10^{-5} = \frac{[CH_3COO^-][H^+]}{[CH_3COOH]} = \frac{x^2}{(1.0 - x)}$$

This gives us the quadratic equation $x^2/(1.0 - x) = 1.8 \times 10^{-5}$. This equation could be solved as a quadratic by the usual method, but we can greatly simplify the arithmetic by noting that, as we are dealing with a weak acid, the amount of dissociation will be only a small fraction of the total acid present. In other words x will be small compared with 1.0, and we may make the approximation that $(1.0 - x) \approx 1.0$. Simplifying our quadratic equation gives

$$\frac{x^2}{(1.0 - x)} \approx \frac{x^2}{1.0} = 1.8 \times 10^{-5}$$

$$x^2 = 1.8 \times 10^{-5} = 18 \times 10^{-6}$$

$$x = \sqrt{18} \times 10^{-3} = 4.2 \times 10^{-3}$$

Was our simplifying assumption justified? x is clearly small beside 1.0, amounting to only 0.42%. As the original data were only given to two figures, an error of 0.42% is clearly negligible, and our approximation was justified.

Referring back to the original question, we see that x is the concentration of hydrogen ion, so $[H^+] = 4.2 \times 10^{-3} M$.

The percentage dissociation of the acid is defined as the ratio of the amount of acid dissociated to the total amount of acid *originally present* before dissociation. It is *not* the ratio of the concentrations of dissociated to undissociated acid at equilibrium. Although the numerical difference between these concepts is, in this case, negligible, it is important to understand the meaning of the term. In this case,

$$\text{percentage dissociation} = \frac{\text{amount of acid dissociated}}{\text{amount of acid originally present}} \times 100$$

$$= \frac{4.2 \times 10^{-3}}{1.0} \times 100 = 0.42\%$$

(b) In $1.0 \times 10^{-2} M$ CH_3COOH, again let x mole liter^{-1} dissociate, then $(1.0 \times 10^{-2}) - x$ mole liter^{-1} remain undissociated. Our equilibrium expression will be

$$\frac{x^2}{(1.0 \times 10^{-2}) - x} = K_a = 1.8 \times 10^{-5}$$

May we use the same simplification as before to solve this quadratic equation? The only way to find out is to try it. Assuming that x is negligible beside 1.0×10^{-2}, then $(1.0 \times 10^{-2}) - x \approx (1.0 \times 10^{-2})$ and

$$\frac{x^2}{(1.0 \times 10^{-2}) - x} \approx \frac{x^2}{1.0 \times 10^{-2}} = 1.8 \times 10^{-5}$$

$$x^2 = 1.8 \times 10^{-7}$$

$$= 18 \times 10^{-8}$$

$$x = \sqrt{18} \times 10^{-4}$$

$$= 4.2 \times 10^{-4} = [H^+]$$

Is the assumption still justified? 4.2×10^{-4} is 4.2% of 1.0×10^{-2}, so our dismissal of x as negligible beside 1.0×10^{-2} introduced about a 2% error in x (the error is reduced when the square root is taken). A better result may be obtained by feeding this answer back into the denominator of the original equation, that is, substituting 4.2×10^{-4} for x in $(1.0 \times 10^{-2}) - x$. Our equation now becomes

$$\frac{x^2}{(1.0 \times 10^{-2}) - (4.2 \times 10^{-4})} = 1.8 \times 10^{-5}$$

or

$$\frac{x^2}{0.96 \times 10^{-2}} = 1.8 \times 10^{-5}$$

from which

$$x = 4.16 \times 10^{-4}$$

Mathematically speaking, this is a more accurate answer than that obtained previously. However, within the accuracy of the figures given, it is not significantly different, and the previous value of 4.2×10^{-4} was quite good enough.

Students often wonder whether or not they should use approximations in solving quadratic equations of this nature. Rather than attempt to judge the meaning of the term "negligible" in each case, we may establish an arbitrary limit of 5%, above which more accurate methods should be used. The procedure then becomes:

1. Set up the equilibrium expression.
2. If dealing with a weak acid, *assume* that the amount of dissociation makes a negligible difference to the amount of acid originally present.
3. Using the simplifying assumption, solve the equation.
4. If the amount of dissociation turns out to be less than 5%, the answer will generally be acceptable.
5. If the dissociation is more than 5%, a more accurate calculation should be performed, either by successive approximations, or by solving the original quadratic rigorously.

(c) The above considerations may be applied to the more dilute solution of CH_3COOH, $1.0 \times 10^{-3} M$. Again, let x mole liter^{-1} dissociate, then,

$$CH_3COOH \rightleftharpoons CH_3COO^- + H^+$$

$$(1.0 \times 10^{-3}) - x \qquad\qquad x \qquad\quad x$$

$$\frac{x^2}{(1.0 \times 10^{-3}) - x} = K_a = 1.8 \times 10^{-5}$$

Assume x is negligible beside 1.0×10^{-3}:

$$\frac{x^2}{(1.0 \times 10^{-3}) - x} \approx \frac{x^2}{1.0 \times 10^{-3}} = 1.8 \times 10^{-5}$$

$$x^2 = 1.8 \times 10^{-8}$$

$$x = 1.34 \times 10^{-4}$$

This gives a percentage dissociation of

$$\frac{1.34 \times 10^{-4}}{1.0 \times 10^{-3}} \times 100 = 13.4\%$$

In this case, our assumption was *not* justified, and neglect of x beside 1.0×10^{-3} has introduced appreciable error. Feeding back this rough value for a better answer gives

$$\frac{x^2}{(1.0 \times 10^{-3}) - (1.34 \times 10^{-4})} = 1.8 \times 10^{-5}$$

$$x^2 = (1.8 \times 10^{-5}) \times (8.66 \times 10^{-4}) = 1.6 \times 10^{-8}$$

from which

$$x = [H^+] = 1.26 \times 10^{-4} M$$

(This is appreciably different from the rough value obtained in the previous calculation.)

dissociation: $\dfrac{1.26 \times 10^{-4}}{1.0 \times 10^{-2}} \times 100 = 12.6\%$

Table 13.2 shows the results of the above calculations, together with other values obtained at different initial concentrations of acetic acid. As the concentration of acid

TABLE 13.2   The Dissociation of Acetic Acid in Water

Original value of [CH_3COOH] M	[CH_3COOH] left at equilibrium M	[H^+] = [CH_3COO^-] at equilibrium M	Dissociation of acid $\%$
1.0	1.0	4.2×10^{-3}	0.42
0.10	0.099	1.3×10^{-3}	1.3
1.0×10^{-2}	0.96×10^{-2}	4.2×10^{-4}	4.2
1.0×10^{-3}	0.87×10^{-3}	1.3×10^{-4}	13
1.0×10^{-4}	6.6×10^{-5}	3.4×10^{-5}	34

decreases, the hydrogen ion concentration *decreases*, but the degree of dissociation *increases*. With a weaker acid, the degree of dissociation would be even less, as the following calculation on hydrocyanic acid (HCN) shows.

Example 13.2   Calculate the hydrogen ion concentration and percentage dissociation in a $2.0 \times 10^{-2} M$ solution of HCN, for which $K_a = 4.9 \times 10^{-10}$.

SOLUTION   Let x mole liter^{-1} dissociate:

$$HCN \rightleftharpoons CN^- + H^+$$
$$(2.0 \times 10^{-2}) - x \qquad x \qquad x \text{ mole liter}^{-1}$$

With such a weak acid, the degree of dissociation will be very small and $(2.0 \times 10^{-2}) - x \approx 2.0 \times 10^{-2}$.

The equilibrium expression will be

$$K_a = \frac{[CN^-][H^+]}{[HCN]} = \frac{x^2}{(2.0 \times 10^{-2}) - x} \approx \frac{x^2}{2.0 \times 10^{-2}} = 4.9 \times 10^{-10}$$

$$x^2 = 9.8 \times 10^{-12}$$

$$x = 3.1 \times 10^{-6}$$

It is clear that our simplifying assumption was justified. Hence $[H^+] = x = 3.1 \times 10^{-6} M$.

Percentage dissociation: $\dfrac{3.1 \times 10^{-6}}{2.0 \times 10^{-2}} \times 100 = 1.6 \times 10^{-2} \%$

With a stronger acid, where the degree of dissociation is greater, the concentration of the undissociated acid will be significantly reduced by dissociation and our assumption will not be valid.

Example 13.3   Calculate the hydrogen ion concentration and degree of dissociation in a $1.5 \times 10^{-2} M$ solution of chloroacetic acid, $CH_2ClCOOH$, for which $K_a = 1.4 \times 10^{-3}$.

SOLUTION   Let x mole liter^{-1} dissociate.

$$CH_2ClCOOH \rightleftharpoons CH_2ClCOO^- + H^+$$
$$(1.5 \times 10^{-2}) - x \qquad x \qquad x$$

The equilibrium expression will be

$$K_a = \frac{[CH_2ClCOO^-][H^+]}{[CH_2ClCOOH]} = \frac{x^2}{(1.5 \times 10^{-2}) - x} = 1.4 \times 10^{-3}$$

Suppose we do assume that x is much less than 1.5×10^{-2}. This will give

$$\frac{x^2}{(1.5 \times 10^{-2}) - x} \approx \frac{x^2}{1.5 \times 10^{-2}} = 1.4 \times 10^{-3}$$

$$x^2 = 2.1 \times 10^{-5} = 21 \times 10^{-6}$$

$$x = 4.6 \times 10^{-3}$$

This value of x is over 30% of our original acid concentration, and our assumption was clearly not justified. Without the simplification, the equilibrium equation is

$$\frac{x^2}{(1.5 \times 10^{-2}) - x} = 1.4 \times 10^{-3}$$

$$x^2 + (1.4 \times 10^{-3})x - (2.1 \times 10^{-5}) = 0$$

from which

$$x = [H^+] = 3.9 \times 10^{-3}$$

dissociation: $\dfrac{3.9 \times 10^{-3}}{1.5 \times 10^{-2}} \times 100 = 26\%$

Problems of this type may of course be approached from the opposite direction, where K_a is the unknown quantity.

Example 13.4   A certain acid HA is 1.0% dissociated in 0.10M solution. What is its ionization constant?

SOLUTION   If 1.0% of the acid is dissociated, the concentrations of A^- and H^+ will be 1.0% of 0.10M, i.e. $1.0 \times 10^{-3}M$. The equilibrium will be

$$\text{HA} \; \rightleftharpoons \; \text{A}^- \; + \; \text{H}^+$$

$(0.10 - 10^{-3})$ 1.0×10^{-3} 1.0×10^{-3} mole liter^{-1}

$$K_a = \frac{[A^-][H^+]}{[HA]} = \frac{(1.0 \times 10^{-3})^2}{(0.10 - 10^{-3})} = \frac{1.0 \times 10^{-6}}{0.099} = 1.01 \times 10^{-5}$$

Example 13.5   A 1.0M solution of benzoic acid has a hydrogen ion concentration of $8.0 \times 10^{-3}M$. Calculate

(a) the ionization constant
(b) the concentration at which the degree of dissociation will have increased to 10%

SOLUTION

(a) The degree of ionization is so small that the concentration of undissociated acid in the solution will still be 1.0M. The amount of benzoate ion produced on ionization must of course equal the amount of hydrogen ion, so the equilibrium will be (representing benzoic acid as HBz)

$$\text{HBz} \; \rightleftharpoons \; \text{Bz}^- \; + \; \text{H}^+$$

1.0 8.0×10^{-3} 8.0×10^{-3} mole liter^{-1}

$$K_a = \frac{[Bz^-][H^+]}{[HBz]} = \frac{(8.0 \times 10^{-3})^2}{1.0} = 6.4 \times 10^{-5}$$

(b) Suppose that an original concentration of y mole liter^{-1} of undissociated acid results in 10 % dissociation. Then at equilibrium $0.9y$ mole of undissociated acid remain, while $[H^+] = [Bz^-] = 0.1y$ mole liter^{-1}.

$$K_a = \frac{[Bz^-][H^+]}{[HBz]} = \frac{(0.1y)^2}{0.9y} = 6.4 \times 10^{-5}$$

$$\frac{(1.0 \times 10^{-2})y}{0.9} = 6.4 \times 10^{-5}$$

$$y = \frac{0.9 \times 6.4 \times 10^{-5}}{1.0 \times 10^{-2}} = 5.8 \times 10^{-3} M$$

13.3 Solutions of Bases

We have defined a base as a substance which decreases the hydrogen ion concentration in solution and increases the hydroxide ion concentration. (The connection between these two concentrations is discussed in section 13.4.) Bases are of three general types: (1) ionic hydroxides of metals, (2) neutral molecules that react with water to give OH^-, and (3) ions that react with water to give OH^-. We will discuss classes (1) and (2) here, and class (3) in the first section of Chapter 14.

1. The first group, the *ionic metal hydroxides*, are compounds that contain hydroxide ions in the solid state. On dissolving, all the hydroxide ions go into solution (as with any other ionic compound, we neglect the possibility of undissociated "molecules" in solution), so the calculation of the concentration of OH^- is simple.

There are a great many metal hydroxides, but the majority of them have only a very limited solubility in water. The only freely soluble hydroxides are the familiar Group I alkali metal compounds LiOH, NaOH, KOH, RbOH, and CsOH; and barium hydroxide, $Ba(OH)_2$. Other group II metal hydroxides are sufficiently soluble to make useful basic solutions such as "lime water" (saturated $Ca(OH)_2$: a liter of water dissolves 0.025 mole at 0°C and the solubility *decreases* on warming.)

Example 13.6 Calculate $[OH^-]$ in the following solutions:

(a) 1.48 g NaOH in 100 ml solution
(b) 2.23 g $Ba(OH)_2 \cdot 8H_2O$ in 50.0 ml solution

SOLUTION

(a) 1.48 g in 100 ml is 14.8 g liter^{-1} and 14.8 g NaOH is $14.8/40.0 = 0.370$ mole.

$$NaOH \longrightarrow Na^+ + OH^-$$

Each mole of NaOH gives one mole OH^- in solution, so 0.370 mole NaOH gives $[OH^-] = 0.370M$.

(b) 2.23 g in 50.0 ml is 44.6 g liter^{-1} and 44.6 g $Ba(OH)_2 \cdot 8H_2O$ is $44.6/315.5 = 0.141$ mole. (*Note*: Barium hydroxide crystallizes as a hydrate.)

$$Ba(OH)_2 \longrightarrow Ba^{2+} + 2OH^-$$

Since each mole $Ba(OH)_2$ gives 2 mole OH^- in solution, 0.141 mole $Ba(OH)_2$ gives

$$[OH^-] = 2 \times 0.141 = 0.282M$$

Obviously, the above calculations are the same as those for any other ionic salt going into solution. Because these hydroxides are completely ionized, they are often referred to as "strong" bases.

2. In the second class of bases, *neutral molecules reacting with water*, one compound completely dominates the group; the ammonia molecule, NH_3. Other molecules that react in a similar way are almost always related compounds such as methylamine, CH_3NH_2; triethylamine, $(C_2H_5)_3N$; etc.

Ammonia is a pungent-smelling gas (bp $-33°C$), which is extremely soluble in water (one liter of water dissolves about 118 liters of NH_3 at STP). Such a high solubility indicates a strong interaction between NH_3 and H_2O molecules, which undoubtedly occurs through hydrogen bonding. However, a *small* amount of dissolved NH_3 reacts with water by transfer of a proton to give a basic solution:

$$NH_3(aq) + H_2O \rightleftharpoons NH_4^+ + OH^-$$

It should be realized that *the great majority of dissolved NH_3 is still in the form of the neutral molecule*, so we refer to the solution as "aqueous ammonia," $NH_3(aq)$. However, common practice is to describe this solution as "ammonium hydroxide," and this label is often seen on stock bottles in the laboratory. We will keep to the name "aqueous ammonia" in this book, but you should bear in mind the meaning of the alternative name when you meet it.

The above ionization has an equilibrium constant with the symbol K_b (b for base). Neglecting $[H_2O]$, we write

$$K_b = \frac{[NH_4^+][OH^-]}{[NH_3(aq)]} = 1.8 \times 10^{-5} \qquad \text{(at } 25°C\text{)}$$

(*Note*: Fortuitously, the numerical value is very similar to K_a for acetic acid.)

The small value of K_b tells us at once that the equilibrium will be to the left, and only a small amount of $NH_3(aq)$ will react with water in solutions of reasonable concentration. Ammonia is therefore referred to as a "weak" base.

Knowing K_b, we may calculate the position of the equilibrium.

Example 13.7   Calculate the concentration of hydroxide ion and the percentage ionization in solutions of $NH_3(aq)(K_b = 1.8 \times 10^{-5})$:

(a) $1.0M$
(b) $1.0 \times 10^{-2}M$

SOLUTION

(a) Let x be the number of mole liter^{-1} ionized. Then $(1.0 - x)$ mole liter^{-1} remain and the equilibrium is

$$NH_3(aq) + H_2O \rightleftharpoons NH_4^+ + OH^-$$
$$(1.0 - x) \qquad\qquad\qquad x \qquad x \quad \text{mole liter}^{-1}$$

$$K_b = \frac{[NH_4^+][OH^-]}{[NH_3(aq)]} = \frac{x^2}{(1.0 - x)} = 1.8 \times 10^{-5}$$

As with the weak acid dissociation, we may simplify the solving of this equation by assuming that x is small, i.e., that $(1.0 - x) \approx 1.0$, then

$$\frac{x^2}{1.0 - x} \approx \frac{x^2}{1.0} = 1.8 \times 10^{-5}$$

$$x^2 = 1.8 \times 10^{-5} = 18 \times 10^{-6}$$

$$x = 4.2 \times 10^{-3}$$

Clearly, our assumption was justified, since x is, within the accuracy of the given data, negligible beside 1.0. In this problem, which is otherwise similar to the weak acid case, x is the *hydroxide* ion concentration, that is,

$$[OH^-] = 4.2 \times 10^{-3} M$$

The degree of ionization is

$$\frac{4.2 \times 10^{-3}}{1.0} \times 100 = 0.42\%$$

(b) In the $1.0 \times 10^{-2} M$ solution, suppose x mole liter^{-1} have ionized. The equilibrium is

$$NH_3(aq) + H_2O \rightleftharpoons NH_4^+ + OH^-$$
$$(1.0 \times 10^{-2}) - x \qquad\qquad x \qquad x$$

$$K_b = \frac{x^2}{(1.0 \times 10^{-2}) - x} = 1.8 \times 10^{-5}$$

As before, we assume x is very small, then,

$$\frac{x^2}{(1.0 \times 10^{-2}) - x} \approx \frac{x^2}{(1.0 \times 10^{-2})} = 1.8 \times 10^{-5}$$

$$x^2 = 1.8 \times 10^{-7} = 18 \times 10^{-8}$$

$$x = 4.2 \times 10^{-4}$$

Checking to see that our assumption was justified, we see that x comes out to be 4.2% of our original base concentration, which is within our limit of 5% as a justifiable criterion of "negligible." Hence

$$[OH^-] = 4.2 \times 10^{-4} M$$

degree of ionization: $\dfrac{4.2 \times 10^{-4}}{1.0 \times 10^{-2}} \times 100 = 4.2\%$

As with a weak acid, a reduction in concentration brings a decrease in ion concentrations (OH^- in this case) but an increase in degree of ionization.

Values of K_b for a number of weak, nitrogen-containing, bases are given in Table 13.3.

TABLE 13.3   Base Ionization Constants (near 25°C)

Base*	K_b	Base*	K_b
ammonia, NH_3	1.8×10^{-5}	hydrazine, N_2H_4	1.3×10^{-6}
aniline, $C_6H_5NH_2$	4.0×10^{-10}	pyridine, C_5H_5N	1.7×10^{-9}
ethylamine, $C_2H_5NH_2$	4.3×10^{-4}	triethylamine, $(C_2H_5)_3N$	5.3×10^{-4}

* All these bases are nitrogen compounds. They react with water by transfer of a proton from water to a nitrogen atom, for example,

$$C_5H_5N(aq) + H_2O \rightleftharpoons C_5H_5NH^+ + OH^-$$

$$K_b = \frac{[C_5H_5NH^+][OH^-]}{[C_5H_5N(aq)]}$$

13.4   The Dissociation of Water: pH and pOH

We considered above the concentrations of hydrogen or hydroxide ion in solutions of acids or bases respectively. However, the concentrations of these ions cannot be independent, since they are related by the equilibrium

$$H_2O + H_2O \rightleftharpoons H_3O^+ + OH^-$$

which may be more simply represented as

$$H_2O \rightleftharpoons H^+ + OH^-$$

The equilibrium constant for the latter will be

$$K_{eq} = \frac{[H^+][OH^-]}{[H_2O]}$$

and ignoring $[H_2O]$, which is constant, gives us the important constant K_w known as the *ion product* for water:

$$K_w = [H^+][OH^-] = 1.0 \times 10^{-14} \qquad \text{(at 25°C)}$$

It is most important to realize that this equilibrium is *always* present in an aqueous solution. Whether the solution is acidic, neutral, or basic, the concentrations of H^+ and OH^- are always linked by the above equation and we can never alter one without altering the other.

Rearranging the equation for K_w gives

$$[H^+] = \frac{K_w}{[OH^-]} = \frac{1.0 \times 10^{-14}}{[OH^-]} \qquad [OH^-] = \frac{K_w}{[H^+]} = \frac{1.0 \times 10^{-14}}{[H^+]}$$

Example 13.8   Using the results of Examples 13.1 and 13.7, calculate both $[H^+]$ and $[OH^-]$ in solutions of acetic acid and aqueous ammonia.

SOLUTION   Using $[OH^-] = \dfrac{1.0 \times 10^{-14}}{[H^+]}$ for acetic acid:

Original $[CH_3COOH]$	$[H^+]$	$[OH^-]$
$1.0M$	$4.2 \times 10^{-3}M$	$2.4 \times 10^{-12}M$
$1.0 \times 10^{-2}M$	$4.2 \times 10^{-4}M$	$2.4 \times 10^{-11}M$
$1.0 \times 10^{-3}M$	$1.3 \times 10^{-4}M$	$7.7 \times 10^{-11}M$

Using $[H^+] = \dfrac{1.0 \times 10^{-14}}{[OH^-]}$ for aqueous ammonia:

Original $[NH_3(aq)]$	$[OH^-]$	$[H^+]$
$1.0M$	$4.2 \times 10^{-3}M$	$2.4 \times 10^{-12}M$
$1.0 \times 10^{-2}M$	$4.2 \times 10^{-4}M$	$2.4 \times 10^{-11}M$

As $[H^+]$ decreases, $[OH^-]$ increases and vice versa.

Example 13.9   Calculate both $[H^+]$ and $[OH^-]$ in $0.10M$ solutions of (a) HCl and (b) NaOH.

SOLUTION

(a) Taking $0.10M$ HCl to be completely dissociated gives $[H^+] = 0.10M$. Then

$$[OH^-] = \frac{1.0 \times 10^{-14}}{[H^+]} = \frac{1.0 \times 10^{-14}}{0.10} = 1.0 \times 10^{-13}M$$

(b) Taking $0.10M$ NaOH to be completely dissociated gives $[OH^-] = 0.10M$. Then

$$[H^+] = \frac{1.0 \times 10^{-14}}{[OH^-]} = \frac{1.0 \times 10^{-14}}{0.10} = 1.0 \times 10^{-13}M$$

In the solutions of acids and bases dealt with in these examples, there is always a big difference between H^+ and OH^- concentrations. However, if we start with pure water, the two ions must be present in equal concentrations, since the ionization of one water molecule produces one ion of each type.

Example 13.10   Calculate $[H^+]$ and $[OH^-]$ in pure water, given that $K_w = 1.0 \times 10^{-14}$.

SOLUTION   Let x moles liter^{-1} dissociate:

$$H_2O \rightleftharpoons H^+ + OH^-$$

Then $[H^+] = [OH^-] = x$ and

$$K_w = [H^+][OH^-] = x^2 = 1.0 \times 10^{-14}$$

$$x = [H^+] = [OH^-] = 1.0 \times 10^{-7}$$

Any solution in which $[H^+] = [OH^-] = 1.0 \times 10^{-7}$ is said to be *neutral*. If $[H^+] > 1.0 \times 10^{-7}$ (and $[OH^-] < 1.0 \times 10^{-7}$) then the solution is *acidic*. If $[H^+] < 1.0 \times 10^{-7}$ (and $[OH^-] > 1.0 \times 10^{-7}$) the solution is *basic*.

Since the hydrogen ion concentration is usually small, we have to express it with factors like 10^{-3} or 10^{-8} tacked on. To avoid this, a logarithmic scale for expressing $[H^+]$ is very widely used. We use the symbol pH (meaning roughly "*power of Hydrogen*") to express hydrogen ion concentration, where

$$pH = -\log[H^+]$$

The negative sign before the logarithm means that pH values are almost always positive, since $[H^+]$ is usually less than 1. However, it would in theory be possible, in a solution of a very strong acid above $1M$ concentration, to have a negative pH.

The corresponding expression for the hydroxide ion concentration is pOH, defined as

$$pOH = -\log[OH^-]$$

However, the pOH value does not tell us anything new about the solution, since it is closely related to the pH. The relationship between the two may be seen from the expression for K_w:

$$[H^+][OH^-] = K_w$$

taking logs: $\log[H^+] + \log[OH^-] = \log K_w$
change signs: $-\log[H^+] - \log[OH^-] = -\log K_w$

The quantity $-\log K_w$ is often called pK_w, so this equation may be written:

$$pH + pOH = pK_w$$

Since $K_w = 1.0 \times 10^{-14}$, the value of pK_w will obviously be $-\log(1.0 \times 10^{-14}) = 14.0$. Hence

$$pH + pOH = 14.0$$

Thus, if we can calculate either pH or pOH, the other may immediately be found. This is in fact the chief use of pOH; as an intermediate in calculations of pH, particularly in alkaline solution.

Our definitions of acid, neutral, and basic solutions may now be rewritten in terms of pH and pOH. In an *acid* solution, pH < 7.0 and pOH > 7.0. In a *neutral* solution, pH = pOH = 7.0. In a *basic* solution, pH > 7.0 and pOH < 7.0. Because of the negative sign in the pH expression, a smaller pH means a higher hydrogen ion concentration and vice versa.

(*Note*: Even with the aid of an electronic calculator, you can't escape using and understanding logarithms when doing pH problems. Read the discussion of logarithms, Appendix A.3, if you have any trouble with the following numerical problems.)

Example 13.11   Calculate the pH and pOH of solutions of HCl, in the following concentrations: (a) $1.0M$, (b) $0.20M$, (c) $3.1 \times 10^{-3}M$.

SOLUTION   The dissociation of the HCl may be assumed to be complete in each case.

(a) $[H^+] = 1.0$
    $pH = -\log 1.0 = 0.0$ (remember the log of 1 is zero)
    $pOH = 14.0 - pH = 14.0 - 0.0 = 14.0$
(b) $[H^+] = 0.20 = 2.0 \times 10^{-1}$
    $pH = -\log(2.0 \times 10^{-1}) = -\log 2.0 - \log 10^{-1} = +1 - \log 2.0 = +1 - 0.30 = +0.70$
    $pOH = 14.0 - pH = 14.0 - 0.70 = 13.30$
(c) $[H^+] = 3.1 \times 10^{-3}$
    $pH = -\log(3.1 \times 10^{-3}) = 3 - \log 3.1 = 3 - 0.49 = 2.51$
    $pOH = 14.0 - 2.51 = 11.49$

In calculations of the above type, it is very easy to make a mistake in converting to the logarithm. Always convert $[H^+]$ to two parts, a quantity between 1 and 10 and a power of 10, then take the log of each part separately and add them together. Unless the solution is a very concentrated acid, the pH will be positive, so be very suspicious of error if you find a pH outside the range 0 to $+14$.

The question of significant figures must be applied to pH as to any other quantity, but the figures are not counted in the same manner when a quantity is expressed logarithmically. Thus, in Example 13.11(b) above, it might appear that the pH, 0.70, contains two figures, while the pOH, 13.30, contains four, although both have been derived from the same original concentration, 0.20, given to two figures only. Is this justified? The answer is yes, because when a quantity is expressed logarithmically, the numbers coming before the decimal point only express powers of ten, and are not "significant" in the sense in which we use the word. The "significant" figures occur *after* the decimal point, and in both the above examples there are only two. In all the examples given in this book, we shall calculate pH to two places of decimals, which is roughly equivalent to expressing $[H^+]$ to two significant figures.

Example 13.12   Calculate the pH of solutions of sodium hydroxide: (a) $1.0M$ (b) $0.30M$ (c) $4.7 \times 10^{-3}M$.

SOLUTION   We will assume the NaOH to be completely dissociated, thus giving us $[OH^-]$ directly. From this, we could calculate $[H^+]$ by the method of Example 13.9 and hence find pH. However, it is easier (because subtraction is easier than division) to calculate pOH, then find pH from that.

(a) $[OH^-] = 1.0$
    $pOH = -\log[OH^-] = -\log 1.0 = 0.0$
    $pH = 14.0 - pOH = 14.0 - 0.0 = 14.0$
(b) $[OH^-] = 0.30 = 3.0 \times 10^{-1}$
    $pOH = -\log(3.0 \times 10^{-1}) = 1 - \log 3.0 = 1 - 0.48 = 0.52$
    $pH = 14.0 - 0.52 = 13.48$
(c) $[OH^-] = 4.7 \times 10^{-3}$
    $pOH = -\log(4.7 \times 10^{-3}) = 3 - \log 4.7 = 3 - 0.67 = 2.33$
    $pH = 14 - 2.33 = 11.67$

In calculating the pH of solutions of weak acids and bases, there are two possible approaches. The more obvious approach is to calculate $[H^+]$ by the methods developed earlier in this chapter, then find pH from that.

Example 13.13   Use the results of Example 13.8 to tabulate the pH of various solutions of CH_3COOH and $NH_3(aq)$.

SOLUTION   Knowing the various values of $[H^+]$, the corresponding pH values may be evaluated.

Solution (M)		$[H^+]$	pH
CH_3COOH:	1.0	4.2×10^{-3}	2.38
CH_3COOH:	1.0×10^{-2}	4.2×10^{-4}	3.38
CH_3COOH:	1.0×10^{-3}	1.3×10^{-4}	3.89
$NH_3(aq)$:	1.0	2.4×10^{-12}	11.62
$NH_3(aq)$:	1.0×10^{-2}	2.4×10^{-11}	10.62

As you have probably realized by now, a difference of one unit of pH corresponds to a difference in $[H^+]$ of a factor of ten.

Example 13.14   Calculate the pH of an $0.020M$ solution of a weak acid, HA, for which $K_a = 3.0 \times 10^{-6}$.

SOLUTION   Let the amount of acid dissociated be x mole liter^{-1}. Then the equilibrium is

$$HA \rightleftharpoons A^- + H^+$$
$$(0.020 - x) \qquad x \qquad x \text{ mole liter}^{-1}$$

$$K_a = \frac{[A^-][H^+]}{[HA]} = \frac{x^2}{(0.020 - x)} = 3.0 \times 10^{-6}$$

Making our usual assumption that x is negligible beside 0.020 gives

$$\frac{x^2}{(0.020 - x)} \approx \frac{x^2}{0.020} = 3.0 \times 10^{-6}$$

$$x^2 = 6.0 \times 10^{-8} \qquad x = 2.45 \times 10^{-4}$$

x is about 1% of our original 0.020 mole of acid, so our assumption was clearly justified.

$$[H^+] = x = 2.45 \times 10^{-4}M$$

and

$$pH = -\log[H^+] = -\log(2.45 \times 10^{-4}) = 4 - \log 2.45 = 4 - 0.39 = 3.61$$

13.5   The Concept of pK

An alternative method of calculating the pH of solutions of this type is to work out a general formula for pH in terms of the known K_a value and the concentration of the solution of weak acid. This involves a new term, the pK_a value of the acid. pK_a is defined in a very similar manner to pH and pK_w, that is,

$$pK_a = -\log K_a$$

It is thus not a new constant for an acid, but a handy version of an old one, K_a. It has the advantage of being always positive (except for a few very strong acids for which $K_a > 1$) and of avoiding awkward powers of 10. A very similar constant, pK_b, is used for bases:

$$pK_b = -\log K_b$$

Example 13.15   Calculate pK_a or pK_b values, as appropriate, for the following:

(a) CH_3COOH, $K_a = 1.8 \times 10^{-5}$
(b) HIO_3, $K_a = 1.7 \times 10^{-1}$
(c) $NH_3(aq)$, $K_b = 1.8 \times 10^{-5}$
(d) quinoline, $K_b = 6.3 \times 10^{-10}$

SOLUTION

(a) $K_a = 1.8 \times 10^{-5}$, $pK_a = -\log(1.8 \times 10^{-5}) = 5 - \log 1.8 = 5 - 0.26 = 4.74$
(b) $K_a = 1.7 \times 10^{-1}$, $pK_a = -\log(1.7 \times 10^{-1}) = 1 - \log 1.7 = 1 - 0.23 = 0.77$
(c) $K_b = 1.8 \times 10^{-5}$, $pK_b = -\log(1.8 \times 10^{-5}) = 5 - \log 1.8 = 5 - 0.26 = 4.74$
(d) $K_b = 6.3 \times 10^{-10}$, $pK_b = -\log(6.3 \times 10^{-10}) = 10 - \log 6.3 = 10 - 0.80 = 9.20$

From the form of the pK expressions, it follows that the *weaker* the acid, or base, the *larger* will be the pK value. By coincidence, the pK values of acetic acid and ammonium hydroxide are numerically the same, but their designation as pK_a and pK_b reminds us that they represent different quantities.

The application of pK values may be seen in the following examples, and in others in Chapter 14.

Example 13.16   Develop a general expression for the pH of a solution of a weak acid, ionization contant K_a, in a solution of molarity M.

SOLUTION   Let x mole liter^{-1} of the acid dissociate, then the equilibrium will be

$$HA \rightleftharpoons A^- + H^+$$

$$(M - x) \qquad x \qquad x \text{ mole liter}^{-1}$$

$$K_a = \frac{[A^-][H^+]}{[HA]} = \frac{x^2}{(M - x)}$$

Assume, since this is a weak acid, that the degree of dissociation is small, that is, x is much less than M and $(M - x) \approx M$. Then

$$\frac{x^2}{(M - x)} \approx \frac{x^2}{M} = K_a \quad \text{so} \quad x^2 = K_a M$$

Taking logs, we get

$$2 \log x = \log K_a + \log M$$

$$\log x = \frac{1}{2} (\log K_a + \log M)$$

Now $x = [H^+]$ or $\log x = \log[H^+]$, so

$$pH = -\log x = -\frac{1}{2} (\log K_a + \log M) = \frac{1}{2} (-\log K_a - \log M)$$

But $-\log K_a = pK_a$, so

$$pH = \frac{1}{2} (pK_a - \log M)$$

Example 13.17   Use the result of Example 13.16 to calculate the pH of the solution of weak acid referred to in Example 13.14.

SOLUTION   The weak acid of Example 13.14 had $K_a = 3.0 \times 10^{-6}$ and $M = 2.0 \times 10^{-2}$, so

$$pK_a = -\log(3.0 \times 10^{-6}) = 6 - \log 3.0 = 6 - 0.48 = 5.52$$

$$pH = \frac{1}{2} (pK_a - \log M) = \frac{1}{2} [5.52 - \log(2.0 \times 10^{-2})]$$

$$= \frac{1}{2} (5.52 + 2 - \log 2.0) = \frac{1}{2} (7.52 - 0.30)$$

$$= \frac{1}{2} \times 7.22 = 3.61$$

Not surprisingly, the result of Example 13.17 is the same as that of Example 13.14. The two calculations are of course the same, except that the steps are done in slightly different order. It is important to realize that the "x is negligibly small" assumption is *implicit* in the use of the formula derived in Example 13.16. However, the means of checking the validity of this assumption is not as easy as in the method of Example 13.14, since the actual value of x is never found. Hence this formula should be used very cautiously in calculating pH, but it is useful if a number of repetitive calculations are being done on the same or similar acids under conditions where the degree of dissociation is known to be small.

A similar formula may be calculated for solutions of weak bases:

Example 13.18   Derive a general expression for the pH of a solution of a weak base, ionization constant K_b, in a solution of molarity M. Use your result to calculate the pH of a solution of $NH_3(aq)$, $1.0 \times 10^{-2}M$, for which $K_b = 1.8 \times 10^{-5}$.

SOLUTION   Let the amount of base ionized be x mole liter^{-1}, where x is negligible beside M. The equilibrium is then

$$B \ + H_2O \ \rightleftharpoons \ BH^+ + OH^-$$

$$(M - x) \qquad\qquad x \qquad x \quad \text{mole liter}^{-1}$$

$$K_b = \frac{x^2}{(M - x)} \approx \frac{x^2}{M} \quad \text{since} \quad (M - x) \approx M$$

$$x^2 = K_b M$$

Taking logs:

$$2 \log x = \log K_b + \log M$$

$$\log x = \frac{1}{2}(\log K_b + \log M)$$

Now $x = [OH^-]$, so $pOH = -\log x$ and, since $pK_b = -\log K_b$,

$$-pOH = \frac{1}{2}(-pK_b + \log M)$$

To express this in terms of pH rather than pOH, we add pK_w to each side:

$$pK_w - pOH = pK_w + \frac{1}{2}(-pK_b + \log M)$$

But

$$pK_w - pOH = pH \quad \text{so} \quad pH = \frac{1}{2}(2pK_w - pK_b + \log M)$$

Applying this to the example given, we have

$$pK_b = -\log K_b = -\log(1.8 \times 10^{-5}) = 4.74$$

we know $M = 1.0 \times 10^{-2}$, so $\log M = -2.00$; then, since $pK_w = 14.0$,

$$pH = \frac{1}{2}(28.0 - 4.74 - 2.00) = 10.63$$

This formula, like the previous one for weak acids, depends on the assumption that the degree of dissociation of the base is negligible, so it should be used with caution.

KEY WORDS

acid; base
ionization constant
strong base
weak base
pH; pOH

ionization
strong acid
weak acid
ion product of water

STUDY QUESTIONS

1. With respect to water, what constitutes an acid? A base?

2. What is a proton transfer reaction? Give examples.

3. Give examples of several strong acids and several strong bases. How do they react with water?

4. What do we understand by a "weak" acid? What equilibrium exists when a weak acid is dissolved in water?

5. How do we approach the quantitative solution of the weak acid ionization equilibrium? What does the ionization constant represent?

6. What simplification do we make to facilitate solving the preceding problem? Under what conditions is the simplification justified? How do we determine whether or not we may simplify a particular problem?

7. Given the original concentration of a weak acid and the equilibrium value of $[H^+]$, how would you calculate K_a for the acid?

8. What equilibrium is set up when ammonia, NH_3, is dissolved in water? Give examples of other substances that react similarly. In which direction does an equilibrium of this type lie?

9. How are $[H^+]$ and $[OH^-]$ connected in aqueous solution? Is this relationship always true?

10. What do we mean by pH, pK_w and pOH? How are they connected?

11. What constitutes (a) an acidic solution, (b) a neutral solution, and (c) a basic solution?

PROBLEMS

Refer to Tables 13.1 and 13.3 for the ionization constants needed in solving these problems.

1. Calculate the hydrogen ion concentration, percentage dissociation, and pH in the following solutions:
   (a) formic acid, $1.20M$
   (b) chloracetic acid, $0.200M$
   (c) hydrocyanic acid, $1.6 \times 10^{-2}M$
   (d) nitrous acid, $1.2 \times 10^{-2}M$
   (e) benzoic acid, $0.11M$
   (f) hydrazoic acid, $4.2 \times 10^{-3}M$

2. Calculate the hydroxide ion concentration, percentage ionization and pH in the following aqueous solutions:
   (a) ethylamine, $1.50M$
   (b) hydrazine, $2.6 \times 10^{-4}M$
   (c) ammonia, $0.066M$
   (d) triethylamine, $8.5 \times 10^{-2}M$

3. Acid HA has a molecular weight of 88. When 2.73 g of HA is dissolved in water and made up to 100 ml volume, the pH of the solution is 4.65. What is the ionization constant of HA?

4. Base B has a molecular weight of 160. When 18.4 g of B is made up to a volume of 1.00 liter with water, the solution has a pH of 11.50. What is the ionization constant of B?

5. (a) Calculate the percent ionization in an $0.10M$ solution of formic acid.
   (b) How will this change if 0.020 mole liter^{-1} of HCl is added (neglect volume change)? (*Hint*: Remember that H^+ is contributed by two solutes at the same time.)

6. What is the pH of the solutions made by dissolving, in 100 ml water, 1.00 liter (STP) of the following gases:
   (a) HCl   (b) NH_3?

7. What is the degree of ionization of ethylamine dissolved in $0.200M$ concentration in
   (a) water
   (b) $1.0 \times 10^{-2}M$ sodium hydroxide solution? (*Hint*: This is similar to problem (5) above, but this time OH^- comes from two solutes. You'll have to solve a quadratic equation here!)

SOLUTIONS TO PROBLEMS

1. These problems are all approached in the same way. If x mole liter^{-1} of the acid dissociate, then $(M - x)$ remain, where M is original concentration of acid, so

$$\frac{x^2}{(M - x)} = K_a$$

Making the approximation $M - x \approx M$ gives

$$\frac{x^2}{M} \approx K_a \qquad x^2 = MK_a$$

Each time we do this, we must check that the approximation was justified.
   (a) concentration, $1.20M$; $K_a = 1.8 \times 10^{-4}$
   $x^2 = 1.20 \times 1.8 \times 10^{-4} = 2.2 \times 10^{-4}$
   $x = 1.5 \times 10^{-2} = [H^+]$

   percent dissociation: $\dfrac{1.5 \times 10^{-2}}{1.20} \times 100 = 1.2\%$

   so the simplifying assumption is justified
   pH: $-\log(1.5 \times 10^{-2}) = 1.82$

(b)   concentration, $0.200M$; $K_a = 1.4 \times 10^{-3}$
$x^2 = 0.200 \times 1.4 \times 10^{-3} = 2.8 \times 10^{-4}$
$x = 1.7 \times 10^{-2}$

percent dissociation: $\dfrac{1.7 \times 10^{-2}}{0.200} \times 100 = 8.5\%$

which is *not* negligible
For a more accurate answer, say

$$M - x = 0.200 - (1.7 \times 10^{-2}) = 0.183$$

$x^2 = 0.183 \times 1.4 \times 10^{-3} = 2.6 \times 10^{-4}$
$x = 1.6 \times 10^{-2} = [H^+]$

percent dissociation: $\dfrac{1.6 \times 10^{-2}}{0.200} \times 100 = 8.0\%$

pH: $-\log(1.6 \times 10^{-2}) = 1.80$

(c)   concentration, $1.6 \times 10^{-2}M$; $K_a = 4.9 \times 10^{-10}$
$x^2 = 1.6 \times 10^{-2} \times 4.9 \times 10^{-10} = 7.8 \times 10^{-12}$
$x = 2.8 \times 10^{-6} = [H^+]$

percent dissociation $= \dfrac{2.8 \times 10^{-6}}{1.6 \times 10^{-2}} \times 100 = 1.8 \times 10^{-2}\%$

so the simplifying assumption is justified
pH $= -\log(2.8 \times 10^{-6}) = 5.55$

(d)   concentration, $1.2 \times 10^{-2}M$; $K_a = 4.5 \times 10^{-4}$
$x^2 = 1.2 \times 10^{-2} \times 4.5 \times 10^{-4} = 5.4 \times 10^{-6}$
$x = 2.3 \times 10^{-3}$

percent dissociation $= \dfrac{2.3 \times 10^{-3}}{1.2 \times 10^{-2}} \times 100 = 19\%$

which is *not* negligible
For a more accurate answer, say
$M - x = 1.2 \times 10^{-2} - (2.3 \times 10^{-3}) = 9.7 \times 10^{-3}$
$x^2 = 9.7 \times 10^{-3} \times 4.5 \times 10^{-4} = 4.4 \times 10^{-6}$
$x = 2.1 \times 10^{-3} = [H^+]$

percent dissociation $= \dfrac{2.1 \times 10^{-3}}{1.2 \times 10^{-2}} \times 100 = 17\%$

pH: $-\log(2.1 \times 10^{-3}) = 2.68$

(e)   concentration $0.11M$; $K_a = 6.4 \times 10^{-5}$
$x^2 = 0.11 \times 6.4 \times 10^{-5} = 7.0 \times 10^{-6}$
$x = 2.7 \times 10^{-3} = [H^+]$

percent dissociation $= \dfrac{2.7 \times 10^{-3}}{0.11} \times 100 = 2.4\%$

so the simplifying assumption is justified
pH $= -\log(2.7 \times 10^{-3}) = 2.57$

(f) concentration, $4.2 \times 10^{-3}M$; $K_a = 1.9 \times 10^{-5}$
$$x^2 = 4.2 \times 10^{-3} \times 1.9 \times 10^{-5} = 8.0 \times 10^{-8}$$
$$x = 2.8 \times 10^{-4}$$

$$\text{percent dissociation} = \frac{2.8 \times 10^{-4}}{4.2 \times 10^{-3}} \times 100 = 6.7\%$$

which is *not* negligible
For a more accurate answer, say
$$M - x = 4.2 \times 10^{-3} - (2.8 \times 10^{-4}) = 3.9 \times 10^{-3}$$
$$x^2 = 3.9 \times 10^{-3} \times 1.9 \times 10^{-5} = 7.4 \times 10^{-8}$$
$$x = 2.7 \times 10^{-4} = [H^+]$$

$$\text{percent dissociation} = \frac{2.7 \times 10^{-4}}{4.2 \times 10^{-3}} \times 100 = 6.5\%$$

$$\text{pH} = -\log(2.7 \times 10^{-4}) = 3.56$$

2. The logic here is very similar. If x mole liter^{-1} of the base B react with water, we have

$$\text{B} + \text{H}_2\text{O} \rightleftharpoons \text{BH}^+ + \text{OH}^-$$
$$(M - x) \qquad\qquad\qquad x \qquad x$$

$$\frac{x^2}{M - x} = K_b$$

If x is negligible beside M,

$$M - x \approx M \qquad \frac{x^2}{M} \approx K_b \qquad x^2 \approx MK_b$$

Every time we use this, we must check the accuracy of the approximation and recalculate if necessary.
(a) concentration, $1.50M$; $K_b = 4.3 \times 10^{-4}$
$$x^2 = 1.50 \times 4.3 \times 10^{-4} = 6.5 \times 10^{-4}$$
$$x = 2.5 \times 10^{-2} = [OH^-]$$

$$\text{percent ionization:} \quad \frac{2.5 \times 10^{-2}}{1.50} \times 100 = 1.7\%$$

so the simplifying assumption is justified
$$\text{pOH} = -\log(2.5 \times 10^{-2}) = 1.60$$
$$\text{pH:} \quad 14.00 - 1.60 = 12.40$$
(b) concentration, $2.6 \times 10^{-4}M$; $K_b = 1.3 \times 10^{-6}$
$$x^2 = 2.6 \times 10^{-4} \times 1.3 \times 10^{-6} = 3.4 \times 10^{-10}$$
$$x = 1.8 \times 10^{-5}$$

$$\text{percent ionization:} \quad \frac{1.8 \times 10^{-5}}{2.6 \times 10^{-4}} \times 100 = 7.1\%$$

which is *not* negligible
For a more accurate answer, say
$$M - x = 2.6 \times 10^{-4} - (1.8 \times 10^{-5}) = 2.4 \times 10^{-4}$$
$$x^2 = 2.4 \times 10^{-4} \times 1.3 \times 10^{-6} = 3.1 \times 10^{-10}$$
$$x = 1.8 \times 10^{-5} = [OH^-]$$
(*Note*: Rounded to two significant figures, this is the same answer we had previously.)
percent ionization: 7.1%
$$\text{pOH:} \quad -\log(1.8 \times 10^{-5}) = 4.74$$
$$\text{pH:} \quad 14.00 - 4.74 = 9.26$$

(c)   concentration: $0.066M$; $K_b = 1.8 \times 10^{-5}$
$x^2 = 0.066 \times 1.8 \times 10^{-5} = 1.2 \times 10^{-6}$
$x = 1.1 \times 10^{-3} = [OH^-]$

$$\text{percent ionization} = \frac{1.1 \times 10^{-3}}{0.066} \times 100 = 1.7\%$$

so the simplifying assumption is justified
$pOH = -\log(1.1 \times 10^{-3}) = 2.96$
$pH = 14.00 - 2.96 = 11.04$

(d)   concentration, 8.5×10^{-2}; $K_b = 5.3 \times 10^{-4}$
$x^2 = 8.5 \times 10^{-2} \times 5.3 \times 10^{-4} = 4.5 \times 10^{-5}$
$x = 6.7 \times 10^{-3}$

$$\text{percent ionization} = \frac{6.7 \times 10^{-3}}{8.5 \times 10^{-2}} \times 100 = 7.9\%$$

which is *not* negligible
For a more accurate answer, say
$M - x = 8.5 \times 10^{-2} - (6.7 \times 10^{-3}) = 7.8 \times 10^{-2}$
$x^2 = 7.8 \times 10^{-2} \times 5.3 \times 10^{-4} = 4.2 \times 10^{-5}$
$x = 6.4 \times 10^{-3} = [OH^-]$

$$\text{percent ionization} = \frac{6.4 \times 10^{-3}}{8.5 \times 10^{-2}} \times 100 = 7.6\%$$

$pOH = -\log(6.4 \times 10^{-3}) = 2.19$
$pH = 14.00 - 2.19 = 11.81$

3.   2.73 g of HA is $2.73/88 = 3.1 \times 10^{-2}$ mole

concentration is initially $\dfrac{3.1 \times 10^{-2} \text{ mole}}{0.100 \text{ liter}} = 0.31M$

at equilibrium, $pH = 4.65$, so
$[H^+] = \text{antilog}(-4.65) = \text{antilog}(-5 + 0.35) = 10^{-5} \times \text{antilog}(0.35) = 2.2 \times 10^{-5}$
This is also the value of $[A^-]$, so the degree of dissociation is very small.

$$\text{dissociation} = \frac{2.2 \times 10^{-5}}{0.31} \times 100 = 7.2 \times 10^{-3}\%$$

substituting to find K_a gives

$$K_a = \frac{[H^+][A^-]}{[HA]} = \frac{(2.2 \times 10^{-5})^2}{0.31} = 1.6 \times 10^{-9}$$

4.   18.4 g of B is $18.4/160 = 0.115$ mole of B
This is in 1.00 liter, so $[B] = 0.115M$.
The pH is 11.50, so $pOH = 2.50$
$[OH^-] = \text{antilog}(-2.50) = \text{antilog}(-3 + 0.50) = 10^{-3} \times \text{antilog}(0.50) = 3.2 \times 10^{-3}$
This is also the value of BH^+, so again the degree of dissociation is small.

$$\text{dissociation:} \quad \frac{3.2 \times 10^{-3}}{0.115} \times 100 = 2.8\%$$

substituting to find K_b gives

$$K_b = \frac{[BH^+][OH^-]}{[B]} = \frac{(3.2 \times 10^{-3})^2}{0.115} = 8.7 \times 10^{-5}$$

5. (a) For formic acid alone, if x mole liter^{-1} dissociate:
        concentration $= 0.10M$; $K_a = 1.8 \times 10^{-4}$
        $x^2 = 0.10 \times 1.8 \times 10^{-4} = 1.8 \times 10^{-5}$
        $x = 4.2 \times 10^{-3} = [H^+]$

        percent dissociation: $\dfrac{4.2 \times 10^{-3}}{0.10} \times 100 = 4.2\%$

        so the assumption of negligible dissociation is justified

    (b) If we start out with 0.020 mole liter^{-1} of HCl (a strong acid) present, the HCl will be fully dissociated and give 0.020 mole liter^{-1} of H^+. The H^+ produced by the dissociation of formic acid will be additional to this so, if x mole liter^{-1} come from formic acid, the total $[H^+]$ will be $0.020 + x$. The equilibrium is then

$$\text{HCOOH} \rightleftharpoons \text{HCOO}^- + \text{H}^+$$
$$(0.10 - x) \qquad\qquad x \qquad (0.020 + x)$$

This must still fit the formula for K_a, so

$$K_a = \frac{[H^+][\text{HCOO}^-]}{[\text{HCOOH}]} = \frac{(0.020 + x)x}{(0.10 - x)} = 1.8 \times 10^{-4}$$

Assuming x is negligible beside 0.020 and 0.10, we simplify this quadratic equation to

$$\frac{0.020x}{0.10} = 1.8 \times 10^{-4} \qquad x = \frac{0.10 \times 1.8 \times 10^{-4}}{0.020} = 9.0 \times 10^{-4}$$

Obviously, the assumption was justified.

percent dissociation: $\dfrac{9.0 \times 10^{-4}}{0.10} \times 100 = 0.90\%$

6. 1.00 liter at STP is $1.00/22.4 = 4.46 \times 10^{-2}$ mole

    initial concentration: $\dfrac{4.46 \times 10^{-2} \text{ mole}}{0.100 \text{ liter}} = 0.446M$

    (a) $0.446M$ HCl dissociates completely to give
        $[H^+] = 0.446M$
        $\text{pH} = -\log(0.446) = -\log(4.46 \times 10^{-1})$
        $\quad = -(0.65 - 1) = +0.35$

    (b) $0.446M$ aqueous ammonia; $K_b = 1.8 \times 10^{-5}$
        If x mole liter^{-1} are ionized,
        $x^2 = 0.446 \times 1.8 \times 10^{-5} = 8.0 \times 10^{-6}$
        $x = 2.8 \times 10^{-3} = [\text{OH}^-]$

        percent ionization: $\dfrac{2.8 \times 10^{-3}}{0.446} \times 100 = 0.64\%$

        so the assumption of negligible ionization is justified
        $\text{pOH} = -\log(2.8 \times 10^{-3}) = 2.55$
        $\text{pH} = 14.00 - 2.55 = 11.45$

7. (a) In water, concentration $= 0.200M$; $K_b = 4.3 \times 10^{-4}$

If x mole liter^{-1} are ionized,

$$x^2 = 0.200 \times 4.3 \times 10^{-4} = 8.6 \times 10^{-5}$$

$$x = 9.3 \times 10^{-3} = [OH^-]$$

percent dissociation: $\dfrac{9.3 \times 10^{-3}}{0.200} \times 100 = 4.6\%$

so the assumption of negligible ionization is justified.

(b) The initial contribution of OH^- from NaOH is $1.0 \times 10^{-2}M$. Suppose the dissolved ethylamine contributes an additional y mole liter^{-1}, then at equilibrium we have

$$C_2H_5NH_2 + H_2O \rightleftharpoons C_2H_5NH_3^+ + OH^-$$

$$(0.200 - y) \qquad\qquad\qquad y \qquad (1.0 \times 10^{-2} + y)$$

$$K_b = \frac{[C_2H_5NH_3^+][OH^-]}{[C_2H_5NH_2]} = \frac{y(1.0 \times 10^{-2} + y)}{0.200 - y} = 4.3 \times 10^{-4}$$

Making the usual assumption, that y is negligible beside 1.0×10^{-2} and 0.200, gives the simplified equation

$$\frac{y \times 1.0 \times 10^{-2}}{0.200} = 4.3 \times 10^{-4}$$

so

$$y = \frac{0.200 \times 4.3 \times 10^{-4}}{1.0 \times 10^{-2}} = 8.6 \times 10^{-3}$$

Obviously our assumption was badly in error, since y is nearly as large as 1.0×10^{-2}. In cases like this, where the simplifying assumption breaks down badly, it is best to solve the quadratic equation properly (see Appendix A.5) rather than use successive approximation:

$$\frac{y(1.0 \times 10^{-2} + y)}{0.200 - y} = 4.3 \times 10^{-4}$$

$$(1.0 \times 10^{-2})y + y^2 = (8.6 \times 10^{-5}) - (4.3 \times 10^{-4})y$$

$$y^2 + (1.0 \times 10^{-2})y - 8.6 \times 10^{-5} = 0$$

from which

$$y = -1.6 \times 10^{-2} \quad \text{or} \quad +5.5 \times 10^{-3}$$

Rejecting the first value (y cannot be negative) gives $y = 5.5 \times 10^{-3}$

percent ionization: $\dfrac{5.5 \times 10^{-3}}{0.200} \times 100 = 2.8\%$

Chapter 14

Equilibrium in Solution: Hydrolysis, Buffers, and Simultaneous Equilibrium

"Well, have you solved it?" I asked.
"Yes. It was the bisulphate of baryta."
"No, No, the mystery!" I cried.   A CASE OF IDENTITY

14.1   Hydrolysis of Ions

In the most general sense, "hydrolysis" simply means "reaction with water." In this section, however, we are thinking of hydrolysis in a specific sense where ionic salts are concerned:

For an ion in solution, an hydrolysis reaction is a reaction of the ion with water in which proton transfer occurs.

Note that we are talking of the reaction of an *ion* with water, not the reaction of an ionic salt. As we know, any ionic salt dissolves in water to give oppositely charged ions, which are hydrated. It is the subsequent reaction of these ions with water that we call *hydrolysis*, and for this process we consider the reaction of the cation and anion separately.

What type of ion undergoes an hydrolysis reaction? It is important to realize that many ions do *not* undergo appreciable hydrolysis in solution. In fact, there are only two types of ion we need consider:

1. An anion (negative ion) which is the conjugate base of a *weak* acid.
2. A cation (positive ion) which is the conjugate acid of a *weak* base.

We can define these terms and give examples of these two cases separately.

1. The *conjugate base* of an acid is the ion produced by transferring a proton from the acid to water:

$$HA + H_2O \rightleftharpoons \underset{\substack{\text{conjugate} \\ \text{base}}}{A^-} + H_3O^+$$

Thus chloride ion is the conjugate base of HCl; the acetate ion is the conjugate base of acetic acid, etc.

In the hydrolysis reaction, the anion (the conjugate base) reacts with water. A proton is transferred from water to the base to leave OH^-, so the solution becomes *basic*.

$$A^- + H_2O \rightleftharpoons HA + OH^-$$

As we might expect, the equilibrium constant for this process is closely related to the dissociation constant for the acid. For the hydrolysis reaction, leaving out $[H_2O]$, we can write

$$K_b = \frac{[HA][OH^-]}{[A^-]}$$

(*Note:* We use K_b as the symbol for this constant because it represents the reaction of a *base* (the ion A^-) with water. Another symbol sometimes used is K_h, the hydrolysis constant, which has the same meaning.) Multiplying top and bottom by $[H^+]$ gives

$$K_b = \frac{[HA][OH^-][H^+]}{[H^+][A^-]} = \frac{[HA]}{[H^+][A^-]} \times [H^+][OH^-]$$

Written in this way, we see that K_b is made up of two terms. The first is $1/K_a$ and the second is K_w, so we have the simple relationship

$$K_b = \frac{K_w}{K_a}$$

Thus, the extent to which hydrolysis of the base A^- occurs is *inversely* proportional to the strength of its conjugate acid HA. Only the anions derived from *weak* acids will undergo any appreciable degree of hydrolysis.

2. When we consider the possibility of a cation (positive ion) undergoing an hydrolysis reaction, the great majority of simple metal ions may be ignored, since they are not related to weak bases. In other words, an ion such as Na^+ in solution cannot influence acid–base equilibrium in any way. The equilibrium

$$Na^+ + H_2O \rightleftharpoons NaOH(aq) + H^+$$

lies completely to the *left*, because NaOH is an ionic substance and does not exist in solution as undissociated NaOH molecules.

(*Note:* The formation of an *insoluble* metal hydroxide can influence acid–base equilibria by removing OH^- from solution as the solid hydroxide. Many metal hydroxides have very low solubilities, so reactions of this type occur:

$$Fe^{3+}(aq) + 3H_2O \rightleftharpoons Fe(OH)_3(s) + 3H^+$$

Although this type of reaction is correctly described as an hydrolysis, we will not consider it here, because we are talking of homogeneous equilibria in solution.)

We have already seen that weak bases are restricted to ammonia and a few related compounds of nitrogen. They react with water by proton transfer to give positive ions, so the ion NH_4^+ is the *conjugate acid* of the base NH_3:

$$\underset{\text{base}}{NH_3} + H_2O \rightleftharpoons \underset{\substack{\text{conjugate} \\ \text{acid}}}{NH_4^+} + OH^-$$

If we take an ammonium salt, such as NH_4Cl, in water, it provides ammonium ions by complete dissociation (the counter-ion, Cl^-, may be ignored). Subsequent hydrolysis of this conjugate acid then produces an *acidic* solution:

$$NH_4^+ + H_2O \rightleftharpoons NH_3 + H_3O^+$$

or, using our abbreviation of H^+ for the aqueous proton,

$$NH_4^+ \rightleftharpoons NH_3 + H^+$$

The equilibrium constant here will be given the symbol K_a, because this is the reaction of an acid, and we have

$$K_a = \frac{[NH_3(aq)][H^+]}{[NH_4^+]} = \frac{K_w}{K_b}$$

where K_b is the base dissociation constant for aqueous ammonia (You should derive the above relationship for yourself).

Again, the symbol K_h (hydrolysis constant) is sometimes used in place of K_a in the above expression. We will use K_a because it conveys more information, reminding us that we are dealing with an acid. Although the concept of the ammonium ion, NH_4^+, as a weak acid may seem strange at first sight, you should realize that it comes under just the same definition as more familiar acids, such as acetic acid, because it reacts with water to increase the concentration of hydrogen ions. The only difference is that NH_4^+ carries a charge, whereas CH_3COOH is a neutral molecule.

So we have exactly the same relationship between K_a, K_b, and K_w for the *conjugate acid–base pair* NH_4^+/NH_3 that we previously had for CH_3COOH/CH_3COO^-, and this is true for any such pair.

If we represent a general weak base as B, where B might be NH_3, CH_3NH_2 (methylamine), C_5H_5N (pyridine), or similar compounds, the two equilibria are written

$$B + H_2O \rightleftharpoons BH^+ + OH^- \qquad K_b = \frac{[BH^+][OH^-]}{[B]}$$
(reaction of base with water)

$$BH^+ + H_2O \rightleftharpoons B + H_3O^+ \qquad K_a = \frac{[B][H_3O^+]}{[BH^+]} = \frac{K_w}{K_b}$$
(reaction of conjugate acid with water)

We can apply this in working out the pH of salts of weak acids or weak bases.

Example 14.1 Calculate the pH and percent hydrolysis occurring in the following solutions (Ac^- is the acetate ion, CH_3COO^-).

(a) $0.100M$ NaAc
(b) $1.00 \times 10^{-3}M$ NaAc
(c) $0.200M$ NH_4Cl
(d) $2.00 \times 10^{-3}M$ NH_4Cl

SOLUTION

(a) In the NaAc solutions, the hydrolysis of Na^+ may be neglected, because the conjugate base, NaOH, is very strong, so the reaction determining the pH of the solution is

$$Ac^- + H_2O \rightleftharpoons HAc + OH^-$$

Suppose the amount of acetate ion which hydrolyzes is x mole liter^{-1}. Then the equilibrium concentrations will be

$$\underset{(0.100 - x)}{Ac^-} + H_2O \rightleftharpoons \underset{x}{HAc} + \underset{x}{OH^-} \quad \text{mole liter}^{-1}$$

$$K_b = \frac{[HAc][OH^-]}{[Ac^-]} = \frac{x^2}{(0.100 - x)} = \frac{K_w}{K_a} = \frac{1.0 \times 10^{-14}}{1.8 \times 10^{-5}} = 5.6 \times 10^{-10}$$

This gives us a quadratic equation in x, but we can simplify it as before by assuming that the degree of hydrolysis will be very small, so that

$$0.100 - x \approx 0.100$$

$$\frac{x^2}{(0.100 - x)} \approx \frac{x^2}{0.100} = 5.6 \times 10^{-10}$$

$$x^2 = 5.6 \times 10^{-11} \qquad x = 7.5 \times 10^{-6}$$

Checking on the value of x, we see that the assumption that it will be small is justified; x is completely negligible compared to 0.100. Since x represents $[OH^-]$, we calculate pOH

$$pOH = -\log x = -\log(7.5 \times 10^{-6}) = 5.13$$

$$pH = 14 - pOH = 14 - 5.13 = 8.87$$

The percentage hydrolysis is

$$\frac{7.5 \times 10^{-6}}{0.100} \times 100 = 7.5 \times 10^{-3}\%$$

(b) For $1.00 \times 10^{-3} M$ NaAc, the calculation follows the same lines. Suppose x mole liter^{-1} hydrolyze:

$$\underset{(1.00 \times 10^{-3}) - x}{Ac^-} + H_2O \rightleftharpoons \underset{x}{HAc} + \underset{x}{OH^-}$$

$$K_b = \frac{[HAc][OH^-]}{[Ac^-]} = \frac{x^2}{(1.00 \times 10^{-3}) - x} = 5.6 \times 10^{-10}$$

Now assume x is negligible beside 1.00×10^{-3}:

$$\frac{x^2}{(1.00 \times 10^{-3}) - x} \approx \frac{x^2}{(1.00 \times 10^{-3})} = 5.6 \times 10^{-10}$$

$$x^2 = 5.6 \times 10^{-13} \qquad x = 7.5 \times 10^{-7}$$

(The simplifying assumption is justified.)

$$pOH = -\log x = -\log(7.5 \times 10^{-7}) = 6.13$$

$$pH = 14 - pOH = 7.87$$

The percentage hydrolysis is

$$\frac{7.5 \times 10^{-7}}{1.00 \times 10^{-3}} \times 100 = 7.5 \times 10^{-2}\%$$

Comparing (a) and (b), we see that the degree of hydrolysis is greater in the dilute solution, although still very small. However, the $[OH^-]$ resulting from the more dilute solution is smaller, so the pH is less.

(c) For NH_4Cl, the ions present in solution are NH_4^+ and Cl^-. No hydrolysis of Cl^- need be considered, so the dominant reaction is

$$NH_4^+ + H_2O \rightleftharpoons NH_3(aq) + H^+$$

In this case, $[NH_4^+]$ is initially $0.200M$. Suppose x mole liter^{-1} undergo hydrolysis. At equilibrium, the concentrations are

$$\begin{array}{ccc} NH_4^+ & + H_2O \rightleftharpoons & NH_3(aq) + H^+ \\ (0.200 - x) & & x \qquad\quad x \end{array}$$

$$K_a = \frac{[NH_3(aq)][H^+]}{[NH_4^+]} = \frac{x^2}{(0.200 - x)} = \frac{K_w}{K_b} = \frac{1.0 \times 10^{-14}}{1.8 \times 10^{-5}} = 5.6 \times 10^{-10}$$

As before, we assume the degree of hydrolysis will be very small, so that $(0.200 - x) \approx 0.200$:

$$\frac{x^2}{(0.200 - x)} \approx \frac{x^2}{0.200} = 5.6 \times 10^{-10}$$

$$x^2 = 1.1 \times 10^{-10} \qquad x = 1.05 \times 10^{-5} = [H^+]$$

(The simplifying assumption is justified.)

$$pH = -\log[H^+] = -\log x = 4.98$$

The percentage hydrolysis is

$$\frac{1.05 \times 10^{-5}}{0.200} \times 100 = 5.3 \times 10^{-3}\%$$

(d) When the initial NH_4Cl concentration is $2.00 \times 10^{-3}M$, the calculation follows a similar course. Suppose x mole liter^{-1} are hydrolyzed:

$$\begin{array}{ccc} NH_4^+ & + H_2O \rightleftharpoons & NH_3(aq) + H^+ \\ (2.00 \times 10^{-3}) - x & & x \qquad\quad x \end{array}$$

assuming x to be negligible,

$$K_a = \frac{x^2}{(2.00 \times 10^{-3}) - x} \approx \frac{x^2}{2.00 \times 10^{-3}} = 5.6 \times 10^{-10}$$

$$x^2 = 1.1 \times 10^{-12} \qquad x = 1.05 \times 10^{-6} \qquad \text{(assumption justified)}$$

$$pH = -\log x = 5.98$$

The percentage hydrolysis is

$$\frac{1.05 \times 10^{-6}}{2.00 \times 10^{-3}} \times 100 = 5.3 \times 10^{-2}\%$$

As in the case of sodium acetate, we see that the degree of hydrolysis is greater in the more dilute solution, but it is very small in each case.

We should briefly consider a further possibility for hydrolysis: when we have a salt obtained by neutralizing a weak acid with a weak base, e.g., ammonium acetate. This is not an important case, and we will not consider the calculation in detail, but there are some points of interest. The two reactions may be written as before

$$NH_4^+ + H_2O \rightleftharpoons NH_3(aq) + H^+$$

$$Ac^- + H_2O \rightleftharpoons HAc + OH^-$$

However, if we produce both H^+ and OH^- at the same time, they will obviously react together to form water:

$$H^+ + OH^- \rightleftharpoons H_2O$$

Adding all *three* equations together gives the overall hydrolysis reaction:

$$NH_4^+ + Ac^- + H_2O \rightleftharpoons NH_3(aq) + HAc$$

The removal of H^+ and OH^- displaces the hydrolysis reactions of the individual ions to the right, so the overall hydrolysis reaction occurs to a much greater extent (about 50%) than was the case with the salt of each ion separately.

Calculation of the pH is quite complicated, since we have now two dissociation constants to consider, but it turns out that there is a simple relationship obtainable by making reasonable approximations:

$$\frac{[H^+]}{[OH^-]} = \frac{K_a}{K_b}$$

where K_a is the ionization constant of the weak acid and K_b that of the weak base. In the case of ammonium acetate, where the two ionization constants happen to be equal, the above equation tells us that $[H^+] = [OH^-]$, giving a neutral solution, as we would intuitively expect. In other cases, if we recall that $[OH^-] = K_w/[H^+]$, we may write

$$\frac{[H^+]}{[OH^-]} = \frac{[H^+]^2}{K_w} = \frac{K_a}{K_b} \qquad [H^+] = \sqrt{\frac{K_w K_a}{K_b}}$$

(*Note*: The molarity of the salt does not enter into this expression. In reasonable concentrations, the pH is independent of the salt concentration, as is the case for a buffer solution—which is discussed in the following section. In fact, a solution of ammonium acetate is an excellent buffer.)

Example 14.2   What is the pH of $0.100M$ ammonium cyanide solution?

K_b for $NH_3(aq)$: 1.8×10^{-5}
K_a for HCN: 4.9×10^{-10}

SOLUTION As we noted above, the concentration of the salt does not matter. Using the data given

$$[H^+] = \sqrt{\frac{K_w K_a}{K_b}} = \sqrt{\frac{(1.0 \times 10^{-14})(4.9 \times 10^{-10})}{1.8 \times 10^{-5}}}$$

$$[H^+] = \sqrt{2.7 \times 10^{-19}} = 5.2 \times 10^{-10}$$

$$pH = -\log(5.2 \times 10^{-10}) = 9.28$$

14.2 Buffer Solutions

A *buffer solution* is one in which the pH is kept within narrow limits, despite the addition to it of acids, bases, or other electrolytes. Since the functioning of any living organism is very dependent on the maintenance of a steady pH in its physiological fluids, we should see how this balance is maintained. We will find that the equilibrium involved is very similar to the acid–base reactions we have already studied in this chapter.

What constitutes a buffer solution? A buffer is a mixture of a solution of a weak acid with one of its salts, *or* a mixture of a solution of a weak base with one of its salts. We will refer to these as "acid buffer" or "base buffer" respectively.

Let's deal with the acid buffer case first, taking as a typical example a solution containing 0.100 moles of acetic acid *and* 0.100 moles of sodium acetate dissolved in water and made up to a total volume of 1.00 liter. How should we calculate the pH of such a solution? We cannot treat the two solutes separately, since the same ion (Ac^-) is common to both. In fact, the calculation is much easier than the one we did previously for the separate cases of HAc and NaAc solutions, provided we realize what is going on in solution.

First, we note that sodium acetate is an ionic salt, and, as such, is considered to be completely dissociated in solution. There is no question of "undissociated" sodium acetate being present. *This fixes the concentration of acetate ion.* In the present case, dissociation of 0.100 moles of sodium acetate gives $[Ac^-] = 0.100M$ (also giving $[Na^+] = 0.100M$, but the sodium ion does not enter into further reaction, and may be ignored).

If we had $0.100M$ acetic acid alone, we would consider its partial dissociation to H^+ and Ac^-. However, the presence in this case of a large amount of Ac^- from the NaAc we added will repress the ionization of acetic acid almost completely. We may regard the HAc as being present, undissociated, in the solution in exactly the concentration we originally added. *This fixes the concentration of acetic acid.* In the present solution, $[HAc] = 0.100M$.

We return to the familiar equation for the dissociation constant of the acid,

$$K_a = \frac{[H^+][Ac^-]}{[HAc]} \quad \text{or} \quad [H^+] = K_a \times \frac{[HAc]}{[Ac^-]}$$

In this buffer solution, we have fixed [HAc] and [Ac^-] by the concentrations in which we made up the solution. It is not necessary to consider any reactions occurring in the solution after it is made up.

Clearly, then, we can calculate [H^+] directly from the known value of the dissociation constant and the two concentrations. Since the concentrations of [HAc] and [Ac^-] are both $0.100M$ in the present case, they cancel out, giving the very simple result

$$[H^+] = K_a = 1.8 \times 10^{-5} \qquad pH = -\log(1.8 \times 10^{-5}) = 4.74$$

The above equation, which may be used for any acid buffer, may readily be converted to a logarithmic form:

$$[H^+] = K_a \times \frac{[HAc]}{[Ac^-]}$$

taking logs of both sides,

$$\log[H^+] = \log K_a + \log \frac{[HAc]}{[Ac^-]}$$

changing sign

$$-\log[H^+] = -\log K_a + \log \frac{[Ac^-]}{[HAc]} \quad \text{or} \quad pH = pK_a + \log \frac{[Ac^-]}{[HAc]}$$

This last equation may be written in the slightly more general form

$$pH = pK_a + \log \frac{[salt]}{[acid]}$$

which applies to any acid buffer, remembering that [salt] refers to the concentration of a salt of the weak acid (usually with sodium or potassium as the inert cation) which gives an equivalent concentration of the acid anion by complete dissociation in solution.

For a basic buffer, for example, $NH_3(aq) + NH_4Cl$, the expression is very similar, with $[OH^-]$ and pK_b instead of $[H^+]$ and pK_a:

$$[OH^-] = K_b \times \frac{[NH_3(aq)]}{[NH_4^+]} \qquad pOH = pK_b + \log \frac{[salt]}{[base]}$$

Example 14.3   What is the pH of a solution $0.200M$ in nitrous acid, HNO_2, and $0.050M$ in sodium nitrite? K_a for HNO_2 is 4.5×10^{-4}.

SOLUTION   Using the equation

$$pH = pK_a + \log \frac{[salt]}{[acid]} = pK_a + \log \frac{[NO_2^-]}{[HNO_2]}$$

we have here

pK_a for HNO_2: $-\log(4.5 \times 10^{-4}) = 3.35$
$[NO_2^-] = 0.050$ (from the sodium nitrite)
$[HNO_2] = 0.200$

$$pH = 3.35 + \log \frac{0.050}{0.200} = 3.35 + \log(0.25)$$

$$= 3.35 + \log(2.5 \times 10^{-1}) = 3.35 + 0.40 - 1.00 = 2.75$$

Example 14.4   Calculate the pH of a solution that is $0.030M$ in $NH_3(aq)$ and $0.070M$ in NH_4NO_3; $K_b = 1.8 \times 10^{-5}$ for $NH_3(aq)$.

SOLUTION   The NH_4NO_3 is an ionic salt, giving by dissociation $[NH_4^+] = 0.070$. (*Note*: The NO_3^- ions do not come into the picture, because HNO_3 is a strong acid.)

$$pK_b \text{ for } NH_3: \quad -\log(1.8 \times 10^{-5}) = 4.74$$

$$pOH = pK_b + \log \frac{[NH_4^+]}{[NH_3(aq)]} = 4.74 + \log \frac{0.070}{0.030}$$

$$= 4.74 + \log 2.33 = 4.74 + 0.37 = 5.11$$

$$pH = 14 - 5.11 = 8.89$$

14.3   Buffering Action

We have considered the method of calculating pH in buffer solutions above, but we have not yet seen how they show resistance to pH changes. This can be very simply explained if we remember what is present in a buffer mixture. Both acidic and basic buffers contain mixtures of a conjugate acid–base pair, e.g.,

acetic acid [acid] + acetate ion   (conjugate base)

ammonia [base] + ammonium ion   (conjugate acid)

If we add additional *acid* to either of these mixtures, it will be neutralized by reaction with the "base" in the mixture; while added *base* will be removed by reaction with the "acid" component. In this way, large changes in pH are not likely to occur.

Example 14.5   A solution is $0.200M$ in acetic acid and $0.200M$ in sodium acetate. What is the pH? What change in pH occurs on adding to a liter of this solution 0.050 mole of (a) HCl and (b) NaOH? (Ignore changes in volume.)

SOLUTION   The equation we need here is

$$pH = pK_a + \log \frac{[Ac^-]}{[HAc]} \qquad (pK_a = 4.74)$$

In the original solution, $[Ac^-] = [HAc] = 0.200M$, so

$$\log \frac{[Ac^-]}{[HAc]} = \log 1.00 = 0.0 \qquad pH = pK_a = 4.74$$

(a) When we add 0.050 mole liter^{-1} of HCl, it reacts *completely* with acetate ion to give acetic acid

$$HCl + Ac^- \longrightarrow HAc + Cl^-$$

This changes the concentrations of Ac^- and HAc:

$$[Ac^-] = 0.200 - 0.050 = 0.150M$$

$$[HAc] = 0.200 + 0.050 = 0.250M$$

The pH can be recalculated from these values

$$pH = pK_a + \log \frac{[Ac^-]}{[HAc]} = 4.74 + \log \frac{0.150}{0.250}$$

$$= 4.74 + \log 0.60 = 4.74 - 0.22 = 4.52$$

The pH has decreased by only 0.22 unit.

(b) When 0.050 mole liter^{-1} of NaOH is added, the reverse process happens. [HAc] is reduced and [Ac$^-$] is increased by the reaction

$$\text{HAc} + \text{NaOH} \longrightarrow \text{Ac}^- + \text{Na}^+ + \text{H}_2\text{O}$$

This reaction is complete, so [HAc] and [Ac$^-$] change:

$$[\text{HAc}] = 0.200 - 0.050 = 0.150$$

$$[\text{Ac}^-] = 0.200 + 0.050 = 0.250$$

$$\text{pH} = \text{p}K_a + \log \frac{[\text{Ac}^-]}{[\text{HAc}]} = 4.74 + \log \frac{0.250}{0.150}$$

$$= 4.74 + \log 1.67 = 4.74 + 0.22 = 4.96$$

Again, the pH has changed by only 0.22 unit, but this time in the reverse direction; it has increased with the addition of base.

We see in the above examples how the pH changed by only a small fraction of a pH unit on the addition of acid or base. If the same amount of either had been added to pure water, the pH would have changed by more than 5 pH units:

1 liter pure water, pH 7.00

addition of 0.050 mole of HCl gives

$[\text{H}^+] = 0.050$ (HCl dissociates completely)

$$\text{pH} = -\log(0.050) = -\log(5.0 \times 10^{-2})$$

$$= -(0.70 - 2) = -(-1.30) = 1.30$$

Similarly, $0.050M$ NaOH has pOH 1.30, pH 12.70. The same effect is seen in a basic buffer solution.

Example 14.6

(a) What is the pH of a solution $0.100M$ in NH$_3$ and $1.50M$ in NH$_4$Cl?
(b) What change will occur if 0.100 mole liter^{-1} of KOH is added?

SOLUTION

(a) The NH$_4$Cl dissociates completely to NH$_4^+$ and Cl$^-$, so $[\text{NH}_4^+] = 1.50$.

$$\text{pOH} = \text{p}K_b = \log \frac{[\text{NH}_4^+]}{[\text{NH}_3(aq)]} = 4.74 + \log \frac{1.50}{0.100}$$

$$= 4.74 + \log 15.0 = 4.74 + 1.18 = 5.92$$

$$\text{pH} = 14.0 - 5.92 = 8.08$$

(b) When KOH is added, it reacts completely with NH$_4^+$, the acid component of the mixture

$$\text{KOH} + \text{NH}_4^+ \longrightarrow \text{K}^+ + \text{NH}_3(aq) + \text{H}_2\text{O}$$

The addition of 0.100 mole KOH removes 0.100 mole NH_4^+ from solution and produces the same amount of NH_3. New concentrations will be:

$[NH_4^+]$: $1.50 - 0.100 = 1.40$
$[NH_3(aq)]$: $0.100 + 0.100 = 0.200$

$$pOH = 4.74 + \log \frac{1.40}{0.200} = 5.59$$

$$pH = 14.0 - 5.59 = 8.41$$

Example 14.7 Using acetic acid and sodium acetate, design a buffer solution having a pH of (a) 5.00 and (b) 7.00.

SOLUTION   Here we shall be using the same equation in the reverse direction. The pH is known, but we have to find from it the concentrations.

$$pH = pK_a + \log \frac{[Ac^-]}{[HAc]} \qquad (pK_a = 4.74)$$

(a) For a solution with pH of 5.00, we write

$$5.00 = 4.74 + \log \frac{[Ac^-]}{[HAc]} \qquad \log \frac{[Ac^-]}{[HAc]} = 5.00 - 4.74 = 0.26 \qquad \frac{[Ac^-]}{[HAc]} = 1.82$$

We only define the *ratio* of $[Ac^-]$ to $[HAc]$ in this equation. Any reasonable solution having this ratio would give the desired pH. For example, we could take

$[Ac^-] = 0.182M$ (i.e., sodium acetate $= 0.182M$)
$[HAc] = 0.100M$

(b) In the case of a pH of 7.00, we can write

$$7.00 = 4.74 + \log \frac{[Ac^-]}{[HAc]} \qquad \log \frac{[Ac^-]}{[HAc]} = 7.00 - 4.74 = 2.26 \qquad \frac{[Ac^-]}{[HAc]} = 182$$

This solution has a much greater disparity between the two concentrations, but, as before, only their ratio is defined. We might choose

$[Ac^-] = 1.82M$ (i.e., sodium acetate $= 1.82M$)
$[HAc] = 0.010M$

The above calculation shows the limitations on the ability of a buffer solution to control pH. Good results are only obtained for values of pH fairly close to the pK value of the weak acid or base being used (say, within one pH unit of this value). Further away from the pK value, the disparity between the two concentrations becomes considerable and the solution does not show good buffering ability.

There must, of course, be a limit to the buffering capacity of a solution. Thus, a solution containing 0.10 mole liter^{-1} of both HAc and Ac^- could absorb added acid or base up to a limit of 0.10 mole liter^{-1} of either. Beyond this, the buffering action would no longer operate, because one component of the acid–base mixture would be removed.

14.4 Mixing Acids and Bases

A lot of students can calculate the pH of solutions of acids, bases, salts, buffers, etc., but come to grief when faced with the data presented in a slightly different way. There is a very simple precept to follow when this type of problem appears: *When acid and base are mixed, reaction occurs.*

All we have to do is to work out, from the amounts of acid and base added, the concentrations of various species present at equilibrium, then calculate the pH accordingly. Let's start with examples from strong-acid, strong-base mixtures.

Example 14.8 50.0 ml of 0.100M HCl is titrated with 0.100M KOH. Calculate the pH when the following volumes of base have been added:

(a) none   (b) 25.0 ml   (c) 49.0 ml   (d) 49.9 ml
(e) 50.0 ml   (f) 50.1 ml   (g) 51.0 ml   (h) 75.0 ml

SOLUTION   We start out with $50.0 \times 0.100 = 5.00$ mmole of HCl. As we add base, this reacts to reduce the concentration of H^+ as the titration proceeds (remember the total volume changes also).

(a) Initially $[H^+] = 0.100$ and $pH = 1.00$
(b) Base added is $25.0 \times 0.100 = 2.50$ mmole. This reacts with 2.50 mmole of acid, leaving 2.50 mmole unreacted. Total volume is 75.0 ml, so

$$[H^+] = \frac{2.50}{75.0} = 3.33 \times 10^{-2}M$$

$$pH = -\log(3.33 \times 10^{-2}) = 2 - \log 3.33 = 2 - 0.52 = 1.48$$

(c) Base added, 4.90 mmole; acid remaining, 0.10 mmole. Total volume is 99.0 ml, so

$$[H^+] = \frac{0.10}{99.0} = 1.01 \times 10^{-3}M$$

$$pH = -\log(1.01 \times 10^{-3}) = 3 - \log 1.01 = 3 - 0.004 = 3.00$$

(d) Base added, 4.99 mmole; acid remaining, 0.01 mmole. Total volume is 99.9 ml, so

$$[H^+] = \frac{0.01}{99.9} = 1.00 \times 10^{-4}M$$

$$pH = -\log(1.00 \times 10^{-4}) = 4.00$$

(e) When 50.0 ml of base have been added, we have completely neutralized the acid, so the only source of H^+ is the dissociation of water

$$[H^+] = 1.0 \times 10^{-7} \qquad pH = 7.00$$

This, of course, represents the end point of the titration.

(f) Once past the end point, we have an excess of base over that required to neutralize the acid (5.00 mmole). This makes the solution alkaline.

50.1 ml 0.100M base contain 5.01 mmole OH^-

5.00 mmole reacts with acid, leaving 0.01 mmole of OH^-

total volume 100.1 ml

$$[OH^-] = \frac{0.01}{100.1} = 1 \times 10^{-4}M$$

$$pOH = 4.0 \qquad pH = 10.0$$

(g) Base added, 5.10 mmole; remaining unreacted, 0.10 mmole; total volume is 101.0 ml, so

$$[OH^-] = \frac{0.10}{101.0} = 1.0 \times 10^{-3}M$$

$$pOH = 3.00 \qquad pH = 11.00$$

(h) Base added, 7.50 mmole; remaining unreacted, 2.50 mmole; total volume is 125.0 ml, so

$$[OH^-] = \frac{2.50}{125.0} = 0.020M$$

$$pOH = -\log(2.0 \times 10^{-2}) = 2 - \log 2.0 = 2 - 0.30 = 1.70$$

$$pH = 14 - 1.70 = 12.30$$

The above calculations give a *titration curve*, that is, a plot of the change in pH as a function of the volumes of acid and base mixed during a titration. Note the sudden change in pH near the end point. With a weak acid, the calculation follows a slightly different course.

Example 14.9 To 50.0 ml of 0.100M NaOH is added 0.100M acetic acid. Calculate the pH after the addition of the following volumes of acid:

(a) 25.0 ml   (b) 50.0 ml   (c) 75.0 ml

SOLUTION   We should first think qualitatively about what happens when a weak acid, such as acetic, is mixed with a strong base (NaOH). We know that the acid is only slightly ionized, so there is only a small amount of H^+ available initially to react with the OH^- in the basic solution. But *removal of this H^+ causes immediate ionization of more weak acid* and further reaction, so the overall process may be written.

$$HAc \quad + \quad OH^- \quad \longrightarrow \quad Ac^- + H_2O$$
$$\text{(un-ionized} \quad \text{(strong}$$
$$\text{weak acid)} \quad \text{base)}$$

How complete is this process? A glance at the above equation shows that it is the exact *reverse* of the equation we previously considered for the hydrolysis of acetate ion, so its equilibrium constant will be the inverse of the K_b value we used in that case:

$$K = \frac{[Ac^-]}{[HAc][OH^-]} = \frac{K_a}{K_w}$$

For acetic acid, the numerical value of this constant is $(1.8 \times 10^{-5})/(1.0 \times 10^{-14})$, or 1.8×10^9. Obviously the equilibrium is very much in favor of the products, so the quantity of unreacted HAc remaining will be negligible. The following conclusions should be carefully borne in mind in systems of this type:

A weak acid only ionizes slightly in water, but *reacts quantitatively* with added strong base (OH^-)

A weak base only ionizes slightly in water, but *reacts quantitatively* with added strong acid (H^+).

Applying this to the present example, we start with 50.0 ml of $0.100M$ NaOH, which is 5.00 mmole of OH^-.

(a) 25.0 ml of $0.100M$ HAc contains 2.50 mmole HAc. This reacts with 2.50 mmole OH^-, leaving 2.50 mmole OH^- unreacted. The total volume is 75.0 ml, so

$$[OH^-] = \frac{2.50}{75.0} = 3.33 \times 10^{-2}$$

$$pOH = 1.48 \qquad pH = 12.52$$

(*Note*: We have a mixture of Ac^- and OH^- ions in solution. This is *not* a buffer, and the ions Ac^- do not affect the pH.)

(b) Addition of 50.0 ml of $0.100M$ HAc takes us to the equivalence point, where all the base has been neutralized. Our 5.00 mmole of OH^- has been converted to 5.00 mmole of Ac^- by reaction with HAc.

In other words, we have a solution of sodium acetate. Since Ac^- is a weak base, hydrolysis will occur, and the solution will not be neutral. We have 5.00 mmole Ac^- in 100.0 ml of solution, so $[Ac^-] = 5.00 \times 10^{-2}$. The equilibrium is

$$
\begin{array}{ccccc}
Ac^- & + H_2O & \rightleftharpoons & HAc & + OH^- \\
(5.00 \times 10^{-2}) - x & & & x & x
\end{array}
$$

$$\frac{[HAc][OH^-]}{[Ac^-]} = \frac{x^2}{(5.00 \times 10^{-2}) - x} = K_b = \frac{K_w}{K_a} = 5.6 \times 10^{-10}$$

Assuming x is negligible beside 5.00×10^{-2} gives

$x^2 = (5.00 \times 10^{-2})(5.6 \times 10^{-10}) = 2.8 \times 10^{-11}$
$x = 5.3 \times 10^{-6}$ (assumption justified)
$x = [OH^-]$

$$pOH = 5.28 \qquad pH = 8.72$$

(*Note*: There is a difference between this case and (a) above. In (a), we did not have to consider hydrolysis of Ac^- because any OH^- from that source would be negligible beside the OH^- present from excess, unneutralized, OH^- from NaOH).

(c) Once we are past the end point, no further reaction of added HAc will occur, so we have a mixture of Ac^- and HAc. *This is a buffer solution.*

Specifically, we have the 5.00 mmole of Ac^- resulting from the neutralization of 5.00 mmole of HAc. Since 75.0 ml of $0.100M$ HAc contains 7.50 mmole of HAc, this leaves 2.50 mmole unreacted. The total volume is 125.0 ml, so

$$[Ac^-] = \frac{5.00}{125.0} = 0.0400M \qquad [HAc] = \frac{2.50}{125.0} = 0.0200M$$

As before in buffer solutions, we substitute these values directly into the K_a expression to find $[H^+]$. No ionization or hydrolysis reaction need be considered.

$$K_a = \frac{[H^+][Ac^-]}{[HAc]} \quad \text{or} \quad [H^+] = \frac{K_a[HAc]}{[Ac^-]}$$

$$[H^+] = (1.8 \times 10^{-5})\frac{0.0200}{0.0400} = 9.0 \times 10^{-6}$$

$$pH = 5.05$$

The calculations of this last example give three points in the titration curve of a strong base–weak acid mixture. Such calculations are very simple provided we remember what is going on. Summarizing:

When a weak acid is added to a strong base (or a weak base added to a strong acid) the solution will be buffered *after* the equivalence point.

When a strong base is added to a weak acid (or a strong acid added to a weak base) the solution will be buffered *before* the equivalence point.

When a weak acid and weak base are mixed, the solution will be buffered throughout. The pH will change gradually throughout; no end-point will be seen, and such titrations are not practicable.

Example 14.10   What is the pH of the solution resulting when 150 ml of 0.200M $NH_3(aq)$ solution ($K_b = 1.8 \times 10^{-5}$) are mixed with 200 ml of 0.100M HCl?

SOLUTION   This is very similar to the previous problem.

150 ml 0.200M $NH_3(aq)$ contain $150 \times 0.200 = 30.0$ mmole NH_3
200 ml 0.100M HCl contain $200 \times 0.100 = 20.0$ mmole HCl (ionized)

When these are mixed, they react to produce NH_4^+ and Cl^-.

$$NH_3 + H^+ + Cl^- \longrightarrow NH_4^+ + Cl^-$$

This reaction is complete: 20.0 mmole of NH_4^+ are produced and 20.0 mmole of NH_3 are removed, leaving $30.0 - 20.0 = 10.0$ mmole of NH_3. Since we have both NH_4^+ and NH_3 present, we have a basic buffer solution.

$$pOH = pK_b + \log\frac{[NH_4^+]}{[NH_3(aq)]} = 4.74 + \log\frac{20.0}{10.0}$$

$$= 4.74 + \log 2.00 = 4.74 + 0.30 = 5.04$$

$$pH = 14.0 - pOH = 14.0 - 5.04 = 8.96$$

(*Note*: We may put numbers of moles into the expression $[NH_4^+]/[NH_3(aq)]$, rather than molarities, since we are interested only in the *ratio* of these quantities. The volume of the solution does not affect this.)

14.5   Simultaneous Equilibria

If two or more equilibria have an ion (or several ions) in common, then they cannot be treated separately. We must consider *both* equilibrium constants in calculations of equilibrium positions, and such reactions are said to be *simultaneous equilibria*. In this section we will be considering several types of process under this heading, including further aspects of acid-base chemistry, solubility equilibria, and complex ion formation.

14.6   Very Dilute Solutions

Let's start with a simple example. What is the pH of $10^{-8}M$ HCl? That seems an easy calculation. Obviously the acid is completely dissociated, so $[H^+] = 10^{-8}M$,

$$pH = -\log(10^{-8}) = +8.0$$

But a pH of 8.0 is an alkaline solution? How can dilute HCl have a *lower* concentration of hydrogen ion than pure water?

The last sentence contains the clue to what has gone wrong in this calculation. We have neglected the H^+ ions produced by the dissociation of water. In any acid-base reaction, the equilibrium

$$H_2O \rightleftharpoons H^+ + OH^-$$

is always present simultaneously with dissociation or hydrolysis reactions that the solute is undergoing. We normally neglect the H^+ produced by the ionization of water, because it is much less than that from the solute, but the very-dilute-HCl paradox draws our attention to it.

More correctly, for any acid, suppose the H^+ contributed by the acid solute is M mole liter^{-1}. At the same time, the dissociation of water produces x mole liter^{-1} of both H^+ and OH^- (they must obviously be produced in equal amounts): Then

total $[H^+] = M + x$

total $[OH^-] = x$

equilibrium is $[H^+][OH^-] = (M + x)x = 10^{-14}$

In reasonable concentrations of acid solute, the value of x is much less than M, and we make the approximation $M + x \approx M$, so

$$[H^+] \approx M \qquad [OH^-] = \frac{10^{-14}}{(M + x)} \approx \frac{10^{-14}}{M}$$

But if $M = 10^{-8}$ (as with this very dilute HCl) the value of x is *not* negligible beside M and we have to solve the quadratic equation

$$(10^{-8} + x)x = 10^{-14}$$

The accurate solution is $x = 0.95 \times 10^{-7}$, so the total $[H^+]$ in the solution is

$$[H^+] = (M + x) = 10^{-8} + (0.95 \times 10^{-7}) = 1.05 \times 10^{-7}$$

We see that this is slightly greater than $[H^+]$ in pure water. If we reduce the solute HCl concentration still further, for example, to $10^{-9}M$, the contribution of H^+ from the dissociation of water will be very much greater than that of the solute, so we can make the approximation $(M + x) \approx x$. The equation $(M + x)x = 10^{-14}$ simplifies to $x^2 = 10^{-14}$; in other words, the value of $[H^+]$ stays at $10^{-7}M$. Obviously, what we are saying is that very dilute solutions approximate to pure water.

The above example may seem rather trivial, since such very dilute solutions are of little practical importance, but it serves to remind us of the approximation we are making when we carry out calculations on solute reactions and ignore the simultaneous equilibrium that the solvent water is always establishing.

Some solubility product calculations illustrate the same point:

Example 14.11   Calculate the concentration of the metal ion and the pH in saturated solutions of the following:

(a) for $Ni(OH)_2$: $K_{sp} = 6.5 \times 10^{-18}$
(b) for $Fe(OH)_3$: $K_{sp} = 1.1 \times 10^{-36}$

SOLUTION

(a) Suppose x mole liter^{-1} of $Ni(OH)_2$ dissolve to give a saturated solution. Then, because one mole of $Ni(OH)_2$ gives *two* moles of OH^- on dissolving, the concentrations produced are $[Ni^{2+}] = x$ and $[OH^-] = 2x$, so

$$K_{sp} = [Ni^{2+}][OH^-]^2 = 4x^3 = 6.5 \times 10^{-18}$$

$$x^3 = 1.63 \times 10^{-18} \qquad x = 1.17 \times 10^{-6}$$

$$[Ni^{2+}] = x = 1.17 \times 10^{-6} \qquad [OH^-] = 2x = 2.34 \times 10^{-6}$$

$$pOH = -\log(2.34 \times 10^{-6}) = 5.63$$

$$pH = 8.37$$

As we would expect, the solution is slightly alkaline.

(b) For $Fe(OH)_3$, suppose x mole liter^{-1} dissolve. Then the concentrations are: $[Fe^{3+}] = x$, $[OH^-] = 3x$ (each mole of $Fe(OH)_3$ now produces *three* moles of OH^- on dissolving).

$$K_{sp} = [Fe^{3+}][OH^-]^3 = 27x^4 = 1.1 \times 10^{-36}$$

$$x^4 = 4.07 \times 10^{-38} \qquad x = 4.5 \times 10^{-10}$$

This leads to $[OH^-] = 3x = 1.35 \times 10^{-9}M$, which gives $pOH = 8.87$ and $pH = 5.13$. At this point, we should start to get suspicious. How does the solution come to be acidic? How can the concentration of OH^- be *lowered* by dissolving $Fe(OH)_3$ in water?

The answer is, of course, that we cannot lower the pH in this way. The above answer is incorrect, because we have neglected the OH^- produced by the dissociation of water. In the example with $Ni(OH)_2$, the contribution of OH^- from the water was negligible beside that from the dissolved hydroxide; but with $Fe(OH)_3$, the reverse is the case. We can neglect the OH^- from the solute and put $[OH^-] = 10^{-7}$ from the water alone. This gives for the metal ion concentration.

$$[Fe^{3+}] = \frac{K_{sp}}{[OH^-]^3} = \frac{1.1 \times 10^{-36}}{(10^{-7})^3} = \frac{1.1 \times 10^{-36}}{10^{-21}} = 1.1 \times 10^{-15}M$$

The pH of the solution is, of course, 7.0.

An equilibrium of this type can be regarded as an example of the common ion effect, discussed in Chapter 12, with the difference that the ion in question (hydroxide) comes from the dissociation of the solvent, rather than from added electrolyte.

14.7 Polyprotic Acids

We spent some time in Chapter 13 considering the dissociation of acids, but we restricted our calculations to acids in which each molecule lost only one hydrogen ion on dissociation. Such acids are called *monoprotic* (= one proton). Many other acids are able to lose two or more hydrogen ions per molecule on ionization, and they are called *polyprotic* (=many protons). An acid that loses two H^+ per molecule is *diprotic*, for example, H_2SO_4, $(COOH)_2$ (oxalic acid). An acid such as phosphoric acid, H_3PO_4, is triprotic. (*Note*: We are not concerned with the *total* number of hydrogen atoms in the molecule, only those that may be lost on ionization. Acetic acid, CH_3COOH, contains four hydrogen atoms in the molecule, but only one may be lost by ionization.)

For a polyprotic acid, each dissociation will have a separate value for K_a. As successive hydrogen ions are removed, the remaining ion acquires an increasing negative charge, so removal of another H^+ is more difficult. The successive values of K_a's for a molecule become smaller as a result. The different K_a values are known as K_1, K_2, etc. For example, for H_3PO_4, there are three K_a values:

$$H_3PO_4 \rightleftharpoons H_2PO_4^- + H^+ \qquad K_1 = 1.1 \times 10^{-2}$$

$$H_2PO_4^- \rightleftharpoons HPO_4^{2-} + H^+ \qquad K_2 = 7.5 \times 10^{-8}$$

$$HPO_4^{2-} \rightleftharpoons PO_4^{3-} + H^+ \qquad K_3 = 4.8 \times 10^{-13}$$

Adding these together gives an overall K

$$H_3PO_4 \rightleftharpoons PO_4^{3-} + 3H^+ \qquad K = K_1 K_2 K_3 = 4.0 \times 10^{-22}$$

In calculating the pH of a solution of a polyprotic acid, we have to take all the equilibria into account.

Example 14.12 Calculate $[H^+]$, the pH, and the concentration of all species present in $1.50M$ selenious acid, H_2SeO_3, for which $K_1 = 3.0 \times 10^{-3}$, $K_2 = 5.0 \times 10^{-8}$. What is the percentage dissociation in each of the two steps?

SOLUTION If we consider the two dissociation steps together, this problem is quite complex, but we can greatly simplify it by first considering only the *first* dissociation step, on the assumption that the *second* dissociation will make negligible differences to the concentrations of the ions. Later on in the calculation, we have to make sure that this assumption is justified.

Suppose x mole liter^{-1} of acid dissociate in the first step. The equilibrium is

$$H_2SeO_3 \rightleftharpoons HSeO_3^- + H^+$$
$$1.50 - x \qquad\qquad x \qquad x$$

$$K_1 = \frac{[HSeO_3^-][H^+]}{[H_2SeO_3]} = \frac{x^2}{1.50 - x} = 3.0 \times 10^{-3}$$

Making the usual assumption that x is small:

$$\frac{x^2}{1.50 - x} \approx \frac{x^2}{1.50} = 3.0 \times 10^{-3}$$

$$x^2 = 4.5 \times 10^{-3} \qquad x = 6.7 \times 10^{-2}$$

This is a degree of dissociation of

$$\frac{6.7 \times 10^{-2}}{1.50} \times 100 = 4.5\% \qquad \text{(assumption justified)}$$

So we can write the concentrations of ions resulting from the *first* stage in the dissociation

$$[H^+] = [HSeO_3^-] = 6.7 \times 10^{-2}M$$

The second stage is the dissociation

$$HSeO_3^- \rightleftharpoons SeO_3^{2-} + H^+$$

for which

$$K_2 = \frac{[SeO_3^{2-}][H^+]}{[HSeO_3^-]} = 5.0 \times 10^{-8}$$

It is easy to go wrong in the calculation here, by saying: "Let x moles dissociate ... etc." This would be wrong, because the second step in the dissociation produces H^+, which is *already present* in the solution as a result of the first dissociation step. (*Note*: This, of course, is why we are considering this process under the heading of simultaneous equilibrium.)

The correct approach is to note that, since the value of K_2 is very small, the degree of dissociation in the second step will be so small that it will make a negligible difference to the concentrations of H^+ and $HSeO_3^-$ already present. We can therefore insert these values into the above equation for K_2:

$$K_2 = \frac{[SeO_3^{2-}][H^+]}{[HSeO_3^-]} = 5.0 \times 10^{-8}$$

But $[H^+] = [HSeO_3^-] = 6.7 \times 10^{-2}$ (they are equal), so $[SeO_3^{2-}] = K_2 = 5.0 \times 10^{-8}$.

The amount of SeO_3^{2-} produced in the second dissociation is equal to the amount of $HSeO_3^-$ removed in this step, and also the additional H^+ produced. Obviously this amount ($5.0 \times 10^{-8}M$) is a negligible fraction of the amount of H^+ and $HSeO_3^-$ present already $(6.7 \times 10^{-2}M)$, so our assumption concerning the second stage equilibrium was completely justified. Summarizing the results gives us,

$$[H_2SeO_3] = 1.50 - 0.07 = 1.43M$$
$$[H^+] = [HSeO_3^-] = 6.7 \times 10^{-2}M$$
$$pH = -\log(6.7 \times 10^{-2}) = 1.17$$

percent dissociation in first stage $= 4.5\%$

percent dissociation in second stage $= \dfrac{5.0 \times 10^{-8}}{6.7 \times 10^{-2}} \times 100 = 7.5 \times 10^{-5}\%$

Example 14.13 Sulfuric acid dissociates in two stages:

$$H_2SO_4 \rightleftharpoons HSO_4^- + H^+$$
$$HSO_4^- \rightleftharpoons SO_4^{2-} + H^+$$

K_1 for the first dissociation is large enough that this may be considered complete; whereas K_2 has the value 1.2×10^{-2}. What will be the concentration of the various species in $0.10M$ sulfuric acid solution? Calculate the pH and the percentage second stage dissociation.

SOLUTION   This problem is similar to the previous example, except that the degree of dissociation is much greater, so we must be careful about the approximations we make.

The complete dissociation in the first stage will give

$$[HSO_4^-] = [H^+] = 0.10M.$$

Suppose the second stage dissociation is x mole liter^{-1}. Then equilibrium concentrations are

$$\begin{array}{ccc} HSO_4^- & \rightleftharpoons & SO_4^{2-} + & H^+ \\ (0.10 - x) & & x & (0.10 + x) \end{array}$$

(*Note*: It is very important to remember that the total H^+ concentration is the *sum* of the H^+ produced in stage 1 *and* stage 2.)

$$K_2 = \frac{x(0.10 + x)}{(0.10 - x)} = 1.2 \times 10^{-2}$$

This is a quadratic equation in x. We can try to simplify it by assuming that x is much smaller than 0.10. (*Note*: this assumption was successful in Example 14.12.) So assume that

$$(0.10 + x) \approx (0.10 - x) \approx 0.10$$

The expression above then simplifies to $K_2 = x = 1.2 \times 10^{-2}$. This shows that x is in fact about 12% of 0.10, in other words our assumption was not justified. A better approximation for x is found by feeding back this rough value and recalculating:

$$\frac{x(0.10 + x)}{(0.10 - x)} \approx \frac{x(0.10 + 0.012)}{(0.10 - 0.012)} = 1.2 \times 10^{-2}$$

$$x = (1.2 \times 10^{-2}) \times \frac{0.088}{0.112} = 9.4 \times 10^{-3}$$

This is a more accurate value of x, and may be fed back for a further improvement:

$$\frac{x(0.10 + 0.0094)}{(0.10 - 0.0094)} = 1.2 \times 10^{-2} \qquad x = (1.2 \times 10^{-2})\frac{0.0906}{0.1094} = 9.9 \times 10^{-3}$$

(The same answer would, of course, be obtained by solving the quadratic rigorously.)

Concentrations are:

$[HSO_4^-]$: $(0.10 - x) = 0.090M$
$[SO_4^{2-}]$: $x = 9.9 \times 10^{-3}M$
$[H^+]$: $(0.10 + x) = 0.11M$

$$pH = -\log(0.11) = -\log(1.1 \times 10^{-1}) = 1 - \log 1.1 = 1 - 0.04 = 0.96$$

$$\text{dissociation in second stage} = \frac{9.9 \times 10^{-3}}{0.10} \times 100 = 9.9\%$$

This calculation shows us that the most abundant species present in $0.10M$ sulfuric acid solution is the bisulfate ion, HSO_4^-, while the amount of the acid which dissociates completely to give the sulfate ion, SO_4^{2-}, is less than 10%.

14.8 Solubility and pH

The solubility of many substances is very dependent on the pH of the solution. For example, a simple experiment will show that barium carbonate is not appreciably soluble in water, but dissolves readily in dilute nitric acid. With barium sulfate, however, there is no change; the compound dissolves neither in water nor in dilute acid. How can we explain such results?

In qualitative terms, it is quite easy to account for the above, and many similar, observations. It will generally be found that a slightly soluble salt of a *weak* acid will show a greater solubility in acid solution, whereas the solubility of a slightly soluble salt of a *strong* acid will be unaffected by the pH of the solution. This is because in the more acid solution the ionization of the weak acid will be repressed, so the concentration of the anion will be lowered. This will reduce the ion product of the slightly soluble salt so that a greater quantity of it will dissolve.

An example should clarify this principle:

Example 14.14 Silver acetate, CH_3COOAg, has a solubility product of 3.7×10^{-3}. What will be the concentration of silver ion in the following solutions in equilibrium with solid silver acetate:

(a) pH 2.00   (b) pH 4.50   (c) pH 7.00

Assume that the dissolving silver acetate is the only source of silver or acetate ions. (K_a for acetic acid is 1.8×10^{-5}.)

SOLUTION There are two equilibria present simultaneously here, that between solid silver acetate and the ions in solution, and that between acetate ion and hydrogen ion in solution. Abbreviating acetate ion to Ac^-, we can write

$$AgAc(s) \rightleftharpoons Ag^+(aq) + Ac^-(aq)$$

$$K_{sp} = [Ag^+][Ac^-] \qquad K_a = \frac{[H^+][Ac^-]}{[HAc]}$$

From the first equation, we can say $[Ac^-] = K_{sp}/[Ag^+]$ and rewriting the second equation gives

$$[HAc] = \frac{[H^+][Ac^-]}{K_a} = \frac{[H^+]K_{sp}}{[Ag^+]K_a}$$

When AgAc goes into solution, Ag^+ and Ac^- ions are produced in equal amounts, so in the absence of any further reaction we would write $[Ag^+] = [Ac^-]$. In this solution, however, we are assuming some further reaction in which Ac^- is converted to HAc, so the total acetate-containing species are divided between Ac^- and HAc. This means that we should write $[Ag^+] = [Ac^-] + [HAc]$. Then substituting the previously obtained expressions for $[Ac^-]$ and $[HAc]$ gives the result

$$[Ag^+] = \frac{K_{sp}}{[Ag^+]} + \frac{[H^+]K_{sp}}{[Ag^+]K_a}$$

which rearranges to

$$[Ag^+]^2 = K_{sp}\left[1 + \frac{[H^+]}{K_a}\right]$$

This very useful and simple result may be applied to any equilibrium of this type, where we have the salt of a weak, monoprotic acid in an equilibrium between solid and solution. Obviously the concentration of the metal ion (Ag^+ in this case) will increase as K_{sp} increases. The effect of changing $[H^+]$ will depend on the value of K_a. If we are dealing with the salt of a weak acid, K_a will be small and $[H^+]/K_a$ will be relatively large; whereas a large value of K_a will reduce $[H^+]/K_a$ so that change of $[H^+]$ has a negligible effect on solubility.

Putting in the specific numbers for this problem (silver acetate) we have

$$K_{sp} = 3.7 \times 10^{-3} \qquad K_a = 1.8 \times 10^{-5}$$

(a) pH = 2.00: $[H^+] = 1.0 \times 10^{-2}$

$$[Ag^+]^2 = 3.7 \times 10^{-3} \left[1 + \frac{1.0 \times 10^{-2}}{1.8 \times 10^{-5}} \right] = 3.7 \times 10^{-3}(1 + 560) = 2.1$$

$$[Ag^+] = 1.4 M$$

(b) at pH = 4.50: $[H^+] = 3.16 \times 10^{-5}$

$$[Ag^+]^2 = 3.7 \times 10^{-3} \left(1 + \frac{3.16 \times 10^{-5}}{1.8 \times 10^{-5}} \right) = 3.7 \times 10^{-3}(1 + 1.8) = 1.0 \times 10^{-2}$$

$$[Ag^+] = 0.10 M$$

(c) at pH = 7.00: $[H^+] = 1.0 \times 10^{-7}$

$$[Ag^+]^2 = 3.7 \times 10^{-3} \left(1 + \frac{1.0 \times 10^{-7}}{1.8 \times 10^{-5}} \right) = 3.7 \times 10^{-3}[1 + (5.6 \times 10^{-3})] = 3.7 \times 10^{-3}$$

$$[Ag^+] = 0.061 M$$

We see that the solubility of silver acetate is much increased in the more acid (lower pH) solution. Remembering that, of the two terms inside the bracket, the first represents $[Ac^-]$, while the second represents [HAc], we see that, in (a), the dissolved acetate is mostly in the form of acetic acid, while in (c), it is mostly ionized to acetate ion. In (b), both acetic acid and acetate ion are present.

Silver acetate is an example of a slightly soluble substance where the two ions have charges of $+1$ and -1. More common are slightly soluble electrolytes where one or both ions has a charge of 2 or more. Where the negative ion has a charge of -2, the salt will be derived from a diprotic acid, so we have to include *two* dissociation constants for the acid in our calculation. This introduces one more constant into the calculation, but the resulting equation is very similar to that found in the previous example.

Example 14.15   Derive an expression for the concentration of barium ion in a solution in equilibrium with solid barium oxalate, $(COO)_2Ba$, as a function of the pH of the solution. Use your expression to work out $[Ba^{2+}]$ in solutions with a pH of 2.00, 4.00, and 7.00. K_{sp} for $(COO)_2Ba$ is 2.3×10^{-8} and dissociation constants for oxalic acid are: K_1, 6.5×10^{-2}; K_2, 6.1×10^{-5}.

SOLUTION As before, the first step is to write down the expressions for all these constants. We'll denote oxalic acid as H_2Ox and the ions as HOx^- and Ox^{2-}:

$$K_{sp} = [Ba^{2+}][Ox^{2-}] \quad \text{or} \quad [Ox^{2-}] = \frac{K_{sp}}{[Ba^{2+}]}$$

$$K_2 = \frac{[H^+][Ox^{2-}]}{[HOx^-]} \quad \text{or} \quad [HOx^-] = \frac{[H^+][Ox^{2-}]}{K_2} = \frac{[H^+]K_{sp}}{[Ba^{2+}]K_2}$$

$$K_1 = \frac{[H^+][HOx^-]}{[H_2Ox]} \quad \text{or} \quad [H_2Ox] = \frac{[H^+][HOx^-]}{K_1} = \frac{[H^+]^2 K_{sp}}{[Ba^{2+}]K_1 K_2}$$

All we have done is to arrange these three definition equations in such a way as to express the three concentrations; $[Ox^{2-}]$, $[HOx^-]$, and $[H_2Ox]$, in terms of $[H^+]$, $[Ba^{2+}]$, and the known constants.

As in the case of the silver acetate, we know that initially, when the salt dissolved, the ions were produced in a 1:1 ratio and $[Ba^{2+}] = [Ox^{2-}]$. However, some of the Ox^{2-} is subsequently converted to HOx^- and H_2Ox, so at equilibrium we know

$$[Ba^{2+}] = [Ox^{2-}] + [HOx^-] + [H_2Ox]$$

Substituting the expressions derived above gives

$$[Ba^{2+}] = \frac{K_{sp}}{[Ba^{2+}]} + \frac{[H^+]K_{sp}}{[Ba^{2+}]K_2} + \frac{[H^+]^2 K_{sp}}{[Ba^{2+}]K_1 K_2}$$

which rearranges to

$$[Ba^{2+}]^2 = K_{sp}\left[1 + \frac{[H^+]}{K_2} + \frac{[H^+]^2}{K_1 K_2}\right]$$

This is the necessary equation for finding $[Ba^{2+}]$ at any given pH. Then we substitute the numbers given.

(a) at pH 2.00: $[H^+] = 1.0 \times 10^{-2}$

$$[Ba^{2+}]^2 = 2.3 \times 10^{-8}\left[1 + \frac{1.0 \times 10^{-2}}{6.1 \times 10^{-5}} + \frac{(1.0 \times 10^{-2})^2}{(6.5 \times 10^{-2})(6.1 \times 10^{-5})}\right]$$

$$= 2.3 \times 10^{-8}(1 + 164 + 25) = 4.4 \times 10^{-6}$$

$$[Ba^{2+}] = 2.1 \times 10^{-3} M$$

(b) at pH 4.00: $[H^+] = 1.0 \times 10^{-4}$

$$[Ba^{2+}]^2 = 2.3 \times 10^{-8}\left[1 + \frac{1.0 \times 10^{-4}}{6.1 \times 10^{-5}} + \frac{(1.0 \times 10^{-4})^2}{(6.5 \times 10^{-2})(6.1 \times 10^{-5})}\right]$$

$$= 2.3 \times 10^{-8}[1 + 1.6 + (2.5 \times 10^{-3})] = 6.1 \times 10^{-8}$$

$$[Ba^{2+}] = 2.5 \times 10^{-4} M$$

(c) at pH 7.00: $[H^+] = 1.0 \times 10^{-7}$

$$[Ba^{2+}]^2 = 2.3 \times 10^{-8}\left[1 + \frac{1.0 \times 10^{-7}}{6.1 \times 10^{-5}} + \frac{(1.0 \times 10^{-7})^2}{(6.5 \times 10^{-2})(6.1 \times 10^{-5})}\right]$$

$$= 2.3 \times 10^{-8}[1 + (1.6 \times 10^{-3}) + (2.5 \times 10^{-9})] = 2.3 \times 10^{-8}$$

$$[Ba^{2+}] = 1.5 \times 10^{-4} M$$

Of the three terms inside the brackets, the first represents $[Ox^{2-}]$, the second $[HOx^-]$, and the third $[H_2Ox]$.

In (a), the majority of the oxalate in solution is in the form of HOx^-, with appreciable amounts of H_2Ox. In (b), the concentration of H_2Ox has fallen to a negligible level, while $[HOx^-]$ and $[Ox^{2-}]$ are of the same order. In (c), virtually all the oxalate in solution is in the form of Ox^{2-}, and the same holds at higher pH. So the acidity of the solution will have an appreciable effect on solubility only when the value of $[H^+]$ is greater than the K_a value for the weak acid involved in the equilibrium.

The above equation for the concentration of $[Ba^{2+}]$ will apply to any slightly soluble salt of the general type $BA \rightleftharpoons B^{2+} + A^{2-}$ where B^{2+} is a divalent metal cation and A^{2-} is the anion of the weak diprotic acid H_2A. An important example of this type of equilibrium is met in the metal sulfides, which are widely used for separating and identifying metals in solution. H_2S is a very weak acid, and any consideration of the solubility of metal sulfides should take this into account.

Example 14.16   For the weak acid H_2S, $K_1 = 9.1 \times 10^{-8}$ and $K_2 = 1.2 \times 10^{-15}$. Qualitative observation shows that PbS will precipitate when H_2S is added to an acidic or neutral solution containing Pb^{2+}, whereas NiS does not precipitate when H_2S is added to acidic solutions of Ni^{2+}, but does precipitate in neutral solution. Account for these observations by calculating $[Pb^{2+}]$ and $[Ni^{2+}]$ in solutions of pH 2.00, 5.00, and 7.00, in equilibrium with each of the solid metal sulfides, given that the K_{sp} values are 3.4×10^{-28} for PbS and 3.0×10^{-19} for NiS.

SOLUTION   We can use the formula developed for Example 14.15, the barium oxalate problem, and write for $[Pb^{2+}]$ in equilibrium with PbS

$$[Pb^{2+}]^2 = K_{sp}\left[1 + \frac{[H^+]}{K_2} + \frac{[H^+]^2}{K_1 K_2}\right]$$

Given the values of K_1 and K_2, we may evaluate this expression for various values of $[H^+]$.

(a) at pH = 2.00: $[H^+] = 10^{-2}$

$$[Pb^{2+}]^2 = 3.4 \times 10^{-28}\left[1 + \frac{10^{-2}}{1.2 \times 10^{-15}} + \frac{(10^{-2})^2}{(9.1 \times 10^{-8})(1.2 \times 10^{-15})}\right]$$

$$= 3.4 \times 10^{-28}[1 + (8.3 \times 10^{12}) + (9.2 \times 10^{17})]$$

$$= (3.4 \times 10^{-28})(9.2 \times 10^{17}) = 3.1 \times 10^{-10}$$

$$[Pb^{2+}] = 1.8 \times 10^{-5} M$$

(b) at pH = 5.00: $[H^+] = 10^{-5}$

$$[Pb^{2+}]^2 = 3.4 \times 10^{-28}[1 + (8.3 \times 10^9) + (9.2 \times 10^{11})]$$

$$= (3.4 \times 10^{-28})(9.2 \times 10^{11}) = 3.1 \times 10^{-16}$$

$$[Pb^{2+}] = 1.8 \times 10^{-8} M$$

(c) at pH = 7.00: $[H^+] = 10^{-7}$

$$[Pb^{2+}]^2 = 3.4 \times 10^{-28}[1 + (8.3 \times 10^7) + (9.2 \times 10^7)]$$

$$= (3.4 \times 10^{-28})(1.7 \times 10^8) = 5.9 \times 10^{-20}$$

$$[Pb^{2+}] = 2.4 \times 10^{-10} M$$

Summarizing these results, we see that the solubility of lead sulfide at any of the above pH values is very small. Although it increases considerably in the more acid solution, a qualitative observation would still classify PbS as "insoluble."

In the case of NiS, the calculation will follow a similar course, with the replacement of K_{sp} for PbS by the corresponding value for NiS, so we need not repeat all the details.

(a) at pH = 2.00: $[H^+] = 10^{-2}$
$[Ni^{2+}]^2 = (3.0 \times 10^{-19})(9.2 \times 10^{17}) = 0.28$
$[Ni^{2+}] = 0.52M$

(b) at pH = 5.00: $[H^+] = 10^{-5}$
$[Ni^{2+}]^2 = (3.0 \times 10^{-19})(9.2 \times 10^{11}) = 2.8 \times 10^{-7}$
$[Ni^{2+}] = 5.3 \times 10^{-4}M$

(c) at pH = 7.00: $[H^+] = 10^{-7}$
$[Ni^{2+}]^2 = (3.0 \times 10^{-19})(1.7 \times 10^8) = 5.1 \times 10^{-11}$
$[Ni^{2+}] = 7.2 \times 10^{-6}M$

The figures for NiS tell us immediately that this compound will be freely soluble in acid solution, but virtually insoluble in neutral solution, in accordance with experimental observation.

We can further see from these figures that, in the acid solutions, virtually all the sulfide in solution is in the form of undissociated H_2S, represented by the third of the terms inside the brackets. In neutral solution, there are present approximately equal concentrations of H_2S and HS^-, but the concentration of S^{2-} is quite negligible.

The above problems may appear rather complex at first sight, but a closer look will show that no new concepts are involved. All we are considering are the familiar solubility and acid–base equilibria, which were covered in Chapters 12 and 13, but we are looking at systems where both types of equilibrium are present simultaneously. Once we realize what is going on, we have only to write down the equations and equilibrium constants for the various reactions, then arrange them to solve for whatever unknown we are seeking.

14.9   Complex Ions and Stability Constants

If we add aqueous ammonia to a solution of copper sulfate, a blue precipitate of copper hydroxide appears immediately. The addition of more ammonia, however, will eventually cause this precipitate to redissolve, as the color of the solution deepens to a much darker blue than was originally present. A similar effect is found when aqueous ammonia is added to solutions of nickel, cobalt, or many other metal ions. How is this explained?

The reaction responsible for the redissolving of the copper hydroxide and the change in color of the solution is the formation of a *complex ion* containing four ammonia molecules around a copper ion, $[Cu(NH_3)_4]^{2+}$. Actually, the copper ion in the original aqueous solution was in a very similar position, being surrounded by four water molecules to form $[Cu(H_2O)_4]^{2+}$.

The ion $[Cu(NH_3)_4]^{2+}$ is in equilibrium with aqueous Cu^{2+} and NH_3 by the reaction

$$Cu^{2+}(aq) + 4NH_3(aq) \rightleftharpoons [Cu(NH_3)_4]^{2+}$$

As for any other reaction, there is an equilibrium constant associated with this process, and it is known as the *stability constant* for the ion $[Cu(NH_3)_4]^{2+}$:

$$K_{stab} = \frac{[Cu(NH_3)_4^{2+}]}{[Cu^{2+}][NH_3]^4} = 1.0 \times 10^{14}$$

We see that this stability constant has a large value, telling us that the equilibrium will be in favor of the formation of the complex ion. In other words, if there is a considerable concentration of NH_3 in the solution, the concentration of Cu^{2+} will be reduced to the extent that the solubility product of $Cu(OH)_2$ will not be exceeded, even in alkaline solution.

Example 14.17 A solution is originally $0.10M$ in Cu^{2+}. NH_3 gas is added until $[NH_3(aq)] = 0.10$. Calculate $[Cu^{2+}]$ and also the pH of the solution. Is the solubility product of $Cu(OH)_2$ exceeded? ($K_{sp} = 1.8 \times 10^{-14}$). The stability constant of $[Cu(NH_3)_4]^{2+}$ is 1.0×10^{14} and K_b for aqueous ammonia is 1.8×10^{-5}.

SOLUTION   The reaction is

$$Cu^{2+} + 4NH_3 \rightleftharpoons Cu(NH_3)_4^{2+}$$

One mole of Cu^{2+} gives one mole of the ammonia complex, so, if we assume that virtually all of the original Cu^{2+} in the solution is converted into the complex $[Cu(NH_3)_4]^{2+}$, the concentration of this ion will be $0.10M$. The equation for the stability constant is

$$\frac{[Cu(NH_3)_4^{2+}]}{[Cu^{2+}][NH_3]^4} = 1.0 \times 10^{14} \quad \text{or} \quad [Cu^{2+}] = \frac{[Cu(NH_3)_4^{2+}]}{[NH_3]^4(1.0 \times 10^{14})}$$

Putting in the given concentration values gives

$$[Cu^{2+}] = \frac{0.10}{(0.10)^4(1.0 \times 10^{14})} = 1.0 \times 10^{-11}M$$

We see at once that our assumption was justified; virtually all the copper is present in solution as the complex ion and hardly any remains as Cu^{2+}.

To evaluate the ion product for $Cu(OH)_2$, we need the concentration of OH^-. We have aqueous NH_3 present in $0.10M$ concentration, so we can calculate the concentration of hydroxide ion as before.

Letting x mole liter^{-1} of NH_3 ionize, then

$$K_b = \frac{[NH_4^+][OH^-]}{[NH_3(aq)]} = \frac{x^2}{0.10 - x} \approx \frac{x^2}{0.10} \quad \text{(assuming } x \text{ is small)}$$

$$\frac{x^2}{0.10} = 1.8 \times 10^{-5} \qquad x^2 = 1.8 \times 10^{-6} \qquad x = 1.3 \times 10^{-3}$$

The value of $[OH^-]$ is $1.3 \times 10^{-3}M$, so our ion product expression for $Cu(OH)_2$ in this solution is

$$[Cu^{2+}][OH^-]^2 = (1.0 \times 10^{-11})(1.3 \times 10^{-3})^2 = 1.8 \times 10^{-17}$$

We see that this is considerably less than K_{sp} for this compound, so $Cu(OH)_2$ initially precipitated on adding NH_3 will redissolve when $[NH_3]$ reaches a sufficiently high value.

Example 14.18   When KOH is first added to a solution of Zn^{2+}, $Zn(OH)_2$ precipitates; but, after an excess of KOH is added, the $Zn(OH)_2$ redissolves to give $Zn(OH)_4^{2-}$. If the initial concentration of Zn^{2+} ion is $2.0 \times 10^{-3} M$, to what value must $[OH^-]$ be raised to redissolve all the $Zn(OH)_2$ initially formed?

K_{sp} for $Zn(OH)_2$: 1.2×10^{-17}
stability constant for $[Zn(OH)_4]^{2-}$: 2×10^{15}

SOLUTION   The two steps in the reaction are

$$Zn^{2+} + 2OH^- \rightleftharpoons Zn(OH)_2(s)$$

$$Zn(OH)_2(s) + 2OH^- \rightleftharpoons Zn(OH)_4^{2-}$$

At the point where the last of the precipitated $Zn(OH)_2$ just redissolves, the solution may be considered saturated with respect to this compound, so we can write

$$[Zn^{2+}][OH^-]^2 = 1.2 \times 10^{-17} \quad \text{or} \quad [Zn^{2+}] = \frac{1.2 \times 10^{-17}}{[OH^-]^2}$$

At the same time, we have the equilibrium for the overall complex ion reaction

$$Zn^{2+} + 4OH^- \rightleftharpoons Zn(OH)_4^{2-}$$

$$K_{stab} = \frac{[Zn(OH)_4^{2-}]}{[Zn^{2+}][OH^-]^4} = 2 \times 10^{15}$$

We can substitute in this equation the previously obtained expression for $[Zn^{2+}]$, giving

$$K_{stab} = \frac{[Zn(OH)_4^{2-}][OH^-]^2}{K_{sp}[OH^-]^4} = 2 \times 10^{15}$$

or

$$[OH^-]^2 = \frac{[Zn(OH)_4^{2-}]}{(1.2 \times 10^{-17})(2 \times 10^{15})} = 42[Zn(OH)_4^{2-}]$$

In our solution, the original concentration of Zn^{2+} was $2.0 \times 10^{-3} M$, and we may assume that all of this is now present as the complex $[Zn(OH)_4]^{2-}$, so the concentration of complex ion is now $2.0 \times 10^{-3} M$. So we can find the hydroxide ion concentration from

$$[OH^-]^2 = 42 \times (2.0 \times 10^{-3}) = 0.084 \qquad [OH^-] = 0.29 M$$

This is the *minimum* hydroxide ion concentration required to dissolve this amount of Zn^{2+} as the complex ion. Below this value, $Zn(OH)_2$ will remain undissolved. Of course, at very low values of $[OH^-]$ (that is, in acid solution), $Zn(OH)_2$ will not be precipitated, because its K_{sp} value is not exceeded. Using the known K_{sp} value and putting $[Zn^{2+}] = 2.0 \times 10^{-3} M$, we can easily calculate the minimum value of $[OH^-]$ needed to precipitate $Zn(OH)_2$:

$$[OH^-]^2 = \frac{K_{sp}}{[Zn^{2+}]} = \frac{1.2 \times 10^{-17}}{2.0 \times 10^{-3}} = 6.0 \times 10^{-15}$$

$$[OH^-] = 7.8 \times 10^{-8} \qquad pOH = 7.1 \qquad pH = 6.9$$

A metal hydroxide that will dissolve in both acid or alkali, as does zinc hydroxide, is known as *amphoteric*. This behavior will be seen for metals where there is a suitable combination of the solubility product of the hydroxide and the stability constant of the complex hydroxy-anion.

KEY WORDS

hydrolysis	conjugate acid–base pair
buffer solution	buffering action
equivalence point of titration	monoprotic acid
polyprotic acid	diprotic acid
stability constant	

STUDY QUESTIONS

1. What types of ion undergo hydrolysis in water? Give a general equation for the hydrolysis of (a) a cation and (b) an anion.

2. An ionic salt is dissolved in pure water. What factors will determine whether the solution (a) becomes acidic, (b) becomes basic, or (c) remains neutral?

3. What do we mean by the following? Give examples:
   (a) the conjugate base of a weak acid
   (b) the conjugate acid of a weak base

4. How is K_a for an acid related to K_b for its conjugate base?

5. How do we approach the quantitative determination of degree of hydrolysis of an ion? Where do we find values of the appropriate equilibrium constants? What simplifying assumption is used?

6. What is a buffer solution? Give examples of acidic buffers and basic buffers.

7. How do we determine the concentrations of the *two* species necessary to make a buffer solution? Do we worry about dissociation or hydrolysis of these species? Why not?

8. How does a buffer work? Give specific equations showing what happens when:
   (a) strong acid is added to an acid buffer
   (b) strong base is added to an acid buffer
   (c) strong acid is added to a basic buffer
   (d) strong base is added to a basic buffer

9. How do we construct a buffer to give a specific pH value?

10. What happens when acid and base are mixed?

11. Give equations showing the *principal* reaction occurring when:
    (a) strong acid is mixed with strong base
    (b) weak acid is mixed with strong base
    (c) strong acid is mixed with weak base

12. What type of solution results when a portion of *strong* base solution is titrated with *strong* acid solution:
    (a) before the equivalence point
    (b) at the equivalence point
    (c) beyond the equivalence point

13. Reconsider the previous question for the case of:
    (a) strong acid titrated with weak base
    (b) strong base titrated with weak acid
    (c) weak acid titrated with strong base
    (d) weak base titrated with strong acid

14. At what time is a buffer mixture present in each of the above titrations?

15. Give examples of acids which are (a) monoprotic, (b) diprotic, and (c) triprotic. How many ionization constants does each have? Write equations for the successive dissociations, with the appropriate expression for the ionization constant for each.

16. How do we go about determining the equilibrium concentrations of various species in a solution of a diprotic acid? What simplifying assumption do we make?

17. In very dilute solutions of acids or bases, we have to modify our approach to calculating pH. Explain why. What contribution to $[H^+]$ or $[OH^-]$ do we usually neglect at reasonably high solute concentrations?

18. Explain qualitatively why the slightly soluble salt of a weak acid will dissolve in dilute nitric acid, whereas the slightly soluble salt of a strong acid will not.

19. Why is it that the formation of a complex ion will cause a precipitate to redissolve?

20. What is an amphoteric hydroxide? Why does it dissolve in base? Give an equation.

PROBLEMS

Refer to Tables 13.1 and 13.2 (Chapter 13) for K_a and K_b values.

1. Calculate the pH and percent hydrolysis in the following solutions:
    (a) sodium formate, $0.60M$
    (b) potassium propionate, $2.3 \times 10^{-2}M$
    (c) ethylammonium chloride, $C_2H_5NH_3^+ \cdot Cl^-$, $0.55M$

2. Calculate the pH of the following solutions:
   (a) 0.200M in chloracetic acid and 0.060M in sodium chloracetate
   (b) 0.015M in propionic acid and 0.100M in potassium propionate
   (c) 0.065M in pyridine, C_5H_5N, and 0.200M in pyridinium chloride, $C_5H_5NH^+ \cdot Cl^-$.

3. What is the pH of the solution produced by mixing the following solutions?
   (a) 50.0 ml of 0.080M NaOH with 40.0 ml of 0.20M hydrofluoric acid
   (b) 100 ml of triethylamine, 0.080M, with 20 ml of HCl, 0.10M
   (c) 50 ml of HCl, 0.15M, with 75 ml of NaOH, 0.12M

4. 20.0 ml of formic acid solution, 0.150M, is titrated with 0.200M sodium hydroxide. Calculate the pH after the addition of the following amounts of base:
   (a) 10.0 ml   (b) 15.0 ml   (c) 20.0 ml

5. 40.0 ml of 0.0500M nitric acid is titrated with 0.0800M aqueous ammonia. Calculate the pH after the addition of the following amounts of base:
   (a) 15.0 ml   (b) 25.0 ml   (c) 50.0 ml

6. A solution is 0.150M in nitrous acid and 0.080M in sodium nitrite. What is the pH?
   (a) originally
   (b) after addition of 0.050 moles HCl to 1.00 liter of solution
   (c) after addition of 0.050 moles NaOH to 1.00 liter of solution

7. A solution is 0.280M in ammonium chloride and 0.160M in aqueous ammonia. What is the pH?
   (a) originally
   (b) after addition of 0.120 moles HCl to 1.00 liter of solution
   (c) after addition of 0.120 moles KOH to 1.00 liter of solution

8. Using formic acid and sodium formate, suggest a recipe for a buffer with a pH of 3.00.

9. Using hydrazine and hydrazinium chloride, $N_2H_5^+ \cdot Cl^-$, suggest a recipe for a buffer with a pH of 7.50.

10. Calculate the $[H^+]$ and pH in $4.0 \times 10^{-7}M$ HCl solution. What fraction of the total $[H^+]$ comes from the solvent?

11. Calculate the concentration of the metal ion and the pH in saturated solutions of the following in water:
    (a) $Mn(OH)_2$, $K_{sp} = 4 \times 10^{-14}$
    (b) $Pt(OH)_2$, $K_{sp} = 1 \times 10^{-35}$
    (c) What is the maximum value of $[Pt^{2+}]$ possible in a solution acidified to pH 1.0?

12. Succinic acid, $C_2H_4(COOH)_2$, is a diprotic acid with $K_1 = 6.2 \times 10^{-5}$ and $K_2 = 2.3 \times 10^{-6}$. Calculate the concentrations of all species present in an 0.10M solution of the acid, also the pH, and the percentage dissociation in the first and second dissociation steps.

13. Sulfurous acid, H_2SO_3, is a diprotic acid with $K_1 = 1.7 \times 10^{-2}$ and $K_2 = 6.5 \times 10^{-8}$. Calculate the concentrations of all species present in $0.020M$ sulfurous acid solution, together with the pH, and the percentage dissociation in the first and second steps.

14. Strontium chromate, $SrCrO_4$, has $K_{sp} = 3.6 \times 10^{-5}$. Calculate the concentration of strontium ions, $[Sr^{2+}]$, in a solution in equilibrium with solid strontium chromate at pH values (a) 2.00, (b) 5.00, and (c) 7.00. Dissociation constants for chromic acid, H_2CrO_4, are $K_1 = 0.18$, $K_2 = 3.3 \times 10^{-6}$.

15. The dissociation constants for carbonic acid, H_2CO_3, are $K_1 = 4.5 \times 10^{-7}$, $K_2 = 4.7 \times 10^{-11}$. What will be the concentration of calcium ion, $[Ca^{2+}]$, in solutions in equilibrium with solid calcium carbonate, $CaCO_3$, at pH values of (a) 4.00, (b) 7.00, and (c) 10.00? For $CaCO_3$, $K_{sp} = 4.7 \times 10^{-9}$.

16. Lithium carbonate, Li_2CO_3, has $K_{sp} = 1.7 \times 10^{-3}$. Work out an equation for the concentration of the lithium ion, $[Li^+]$, as a function of the hydrogen ion concentration and the dissociation constants of H_2CO_3, for solutions in equilibrium with solid Li_2CO_3. Use your equation to find $[Li^+]$ at pH $= 8.00$ and pH $= 10.00$ in such a solution. (K_a values are given in problem 15.) (*Hint*: Write equations for Li_2CO_3 dissolving and for the subsequent reactions of CO_3^{2-} with water. What is the stoichiometry?)

17. Experimental observation shows that cobalt sulfide, CoS, is soluble in acid, but precipitates from neutral or alkaline solution, whereas copper sulfide, CuS, is not dissolved at any pH. Account for this difference by calculating the concentrations of Co^{2+} and Cu^{2+} ions in solutions in equilibrium with the respective solid sulfides at pH values of (a) 1.00, (b) 3.00, and (c) 6.00. The K_{sp} values are CoS, 1×10^{-21}; CuS, 1×10^{-40}. For H_2S, $K_1 = 9.1 \times 10^{-8}$ and $K_2 = 1.2 \times 10^{-15}$.

18. The complex ion $Zn(CN)_4^{2-}$ has a stability constant of 1.6×10^{20}. A solution is originally $0.20M$ in Zn^{2+}, and sufficient potassium cyanide is added to form the complex and give a concentration of free cyanide ion, CN^-, of $0.010M$.
    (a) What will be the concentration of Zn^{2+} ion in this solution?
    (b) Will zinc carbonate, $ZnCO_3(K_{sp} = 1.4 \times 10^{-11})$ precipitate if the solution is made $0.10M$ in carbonate ion?

19. An acidic solution contains Al^{3+} in $0.050M$ concentration. KOH is slowly added and $Al(OH)_3$ begins to precipitate. Eventually, when further KOH has been added, the precipitate redissolves as the complex ion $Al(OH)_4^-$ forms. Calculate the pH at which $Al(OH)_3$ starts to appear and the pH at which the precipitate has just dissolved as the complex ion. Ignore volume changes. K_{sp} for $Al(OH)_3$ is 2×10^{-32} and the stability constant for $Al(OH)_4^-$ is 2×10^{33} for formation from $Al^{3+} + 4OH^-$. (*Hint*: Assume that the final solution is saturated with respect to $Al(OH)_3$ and that essentially all the aluminum in this solution is in the form of the complex ion.)

20. Iron(III) forms a complex with oxalate ion, $C_2O_4^{2-}$ by the reaction

$$Fe^{3+} + 3C_2O_4^{2-} \rightleftharpoons [Fe(C_2O_4)_3]^{3-}$$

and the complex has a stability constant of 1.6×10^{20}. A solution is originally $0.040M$ in

Fe^{3+} and enough potassium oxalate is added to form the complex and give a concentration of free oxalate ion of $0.010M$. What will be the concentration of Fe^{3+} in this solution? If the pH is increased to 7.00, will $Fe(OH)_3$ precipitate? (K_{sp} for $Fe(OH)_3$ is 1.1×10^{-36}.) (*Hint*: Assume that the concentration of the complex in the final solution is the same as the original concentration of Fe^{3+}.)

SOLUTIONS TO PROBLEMS

1. (a) Sodium formate, HCOONa, $0.60M$, ionizes to give, initially, $[HCOO^-] = 0.60$. Let x mole liter^{-1} hydrolyze:

$$HCOO^- + H_2O \rightleftharpoons HCOOH + OH^-$$
$$(0.60 - x) \qquad\qquad x \qquad x$$

$$\frac{[HCOOH][OH^-]}{[HCOO^-]} = \frac{x^2}{0.60 - x} = K_b \quad \text{for } HCOO^-$$

$$K_b = \frac{K_w}{K_a} \quad \text{where} \quad K_a = 1.8 \times 10^{-4} \quad \text{for formic acid}$$

Assuming x is small beside 0.60, we get

$$\frac{x^2}{0.60 - x} \approx \frac{x^2}{0.60} = \frac{K_w}{K_a} = \frac{10^{-14}}{1.8 \times 10^{-4}}$$

$$x^2 = \frac{0.60 \times 10^{-14}}{1.8 \times 10^{-4}} = 3.3 \times 10^{-11}$$

$$x = 5.8 \times 10^{-6} \quad \text{(assumption is justified)}$$
$$pOH = -\log(5.8 \times 10^{-6}) = 5.24 \quad \text{and} \quad pH = 8.76$$

$$\text{percent hydrolysis:} \quad \frac{5.8 \times 10^{-6}}{0.60} \times 100 = 9.7 \times 10^{-4}\%$$

(b) Initially, $[C_2H_5COO^-] = 2.3 \times 10^{-2}M$
$$C_2H_5COO^- + H_2O \rightleftharpoons C_2H_5COOH + OH^-$$
Using the same argument as in (a), gives us

$$\frac{x^2}{(2.3 \times 10^{-2}) - x} \approx \frac{x^2}{2.3 \times 10^{-2}} = \frac{K_w}{K_a} = \frac{10^{-14}}{1.3 \times 10^{-5}}$$

$$x^2 = \frac{10^{-14} \times 2.3 \times 10^{-2}}{1.3 \times 10^{-5}} = 1.8 \times 10^{-11}$$

$$x = 4.2 \times 10^{-6} \quad \text{(assumption is justified)}$$
$$pOH = -\log(4.2 \times 10^{-6}) = 5.38 \quad \text{and} \quad pH = 8.62$$

$$\text{percent hydrolysis:} \quad \frac{4.2 \times 10^{-6}}{2.3 \times 10^{-2}} \times 100 = 1.8 \times 10^{-2}\%$$

(c)   Initially, $[C_2H_5NH_3^+] = 0.55M$

$$C_2H_5NH_3^+ + H_2O \rightleftharpoons C_2H_5NH_2 + H_3O^+$$

Let x mole liter^{-1} hydrolyze, where x is small:

$$\frac{x^2}{0.55 - x} \approx \frac{x^2}{0.55} = K_a = \frac{K_w}{K_b}$$

where K_b for ethylamine $= 4.3 \times 10^{-4}$, so

$$\frac{x^2}{0.55} = \frac{10^{-14}}{4.3 \times 10^{-4}} \quad \text{and} \quad x^2 = \frac{0.55 \times 10^{-14}}{4.3 \times 10^{-4}} = 1.3 \times 10^{-11}$$

$x = 3.6 \times 10^{-6}$ (assumption is justified)

(Remember $x = [H^+]$, since a conjugate acid is reacting.)

pH $= -\log(3.6 \times 10^{-6}) = 5.44$

$$\text{percent hydrolysis:} \quad \frac{3.6 \times 10^{-6}}{0.55} \times 100 = 6.5 \times 10^{-4}\%$$

2.   All these are buffer solutions.

   (a)   Chloracetic acid, $CH_2ClCOOH$, has $K_a = 1.4 \times 10^{-3}$.

   $pK_a = -\log(1.4 \times 10^{-3}) = 2.85$

   acid buffer:

$$pH = pK_a + \log \frac{[CH_2ClCOO^-]}{[CH_2ClCOOH]}$$

$$= 2.85 + \log \frac{0.060}{0.200} = 2.85 + \log(0.30)$$

$$= 2.85 - 0.52 = 2.33$$

   (b)   Propionic acid, C_2H_5COOH, has $K_a = 1.3 \times 10^{-5}$.

   $pK_a = -\log(1.3 \times 10^{-5}) = 4.89$

   acid buffer:

$$pH = pK_a + \log \frac{[C_2H_5COO^-]}{[C_2H_5COOH]}$$

$$= 4.89 + \log \frac{0.100}{0.015} = 4.89 + \log 6.7$$

$$= 4.89 + 0.82 = 5.71$$

   (c)   Pyridine, C_5H_5N, has $K_b = 1.7 \times 10^{-9}$.

   $pK_b = -\log(1.7 \times 10^{-9}) = 8.77$

   Base buffer: $pOH = pK_b + \log \dfrac{[C_5H_5NH^+]}{[C_5H_5N]}$

$$= 8.77 + \log \frac{0.200}{0.065} = 8.77 + \log 3.08$$

$$= 8.77 + 0.49 = 9.26$$

   pH $= 4.74$

   (*Note*: Pyridine is such a weak base that it buffers at pH values below 7, i.e., buffer solutions are acidic.)

3. It is essential in "mixing" problems to recognize what *type* of solution is present.
   (a) 50.0 ml of $0.080M$ NaOH contain $50.0 \times 0.080 = 4.0$ mmole OH^-.
   40.0 ml of $0.20M$ HF contain $40.0 \times 0.20 = 8.0$ mmole HF.
   Reaction removes 4.0 mmole of OH^- and 4.0 mmole of HF to produce 4.0 mmole of F^-.
   4.0 mmole HF remain unreacted.
   HF and F^- are present together; this is an acid buffer solution.
   HF has $K_a = 6.7 \times 10^{-4}$, so $pK_a = 3.17$.

   $$pH = 3.17 + \log \frac{[F^-]}{[HF]} = 3.17 + \log \frac{4.0}{4.0} = 3.17 + \log 1 = 3.17$$

   (b) 100 ml triethylamine, $0.080M$, contain $100 \times 0.080 = 8.0$ mmole $(C_2H_5)_3N$.
   20 ml HCl, $0.10M$, contain $20 \times 0.10 = 2.0$ mmole H^+.
   Reaction removes 2.0 mmole of $(C_2H_5)_3N$.
   $(C_2H_5)_3N + H^+ \longrightarrow (C_2H_5)_3NH^+$
   6.0 mmole $(C_2H_5)_3N$ remain unreacted; this is a base buffer solution.
   $(C_2H_5)_3N$ has $K_b = 5.3 \times 10^{-4}$, so $pK_b = 3.28$.

   $$pOH = pK_b + \log \frac{[(C_2H_5)_3NH^+]}{[(C_2H_5)_3N]} = 3.28 + \log \frac{2.0}{6.0}$$

   $$= 3.28 + \log(0.33) = 3.28 - 0.48 = 2.80$$
   $$pH = 14.00 - 2.80 = 11.20$$

   (c) 50 ml of HCl, $0.15M$ contain $50 \times 0.15 = 7.5$ mmole H^+.
   75 ml of NaOH, $0.12M$, contain $75 \times 0.12 = 9.0$ mmole OH^-.
   Reaction removes 7.5 mmole of each as H_2O, leaving 1.5 mmole of OH^- unreacted.
   This is *not* a buffer solution (strong-acid–strong-base mixture, no buffering action).

   $$\text{total volume 125 ml;} \quad [OH^-] = \frac{1.5 \text{ mmole}}{125 \text{ ml}} = 1.2 \times 10^{-2}M$$

   $$pOH = -\log(1.2 \times 10^{-2}) = 1.92$$
   $$pH = 12.08$$

4. We start with $20.0 \text{ ml} \times 0.150M = 3.00$ mmole formic acid.
   (a) 10.0 ml $0.200M$ NaOH contain $10.0 \times 0.200 = 2.00$ mmole OH^-.
   This reacts with 2.00 mmole formic acid, HCOOH, to give 2.00 mmole $HCOO^-$,
   which leaves unreacted 1.00 mmole HCOOH.
   This is an acid buffer solution.
   K_a for HCOOH $= 1.8 \times 10^{-4}$, so $pK_a = 3.74$

   $$pH = pK_a + \log \frac{[HCOO^-]}{[HCOOH]} = 3.74 + \log \frac{2.00}{1.00}$$

   $$= 3.74 + \log 2.00 = 3.74 + 0.30 = 4.04$$

   (b) 15.0 ml $0.200M$ NaOH contain $15.0 \times 0.200 = 3.00$ mmole OH^-.
   This exactly neutralizes 3.00 mmole HCOOH to give
   3.00 mmole $HCOO^-$ (sodium formate solution).

   $$[HCOO^-] = \frac{3.00 \text{ mmole}}{35.0 \text{ ml}} = 0.0857M$$

Hydrolysis occurs to the extent of x mole liter^{-1}.

$$\frac{x^2}{0.0857 - x} \approx \frac{x^2}{0.0857} = \frac{K_w}{K_a} = \frac{10^{-14}}{1.8 \times 10^{-4}}$$

$$x^2 = \frac{0.0857 \times 10^{-14}}{1.8 \times 10^{-4}} = 4.8 \times 10^{-12} \quad \text{so} \quad x = 2.2 \times 10^{-6} = [OH^-]$$

$$pOH = -\log(2.2 \times 10^{-6}) = 5.66$$
$$pH = 8.34$$

(c) 20.0 ml 0.200M NaOH contain $20.0 \times 0.200 = 4.00$ mmole OH$^-$
This converts 3.00 mmole HCOOH completely to HCOO$^-$, which leaves 1.00 mmole OH$^-$ in excess.
This is *not* a buffer solution.

total volume 40.0 ml: $[OH^-] = \dfrac{1.00 \text{ mmole}}{40.0 \text{ ml}} = 2.50 \times 10^{-2} M$

$$pOH = -\log(2.50 \times 10^{-2}) = 1.60$$
$$pH = 12.40$$

5. Initially, we have $40.0 \times 0.0500 = 2.00$ mmole H$^+$.
(a) 15.0 ml 0.0800M NH$_3$ contain $15.0 \times 0.0800 = 1.20$ mmole NH$_3$.
This reacts 1.20 mmole H$^+$, giving 1.20 mmole NH$_4^+$.
$(2.00 - 1.20) = 0.80$ mmole H$^+$ remain.
This is *not* a buffer solution.

$$[H^+] = \frac{0.80 \text{ mmole}}{55.0 \text{ ml}} = 1.45 \times 10^{-2} M$$

$$pH = -\log(1.45 \times 10^{-2}) = 1.84$$

(b) 25.0 ml 0.0800 NH$_3$ contain $25.0 \times 0.0800 = 2.00$ mmole NH$_3$.
This exactly neutralizes the H$^+$ to give 2.00 mmole of NH$_4^+$ (ammonium nitrate solution).

$$[NH_4^+] = \frac{2.00 \text{ mmole}}{65.0 \text{ ml}} = 3.08 \times 10^{-2} M$$

This undergoes hydrolysis to a degree x mole liter^{-1}:

$$\frac{x^2}{(3.08 \times 10^{-2}) - x} \approx \frac{x^2}{3.08 \times 10^{-2}} = \frac{K_w}{K_b} = \frac{10^{-14}}{1.8 \times 10^{-5}}$$

$$x^2 = \frac{(3.08 \times 10^{-2}) \times 10^{-14}}{1.8 \times 10^{-5}} = 1.7 \times 10^{-11} \quad \text{so} \quad x = 4.1 \times 10^{-6} = [H^+]$$

$$pH = 5.38$$

(c) 50.0 ml 0.0800M NH$_3$ contain $50.0 \times 0.0800 = 4.00$ mmole NH$_3$.
This reacts with 2.00 mmole H$^+$ to give 2.00 mmole NH$_4^+$, which leaves 2.00 mmole NH$_3$ remaining.
Both NH$_3$ and NH$_4^+$ are present; this is a base buffer.
pK_b for NH$_3(aq) = -\log(1.8 \times 10^{-5}) = 4.74$

$$pOH = 4.74 + \log\frac{[NH_4^+]}{[NH_3]} = 4.74 + \log\frac{2.00}{2.00} = 4.74$$

$$pH = 9.26$$

6. $[HNO_2] = 0.150M$ and $[NO_2^-] = 0.080M$.

   (a) This is an acid buffer solution.

   $pK_a = -\log(4.5 \times 10^{-4}) = 3.35$

   $$pH = pK_a + \log\frac{[NO_2^-]}{[HNO_2]} = 3.35 + \log\frac{0.080}{0.150} = 3.35 + \log(0.533) = 3.35 - 0.27 = 3.07$$

   (b) After adding to 1.00 liter 0.050 mole HCl: $H^+ + NO_2^- \rightarrow HNO_2$

   $[NO_2^-] = 0.080 - 0.050 = 0.030$ and $[HNO_2] = 0.150 + 0.050 = 0.200$

   $$pH = 3.35 + \log\frac{0.030}{0.200} = 3.35 + \log(0.150) = 3.35 - 0.82 = 2.53$$

   (c) After adding to 1.00 liter 0.050 mole NaOH: $OH^- + HNO_2 \rightarrow NO_2^- + H_2O$.

   $[NO_2^-] = 0.080 + 0.050 = 0.130$ and $[HNO_2] = 0.150 - 0.050 = 0.100$

   $$pH = 3.35 + \log\frac{0.130}{0.100} = 3.35 + \log 1.30 = 3.35 + 0.11 = 3.46$$

7. $[NH_4^+] = 0.280M$ and $[NH_3(aq)] = 0.160M$

   (a) This is a base buffer solution.

   $pK_b = 4.74$.

   $$pOH = pK_b + \log\frac{[NH_4^+]}{[NH_3(aq)]} = 4.74 + \log\frac{0.280}{0.160} = 4.74 + \log 1.75 = 4.74 + 0.24 = 4.98$$

   $pH = 14.00 - 4.98 = 9.02$

   (b) After adding to 1.00 liter 0.120 mole HCl: $H^+ + NH_3 \rightarrow NH_4^+$.

   $[NH_4^+] = 0.280 + 0.120 = 0.400$ and $[NH_3(aq)] = 0.160 - 0.120 = 0.040$

   $$pOH = 4.74 + \log\frac{0.400}{0.040} = 4.74 + \log 10 = 4.74 + 1.00 = 5.74$$

   $pH = 14.00 - 5.74 = 8.26$

   (c) After adding to 1.00 liter 0.120 mole KOH: $OH^- + NH_4^+ \rightarrow NH_3 + H_2O$.

   $[NH_4^+] = 0.280 - 0.120 = 0.160$ and $[NH_3(aq)] = 0.160 + 0.120 = 0.280$

   $$pOH = 4.74 + \log\frac{0.160}{0.280} = 4.74 + \log(0.57) = 4.74 - 0.24 = 4.50$$

   $pH = 14.00 - 4.50 = 9.50$

8. Formic acid, HCOOH has $K_a = 1.8 \times 10^{-4}$, so $pK_a = 3.74$.

   $$pH = 3.74 + \log\frac{[HCOO^-]}{[HCOOH]}$$

   For a pH of 3.00: $3.00 = 3.74 + \log\dfrac{[HCOO^-]}{[HCOOH]}$

   $$\log\frac{[HCOO^-]}{[HCOOH]} = 3.00 - 3.74 = -0.74 = -1 + 0.26$$

   $$\frac{[HCOO^-]}{[HCOOH]} = \text{antilog}(-1 + 0.26) = 10^{-1} \times 1.8 = 0.18$$

   Any reasonable solution with this ratio of salt to acid would have the desired pH, e.g.,
   $[HCOO^-] = 0.18M$ (sodium formate, $0.18M$)
   $[HCOOH] = 1.00M$

9. Hydrazine has $K_b = 1.3 \times 10^{-6}$, so $pK_b = 5.89$.

$$pOH = 5.89 + \log \frac{[N_2H_5^+]}{[N_2H_4]}$$

For a pH of 7.50: $pOH = 14.00 - 7.50 = 6.50 = 5.89 + \log \frac{[N_2H_5^+]}{[N_2H_4]}$

$$\log \frac{[N_2H_5^+]}{[N_2H_4]} = 6.50 - 5.89 = 0.61$$

$$\frac{[N_2H_5^+]}{[N_2H_4]} = \text{antilog}(0.61) = 4.1$$

Any reasonable solution with this ratio would have the desired pH, e.g.,
$[N_2H_5^+] = 0.41M$ (as $N_2H_5^+ \cdot Cl^-$, $0.41M$)
$[N_2H_4] = 0.10M$

10. Let x mole liter of H^+ and OH^- come from dissociation of water.
$[H^+] = (4.0 \times 10^{-7}) + x$ and $[OH^-] = x$
$[(4.0 \times 10^{-7}) + x]x = K_w = 1.0 \times 10^{-14}$
$x^2 + (4.0 \times 10^{-7})x - 10^{-14} = 0$
Solving the quadratic, $x = 2.4 \times 10^{-8}$

total $[H^+] = (4.0 \times 10^{-7}) + x = 4.2 \times 10^{-7}$
$pH = -\log(4.2 \times 10^{-7}) = 6.37$

percent H^+ from solvent: $\dfrac{2.4 \times 10^{-8}}{4.2 \times 10^{-7}} \times 100 = 5.6\%$

11. (a)   Let x mole liter^{-1} $Mn(OH)_2$ dissolve.
$[Mn^{2+}] = x$ and $[OH^-] = 2x$
$K_{sp} = [Mn^{2+}][OH^-]^2 = 4x^3$
$4x^3 = 4 \times 10^{-14}$ $x^3 = 1 \times 10^{-14}$ $x = 2.2 \times 10^{-5}$
$pOH = -\log(2.2 \times 10^{-5}) = 4.67$ and $pH = 9.33$

(b)   A similar treatment of $Pt(OH)_2$ would give $[OH^-] = 1.4 \times 10^{-12}$ ($pH = 2.1$) which is clearly wrong because of neglect of OH^- from water. The correct assumption is that OH^- from *solute* is negligible:
$[OH^-] = 1.0 \times 10^{-7}$, $pH = 7.00$

$$[Pt^{2+}] = \frac{K_{sp}}{[OH^-]^2} = \frac{1 \times 10^{-35}}{(1.0 \times 10^{-7})^2} = 1 \times 10^{-21}$$

(c)   at pH 1.0: $[H^+] = 10^{-1}$, $[OH^-] = 10^{-13}$

$$[Pt^{2+}] = \frac{1 \times 10^{-35}}{(10^{-13})^2} = 1 \times 10^{-9}$$

(*Conclusion*: Even in highly acidic solution, an aqueous solution of Pt^{2+} cannot be prepared in appreciable concentration.)

12.   Call the acid H_2Su. Let the amount dissociating in the first stage be x mole liter^{-1}:

$$H_2Su \rightleftharpoons HSu^- \quad H^+$$
$$(0.10 - x) \qquad\qquad x \qquad x$$

Assuming x is small:

$$\frac{x^2}{0.10 - x} \approx \frac{x^2}{0.10} = K_1 = 6.2 \times 10^{-5}$$

$$x^2 = 6.2 \times 10^{-6} \qquad x = 2.5 \times 10^{-3} = [H^+] = [HSu^-]$$

$$\text{degree of dissociation} = \frac{2.5 \times 10^{-3}}{0.10} \times 100 = 2.5\% \text{ (The assumption is justified.)}$$

$$pH = -\log(2.5 \times 10^{-3}) = 2.60$$

In the second stage of dissociation, assume that degree of dissociation makes negligible difference to $[H^+]$ and $[HSu^-]$

$$HSu^- \rightleftharpoons Su^{2-} + H^+$$

$$\frac{[Su^{2-}][H^+]}{[HSu^-]} = K_2 = 2.3 \times 10^{-6}$$

But

$$[HSu^-] = [H^+] = 2.5 \times 10^{-3} \qquad [Su^{2-}] = 2.3 \times 10^{-6}$$

$$\text{percent dissociation in second stage:} \quad \frac{2.3 \times 10^{-6}}{2.5 \times 10^{-3}} \times 100 = 0.092\%$$

13.   Let the amount dissociating be x mole liter^{-1} in the first stage

$$H_2SO_3 \rightleftharpoons HSO_3^- + H^+$$
$$(0.020 - x) \qquad\quad x \qquad x$$

$$\frac{x^2}{0.020 - x} = K_1 = 1.7 \times 10^{-2}$$

x cannot be assumed small, as K_1 is quite large, so we solve the quadratic:

$$x^2 + (1.7 \times 10^{-2})x - (3.4 \times 10^{-4}) = 0 \qquad x = 1.2 \times 10^{-2} = [H^+] = [HSO_3^-]$$

$$\text{percent dissociation in first stage:} \quad \frac{1.2 \times 10^{-2}}{0.020} \times 100 = 60\%$$

In the second stage,

$$HSO_3^- \rightleftharpoons SO_3^{2-} + H^+$$

$$\frac{[SO_3^{2-}][H^+]}{[HSO_3^-]} = K_2 = 6.5 \times 10^{-8}$$

K_2 is very small, so this dissociation makes negligible difference to $[H^+]$ and $[HSO_3^-]$. These remain equal, so

$$[SO_3^{2-}] = 6.5 \times 10^{-8}$$

$$\text{percent dissociation in second stage} = \frac{6.5 \times 10^{-8}}{1.2 \times 10^{-2}} \times 100 = 5.4 \times 10^{-4}\%$$

$$pH = -\log(1.2 \times 10^{-2}) = 1.93$$

14. Use the expression developed in Example 14.15:

$$[Sr^{2+}]^2 = K_{sp}\left(1 + \frac{[H^+]}{K_2} + \frac{[H^+]^2}{K_1 K_2}\right)$$

where $K_{sp} = 3.6 \times 10^{-5}$, $K_2 = 3.3 \times 10^{-6}$, and $K_1 = 0.18$.

(a) at pH 2.00: $[H^+] = 1.0 \times 10^{-2}$
$[Sr^{2+}]^2 = 3.6 \times 10^{-5}[1 + (3.0 \times 10^3) + 170] = 0.12$
$[Sr^{2+}] = 0.34M$ (freely soluble)

(b) at pH 5.00: $[H^+] = 1.0 \times 10^{-5}$
$[Sr^{2+}]^2 = 3.6 \times 10^{-5}[1 + 3.0 + (1.7 \times 10^{-4})] = 1.4 \times 10^{-4}$
$[Sr^{2+}] = 0.012M$ (slightly soluble)

(c) at pH 7.00: $[H^+] = 1.0 \times 10^{-7}$
$[Sr^{2+}]^2 = 3.6 \times 10^{-5}[1 + (3.0 \times 10^{-2}) + (1.7 \times 10^{-8})] = 3.7 \times 10^{-5}$
$[Sr^{2+}] = 6.1 \times 10^{-3}M$ (very limited solubility)

15. The above type of expression applies here also:

$$[Ca^{2+}]^2 = K_{sp}\left(1 + \frac{[H^+]}{K_2} + \frac{[H^+]^2}{K_1 K_2}\right)$$

where $K_{sp} = 4.7 \times 10^{-9}$, $K_2 = 4.7 \times 10^{-11}$, and $K_1 = 4.5 \times 10^{-7}$.

(a) at pH 4.00: $[H^+] = 1.0 \times 10^{-4}$
$[Ca^{2+}]^2 = 4.7 \times 10^{-9}[1 + (2.1 \times 10^6) + (4.7 \times 10^8)] = 2.2$
$[Ca^{2+}] = 1.5M$ (freely soluble)

(b) at pH 7.00: $[H^+] = 1.0 \times 10^{-7}$
$[Ca^{2+}]^2 = 4.7 \times 10^{-9}[1 + (2.1 \times 10^3) + (4.7 \times 10^2)] = 1.2 \times 10^{-5}$
$[Ca^{2+}] = 3.5 \times 10^{-3}M$ (slightly soluble)

(c) at pH 10.00: $[H^+] = 1.0 \times 10^{-10}$
$[Ca^{2+}]^2 = 4.7 \times 10^{-9}[1 + 2.1 + (4.7 \times 10^{-4})] = 1.5 \times 10^{-8}$
$[Ca^{2+}] = 1.2 \times 10^{-4}$ (very limited solubility)

16.
$$Li_2CO_3(s) \;\rightleftharpoons\; 2Li^+ + CO_3^{2-}$$

$$K_{sp} = [Li^+]^2[CO_3^{2-}] \quad \text{and} \quad [CO_3^{2-}] = \frac{K_{sp}}{[Li^+]^2}$$

Now, initially, 1 mole Li_2CO_3 dissolves to give 2 moles Li^+ and 1 mole CO_3^{2-}, so we have

$$[Li^+] = 2[CO_3^{2-}]$$

Subsequently, equilibria are set up:

$$HCO_3^- \;\rightleftharpoons\; H^+ + CO_3^{2-}$$

$$\frac{[H^+][CO_3^{2-}]}{[HCO_3^-]} = K_2 \qquad [HCO_3^-] = \frac{[H^+][CO_3^{2-}]}{K_2}$$

$$H_2CO_3 \;\rightleftharpoons\; H^+ + HCO_3^-$$

$$\frac{[H^+][HCO_3^-]}{[H_2CO_3]} = K_1 \qquad [H_2CO_3] = \frac{[H^+][HCO_3^-]}{K_1}$$

or

$$[H_2CO_3] = \frac{[H^+]^2[CO_3^{2-}]}{K_2 K_1}$$

At equilibrium, CO_3^{2-}, HCO_3^-, and H_2CO_3 have been produced from the original CO_3^{2-}, so the stoichiometry of the solution is:

$$[Li^+] = 2([CO_3^{2-}] + [HCO_3^-] + [H_2CO_3]) = 2\left([CO_3^{2-}] + \frac{[H^+][CO_3^{2-}]}{K_2} + \frac{[H^+]^2[CO_3^{2-}]}{K_2 K_1}\right)$$

Substituting $[CO_3^{2-}] = \dfrac{K_{sp}}{[Li^+]^2}$ gives

$$[Li^+] = \frac{2K_{sp}}{[Li^+]^2}\left(1 + \frac{[H^+]}{K_2} + \frac{[H^+]^2}{K_2 K_1}\right) \qquad [Li^+]^3 = 2K_{sp}\left(1 + \frac{[H^+]}{K_2} + \frac{[H^+]^2}{K_2 K_1}\right)$$

The desired equation is on the right.
at pH 8.00: $[H^+] = 1.0 \times 10^{-8}$:
$[Li^+]^3 = 2 \times 1.7 \times 10^{-3}(1 + 213 + 4.7) = 0.74$
   $[Li^+] = 0.91 M$
at pH 10.00: $[H^+] = 1.0 \times 10^{-10}$:
$[Li^+]^3 = 2 \times 1.7 \times 10^{-3}[1 + 2.13 + (4.7 \times 10^{-4})] = 0.011$
   $[Li^+] = 0.22 M$

17. The general expression for the sulfide of a metal 2+ ion, MS, was treated in Example 14.16, which supplied the equation

$$[M^{2+}]^2 = K_{sp}\left(1 + \frac{[H^+]}{K_2} + \frac{[H^+]^2}{K_2 K_1}\right)$$

Inserting the given data into this expression gives
(a) at pH 1.00: $[Co^{2+}] = 0.30 M$, $[Cu^{2+}] = 9.6 \times 10^{-11} M$
(b) at pH 3.00: $[Co^{2+}] = 3.0 \times 10^{-3} M$, $[Cu^{2+}] = 9.6 \times 10^{-13} M$
(c) at pH 6.00: $[Co^{2+}] = 3.2 \times 10^{-6} M$, $[Cu^{2+}] = 1.0 \times 10^{-15} M$

This is consistent with experiment, since it shows that CoS dissolves in the acidic solution only (pH 1.0) whereas CuS does not dissolve at any pH.

18. Assuming all Zn^{2+} is converted to the complex ion, then

$$[Zn(CN)_4^{2-}] = 0.20 M$$

The stability constant expression is

$$\frac{[Zn(CN)_4^{2-}]}{[Zn^{2+}][CN^-]^4} = K_{stab} = 1.6 \times 10^{20} \quad so \quad [Zn^{2+}] = \frac{[Zn(CN)_4^{2-}]}{1.6 \times 10^{20}[CN^-]^4}$$

(a) If $[CN^-] = 0.010 M$,

$$[Zn^{2+}] = \frac{0.20}{1.6 \times 10^{20} \times (0.010)^4} = 1.3 \times 10^{-13}$$

(b) If $[CO_3^{2-}] = 0.10 M$, the ion product for $ZnCO_3$ is

$$[Zn^{2+}][CO_3^{2-}] = 1.3 \times 10^{-13} \times 0.10 = 1.3 \times 10^{-14}$$

This is less than K_{sp}, so precipitation does not occur.

19. If $[Al^{3+}] = 0.050$, $Al(OH)_3$ starts to precipitate when

$$[OH^-]^3 = \frac{K_{sp}}{[Al^{3+}]} = \frac{2 \times 10^{-32}}{0.050} = 4 \times 10^{-31}$$

$$[OH^-] = 7.4 \times 10^{-11} \qquad pOH = 10.1 \qquad pH = 3.9$$

Assuming that, when the precipitate redissolves,
$[Al(OH)_4^-] = 0.050$ (corresponding to all of the original Al^{3+})

$$K_{stab} = \frac{[Al(OH)_4^-]}{[Al^{3+}][OH^-]^4} = 2 \times 10^{33}$$

But if the solution is saturated in $Al(OH)_3$,

$$[Al^{3+}] = \frac{K_{sp}}{[OH^-]^3} = \frac{2 \times 10^{-32}}{[OH^-]^3}$$

Inserting this into the stability constant equation gives

$$\frac{[Al(OH)_4^-]}{2 \times 10^{-32}[OH^-]} = 2 \times 10^{33} \qquad [OH^-] = \frac{[Al(OH)_4^-]}{(2 \times 10^{-32})(2 \times 10^{33})} = \frac{[Al(OH)_4^-]}{40}$$

If $[Al(OH)_4^-] = 0.050$ then $[OH^-] = 0.050/40 = 1.3 \times 10^{-3}$, and

$$pOH = 2.9 \qquad pH = 11.1$$

Another amphoteric hydroxide, precipitating at pH 3.9 and redissolving at pH 11.1.

20. Conversion of essentially all Fe^{3+} to the complex gives

$$[Fe(C_2O_4)_3^{3-}] = 0.040$$

Using the stability constant expression

$$\frac{[Fe(C_2O_4)_3^{3-}]}{[Fe^{3+}][C_2O_4^{2-}]^3} = K_{stab} = 1.6 \times 10^{20}$$

If free oxalate, $[C_2O_4^{2-}]$, is $0.010 M$

$$[Fe^{3+}] = \frac{[Fe(C_2O_4)_3^{3-}]}{1.6 \times 10^{20} \times (0.010)^3} = \frac{0.040}{1.6 \times 10^{14}} = 2.5 \times 10^{-16}$$

At pH 7.0, $[OH^-] = 1.0 \times 10^{-7}$
Ion product for $Fe(OH)_3$ is

$$[Fe^{3+}][OH^-]^3 = (2.5 \times 10^{-16}) \times (1.0 \times 10^{-7})^3 = 2.5 \times 10^{-37}$$

No precipitation will occur, as this is below K_{sp}.

Chapter 15

Electrochemistry: Oxidation and Reduction

Since I ran down that coiner by the zinc and copper filings in the seam of his cuff, they have begun to realise the importance of the microscope. THE ADVENTURE OF SHOSCOMBE OLD PLACE

If a difference in electrical potential is applied to an ionic crystal through electrodes, nothing will happen. The charged ions experience an attraction to the electrode of opposite charge, but they are held in position in the crystal by the strong electrostatic forces within the lattice, and cannot move. But the result will be very different if the lattice is broken down by melting the crystal or dissolving it in a solvent, so that the ions are free to move about. Under these conditions, the ions will migrate toward the electrodes under the influence of the externally applied potential.

For a current to flow through a molten salt or a solution, a chemical reaction must occur at each electrode. These reactions are of two types, according to the direction of electron transfer between the solution and the electrode. If electrons are transferred *from* the electrode *to* the solution then a species in solution (either an ion or a solvent molecule) will gain one or more electrons. This gain of electrons is called *reduction*, and the electrode where it occurs is called the *cathode*.

If electrons are transferred from the solution to the electrode, then a species in solution is *losing* one or more electrons. This loss of electrons is called *oxidation*, and the electrode where it occurs is called the *anode*. Oxidation and reduction always occur together; in other words, if there are two electrodes in a solution, one must be an anode and the other a cathode.

As a simple illustration, consider two inert electrodes carrying a current through molten sodium chloride. The reactions occurring will be

at the anode: $$Cl^- \longrightarrow \frac{1}{2} Cl_2 + e^-$$

The chloride ion (an *anion*) loses an electron. This is an oxidation.

at the cathode: $$Na^+ + e^- \longrightarrow Na$$

The sodium ion (a *cation*) gains an electron. This is a reduction.

If we add these two together, the electron will cancel out to give the overall reaction occurring as

$$Na^+ + Cl^- \longrightarrow Na + \frac{1}{2}Cl_2$$

The physical meaning of the disappearance of the electron from these equations is that it has gone from the solution into the anode, then through the external circuit to the cathode, from which it goes back into the solution. This process consumes energy, which is supplied by a battery or other source of electric potential difference external to the solution.

An electron transfer process represents the sum of two electrode reactions, which are called "half-cell" reactions. Since one is a reduction and the other an oxidation, the overall reaction ("cell reaction") is called a *redox* or *oxidation-reduction* reaction. We have met the concept of half-reactions before, in Chapter 3, when we used it as an aid to balancing redox equations. Although we have been speaking of a specific case involving actual electrodes, the concept is quite general, and it is not necessary for actual electrodes to be present.

Any redox reaction may be split into two half-cell reactions, one a reduction and the other an oxidation. For a balanced overall reaction, the number of electrons lost in the oxidation (anode) reaction must equal the number of electrons gained in the reduction (cathode) reaction.

(*Note*: There are various mnemonics for remembering which process is which. Try: "Gain a red cat, but lose an odd ox"—(translated, *gain* electrons, *red*uction, *cat*hode; *lose* electrons at the *an*ode, *ox*idation.)

15.1 Quantitative Electrolysis and Electrolytic Cells

One important aspect of electrolysis reactions is their stoichiometry. Suppose we pass a current through a solution or a molten salt, how much substance will be liberated from solution at each electrode? As with any problem of this type, we will work on a mole basis. Returning to the reaction

$$Na^+ + e^- \longrightarrow Na$$

we see that one electron reduces one sodium ion to an atom, so if we want to produce one mole of sodium, we will have to pass one mole (Avogadro's number) of electrons through the electrode. Knowing the charge on the electron and the value of Avogadro's number, we can easily work out what quantity of electricity this is:

$$(1.602 \times 10^{-19} \text{ coulomb}) \times (6.023 \times 10^{23} \text{ mole}^{-1}) = 96,490 \text{ coulomb mole}^{-1}$$

A coulomb (symbol C) is defined as the quantity of electricity that passes when a current of 1 ampere (symbol A) flows for 1 second, so to pass one mole of electrons through an electrode we would have to pass 1 ampere for 96,490 seconds, or 26.80 hours.

This quantity of electricity is known as the *Faraday constant*, in commemoration of Michael Faraday, whose laws of electrolysis laid the foundation of electrochemistry. For most purposes, we take the Faraday constant (symbol $\mathcal{F}$) as 96,500 C mole^{-1}; that is, to three significant figures.

The reduction of Na^+ is a one-electron process, that is to say, one electron is transferred for each ion that reacts. Other oxidation or reduction processes may involve the transfer of more than one electron.

Two-electron processes:

$$Cu^{2+} + 2e^- \longrightarrow Cu$$

$$Fe \longrightarrow Fe^{2+} + 2e^-$$

$$(COO)_2^{2-} \longrightarrow 2CO_2 + 2e^-$$

$$H_2O \longrightarrow 2H^+ + \frac{1}{2}O_2 + 2e^-$$

Three-electron processes:

$$Al^{3+} + 3e^- \longrightarrow Al$$

$$Fe \longrightarrow Fe^{3+} + 3e^-$$

$$CrO_4^{2-} + 8H^+ + 3e^- \longrightarrow Cr^{3+} + 4H_2O$$

Four-electron processes:

$$Ce \longrightarrow Ce^{4+} + 4e^-$$

$$SO_3^{2-} + 6H^+ + 4e^- \longrightarrow S + 3H_2O$$

Five-electron processes:

$$MnO_4^- + 8H^+ + 5e^- \longrightarrow Mn^{2+} + 4H_2O$$

etc.

Obviously, the relationship between the amount of substance undergoing gain or loss of electrons and the quantity of electricity passed will depend on the number of electrons required to effect the change on each ion or molecule. If it requires two electrons to reduce an ion of Cu^{2+} to Cu, then it will require two moles of electrons to reduce one mole of Cu^{2+}. In general,

---

The number of moles of a substance oxidized or reduced at an electrode is equal to the number of moles of electrons passed, divided by the number of electrons involved in the electrode reaction for each ion or molecule of the substance.

So if we pass a current of i amperes for a period of t seconds through an electrode at which an n electron change is occurring, the amount of substance undergoing reaction at that electrode will be

$$\frac{it}{n\mathscr{F}} = \frac{it}{96,500n} \text{ moles}$$

The quantity of substance undergoing reaction when one mole of electrons passes will obviously be $1/n$ moles, and this is often referred to as one *equivalent* of the substance. However, the value of the equivalent will depend on the value of n, that is, on the reaction that is occurring, so it is not fixed for a given substance (*Note*: In the above equations, iron may enter into reactions where $n = 2$ or $n = 3$.) It is a mistake to speak of an equivalent of a substance without specifying the reaction it is to undergo. For this reason, the concept will not be used in this discussion, and we will use instead the mole concept coupled with the value of n derived from the half-cell reaction equation.

Example 15.1 How much Br_2 is liberated by electrolysis of molten KBr for 30 minutes at a current of 2.0 A?

SOLUTION 30 minutes $= 30 \times 60 = 1800$ s. At a current of 2.0 A (2.0 C s^{-1}), the amount of electricity passed is

$$2.0 \times 1800 = 3600 \text{ C} = \frac{3600}{96,500} = 0.0373 \text{ mole of electrons}$$

The anode reaction is

$$Br^- \longrightarrow \frac{1}{2} Br_2 + e^-$$

In a one-electron process, 1/2 mole of Br_2 is liberated:

$$\frac{1}{2} \text{ mole } Br_2 = 79.9 \text{ g liberated by 1 mole of electrons}$$

so 0.0373 mole of electrons liberates $0.0373 \times 79.9 = 3.0$ g Br_2.

Example 15.2 How long must a current of 20.0 A be passed through molten $CaCl_2$ to produce 15.0 g of calcium?

SOLUTION The reaction is

$$Ca^{2+} + 2e^- \longrightarrow Ca$$

1 mole of calcium (40.0 g) is produced by passing 2 moles of electrons

$$15.0 \text{ g calcium require } \frac{15.0 \times 2}{40.0} = 0.75 \text{ mole of electrons}$$

0.75 mole of electrons is $0.75 \times 96,500 = 72,400$ C

At a current flow of 20.0 A, this will take

$$\frac{72,400}{20.0} = 3620 \text{ s} = 1 \text{ h } 20 \text{ s}$$

Example 15.3 Water is electrolyzed by passing a current through dilute sodium sulfate solution. The reactions are

cathode: $\qquad\qquad 2H^+ + 2e^- \longrightarrow H_2$

anode: $\qquad\qquad H_2O \longrightarrow \frac{1}{2} O_2 + 2H^+ + 2e^-$

What volumes of gas will be liberated at each electrode if a current of 2.00 A passes for 40.0 minutes at STP?

SOLUTION   The amount of electricity passed is

$$2.00 \times 40.0 \times 60 = 4800 \text{ C} = \frac{4800}{96,500} = 0.0497 \text{ mole of electrons}$$

at the cathode, 2 moles of electrons give 1 mole H_2 and 0.0497 mole of electrons give

$$\frac{0.0497}{2} = 0.0249 \text{ mole } H_2 = 0.0249 \times 22.4 \text{ liter}$$

$$= 0.557 \text{ liter } H_2$$

At the anode, 2 moles of electrons give 0.5 mole O_2 and 0.0497 mole of electrons gives

$$\frac{0.5 \times 0.0497}{2} = 0.0124 \text{ mole } O_2$$

$$= 0.0124 \times 22.4 = 0.279 \text{ liter } O_2$$

As we would expect, the volumes of gas produced by the electrolysis of water are in a 2:1 ratio.

Example 15.4   A "silver coulometer" is employed as an accurate means of measuring current flow, using the change in weight of a silver electrode. The reaction is

$$Ag^+ + e^- \longrightarrow Ag$$

If a silver electrode gains 0.3482 g in weight over a period of 4250 s, what is the average current flowing during this time?

SOLUTION   A mole of silver is 107.9 g, so the quantity deposited in this electrolysis is

$$\frac{0.3482}{107.9} = 3.227 \times 10^{-3} \text{ mole}$$

The electrode process is a one-electron change, so this means that 3.227×10^{-3} mole of electrons passed:

$$3.227 \times 10^{-3} \text{ mole of electrons} = 3.227 \times 10^{-3} \times 96,490 = 311.4 \text{ C}$$

If this took 4250 s to pass, the current was

$$\frac{311.4 \text{ C}}{4250 \text{ s}} = 7.327 \times 10^{-2} \text{ A}$$

So far, we have been discussing electrolysis reactions occurring in cells called *electrolytic cells*. The reaction is carried out by using an external potential difference to supply the necessary energy, since the electrolysis reaction will not occur spontaneously. For example, the reaction

$$Na^+ + Cl^- \longrightarrow Na(l) + \frac{1}{2} Cl_2(g)$$

(electrolysis of molten NaCl) requires considerable energy input to go as written; indeed, the products of reaction at each electrode must be kept apart to prevent spontaneous reaction occurring in the *reverse* direction.

15.2 Voltaic Cells and Electrode Potentials

If we approach the problem of redox reactions in a slightly different manner, we can assemble a cell in which the reaction is allowed to proceed by itself (*spontaneously*) at two electrodes while we measure the potential difference between the electrodes. We could use this cell as a source of electrical energy to power lamps, radios, motors, etc., in which case we would commonly refer to it as a battery. In chemistry, however, we use the term *voltaic cell* (or *galvanic cell*) to describe this arrangement and to distinguish it from an electrolytic cell.

In a voltaic cell, a difference of electrical potential between the electrodes is produced by chemical reactions occurring spontaneously at each electrode.

Clearly, the two electrodes in a voltaic cell must differ in some way in order to produce a difference in potential. But how can we divide the observed potential between the two electrodes? Suppose we observe an overall potential of one volt. Can we say that 0.3 volt is produced at one electrode and 0.7 volt at the other?

The answer to the above question is that we cannot split up the observed potential in this way. There is no way of observing the potential at a single electrode, we must always take them in pairs. For example, suppose we immerse a copper rod in a solution of copper sulfate. A potential difference will appear at the interface between the metal and the solution, but the only way we can attempt to measure it is to connect one terminal of a voltmeter to the copper metal and the other terminal to a wire which we dip into the solution. The action of placing the wire into the solution creates a second metal–solution interface, and the potential difference registered on the voltmeter will be the *difference* between the two individual electrode potentials.

Note that the cell potential must be the difference between, rather than the sum of, the two electrode potentials. This is because they must be connected in opposition, so that the complete circuit goes voltmeter–electrode, electrode–solution, solution–electrode, electrode–voltmeter.

For convenience in working with cell potentials, there is a universally accepted convention for measuring cell potentials. We construct a *standard hydrogen electrode*, which is an electrode in which gaseous hydrogen at one atmosphere pressure is in equilibrium with hydrogen ions in solution, concentration $1M$, at 25°C, over a platinum surface. The potential difference at this electrode is then arbitrarily taken to be zero volts, and other electrode potentials are measured with reference to the standard hydrogen electrode.

For example, suppose we assemble a cell by combining a standard hydrogen electrode with a zinc metal electrode immersed in $1M$ $ZnSO_4$ solution. When we connect a voltmeter to this cell, it registers 0.76 V; and we say that this is due solely to the potential difference at the zinc electrode, based on the assignment of zero potential to the hydrogen electrode. (*Note*: We ignore the possibility of any potential difference at the interface between the $ZnSO_4$ solution and the H^+ solution. In practice, there are various techniques for arranging electrical contact without mixing between two electrode solutions.)

When we connect a wire across the above cell, current flows and chemical reaction occurs at each electrode. We write down the equation for each half-cell reaction with its accompanying *standard electrode potential*, which is given the symbol $\mathcal{E}^0$:

cathode:	$2H^+ + 2e^-$	$\longrightarrow H_2$	$\mathcal{E}^0 = \quad 0.00$ V
anode:	Zn	$\longrightarrow Zn^{2+} + 2e^-$	$\mathcal{E}^0 = +0.76$ V

overall:	$Zn + 2H^+$	$\longrightarrow Zn^{2+} + H_2$	$\mathcal{E}^0 = +0.76$ V

If we reverse each of the half-reaction equations, we must change the sign for each, so that the overall reaction has the reverse sign:

cathode:	$Zn^{2+} + 2e^-$	$\longrightarrow Zn$	$\mathcal{E}^0 = -0.76$ V
anode:	H_2	$\longrightarrow 2H^+ + 2e^-$	$\mathcal{E}^0 = \quad 0.0$ V

overall:	$Zn^{2+} + H_2$	$\longrightarrow Zn + 2H^+$	$\mathcal{E}^0 = -0.76$ V

Both of the above equations represent balanced redox processes, but there is an important difference between them. Experiment shows that, when the cell is assembled and the circuit complete, *the direction of current flow is such that the first reaction proceeds spontaneously* (zinc dissolves in acid solution, giving hydrogen and Zn^{2+} ions).

The overall cell potential is positive for this reaction, which is an example of a general result of great importance:

Reaction will spontaneously proceed in the direction for which the overall cell potential is positive.

Suppose we assemble a slightly different cell with a copper electrode in $1M$ $CuSO_4$ solution and a standard hydrogen electrode. The half-cell reactions, with their $\mathcal{E}^0$ values, are

anode:	H_2	$\longrightarrow 2H^+ + 2e^-$	$\mathcal{E}^0 = \quad 0.00$ V
cathode:	$Cu^{2+} + 2e^-$	$\longrightarrow Cu$	$\mathcal{E}^0 = +0.34$ V

overall:	$Cu^{2+} + H_2$	$\longrightarrow Cu + 2H^+$	$\mathcal{E}^0 = +0.34$ V

We see that this reaction goes spontaneously in the direction that causes reduction of the metal ion (Cu^{2+}) by hydrogen, in contrast to the zinc reaction.

If we now assemble a cell by taking one Cu/Cu^{2+} electrode and one Zn/Zn^{2+} electrode, we can use the above half-cell potentials to forecast the overall cell potential and the direction in which reaction will spontaneously proceed. Of course, we must always write one half-reaction as an oxidation and the other as a reduction when we add them together:

anode:	Zn	$\longrightarrow Zn^{2+} + 2e^-$	$\mathcal{E}^0 = +0.76$ V
cathode:	$Cu^{2+} + 2e^-$	$\longrightarrow Cu$	$\mathcal{E}^0 = +0.34$ V

overall:	$Zn + Cu^{2+}$	$\longrightarrow Zn^{2+} + Cu$	$\mathcal{E}^0 = +0.76 + 0.34 = 1.10$ V

For the reverse reaction, $\mathcal{E}^0$ is negative:

$$Zn^{2+} + Cu \longrightarrow Zn + Cu^{2+} \qquad \mathcal{E}^0 = -1.10 \text{ V}$$

The first reaction will be the one that occurs spontaneously, since $\mathcal{E}^0$ is positive.

It is most important to realize that, although the half-cell potentials depend on our arbitrary choice of the hydrogen reference electrode, the overall cell potential is a real, experimentally verifiable quantity that is independent of our choice of standard. So we can measure the half-cell potentials of a whole variety of electrodes, relative to the standard hydrogen electrode, and work out the overall cell potential of any pair of these in combination by taking the necessary data from this list. Such a tabulation is given in Table 15.1. These are $\mathcal{E}^0$ values, that is, they are measured under standard conditions, which we may define as:

1. Pure, solid (except mercury!) elements for metal electrodes (Cu, Zn, Mg, etc.)
2. Gaseous components at partial pressures of one atmosphere (H_2, O_2, Cl_2, etc.)
3. Soluble ionic species at 1 molar concentration (Cu^{2+}, H^+, Cl^-, etc.)
4. A temperature of 25°C.

(*Note*: In advanced work, a slightly more complicated definition of standard conditions is needed, but the above is sufficient for our purposes.)

As you will see, the half-cell potentials are all given as reductions (cathode processes). This is usual practice, and we refer to the table as "reduction potentials." Whenever we have to use one of these potentials in an oxidation half-reaction (anode process), we must remember to reverse the sign of $\mathcal{E}^0$ as we write the equation the other way round.

We can use this table to work out $\mathcal{E}^0$ values for various redox processes. Remembering the significance of a positive overall value, we can then forecast the direction in which the reactions will spontaneously go.

Example 15.6   Calculate $\mathcal{E}^0$ for the following reactions:

(a) $Sn + 2H^+ \longrightarrow Sn^{2+} + H_2$

(b) $Ag + H^+ \longrightarrow Ag^+ + \frac{1}{2} H_2$

Will either of the above metals, Sn or Ag, spontaneously dissolve in acid solution?

SOLUTION

(a) anode: $Sn \longrightarrow Sn^{2+} + 2e^-$ $\mathcal{E}^0 = +0.14$ V

   (tin is oxidized, so we reverse the given reduction potential of -0.14 V)

   cathode: $2H^+ + 2e^- \longrightarrow H_2$ $\mathcal{E}^0 = 0.00$ V

   overall: $Sn + 2H^+ \longrightarrow Sn^{2+} + H_2$ $\mathcal{E}^0 = +0.14$ V

(b) anode: $Ag \longrightarrow Ag^+ + e^-$ $\mathcal{E}^0 = -0.80$ V

   cathode: $H^+ + e^- \longrightarrow \frac{1}{2} H_2$ $\mathcal{E}^0 = 0.00$ V

   overall: $Ag + H^+ \longrightarrow Ag^+ + \frac{1}{2} H_2$ $\mathcal{E}^0 = -0.80$ V

Since the overall potential is positive for Sn but negative for Ag, we conclude that Sn will spontaneously dissolve in acid solution, whereas Ag will not.

TABLE 15.1 Standard Electrode Potentials in Aqueous Solution at 25°C

Reaction	$\mathcal{E}^0$, volts
In Acid Solution	
$Li^+ + e^- \longrightarrow Li(s)$	-3.05
$Na^+ + e^- \longrightarrow Na(s)$	-2.71
$Mg^{2+} + 2e^- \longrightarrow Mg(s)$	-2.36
$Al^{3+} + 3e^- \longrightarrow Al(s)$	-1.66
$Mn^{2+} + 2e^- \longrightarrow Mn(s)$	-1.18
$Zn^{2+} + 2e^- \longrightarrow Zn(s)$	-0.76
$Cr^{3+} + 3e^- \longrightarrow Cr(s)$	-0.74
$Fe^{2+} + 2e^- \longrightarrow Fe(s)$	-0.44
$Cr^{3+} + e^- \longrightarrow Cr^{2+}$	-0.41
$Tl^+ + e^- \longrightarrow Tl(s)$	-0.34
$Ni^{2+} + 2e^- \longrightarrow Ni(s)$	-0.25
$H_3PO_4 + 2H^+ + 2e^- \longrightarrow H_3PO_3 + H_2O$	-0.28
$Sn^{2+} + 2e^- \longrightarrow Sn(s)$	-0.14
$Pb^{2+} + 2e^- \longrightarrow Pb(s)$	-0.13
$H^+ + e^- \longrightarrow \frac{1}{2}H_2(g)$	0.00
$S(s) + 2H^+ + 2e^- \longrightarrow H_2S(aq)$	$+0.14$
$Sn^{4+} + 2e^- \longrightarrow Sn^{2+}$	$+0.15$
$SO_4^{2-} + 4H^+ + 2e^- \longrightarrow H_2SO_3 + H_2O$	$+0.17$
$Hg_2Cl_2(s) + 2e^- \longrightarrow 2Hg(l) + 2Cl^-$	$+0.24$
$Cu^{2+} + 2e^- \longrightarrow Cu(s)$	$+0.34$
$\frac{1}{2}I_2(s) + e^- \longrightarrow I^-$	$+0.54$
$O_2(g) + 2H^+ + 2e^- \longrightarrow H_2O_2$	$+0.68$
$Fe^{3+} + e^- \longrightarrow Fe^{2+}$	$+0.77$
$Ag^+ + e^- \longrightarrow Ag(s)$	$+0.80$
$NO_3^- + 3H^+ + 2e^- \longrightarrow HNO_2 + H_2O$	$+0.96$
$IO_3^- + 6H^+ + 5e^- \longrightarrow \frac{1}{2}I_2 + 3H_2O$	$+1.20$
$\frac{1}{2}O_2(g) + 2H^+ + 2e^- \longrightarrow H_2O$	$+1.23$
$\frac{1}{2}Cl_2(g) + e^- \longrightarrow Cl^-$	$+1.36$
$Cr_2O_7^{2-} + 14H^+ + 6e^- \longrightarrow 2Cr^{3+} + 7H_2O$	$+1.33$
$HClO + H^+ + 2e^- \longrightarrow Cl^- + H_2O$	$+1.49$
$MnO_4^- + 8H^+ + 5e^- \longrightarrow Mn^{2+} + 4H_2O$	$+1.51$
$Ce^{4+} + e^- \longrightarrow Ce^{3+}$	$+1.61$
$\frac{1}{2}F_2(g) + H^+ \longrightarrow HF$	$+2.87$
In Alkaline Solution	
$Al(OH)_4^- + 3e^- \longrightarrow Al(s) + 4OH^-$	-2.33
$HPO_3^{2-} + 2H_2O + 2e^- \longrightarrow H_2PO_2^- + 3OH^-$	-1.57
$2H_2O + 2e^- \longrightarrow H_2(g) + 2OH^-$	-0.83
$Fe(OH)_3(s) + e^- \longrightarrow Fe(OH)_2(s) + OH^-$	-0.56
$S(s) + 2e^- \longrightarrow S^{2-}$	-0.45
$Ag_2O(s) + H_2O + 2e^- \longrightarrow 2Ag(s) + 2OH^-$	$+0.34$
$O_2(g) + 2H_2O + 4e^- \longrightarrow 4OH^-$	$+0.40$
$O_3(ozone)(g) + H_2O + 2e^- \longrightarrow O_2(g) + 2OH^-$	$+1.24$

Before going on to more complicated examples, we should consider one important point. It often happens, as we balance a redox equation, that one or both of the half-reaction equations must be multiplied throughout by some coefficient (2, 3, etc.) in order to make the electrons cancel out in the overall equation. When we do this, we do *not* make any change in the accompanying $\mathcal{E}^0$ value.

In other words, an $\mathcal{E}^0$ value is *not* measured on a "per mole" basis. It is simply an expression of the potential difference between an electrode and a solution, and this will not vary according to the amount of substance which is being considered.

Example 15.7 Taking the appropriate $\mathcal{E}^0$ values from Table 15.1, answer the following questions:

(a) What is $\mathcal{E}^0$ for the overall reaction

$$Zn + Sn^{4+} \longrightarrow Zn^{2+} + Sn^{2+}$$

(b) Will metallic Ni reduce
(i) Fe^{3+} to Fe^{2+}
(ii) Zn^{2+} to Zn

(c) In which direction can the reaction

$$Sn^{2+} + 2Fe^{3+} \longrightarrow Sn^{4+} + 2Fe^{2+}$$

go spontaneously?

SOLUTION

(a) The Zn reduction reaction must be reversed to make an oxidation, giving

anode reaction:	$Zn \longrightarrow Zn^{2+} + 2e^-$	$\mathcal{E}^0 = +0.76$ V
cathode reaction:	$Sn^{4+} + 2e^- \longrightarrow Sn^{2+}$	$\mathcal{E}^0 = +0.15$ V
overall reaction:	$Zn + Sn^{4+} \longrightarrow Zn^{2+} + Sn^{2+}$	$\mathcal{E}^0 = 0.76 + 0.15$
		$= +0.91$ V

The reaction can go spontaneously in the forward direction as written.

(b) The anode reaction in each case is

$$Ni \longrightarrow Ni^{2+} + 2e^- \qquad \mathcal{E}^0 = +0.25 \text{ V}$$

since Ni is to be used as a reducing agent.

(i) For the reduction of

$$Fe^{3+} + e^- \longrightarrow Fe^{2+} \qquad \mathcal{E}^0 = +0.77 \text{ V}$$

To obtain a balanced overall reaction, we must double the coefficients in this equation.

$$2Fe^{3+} + 2e^- \longrightarrow 2Fe^{2+} \qquad \mathcal{E}^0 = +0.77 \text{ V}$$

Note that this has no effect on the value of $\mathcal{E}^0$. Then, adding to the nickel equation gives

$$Ni + 2Fe^{3+} \longrightarrow Ni^{2+} + 2Fe^{2+} \qquad \mathcal{E}^0 = 0.25 + 0.77 = +1.02 \text{ V}$$

Yes, this reaction can go, since $\mathcal{E}^0$ is positive. Ni will reduce Fe^{3+} to Fe^{2+}.

(ii) For the reduction of Zn^{2+}, the equation is

$$Zn^{2+} + 2e^- \longrightarrow Zn \qquad \mathcal{E}^0 = -0.76 \text{ V}$$

Adding this to the anode reaction gives

$$Ni + Zn^{2+} \longrightarrow Ni^{2+} + Zn \qquad \mathcal{E}^0 = +0.25 - 0.76 = -0.51 \text{ V}$$

This reaction will *not* spontaneously go as written. It can spontaneously proceed in the *reverse* direction.

(c) The anode reaction is

$$Sn^{2+} \longrightarrow Sn^{4+} + 2e^- \qquad \mathcal{E}^0 = -0.15 \text{ V}$$

For the cathode reaction, we reverse the iron reaction and double the coefficients:

$$2Fe^{3+} + 2e^- \longrightarrow 2Fe^{2+} \qquad \mathcal{E}^0 = +0.77 \text{ V}$$

(*Note*: Again, $\mathcal{E}^0$ did not change when the coefficients in the equation were doubled.) Adding for the overall reaction

$$Sn^{2+} + 2Fe^{3+} \longrightarrow Sn^{4+} + 2Fe^{2+} \qquad \mathcal{E}^0 = -0.15 + 0.77 = +0.62 \text{ V}$$

The reaction can go in the forward direction as written.

A study of $\mathcal{E}^0$ values is very useful in telling us the direction in which a reaction *may* spontaneously go, but we cannot assume that the equation we write will accurately represent what actually happens in practice. The reason for this is that the $\mathcal{E}^0$ value tells us nothing of the *rate* at which a reaction will occur, and it may be that a process that appears very favorable in the equation will only proceed very slowly in practice (remembering the chapter on kinetics, you'll realize that this indicates a large activation energy). Meanwhile there may be a competing reaction proceeding at a very much greater rate, although its $\mathcal{E}^0$ value is lower.

To take specific example, consider the following reaction involving perchloric acid, $HClO_4$:

cathode:	$ClO_4^- + 2H^+ + 2e^- \longrightarrow ClO_3^- + H_2O$	$\mathcal{E}^0 = +1.23$ V
anode:	$Zn \longrightarrow Zn^{2+} + 2e^-$	$\mathcal{E}^0 = +0.76$ V
overall:	$Zn + ClO_4^- + 2H^+ \longrightarrow Zn^{2+} + ClO_3^- + H_2O$	$\mathcal{E}^0 = 1.23 + 0.76$
		$= +1.99$ V

The overall potential of $+1.99$ V tells us the reaction may spontaneously occur as written. If, however, we try the experiment of dissolving zinc in perchloric acid, the ion ClO_4^- takes no part in the reaction and all that happens is

$$Zn + 2H^+ \longrightarrow Zn^{2+} + H_2 \qquad \mathcal{E}^0 = +0.76 \text{ V}$$

The reduction of ClO_4^- is favorable from a long-term point of view, but it is very slow in practice, so the much faster reaction of Zn with H^+ predominates, even though its $\mathcal{E}^0$ is much lower.

Before leaving the topic of overall $\mathcal{E}^0$ values, we should note that it is only correct to sum the half-cell $\mathcal{E}^0$ values into an overall reaction *if* we have ensured that the electrons in the anode and cathode half-cell reactions cancel out, that is, that the number of electrons given up in the oxidation half-reaction exactly equals the number gained in the reduction half-reaction. If we have an overall reaction in which electrons appear, its $\mathcal{E}^0$ value *cannot* be obtained by the simple summation of $\mathcal{E}^0$ values of other processes. Thus, for the iron reactions, we have

(i)	$Fe^{2+} + 2e^- \longrightarrow Fe$	$\mathcal{E}^0 = -0.44$ V
(ii)	$Fe^{3+} + e^- \longrightarrow Fe^{2+}$	$\mathcal{E}^0 = +0.77$ V
(iii)	$Fe^{3+} + 3e^- \longrightarrow Fe$	$\mathcal{E}^0 = -0.036$ V

Although reaction (iii) is obtained by summing the equations for reactions (i) and (ii), its $\mathcal{E}^0$ value is *not* the sum of the two $\mathcal{E}^0$ values, because the electrons do not cancel out. (*Note*: Overall $\mathcal{E}^0$ values in cases like this can be obtained from $\mathcal{E}^0$ values of component reactions by a slightly different procedure, which is described at the end of Chapter 16.)

15.3   Variations in Electrode Potentials: The Nernst Equation

Up to this point we have used only standard electrode potentials to determine the direction in which a given reaction will proceed. Because the concentration of soluble ionic species and the pressure of gaseous components had been fixed by definition of the standard conditions, it was implied that the reactions were also run under the standard conditions. However, there are many instances in which a concentration less than $1M$ would be desirable for an ionic redox reaction, or solubility considerations may limit the concentration of some reactant to a lower value.

In this section, we will consider the effect of a change in the concentration of the soluble ions in a half-cell on the potential of given half-cells and, hence, upon the overall potential of a redox reaction.

For a general overall redox reaction,

$$aA + bB \longrightarrow cC + dD$$

the potential for this reaction is given by the *Nernst equation*

$$\mathcal{E}_{cell} = \mathcal{E}_{cell}^0 - \frac{RT}{n\mathcal{F}} \ln\left\{\frac{[C]^c[D]^d}{[A]^a[B]^b}\right\}$$

where $\mathcal{E}_{cell}$ is the potential for the overall redox reaction, $\mathcal{E}_{cell}^0$ is the standard potential of the overall redox reaction, R is the gas constant (8.314 J mole^{-1} K^{-1}), T is the absolute temperature, $\mathcal{F}$ is the Faraday constant, and n is the number of electrons transferred in the overall redox reaction. The expression within the logarithmic term is the mass-action quotient for the overall cell reaction, obtained by dividing the product of the concentrations of the products by the product of the concentrations of the reactants. Each concentration is raised to the power of its coefficient in the balanced-cell reaction. As in writing expressions for equilibrium constants, pure phases (e.g., solid metals) are omitted in the quotient since the concentration of a pure phase is a constant at a constant temperature. If any species is gaseous, its partial pressure in *atmospheres* is used instead of its molar concentration.

We will be working at a constant temperature of 25°C (298 K), so we can put together several constant terms in the Nernst equation. It's more convenient to work with common logs (base 10) so we put in the usual conversion factor of 2.303 to convert from natural to common logs, giving

$$\mathcal{E}_{cell} = \mathcal{E}_{cell}^0 - \frac{2.303RT}{n\mathcal{F}} \log\left\{\frac{[C]^c[D]^d}{[A]^a[B]^b}\right\}$$

At 298 K,

$$\frac{2.303RT}{\mathcal{F}} = \frac{2.303 \times 8.314 \times 298}{96,500} = 0.0591 \text{ V}$$

$$\mathcal{E}_{cell} = \mathcal{E}_{cell}^0 - \frac{0.0591}{n} \log\left\{\frac{[C]^c[D]^d}{[A]^a[B]^b}\right\}$$

Obviously, if $[C] = [D] = [A] = [B] = 1M$, then the log term becomes zero and $\mathcal{E}_{cell} = \mathcal{E}^0_{cell}$, as we would expect. At other conditions we can work out the value of the log term that must be subtracted from $\mathcal{E}^0_{cell}$ to give the electrode potential, $\mathcal{E}_{cell}$.

The Nernst equation can be used to determine the potential of either a half-cell or a complete redox reaction at nonstandard conditions. Several examples will illustrate how the equation is applied.

Example 15.8 For the process

$$Zn + 2H^+ \longrightarrow Zn^{2+} + H_2$$

the value of $\mathcal{E}^0_{cell}$ is $+0.76$ V. With $p(H_2)$ constant at 1.00 atm, how will $\mathcal{E}_{cell}$ change under the following conditions:

(a) $[Zn^{2+}] = 1.00 \times 10^{-3}M$, $[H^+] = 1.00M$
(b) $[Zn^{2+}] = 1.00M$, $[H^+] = 1.00 \times 10^{-3}M$

SOLUTION The total number of electrons transferred in the balanced equation (n) is two, so the Nernst equation is

$$\mathcal{E}_{cell} = \mathcal{E}^0_{cell} - \frac{0.059}{2} \log\left\{\frac{[Zn^{2+}]p(H_2)}{[H^+]^2}\right\}$$

(*Note*: The concentration of Zn, the pure solid metal, is omitted as being constant.)

We may omit $p(H_2)$, which is constant at 1.00 atm, so putting $\mathcal{E}^0_{cell} = 0.76$ V, we have

$$\mathcal{E}_{cell} = 0.76 - \frac{0.059}{2} \log\left\{\frac{[Zn^{2+}]}{[H^+]^2}\right\}$$

(a) When $[Zn^{2+}] = 1.00 \times 10^{-3}M$ and $[H^+] = 1.00M$,

$$\mathcal{E}_{cell} = 0.76 - \frac{0.059}{2} \log(1.00 \times 10^{-3})$$

$$= 0.76 + \frac{0.059 \times 3}{2} = 0.76 + 0.089 = 0.85 \text{ V}$$

(b) When $[Zn^{2+}] = 1.00M$ and $[H^+] = 1.00 \times 10^{-3}M$,

$$\mathcal{E}_{cell} = 0.76 - \frac{0.059}{2} \log\frac{1}{(1.00 \times 10^{-3})^2}$$

$$= 0.76 - \frac{0.059}{2} \log(1.00 \times 10^6)$$

$$= 0.76 - \frac{0.059 \times 6}{2} = 0.76 - 0.18 = 0.58 \text{ V}$$

Note that $\mathcal{E}_{cell}$ *increases* when we decrease $[Zn^{2+}]$. Remembering that $\mathcal{E}$ is a measure of the tendency the reaction has to proceed spontaneously, this is a reasonable result, since Zn^{2+} is a product of the reaction and its removal will make it easier for the reaction to proceed. Conversely, when $[H^+]$ is reduced, $\mathcal{E}_{cell}$ falls because H^+ is needed to make the reaction go in the forward direction as written.

Since the cell in part (a) still contains a standard hydrogen electrode, its overall potential of $+0.85$ V may be said to be that of a Zn/Zn^{2+} half-cell at $[Zn^{2+}] = 1.00 \times 10^{-3} M$.

In part (b), we remind ourselves that the half-cell potential of the hydrogen electrode is zero *only if* we are at standard conditions (that is, $[H^+] = 1.00 M$.)

Example 15.9   A complete cell is made by combining a standard copper electrode with another copper electrode in which $[Cu^{2+}] = 10^{-2} M$. When this cell functions as a voltaic cell, what will be the overall cell potential? Which electrode will be the anode and which the cathode? What overall reaction occurs?

SOLUTION   Common sense would suggest that as this cell functions, the concentrations of Cu^{2+} around the two electrodes would tend to approach each other; in other words, copper would be removed from solution at the standard electrode (the higher initial concentration) and go into solution at the more dilute electrode. The reactions would then be:

$$\begin{array}{lll}
\text{anode:} & Cu \longrightarrow Cu^{2+}_{dilute} + 2e^- & \mathcal{E}^0 = -0.34 \text{ V} \\
\text{cathode:} & \underline{Cu^{2+}_{std} + 2e^- \longrightarrow Cu} & \mathcal{E}^0 = +0.34 \text{ V} \\
\text{overall:} & Cu^{2+}_{std} + Cu \longrightarrow Cu^{2+}_{dilute} + Cu &
\end{array}$$

Using the Nernst equation, we have $\mathcal{E}^0_{cell} = 0.00$ V and $n = 2$:

$$\mathcal{E}_{cell} = \mathcal{E}^0_{cell} - \frac{0.059}{2} \log \left\{ \frac{[Cu^{2+}_{dilute}]}{[Cu^{2+}_{std}]} \right\}$$

$$= 0.00 - \frac{0.059}{2} \log \frac{10^{-2}}{1.0}$$

$$= 0.00 - \frac{0.059 \times (-2)}{2} = +0.059 \text{ V}$$

The overall cell potential is positive, so the reaction will go in the forward direction as written. Although we had expected this result, the result obtained does not depend on the initial assumption, since, if we had started by assuming that the reaction would go in the reverse direction, we would have obtained a negative cell potential to show that we had the equation the wrong way round.

The above cell is known as a *concentration cell*, since it depends for its overall potential on a difference in concentration between the electrolyte solutions. Only a small potential is obtained in this way.

Example 15.10   A cell is set up where the overall reaction is

$$Pb^{2+} + Sn \longrightarrow Pb + Sn^{2+}$$

Calculate $\mathcal{E}_{cell}$ for this process under the following conditions:

(a) standard conditions at both electrodes
(b) $[Sn^{2+}] = 1.0 \times 10^{-2} M$, $[Pb^{2+}] = 0.20 M$
(c) $[Sn^{2+}] = 0.40 M$, $[Pb^{2+}] = 5.0 \times 10^{-3} M$

In each case, decide the direction in which reaction will spontaneously proceed.

SOLUTION

(a) Under standard conditions:

anode:	$Sn \longrightarrow$	$Sn^{2+} + 2e^-$	$\varepsilon^0 = +0.14$ V
cathode:	$Pb^{2+} + 2e^- \longrightarrow$	Pb	$\varepsilon^0 = -0.13$ V
overall:	$Pb^{2+} + Sn \longrightarrow$	$Pb + Sn^{2+}$	$\varepsilon^0_{cell} = +0.01$ V

Reaction goes forward as written.

(b) Under other conditions, using $\varepsilon^0_{cell} = +0.01$ V and $n = 2$,

$$\varepsilon_{cell} = 0.01 - \frac{0.059}{2} \log\left\{\frac{[Sn^{2+}]}{[Pb^{2+}]}\right\}$$

and when $[Sn^{2+}] = 1.0 \times 10^{-2}$ and $[Pb^{2+}] = 0.20M$,

$$\varepsilon_{cell} = 0.01 - \frac{0.059}{2} \log \frac{1.0 \times 10^{-2}}{0.20}$$

$$= 0.01 - \frac{0.059}{2} \log(5.0 \times 10^{-2})$$

$$= 0.01 - \frac{0.059 \times (-1.30)}{2} = 0.01 + 0.038 = 0.05 \text{ V}$$

Reaction goes forward as written.

(c) When $[Sn^{2+}] = 0.40$ and $[Pb^{2+}] = 5.0 \times 10^{-3}M$,

$$\varepsilon_{cell} = 0.01 - \frac{0.059}{2} \log \frac{0.40}{5.0 \times 10^{-3}}$$

$$= 0.01 - \frac{0.059}{2} \log 80 = 0.01 - \frac{0.059 \times 1.9}{2}$$

$$= 0.01 - 0.056 = -0.05 \text{ V}$$

Reaction will now proceed spontaneously in the *reverse* direction.

For a reaction with a small ε^0 value, changes in electrolyte concentrations can reverse the direction in which reaction spontaneously proceeds.

Example 15.11   A cell is set up where the overall reaction is

$$H_2(g) + Sn^{4+} \longrightarrow 2H^+ + Sn^{2+}$$

The hydrogen electrode is under standard conditions and ε_{cell} is found to be $+0.20$ V. What is the ratio of Sn^{2+} to Sn^{4+} around the other electrode?

SOLUTION   For the reaction as written, $\mathcal{E}^0_{cell} = +0.15$ V and $n = 2$:

$$\mathcal{E}_{cell} = 0.15 - \frac{0.059}{2} \log\left\{\frac{[H^+]^2[Sn^{2+}]}{p(H_2)[Sn^{4+}]}\right\}$$

If the hydrogen electrode is at standard conditions and $\mathcal{E}_{cell} = 0.20$ V, we can write

$$0.20 = 0.15 - \frac{0.059}{2} \log\left\{\frac{[Sn^{2+}]}{[Sn^{4+}]}\right\}$$

$$\frac{0.059}{2} \log\left\{\frac{[Sn^{2+}]}{[Sn^{4+}]}\right\} = -0.20 + 0.15 = -0.05$$

$$\log\left\{\frac{[Sn^{2+}]}{[Sn^{4+}]}\right\} = \frac{-0.05 \times 2}{0.059} = -1.7 \qquad (\bar{2}.3)$$

$$\frac{[Sn^{2+}]}{[Sn^{4+}]} = 2.0 \times 10^{-2}$$

15.4   Equilibrium Constants from Cell Potentials

We see from the above examples that an overall cell potential may be changed by varying the concentrations of the reacting ions around one or both electrodes. In Example 15.10, we actually reversed the direction in which the reaction would spontaneously proceed, and that leads us to a most important point. Suppose we arranged the concentrations so that $\mathcal{E}_{cell}$ was zero, then the reaction would have no tendency to proceed spontaneously in *either* direction. What would the significance of this be? The answer is very simple:

When an overall cell potential is zero, the reaction corresponding to this cell is under conditions of chemical equilibrium.

This is the chemical explanation of the familiar problem of the "dead battery." In any voltaic cell, the overall $\mathcal{E}_{cell}$ will always fall as reaction proceeds and a current flows. As the potential falls to zero, the cell reaches a state of equilibrium. We can use the Nernst equation to express this:

$$\mathcal{E}_{cell} = \mathcal{E}^0_{cell} - \frac{0.059}{n} \log\left\{\frac{[C]^c[D]^d}{[A]^a[B]^b}\right\}$$

When the reaction is at equilibrium,

$$\mathcal{E}_{cell} = 0 \qquad \frac{[C]^c[D]^d}{[A]^a[B]^b} = K$$

where K is the equilibrium constant,

$$0 = \mathcal{E}^0_{cell} - \frac{0.059}{n} \log K \quad \text{and} \quad \log K = \frac{n\mathcal{E}^0_{cell}}{0.059} \qquad \text{(at 298 K)}$$

This very simple relationship enables us to calculate equilibrium constants for any reaction whose $\mathcal{E}^0$ value can be found. The K values often come out to be very large or very small.

Example 15.12   Use $\mathcal{E}^0$ values to evaluate the equilibrium constants for the following processes. In each case, write the expression for K and also work out K' for the reverse process.

(a) $Cu^{2+} + Zn \rightleftharpoons Cu + Zn^{2+}$

(b) $Fe^{2+} + H_2 \rightleftharpoons Fe + 2H^+$

(c) $MnO_4^- + 5Cl^- + 8H^+ \rightleftharpoons Mn^{2+} + \dfrac{5}{2}Cl_2 + 4H_2O$

SOLUTION   In each case, we work out $\mathcal{E}^0$ from the half-cell values (Table 15.1).

(a) $\mathcal{E}^0 = +0.76 - (-0.34) = +1.10$ V

$n = 2$, so $\log K = \dfrac{2 \times 1.10}{0.059} = 37.3$

$K = 2 \times 10^{37}$

Concentrations of solids (Cu and Zn) are omitted in the equilibrium constant expression, so

$$K = \frac{[Zn^{2+}]}{[Cu^{2+}]} = 2 \times 10^{37}$$

Essentially no $[Cu^{2+}]$ remains at equilibrium. For the reverse reaction

$$K' = \frac{[Cu^{2+}]}{[Zn^{2+}]} = \frac{1}{2 \times 10^{37}} = 5 \times 10^{-38}$$

(b) $\mathcal{E}^0 = -0.44$ V and $n = 2$

$$\log K = \frac{2 \times (-0.44)}{0.059} = -14.9 \qquad (\overline{15}.1)$$

$$K = 1.2 \times 10^{-15} = \frac{[H^+]^2}{[Fe^{2+}]p(H_2)}$$

for the reverse reaction,

$$K' = \frac{[Fe^{2+}]p(H_2)}{[H^+]^2} = \frac{1}{1.2 \times 10^{-15}} = 8 \times 10^{14}$$

(c) $\mathcal{E}^0 = +1.51 - 1.36 = +0.15$ V and $n = 5$

$$\log K = \frac{5 \times 0.15}{0.059} = 12.7$$

$$K = 5 \times 10^{12} = \frac{[Mn^{2+}][p(Cl_2)]^{5/2}}{[MnO_4^-][Cl^-]^5[H^+]^8}$$

for the reverse reaction:

$$K' = \frac{[MnO_4^-][Cl^-]^5[H^+]^8}{[Mn^{2+}][p(Cl_2)]^{5/2}} = \frac{1}{5 \times 10^{12}} = 2 \times 10^{-13}$$

The large values of equilibrium constants for redox reactions show us that they usually go to completion, a valuable feature when they are used for analytical work.

15.5   The Hydrogen Electrode and pH

As we have seen, the value of $\mathcal{E}$ for the hydrogen electrode will depend on the hydrogen ion concentration of the solution, that is, on its pH. Suppose we make up a concentration cell using one standard and one nonstandard hydrogen electrode. What will its $\mathcal{E}_{cell}$ value be? If the standard electrode is the cathode, we have

anode: $\qquad \frac{1}{2} H_2(g) \longrightarrow H^+ + e^- \qquad \mathcal{E}^0 = 0.0 \text{ V}$

cathode: $\qquad H^+_{std} + e^- \longrightarrow \frac{1}{2} H_2(g) \qquad \mathcal{E}^0 = 0.0 \text{ V}$

obviously, $\mathcal{E}^0_{cell} = 0.0$ V and $n = 1$:

$$\mathcal{E}_{cell} = 0.0 - \frac{0.059}{1} \log\left\{\frac{[H^+]p(H_2)}{[H^+_{std}]p(H_2)}\right\}$$

If we make $p(H_2)$ one atm at both electrodes, and remember that $[H^+_{std}] = 1.00$,

$$\mathcal{E}_{cell} = -0.059 \log[H^+] = 0.059 \text{ pH}$$

A very simple relationship exists between the pH of the solution around the nonstandard hydrogen electrode and the overall cell potential. Using the above equation, we may evaluate the overall cell potential, for example,

pH = 3.0: $\mathcal{E} = 0.059 \times 3.0 = 0.18$ V
pH = 8.0: $\mathcal{E} = 0.059 \times 8.0 = 0.47$ V

The cell potentials are positive, so our assignment of anode and cathode half-cells is correct. As with the previous concentration cell (Example 15.9), the spontaneous reaction tends to remove ions from the more concentrated solution (the standard electrode) and increase $[H^+]$ in the more dilute solution.

In practice, we would combine our "unknown" solution with some standard electrode in order to measure the pH of a solution. The hydrogen electrode, which is clumsy to assemble, can be replaced with a "glass electrode" (which is sensitive to pH changes) for convenience, but the principles of calculation can be illustrated with the hydrogen electrode.

Example 15.13   A cell is made in which the overall reaction is

$$Ag^+ + \frac{1}{2} H_2 \longrightarrow Ag + H^+$$

What will $\mathcal{E}_{cell}$ be if $[Ag^+] = 1.00M$ and $p(H_2) = 1.00$ atm when the pH of the solution round the hydrogen electrode is (a) 0.0, (b) 5.0, and (c) 9.5?

SOLUTION   The half-cells are

anode: $\qquad \frac{1}{2} H_2 \longrightarrow H^+ + e^- \qquad \mathcal{E}^0 = 0.0 \text{ V}$

cathode: $\qquad Ag^+ + e^- \longrightarrow Ag \qquad \mathcal{E}^0 = +0.80 \text{ V}$

$$\mathcal{E}^0_{cell} = +0.80 \text{ V} \quad \text{and} \quad n = 1$$

$$\mathcal{E}_{cell} = 0.80 - \frac{0.059}{n} \log\left\{\frac{[H^+]}{[Ag^+]}\right\}$$

If $[Ag^+] = 1.00$, changes in $\mathcal{E}_{cell}$ will be due solely to changes in $[H^+]$:

$$\mathcal{E}_{cell} = 0.80 - 0.059 \log[H^+] = 0.80 + 0.059 \text{ pH}$$

(a) pH = 0.0 and $\mathcal{E}_{cell}$ = 0.80 V. This, of course, is $\mathcal{E}^0_{cell}$, because a pH of zero means standard conditions ([H$^+$] = 1.00) for the hydrogen electrode.

(b) pH = 5.0 and $\mathcal{E}_{cell}$ = 0.80 + (0.059 × 5.0) = 0.80 + 0.30 = 1.10 V

(c) pH = 9.5 and $\mathcal{E}_{cell}$ = 0.80 + (0.059 × 9.5) = 0.80 + 0.56 = 1.36 V

The result obtained in the above problem may be used in a useful general equation. If we combine a hydrogen electrode (where the hydrogen pressure is 1.00 atm) with an electrode of some other type, the second electrode being under standard conditions, the overall cell potential is given by the equation

$$\mathcal{E}_{cell} = \mathcal{E}^0_{cell} + 0.059 \text{ pH}$$

provided that $\mathcal{E}_{cell}$ is evaluated for the reaction in which the hydrogen half-cell forms the anode (oxidation) half-reaction.

Example 15.14 A hydrogen electrode having $p(H_2)$ = 1.00 atm is combined with a standard calomel electrode, whose half-reaction is

$$Hg_2Cl_2(s) + 2e^- \longrightarrow 2Hg(l) + 2Cl^- \qquad \mathcal{E}^0 = 0.242 \text{ V}$$

If the cell potential is 0.800 V, what is the pH of the solution around the hydrogen electrode? What will the cell potential be if the hydrogen electrode is immersed in a neutral solution?

SOLUTION   For the overall reaction

$$Hg_2Cl_2(s) + H_2(g) \longrightarrow 2Hg(l) + 2Cl^- + 2H^+$$

we know that $\mathcal{E}^0_{cell}$ = 0.242 V. Using $\mathcal{E}_{cell} = \mathcal{E}^0_{cell} + 0.059$ pH gives

$$0.059 \text{ pH} = \mathcal{E}_{cell} - \mathcal{E}^0_{cell}$$

If $\mathcal{E}_{cell}$ = 0.800 V, then

$$0.059 \text{ pH} = 0.800 - 0.242 = 0.558$$

$$\text{pH} = \frac{0.558}{0.059} = 9.5$$

If the hydrogen electrode is in neutral solution (pH = 7.00)

$$\mathcal{E}_{cell} = 0.242 + (0.059 \times 7.00) = 0.242 + 0.41 = 0.65 \text{ V}$$

The commercial pH meter contains circuits to compensate for $\mathcal{E}^0_{cell}$ and displays the resulting residual voltage on a dial calibrated directly in pH units.

KEY WORDS

anode	oxidation
cathode	reduction
Faraday constant	electrolytic cell
voltaic (galvanic) cell	standard hydrogen electrode
half-cell	standard electrode potential
Nernst equation	concentration cell

STUDY QUESTIONS

1. What type of chemical reaction occurs at (a) an anode and (b) a cathode?

2. Why do oxidation and reduction occur simultaneously?

3. What determines the number of moles of electrons associated with the oxidation or reduction of a mole of a substance?

4. How many electrons are associated with the Faraday constant? How is this connected with current flow and time?

5. How does an electrochemical cell differ from a voltaic cell?

6. Under standard conditions (how are these defined?), what factors affect the overall emf produced by a voltaic cell?

7. Can the overall emf of a voltaic cell be broken down into absolute potential differences at the two electrode–solution interfaces?

8. Why is the standard hydrogen electrode of particular significance in electrochemistry? What chemical reaction is occurring at this electrode?

9. A cell reaction is constructed from a reduction equation and an oxidation equation. What condition must be fulfilled if we want to calculate the overall cell potential from the two standard half-cell potentials?

10. What is the significance of an overall cell potential having a value that is (a) positive or (b) negative? Suppose we write the equation for the reverse overall reaction, how do we change the cell potential?

11. How is the half-cell potential for the reaction

$$A \longrightarrow A^{n+} + ne^-$$

related to that for the reaction

$$mA \longrightarrow mA^{n+} + mne^-$$

where m and n are integers?

12. Why does a half-cell potential change when we go from standard to nonstandard conditions?

13. What is a "concentration cell"? If current is allowed to flow spontaneously in such a cell by completion of the external circuit, what is the effect on solution concentrations around each electrode?

14. What has happened, chemically speaking, when the overall emf of a voltaic cell has dropped to zero?

15. A Voltaic cell is made up with both electrodes under standard conditions, and current is allowed to flow spontaneously until the emf has dropped to zero. What factors determine:
    (a) the quantity of electricity the cell provides during this process
    (b) the amount of electrical energy the cell provides during this process
    (c) the final solute concentrations around each electrode when current flow ceases

16. How can electrode potentials be connected with the pH of a solution? Is one hydrogen electrode alone sufficient to measure the pH of a solution? Why not? Could you manage it with *two* hydrogen electrodes? How?

PROBLEMS

Refer to Table 15.1 for half-cell potentials needed. Take Faraday's constant = 96,500 coulomb mole^{-1}

1. What mass of metal is deposited in each of the following cathode processes when 0.150 mole of electrons is passed?
   (a) $Li^+ + e^- \longrightarrow Li$ (b) $Cr^{3+} + 3e^- \longrightarrow Cr$
   (c) $Mn^{2+} + 2e^- \longrightarrow Mn$

2. How long must a current of 5.00 A be passed through each of the electrodes in problem 1 to deposit 1.00 g of the respective metals?

3. Water is electrolyzed using a current of 12.0 A for a period of 45.0 minutes. What volume of gas (STP) is produced at each electrode?

4. Elementary fluorine is produced by an electrolytic process using molten KHF_2. The reaction is

$$HF_2^- \longrightarrow HF + \frac{1}{2} F_2 + e^-$$

   At a current of 20.0 A, how long would it take to produce 15.0 liters of F_2 (at STP)?

5. Current is passed through a cathode where the reaction is

$$5e^- + MnO_4^- + 8H^+ \longrightarrow Mn^{2+} + 4H_2O$$

   All the permanganate present in 15.0 ml of solution has been reduced after a current of 0.600 A has passed for 603 seconds. What was the original concentration of permanganate?

6. A meter designed for use in mass spectrometry is advertised as being able to measure a current flow of 63 electrons per second.
   (a) What would this be in ampere?
   (b) If this current had been passing through aqueous sulfuric acid since the creation of the earth (4.5×10^9 years ago), what volume of hydrogen would have been liberated by now? (Charge on electron = 1.60×10^{-19} C.)

7. Calculate $\mathcal{E}^0$ values for the following reactions
   (a) $Fe + Pb^{2+} \longrightarrow Fe^{2+} + Pb$ (acid)
   (b) $HClO + Sn^{2+} + H^+ \longrightarrow Cl^- + Sn^{4+} + H_2O$ (acid)
   (c) $4Fe(OH)_2 + O_2 + 2H_2O \longrightarrow 4Fe(OH)_3$ (alkaline)

8. Use $\mathcal{E}^0$ values to answer the following:
   (a) Will I_2 oxidize H_2SO_3 to SO_4^{2-} in acid solution?
   (b) Will Sn^{4+} oxidize H_2O_2 (giving also Sn^{2+}) in acid solution?
   (c) Will Cl_2 oxidize water to O_2 in acid solution?
   (d) Will Fe^{2+} be oxidized by I_2 to Fe^{3+} (giving I^-) in acid solution?

9. A student prepares a solution of Cr^{2+} by reducing Cr^{3+} with metallic zinc under acid conditions. Although he takes care to keep air or other known oxidizing agents out of the apparatus, he finds the Cr^{2+} always decomposes in a few hours. Can you suggest a reason? (*Hint*: Maybe $\mathcal{E}^0$ values have something to do with it!)

10. Calculate equilibrium constants for the reactions in problem 8. In each case, write the concentration expression for K.

11. Calculate the cell potential for a voltaic cell composed of a piece of chromium metal immersed in a $0.500M$ solution of Cr^{3+} and a piece of zinc in a solution of $0.0500M$ Zn^{2+}. Which electrode is the cathode?

12. If the cell in problem 11 is modified so that the $[Cr^{3+}]$ is now $1.00 \times 10^{-4}M$ and the $[Zn^{2+}]$ is $1.00M$, what is the cell potential? Which electrode is now the cathode in the voltaic cell?

13. What is the potential for the hydrogen electrode

$$\frac{1}{2} H_2 \longrightarrow H^+ + e^-$$

in solutions of pH (a) 7.0 and (b) 12.0?

14. A cell is made up by combining an Fe/Fe^{2+} electrode with a Cu/Cu^{2+} electrode. The overall potential is $+0.80$ V and the concentration of Cu^{2+} is $0.050M$. What is the concentration of Fe^{2+}?

15. A cell is made up using a hydrogen electrode in an unknown solution, plus a silver electrode in $0.100M$ Ag^+ solution. The overall potential is $+1.24$ V. What is the pH of the unknown solution, assuming that the hydrogen gas pressure is 1 atm?

16. For the half-cell reaction

$$Cr_2O_7^{2-} + 14H^+ + 6e^- \longrightarrow 2Cr^{3+} + 7H_2O$$

$\mathcal{E}^0$ is $+1.33$ V. What would be the potential if the concentration of H^+ were reduced to $0.100M$ while the remaining species were maintained at a concentration of $1.00M$?

SOLUTIONS TO PROBLEMS

1. In each case, when 0.150 mole e^- passes:

$$\text{mass of metal} = \frac{0.150\,M}{n}$$

where n = number of electrons in change and M is the molar mass of the metal.

(a) $M = 6.94$ g, $n = 1$, mass $= \dfrac{0.150 \times 6.94}{1} = 1.04$ g

(b) $M = 52.0$ g, $n = 3$, mass $= \dfrac{0.150 \times 52.0}{3} = 2.60$ g

(c) $M = 54.9$ g, $n = 2$, mass $= \dfrac{0.150 \times 54.9}{2} = 4.12$ g

2. 1.00 g is $1.00/M$ mole of metal. This is deposited in t seconds, where

$$\frac{1.00}{M} = \frac{it}{n\mathcal{F}}$$

where i is the current (5.00 A here) and $\mathcal{F} = 96{,}500$ coulomb mole^{-1}

$$t = \frac{96{,}500\,n}{5.00\,M} = \frac{19{,}300\,n}{M}\ \text{s}$$

(a) $M = 6.94$ g and $n = 1$, so $t = \dfrac{19{,}300}{6.94} = 2.78 \times 10^3$ s

(b) $M = 52.0$ g and $n = 3$, so $t = \dfrac{19{,}300 \times 3}{52.0} = 1.11 \times 10^3$ s

(c) $M = 54.9$ g and $n = 2$, so $t = \dfrac{19{,}300 \times 2}{54.9} = 703$ s

3. 12.0 A for 45.0 min is $12.0 \times 45.0 \times 60 = 3.24 \times 10^4$ C $= \dfrac{3.24 \times 10^4}{96{,}500} = 0.336\ \mathcal{F}$.

Cathode reaction: $2H_2O + 2e^- \longrightarrow H_2 + 2OH^-$
This is a two-electron reduction, so 1 $\mathcal{F}$ gives 0.5 mole H_2.
0.336 $\mathcal{F}$ gives $0.336 \times 0.5 = 0.168$ mole H_2, which is $0.168 \times 22.4 = 3.76$ liter H_2.
Anode reaction: $2H_2O \longrightarrow O_2 + 4H^+ + 4e^-$
This is a four-electron oxidation, so 1 $\mathcal{F}$ gives 0.25 mole O_2.
0.336 $\mathcal{F}$ gives $0.336 \times 0.25 = 8.40 \times 10^{-2}$ mole O_2, which is $8.40 \times 10^{-2} \times 22.4 = 1.88$ liter O_2.

4. 0.5 mole F_2 is obtained from 1 $\mathcal{F}$.
We need $15.0/22.4 = 0.670$ mole F_2,

which needs $\dfrac{0.670}{0.5} = 1.34\ \mathcal{F} = 1.34 \times 96{,}500 = 1.29 \times 10^5$ C.

At 20.0 A (20.0 C s^{-1}), this takes $\dfrac{1.29 \times 10^5}{20.0} = 6.46 \times 10^3$ s.

$$t = \frac{6.46 \times 10^3}{3600} = 1.80\ \text{h}$$

5. 0.600 A for 603 s is $0.600 \times 603 = 362$ C $= \dfrac{362}{96{,}500} = 3.75 \times 10^{-3}$ $\mathcal{F}$.

To reduce 1 mole MnO_4^- takes 5 mole e^- (5 $\mathcal{F}$ of electricity must pass).

3.75×10^{-3} $\mathcal{F}$ reduces $\dfrac{3.75 \times 10^{-3}}{5} = 7.50 \times 10^{-4}$ mole MnO_4^-

$$\text{concentration} = \dfrac{7.50 \times 10^{-4} \text{ mole}}{0.0150 \text{ liter}} = 5.00 \times 10^{-2} M$$

6. Flow of electrical charge per second:
$63 \times 1.60 \times 10^{-19} = 1.01 \times 10^{-17} \text{ C s}^{-1}$ $(= A)$
In 4.5×10^9 years, pass $4.5 \times 10^9 \times 365 \times 24 \times 3600 \times 1.01 \times 10^{-17} = 1.43$ C

This is $\dfrac{1.43}{96{,}500} = 1.48 \times 10^{-5}$ $\mathcal{F}$.

1 $\mathcal{F}$ liberates 0.5 mole (11.2 liter) of H_2.
1.48×10^{-5} $\mathcal{F}$ liberates $1.48 \times 10^{-5} \times 11.2 = 1.7 \times 10^{-4}$ liter $= 0.17$ ml H_2

7. (a) anode: $Fe \longrightarrow Fe^{2+} + 2e^-$ $\mathcal{E}^0 = +0.44$ V
    cathode: $Pb^{2+} + 2e^- \longrightarrow Pb$ $\mathcal{E}^0 = -0.13$ V
    overall: $\mathcal{E}^0 = +0.44 - 0.13 = +0.31$ V
    (b) anode: $Sn^{2+} \longrightarrow Sn^{4+} + 2e^-$ $\mathcal{E}^0 = -0.15$ V
    cathode: $HClO + H^+ + 2e^- \longrightarrow Cl^- + H_2O$ $\mathcal{E}^0 = +1.49$ V
    overall: $\mathcal{E}^0 = +1.49 - 0.15 = +1.34$ V
    (c) anode: $4Fe(OH)_2 + 4OH^- \longrightarrow 4Fe(OH)_3 + 4e^-$ $\mathcal{E}^0 = +0.56$ V
    cathode: $O_2 + 2H_2O + 4e^- \longrightarrow 4OH^-$ $\mathcal{E}^0 = +0.40$ V
    overall: $\mathcal{E}^0 = +0.56 + 0.40 = +0.96$ V

8. We assume that the desired reaction will go, and work out $\mathcal{E}^0_{cell}$ on this basis:
    (a) anode: $H_2SO_3 + H_2O \longrightarrow SO_4^{2-} + 4H^+ + 2e^-$ $\mathcal{E}^0 = -0.17$ V
    cathode: $I_2 + 2e^- \longrightarrow 2I^-$ $\mathcal{E}^0 = +0.54$ V
    overall: $\mathcal{E}^0 = -0.17 + 0.54 = +0.37$ V
    Yes, this will go in this direction.
    (b) anode: $H_2O_2 \longrightarrow O_2 + 2H^+ + 2e^-$ $\mathcal{E}^0 = -0.68$ V
    cathode: $Sn^{4+} + 2e^- \longrightarrow Sn^{2+}$ $\mathcal{E}^0 = +0.15$ V
    overall: $\mathcal{E}^0 = -0.68 + 0.15 = -0.53$ V
    This reaction will *not* spontaneously go in this direction.

    (c) anode: $H_2O \longrightarrow \frac{1}{2} O_2 + 2H^+ + 2e^-$ $\mathcal{E}^0 = -1.23$ V

    cathode: $Cl_2 + 2e^- \longrightarrow 2Cl^-$ $\mathcal{E}^0 = +1.36$ V
    overall: $\mathcal{E}^0 = -1.23 + 1.36 = +0.13$ V
    Yes, this reaction will go.
    (d) anode: $Fe^{2+} \longrightarrow Fe^{3+} + e^-$ $\mathcal{E}^0 = -0.77$ V

    cathode: $\frac{1}{2} I_2(s) + e^- \longrightarrow I^-$ $\mathcal{E}^0 = +0.54$ V

    overall: $\mathcal{E}^0 = -0.77 + 0.54 = -0.23$ V
    No, reaction will not go as written.

9. For the half-reaction

$$Cr^{2+} \longrightarrow Cr^{3+} + e^-$$

the value of $\mathcal{E}^0$ is $+0.41$. Combining this with a standard hydrogen electrode gives $\mathcal{E}^0_{cell} = +0.41$ V for the process

$$Cr^{2+} + H^+ \longrightarrow Cr^{3+} + \frac{1}{2} H_2$$

So the acid solution of Cr^{2+} is liable to decompose with evolution of hydrogen (of course, we can't forecast the *rate* at which this will occur).

10. In each case, we use $\log K = \dfrac{n\mathcal{E}^0_{cell}}{0.059}$

  (a) $n = 2$ and $\mathcal{E}^0 = +0.37$ V, so $\log K = \dfrac{2 \times 0.37}{0.059} = 12.5$

  $$K = \frac{[SO_4^{2-}][I^-]^2[H^+]^4}{[H_2SO_3][I_2]} = 3.5 \times 10^{12}$$

  (b) $n = 2$ and $\mathcal{E}^0 = -0.53$ V, so $\log K = \dfrac{2 \times (-0.53)}{0.059} = -18.0$

  $$K = \frac{[Sn^{2+}][H^+]^2 p(O_2)}{[Sn^{4+}][H_2O_2]} = 1 \times 10^{-18}$$

  (c) $n = 2$ and $\mathcal{E}^0 = +0.13$ V, so $\log K = \dfrac{2 \times 0.13}{0.059} = 4.4$

  $$K = \frac{[Cl^-]^2[H^+]^2[p(O_2)]^{1/2}}{p(Cl_2)} = 2.6 \times 10^4$$

  (d) $n = 1$ and $\mathcal{E}^0 = -0.23$ V, so $\log K = \dfrac{-0.23}{0.059} = -3.9$ $\quad(\overline{4}.1)$

  $$K = \frac{[I^-][Fe^{3+}]}{[Fe^{2+}]} = 1.3 \times 10^{-4}$$

11. If zinc is the cathode, reactions are:

  anode: $\qquad\qquad 2Cr \longrightarrow 2Cr^{3+} + 6e^- \qquad \mathcal{E}^0 = +0.74$ V

  cathode: $\qquad 3Zn^{2+} + 6e^- \longrightarrow 3Zn \qquad\quad \mathcal{E}^0 = -0.76$ V

  overall: $\qquad 2Cr + 3Zn^{2+} \longrightarrow 2Cr^{3+} + 3Zn$

  $\mathcal{E}^0_{cell} = +0.74 - 0.76 = -0.02$ V $\quad$ and $\quad n = 6$

  $$\mathcal{E}_{cell} = \mathcal{E}^0_{cell} - \frac{0.059}{6} \log\left\{\frac{[Cr^{3+}]^2}{[Zn^{2+}]^3}\right\}$$

  If $[Cr^{3+}] = 0.500 M$ and $[Zn^{2+}] = 0.0500 M$

  $$\mathcal{E}_{cell} = -0.02 - \frac{0.059}{6} \log \frac{(0.500)^2}{(0.0500)^3}$$

  $$= -0.02 - \frac{0.059}{6} \log(2.0 \times 10^3) = -0.02 - 0.033$$

  $$= -0.05 \text{ V}$$

Since this is negative, our initial assumption was wrong. For spontaneous cell reaction, Cr is the cathode and the reaction goes in the reverse direction from that written above, with $\mathcal{E}_{cell} = +0.05$ V.

12. Again, suppose Zn is the cathode:

$$\mathcal{E}_{cell} = -0.02 - \frac{0.059}{6} \log\left\{\frac{[Cr^{3+}]^2}{[Zn^{2+}]^3}\right\}$$

If $[Cr^{3+}] = 1.00 \times 10^{-4}$ and $[Zn^{2+}] = 1.00 M$,

$$\mathcal{E}_{cell} = -0.02 - \frac{0.059}{6} \log \frac{(1.00 \times 10^{-4})^2}{(1.00)^3}$$

$$= -0.02 - \frac{0.059}{6} \log(1.00 \times 10^{-8})$$

$$= -0.02 + \frac{0.059 \times 8}{6} = -0.02 + 0.08 = +0.06 \text{ V}$$

Now the value of $\mathcal{E}_{cell}$ is positive, so the reaction spontaneously goes as first written, with Zn as the cathode.

13. Using $\mathcal{E} = 0.059$ pH:
    (a) pH = 7.0: $\mathcal{E} = 0.059 \times 7.0 = 0.41$ V
    (b) pH = 12.0: $\mathcal{E} = 0.059 \times 12.0 = 0.71$ V

14. The cell reaction is $Fe + Cu^{2+} \longrightarrow Fe^{2+} + Cu$
    $\mathcal{E}^0_{cell} = +0.44 + 0.34 = +0.78$ V $\qquad n = 2$

$$\mathcal{E}_{cell} = \mathcal{E}^0_{cell} - \frac{0.059}{n} \log\left\{\frac{[Fe^{2+}]}{[Cu^{2+}]}\right\}$$

Putting $\mathcal{E}_{cell} = 0.80$ and $\mathcal{E}^0_{cell} = 0.78$

$$0.80 = 0.78 - \frac{0.059}{2} \log\left\{\frac{[Fe^{2+}]}{[Cu^{2+}]}\right\}$$

$$-\frac{0.059}{2} \log\left\{\frac{[Fe^{2+}]}{[Cu^{2+}]}\right\} = 0.80 - 0.78 = 0.02$$

$$\log\left\{\frac{[Fe^{2+}]}{[Cu^{2+}]}\right\} = \frac{-2 \times 0.02}{0.059} = -0.7 \qquad (\bar{1}.3)$$

$$\frac{[Fe^{2+}]}{[Cu^{2+}]} = 0.2 \qquad [Fe^{2+}] = 0.2[Cu^{2+}]$$

If $[Cu^{2+}] = 0.050 M$, $[Fe^{2+}] = 0.050 \times 0.2 = 0.01 M$

15. The overall reaction is $\frac{1}{2} H_2 + Ag^+ \longrightarrow H^+ + Ag$

$$\mathcal{E}^0_{cell} = 0.00 + 0.80 = 0.80 \text{ V} \qquad n = 1$$

$$\mathcal{E}_{cell} = \mathcal{E}^0_{cell} - \frac{0.059}{n} \log\left\{\frac{[H^+]}{[Ag^+]}\right\}$$

If $\varepsilon_{cell} = 1.24$ V,

$$0.059 \log\left\{\frac{[H^+]}{[Ag^+]}\right\} = \varepsilon^0_{cell} - \varepsilon_{cell} = 0.80 - 1.24 = -0.44$$

$$\log\left\{\frac{[H^+]}{[Ag^+]}\right\} = \frac{-0.44}{0.059} = -7.5 \qquad (\bar{8}.5)$$

$$\frac{[H^+]}{[Ag^+]} = 3 \times 10^{-8}$$

If $[Ag^+] = 0.100M$, $[H^+] = (3 \times 10^{-8}) \times 0.100 = 3 \times 10^{-9}M$
$$pH = -\log(3 \times 10^{-9}) = 8.5$$

16. $\varepsilon = \varepsilon^0 - \dfrac{0.059}{n} \log\left\{\dfrac{[Cr^{3+}]^2}{[Cr_2O_7^{2-}][H^+]^{14}}\right\}$

If $n = 6$ and $[Cr^{3+}] = [Cr_2O_7^{2-}] = 1.00M$

$$\varepsilon = 1.33 - \frac{0.059}{6} \log\left\{\frac{1}{[H^+]^{14}}\right\}$$

If $[H^+] = 0.100$,

$$\varepsilon = 1.33 - \frac{0.059}{6} \log\frac{1}{(0.100)^{14}}$$

$$= 1.33 - \frac{0.059}{5} \log(10^{14}) = 1.33 - \frac{0.059 \times 14}{6}$$

$$= 1.33 - 0.14 = 1.19 \text{ V}$$

Chapter 16

Free Energy and Equilibrium

It is of the highest importance in the art of detection to be able to recognize, out of a number of facts, which are incidental and which vital. THE REIGATE PUZZLE

In previous chapters, we have considered the enthalpy change occurring in a reaction (Chapter 4), the equilibrium constant (Chapter 11), and the standard cell potential (Chapter 15). In this final chapter, we will try to connect these concepts and consider in general terms the factors affecting chemical equilibrium.

16.1 Free Energy

Why does a chemical reaction go? What provides the driving force for a process to occur spontaneously? A reasonable suggestion would be that the enthalpy change has something to do with it. You might think that an exothermic process is more likely to occur than an endothermic process, since common observation shows that many reactions (combustion, etc.) give out heat as they proceed.

However, a simple experiment shows that this idea is not always correct. Put some solid ammonium nitrate or ammonium chloride in a beaker, add a little water, and stir. The ammonium salt dissolves and the temperature *falls*, often to the point where frost forms on the beaker. Obviously the process is endothermic, yet it occurs spontaneously. Another example is seen when an ice cube melts. ΔH is positive and the system ends up with a higher enthalpy content, yet the process will occur spontaneously if heat is supplied at 0°C. Why?

The answer is that the enthalpy change is *not* the factor determining the direction in which a reaction will spontaneously proceed. The important factor is the change in a quantity called, in general terms, "free energy." This may be defined in several ways, but the most useful one in chemistry is that of *Gibbs free energy*, given the symbol G. Throughout this chapter, it should be understood that this is the quantity we have in mind when we refer to free energy.

The change in free energy, ΔG, is related to the enthalpy change by the equation

$$\Delta G = \Delta H - T\Delta S$$

where T is the absolute temperature and ΔS is the change in the *entropy* of the system. Entropy is a measure of the disorder or randomness of the system; the more disordered it is, the higher its entropy content. A highly ordered arrangement, such as a crystalline solid, has a low entropy content, whereas a disordered liquid will have more entropy and a gas, where disorder is at a maximum, will have the greatest entropy.

The importance of the free energy lies in the following principle:

Change will spontaneously proceed in the direction in which ΔG is negative.

Looking at the equation relating ΔG to ΔH and ΔS, we see at once that a large negative value of ΔH is likely to result in a negative value of ΔG, so our initial assumption was partly right. The enthalpy change in a process is certainly one of the factors determining whether or not it will go spontaneously, and in many cases it is the dominant factor.

We also see from the form of the equation that a positive value of ΔS will contribute to a negative value of ΔG, in other words, a process will tend to go spontaneously from an ordered state to a disordered state so that the entropy content of the system increases.

Here is the explanation of what happens when ammonium nitrate dissolves in water. The enthalpy change, ΔH, is positive (as shown by the fall in temperature), but the disorder increases as solid crystals go into solution, so ΔS is positive also. On balance, $T\Delta S$ is greater than ΔH, so the value of ΔG is negative and the process goes spontaneously.

Notice also the appearance of T in the equation for ΔG. As T increases, the $T\Delta S$ term will become larger, so the effect of disorder will be more important at higher temperatures.

In Chapter 4, we learned that ΔH values for a change were independent of the path taken. Exactly the same applies to ΔG and ΔS (H, G, and S are said to be "state functions"), which is a help in determining the changes in a reaction.

Just as we used the difference of total ΔH_f^0 values between products and reactants to find ΔH^0 for a reaction, so we can use total free energies of formation to find ΔG^0 for a reaction. The standard molar free energy of formation of a compound, ΔG_f^0, is defined as the change in free energy when a mole of the compound is formed from its constituent elements in their standard states. For an element in its standard state, therefore, ΔG_f^0 will be zero, as is ΔH_f^0.

Entropy is recorded a little differently. The entropy of a perfect crystal at absolute zero is taken to be zero, so we can work out the absolute entropy of substances at standard conditions (25°C and 1 atmosphere) by measuring the amount of heat required to take a mole of substance from 0 K to 298 K. This gives us values of absolute entropies, S^0, measured under standard conditions in units of $J\ K^{-1}\ mole^{-1}$ (or $cal\ K^{-1}\ mole^{-1}$). For example, the standard absolute entropy of $CO_2(g)$ is $214\ J\ K^{-1}\ mole^{-1}$, while that of $Fe(s)$ is $27.2\ J\ K^{-1}\ mole^{-1}$. Note that S^0 values for elements are *not* zero.

We can use these values to calculate ΔG^0 for reactions occurring between substances in their standard states.

Example 16.1 For the reaction

$$C(s) + O_2(g) \longrightarrow CO_2(g)$$

$\Delta H^0 = -394\ kJ\ mole^{-1}$. If S^0 values are $C(s)$, 5.69; $O_2(g)$, 205.0; and $CO_2(g)$, 213.6 $J\ K^{-1}\ mole^{-1}$, calculate ΔG^0 at 298 K.

SOLUTION   We can calculate ΔS^0 from the difference between the total S^0 of products and the total S^0 of the reactants

$$\Delta S^0 = 213.6 - (205.0 + 5.69) = 2.9 \text{ J K}^{-1}$$

at 298 K,

$$TAS^0 = 298 \times 2.9 = 864 \text{ J} = 0.864 \text{ kJ}$$
$$\Delta G^0 = \Delta H^0 - TAS^0 = -394 - 0.864$$
$$= -395 \text{ kJ mole}^{-1}$$

We see that ΔG^0 and ΔH^0 are very similar, for the reaction at this temperature. In other words, the enthalpy change is dominant. We might expect this result, because in this process we start and end with one mole of gas, so there is not much change in disorder.

Example 16.2   If ΔH_f^0 for $H_2O(l)$ is -286 kJ mole^{-1}, calculate ΔG_f^0 at 298 K, given that values of S^0 are $H_2(g)$, 130.6; $O_2(g)$, 205.0; and $H_2O(l)$, 69.96 J K^{-1}mole^{-1}

SOLUTION   For the process

$$H_2(g) + \frac{1}{2} O_2(g) \longrightarrow H_2O(l)$$

$$\Delta S^0 = 69.96 - \left(130.6 + \frac{205.0}{2}\right) = -163.1 \text{ J K}^{-1}$$

At 298 K,

$$TAS^0 = -(298 \times 163.1) = -4.86 \times 10^4 \text{ J} = -48.6 \text{ kJ mole}^{-1}$$

The above reaction is the formation process, so the values of ΔG^0 and ΔH^0 are equal to ΔG_f^0 and ΔH_f^0 respectively for $H_2O(l)$.

$$\Delta G^0 = \Delta H^0 - TAS^0$$
$$= -286 + 48.6 = -237 \text{ kJ mole}^{-1}$$

We see that ΔS^0 is much larger than it was for the formation of CO_2, and is negative. We have gone to a more ordered (lower entropy) state by forming a mole of liquid from 1.5 mole of gas.

Example 16.3   Calculate ΔH^0, ΔG^0, and ΔS^0 for the process

$$H_2O(l) \longrightarrow H_2O(g) \quad \text{(at 298 K)}$$

from the following data: ΔH_f^0 values are -286 kJ mole^{-1} for $H_2O(l)$ and -242 kJ mole^{-1} for $H_2O(g)$: S^0 values are 69.96 J K^{-1} mole^{-1} for $H_2O(l)$ and 188.7 J K^{-1} mole^{-1} for $H_2O(g)$.

SOLUTION   We first calculate ΔH^0 as usual:

$$\Delta H^0 = -242 - (-286) = +44 \text{ kJ mole}^{-1}$$

Similarly, we find

$$\Delta S^0 = 188.7 - 69.96 = 118.7 \text{ J K}^{-1} \text{ mole}^{-1}$$

So, at 298 K,

$$\Delta G^0 = 44 - (298 \times 0.119) = 44 - 35.5 = 8.5 \text{ kJ mole}^{-1}$$

We find that ΔG^0, although small, is positive, in accord with our observation that, at 298 K, the standard state of water is the liquid rather than the gas.

Example 16.4   For the reaction

$$2Al(s) + \frac{3}{2} O_2(g) \longrightarrow Al_2O_3(s)$$

$\Delta G^0 = -1576$ and $\Delta H^0 = -1669$ kJ mole^{-1} at 298 K. What is ΔS^0?

SOLUTION   Our equation may be written
$$T\Delta S^0 = \Delta H^0 - \Delta G^0$$
$$= -1669 + 1576 = -93 \text{ kJ mole}^{-1}$$

at 298 K

$$\Delta S^0 = \frac{-93}{298} = -0.31 \text{ kJ K}^{-1} \text{ mole}^{-1}$$

As we would expect, ΔS^0 is negative in this reaction where gas forms solid.

16.2   Equilibrium Constants

We stated above that reaction will go spontaneously *only* in the direction in which ΔG^0 is negative. However, we know that many reactions are reversible, and our concept of equilibrium is a situation in which reaction is going on in both forward and reverse direction at the same rate. Isn't there a contradiction here? Surely ΔG^0 has opposite signs for the forward and reverse process, so one should be more favorable than the other.

The answer is that there is no contradiction here. When we say a reaction will spontaneously "go," we do not necessarily mean that it will go to completion. Reaction continues until the total free energy of the system has reached a minimum, which we call an *equilibrium state*, and this often corresponds to a mixture of both reactants and products.

To take a specific example, for the reaction

$$H_2(g) + I_2(g) \rightleftharpoons 2HI(g)$$

if we start with a mixture of pure H_2 and pure I_2, reaction will spontaneously occur to produce some HI. The process is favorable because there is initially no HI present. On the other hand, if we start with pure HI, reaction will initially proceed in the opposite direction to give H_2 and I_2. In each case, a small initial degree of reaction is accompanied by a decrease in the total free energy of the system. The diagram of Figure 16.1 will make this clearer.

The total free energy change at 763 K when one mole of HI is produced from pure H_2 and I_2 is -12.1 kJ. However, the total free energy of the system, G, goes through a minimum value at an intermediate point at which the slope of the line is zero. This line represents the change in G as reaction proceeds, and its minimum value occurs at the position of chemical equilibrium for the system. The diagram illustrates an important principle:

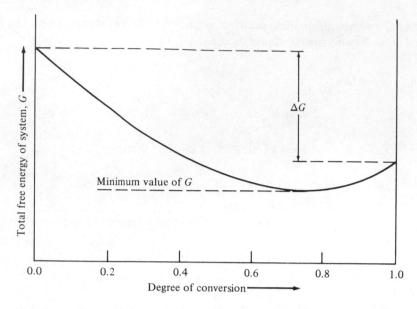

Figure 16.1   Free energy of the system $H_2 + I_2 \rightleftharpoons 2HI$ as a function of the degree of conversion.

For a reaction occurring under conditions of chemical equilibrium, ΔG for both the forward reaction and the reverse reaction is zero.

Note carefully that this ΔG refers to the change in free energy when reactants are converted into products *under the conditions prevailing at equilibrium.* Its value is *not* the same as ΔG^0, which refers to changes between reactants and products in their standard states (i.e., 1 atmosphere pressure for each gas.)

We see now the explanation of the apparent contradiction we noted earlier. In an equilibrium situation, ΔG values will not favor reaction in one direction or the other, so we have a stable situation.

Returning briefly to Example 16.3, we can recalculate ΔG^0 for the process

$$H_2O(l) \longrightarrow H_2O(g)$$

at a temperature of 100°C (373 K). Assuming that the values of ΔH^0 and ΔS^0 are the same as they were at 25°C (a reasonably good assumption, but not perfectly correct) we obtain

$$\Delta G^0 = \Delta H^0 - T\Delta S^0$$
$$= 44 - (373 \times 0.119) = 44 - 44.4$$

Within the accuracy of our calculation, ΔG^0 is zero, consistent with our observation that liquid and gaseous water exist in equilibrium at 373 K and 1 atmosphere pressure. At a higher temperature, $T\Delta S$ would be greater and ΔG^0 would be negative, so reaction would spontaneously go towards the gaseous state.

If we choose a temperature at which $\Delta G^0 = 0$, then a situation of chemical equilibrium will prevail when all substances involved are in their standard states, which we will take as 1 atmosphere pressure for gases and 1 molar concentration for solutions (*Note*: Strictly speaking, this should be "at unit activities," but we will keep to the simpler picture of considering concentrations rather than activities.)

There is a close connection between the overall ΔG^0 value for a reaction and the value of its equilibrium constant. Based on the above principles, it can be shown that

$$\Delta G^0 = -RT \ln K$$

where R is the gas constant, T the absolute temperature, and K the equilibrium constant. Converting from natural to common logarithms and rearranging gives

$$\log K = \frac{-\Delta G^0}{2.303RT}$$

Applying this to the equilibrium considered above:

$$H_2 + I_2 \rightleftharpoons 2HI$$

we know that $\Delta G^0 = -24.2$ kJ (because we are making *two* moles of HI), while $R = 8.314$ J K^{-1} and $T = 490°C$ (763 K). So we can find K:

$$\log K_p = \frac{+24.2 \times 10^3}{2.303 \times 8.314 \times 763} = 1.66$$

$$K_p = 46$$

(*Note*: The factor of 10^3 is needed to convert ΔG^0 from kJ to J.)

This value of K is, of course, the same as that we used previously (Example 11.2). Solving for that equilibrium, we found that the degree of conversion from $H_2 + I_2$ to 2HI was 0.772 (77.2%). We see now that this mixture has the lowest free energy content for the system at this temperature.

Note that, because of the definition of standard states for gases as 1 atmosphere pressure, the values of K obtained from the above approach have gas concentrations as partial pressures in units of atmospheres, rather than in mole liter^{-1}. In other words, they are K_p values. This does not make any numerical difference in the hydrogen iodide equilibrium, which does not involve a change in the number of moles, but it should be kept in mind when the number of moles changes, so we will write K_p whenever we are working with gases.

Example 16.5 For the ammonia equilibrium

$$N_2(g) + 3H_2(g) \rightleftharpoons 2NH_3(g)$$

$\Delta H^0 = -92.4$ kJ. Values of S^0 are NH$_3$, 192.5; N$_2$, 191.5; and H$_2$, 130.6 J K^{-1} mole^{-1}.

Calculate ΔG^0 at 25°C and at 300°C, and the corresponding values of K_p. At what temperature does $\Delta G^0 = 0$?

SOLUTION   For the equilibrium written above, we can calculate ΔS^0 from the given data:

$$\Delta S^0 = (2 \times 192.5) - 191.5 - (3 \times 130.6)$$
$$= -198.3 \text{ J K}^{-1} \quad \text{(for two moles NH}_3\text{)}$$
$$= -0.1983 \text{ kJ K}^{-1}$$

Assuming ΔH^0 and ΔS^0 are constant,

$$\Delta G^0 = \Delta H^0 - T\Delta S^0$$
$$= -92.4 + 0.1983T$$

This equation enables us to calculate ΔG^0 at any desired temperature

at 25°C (298 K): $\Delta G^0 = -92.4 + (0.1983 \times 298)$
$$= -92.4 + 59.1 = -33.3 \text{ kJ.}$$
at 300°C (573 K): $\Delta G^0 = -92.4 + (0.1983 \times 573)$
$$= -92.4 + 113.6 = +21.2 \text{ kJ.}$$

When $\Delta G^0 = 0$, we use the same equation

$$0 = -92.4 + 0.1983T \qquad T = \frac{92.4}{0.1983} = 466 \text{ K } (193°\text{C})$$

The values of K_p are found from

$$\log K_p = \frac{-\Delta G^0}{2.303RT}$$

at 25°C: $\Delta G^0 = -33.3 \text{ kJ}$

$$\log K_p = \frac{+33.3 \times 10^3}{2.303 \times 8.314 \times 298} = 5.84$$

$$K_p = 6.86 \times 10^5 \text{ atm}^{-2}$$
at 300°C: $\Delta G^0 = +21.2 \text{ kJ}$

$$\log K_p = \frac{-21.2 \times 10^3}{2.303 \times 8.314 \times 573} = -1.93$$

$$K_p = 1.17 \times 10^{-2} \text{ atm}^{-2}$$
at 193°C: $\Delta G^0 = 0$
$$K_p = 1.00 \text{ atm}^{-2}$$

Example 16.6   In one of the classic experiments of all time, Lavoisier heated mercury in air at a moderate temperature to give mercury(II) oxide, HgO. On strong heating, this decomposed to give mercury and oxygen. Account for Lavoisier's results by considering the equilibrium

$$HgO(s) \rightleftharpoons Hg(g) + \frac{1}{2} O_2(g)$$

(a) Calculate ΔG^0 at 25°C and 200°C
(b) Calculate K_p at 200°C
(c) At what temperature does $\Delta G^0 = 0$?

	ΔH_f^0 (kJ mole^{-1})	S^0 (J K^{-1} mole^{-1})
HgO(s):	−90.7	72.0
Hg(g):	+60.8	174.9
O$_2$(g):	—	205

SOLUTION

(a) For the reaction as written above,

$$\Delta H^0 = +60.8 - (-90.7) = +151.5 \text{ kJ}$$

$$\Delta S^0 = 174.9 + \frac{205}{2} - 72.0 = 205.4 \text{ J K}^{-1} = 0.2054 \text{ kJ K}^{-1}$$

$$\Delta G^0 = 151.5 - 0.2054T$$

When $T = 298$ K (25°C), $\Delta G^0 = 151.5 - (298 \times 0.2054) = +90.3$ kJ.
When $T = 473$ K (200°C), $\Delta G^0 = 151.5 - (473 \times 0.2054) = +54.3$ kJ.

We see that ΔG^0 is positive at room temperature and at 200°C. Decomposition of HgO into the elements is not favorable. For the reverse reaction, ΔG^0 is negative, so the reverse reaction does occur on moderate heating (at room temperature, although ΔG^0 is favorable, the *rate* of reaction is very slow.)

(b) At 200°C:

$$\Delta G^0 = +54.3 \text{ kJ}$$

$$\log K_p = \frac{-54.3 \times 10^3}{2.303 \times 8.314 \times 473} = -6.00$$

$$K_p = 1.0 \times 10^{-6} \text{ atm}^{1.5}$$

At equilibrium, the partial pressures of Hg(g) and O$_2$(g) will be very low. In other words, heating mercury at 200°C in air or oxygen will cause HgO to form.

(c) When $\Delta G^0 = 0$, we can write

$$\Delta G^0 = 0 = 151.5 - 0.2054T$$

$$T = \frac{151.5}{0.2054} = 737 \text{ K} \qquad (464°C)$$

At this temperature, we have equilibrium between the three compounds in their standard states, i.e., solid HgO is in equilibrium with Hg(g) and O$_2$(g) each at 1 atm pressure

$(K_p = 1.00)$. Heating HgO above this temperature would cause ready decomposition to the elements.

Example 16.7   Experiment shows that solid ammonium chloride dissociates on heating in the following manner:

$$NH_4Cl(s) \rightleftharpoons NH_3(g) + HCl(g)$$

Explain this result by calculating ΔG^0 and K_p at 25°C, then calculating the temperature at which $\Delta G^0 = 0$.

	ΔH_f^0 (kJ mole^{-1})	S^0 (J K^{-1} mole^{-1})
$NH_4Cl(s)$:	−315.4	22.6
$NH_3(g)$:	−46.2	192.5
$HCl(g)$:	−92.3	186.7

SOLUTION   As usual, we calculate ΔH^0 and ΔS^0:

$$\Delta H^0 = -46.2 - 92.3 - (-315.4) = +176.9 \text{ kJ}$$
$$\Delta S^0 = 186.7 + 192.5 - 22.6 = 356.6 \text{ J K}^{-1} = 0.3566 \text{ kJ K}^{-1}$$
at 25°C (298 K): $\qquad \Delta G^0 = 176.9 - (298 \times 0.3566) = +70.6 \text{ kJ}$

$$\log K_p = \frac{-70.6 \times 10^3}{2.303 \times 8.314 \times 298} = -12.4$$

$$K_p = 4 \times 10^{-13} \text{ atm}^2$$

Decomposition is obviously unfavorable, or in other words NH_3 and HCl will readily combine to form NH_4Cl at this temperature.

When $\Delta G^0 = 0$, $\qquad 0 = 176.9 - 0.3566T$, and $T = \dfrac{176.9}{0.3566} = 496 \text{ K}$ $\qquad$ (223°C)

At or above this temperature, decomposition to NH_3 and HCl will occur readily.

Although ΔG^0 is an important and useful quantity, it is convenient to be able to do calculations of the above type in one step rather than two, i.e. to have a direct connection between ΔH^0, ΔS^0, and the equilibrium constant.

If we combine the two equations

$$\log K = \frac{-\Delta G^0}{2.303RT} \quad \text{and} \quad \Delta G^0 = \Delta H^0 - T\Delta S^0$$

we immediately obtain

$$\log K = \frac{1}{2.303R} \left(\Delta S^0 - \frac{\Delta H^0}{T} \right)$$

and we could have used this equation in Example 16.5.

Now suppose we compare the values of K at two different temperatures, T_1, and T_2. The values of K will be

$$\log K_1 = \frac{1}{2.303R}\left(\Delta S^0 - \frac{\Delta H^0}{T_1}\right) \quad \text{and} \quad \log K_2 = \frac{1}{2.303R}\left(\Delta S^0 - \frac{\Delta H^0}{T_2}\right)$$

Combining these two equations by subtraction removes ΔS^0, giving

$$\log\left(\frac{K_1}{K_2}\right) = \frac{\Delta H^0}{2.303R}\left(\frac{1}{T_2} - \frac{1}{T_1}\right)$$

This equation is known as the *van't Hoff equation*, and we see its form is similar to that of the Arrhenius equation (Chapter 10). Its importance is that it enables us to compare equilibrium constants at different temperatures, knowing the value of ΔH^0. Alternatively, if we measure two values of K at different temperatures, we can find ΔH^0 directly. (To avoid confusion, we will consistently call the lower temperature T_1 and the higher temperature T_2).

Example 16.8   For the ammonia equilibrium

$$N_2(g) + 3H_2(g) \rightleftharpoons 2NH_3(g)$$

ΔH^0 is -92.4 kJ for the formation of two moles of NH_3. By what factor will K_p at 25°C differ from K_p at 300°C? Compare your answer with that contained in Example 16.5.

SOLUTION   If the value of K_p is K_1 at 25°C (298 K) and K_2 at 300°C (573 K) then

$$\log\left(\frac{K_1}{K_2}\right) = \frac{-92.4 \times 10^3}{2.303 \times 8.314}\left(\frac{1}{573} - \frac{1}{298}\right)$$

$$= (-4826)(-1.61 \times 10^{-3}) = 7.77$$

$$\frac{K_1}{K_2} = 5.91 \times 10^7$$

Using the data obtained previously,

$$\frac{K_1}{K_2} = \frac{6.86 \times 10^5}{1.17 \times 10^{-2}} = 5.9 \times 10^7$$

As expected, the two methods of calculation give the same result. Note that, in Example 16.8, we did not need to know ΔS^0.

Example 16.9   For the equilibrium

$$N_2O_4(g) \rightleftharpoons 2NO_2(g)$$

measurements of K_p give values of 0.168 atm at 30°C and 0.701 atm at 50°C. Calculate ΔH^0 and ΔS^0 for this process.

SOLUTION   To find ΔH^0, we write

$$\log\left(\frac{0.168}{0.701}\right) = \frac{\Delta H^0}{2.303 \times 8.314}\left(\frac{1}{323} - \frac{1}{303}\right)$$

$$\log 0.240 = \frac{\Delta H^0}{19.15} \times (3.096 - 3.300) \times 10^{-3}$$

$$-0.620 = -\Delta H^0 \times \frac{0.204 \times 10^{-3}}{19.15}$$

$$\Delta H^0 = \frac{19.15 \times 0.620}{0.204 \times 10^{-3}} = 5.82 \times 10^4 \text{ J}$$

$$= 58.2 \text{ kJ for the reaction as written}$$

To find ΔS^0, we use the equation connecting one K_p value with ΔH^0 and ΔS^0

$$\log K_p = \frac{1}{2.303R}\left(\Delta S^0 - \frac{\Delta H^0}{T}\right)$$

$$\log 0.168 = \frac{1}{2.303 \times 8.314}\left(\Delta S^0 - \frac{58.2 \times 10^3}{303}\right)$$

$$-0.775 = \frac{1}{19.15}(\Delta S^0 - 192.1)$$

$$\Delta S^0 = (-0.775 \times 19.15) + 192.1$$
$$= 192.1 - 14.8 = 177 \text{ J}$$

Note that we needed the results of *two* experiments to give us the values of both ΔH^0 and ΔS^0. A single value of K_p at one temperature would have given us one equation with two unknowns, which we could not have solved.

Example 16.10   A reaction has $\Delta H^0 = -100$ kJ. How will the equilibrium constant change as the temperature is increased from 50°C to 100°C?

SOLUTION   We do not have enough information to calculate the actual value of K at any temperature, but we can evaluate the ratio of the two K values.

Let the equilibrium constant be K_1 at 50°C (323 K) and K_2 at 100°C (373 K). Then

$$\log\left(\frac{K_1}{K_2}\right) = \frac{-100 \times 10^3}{2.303 \times 8.314}\left(\frac{1}{373} - \frac{1}{323}\right)$$

$$= (-5.22 \times 10^3)(-4.15 \times 10^{-4}) = +2.17$$

$$\frac{K_1}{K_2} = 147$$

The equilibrium constant is 147 times larger at the lower temperature; in other words, the equilibrium shifts to the left (favoring the reactants) when we raise the temperature.

Example 16.11   The equilibrium constant for a reaction is found to increase by a factor of 250 when the temperature is raised from 100°C to 200°C. Calculate ΔH^0 for the reaction.

SOLUTION   This is the reverse of the previous example. Using the same equation:

$$\log\left(\frac{K_1}{K_2}\right) = \log\left(\frac{1}{250}\right) = \log(4.00 \times 10^{-3}) = -2.40$$

(*Note*: Be careful to get the fraction the right way up! K_2 is the constant at the *higher* temperature, T_2.)

$$-2.40 = \frac{\Delta H^0}{2.303 \times 8.314}\left(\frac{1}{473} - \frac{1}{373}\right) = \frac{\Delta H^0}{19.14}(2.114 - 2.681) \times 10^{-3}$$

$$\Delta H^0 = \frac{-2.40 \times 19.14}{-5.67 \times 10^{-4}} = +8.10 \times 10^4 \text{ J} = 81.0 \text{ kJ}$$

The last four examples bring out an important point. In Examples 16.8 and 16.10, the value of ΔH^0 was negative and the value of K was found to be larger at the lower temperature ($K_1 > K_2$). In Examples 16.9 and 16.11, the situation was reversed, since ΔH^0 values were positive and the equilibrium constant was greater at the higher temperature.

Putting this another way, for the equilibrium

$$A + B \rightleftharpoons C + D$$

if the reaction going from left to right is exothermic (ΔH^0 negative), then an increase in temperature will always shift the equilibrium to the left (smaller K); whereas if the reaction is endothermic (ΔH^0 positive), an increase in temperature will shift the equilibrium to the right (larger K).

This effect is often cited as an example of Le Chatelier's principle (when a stress is applied to a system, the system reacts so as to minimize the effect of that stress), but we can see from the discussion in this chapter that it is an inevitable result of the type of relationship existing between the various quantities connected with an equilibrium.

16.3   Free Energy and Cell Potentials

Finally in this chapter, we relate the free energy change occurring in a reaction to the standard cell potential for that reaction. We have already used the Nernst equation (Chapter 15) to work out equilibrium constants from cell potentials, so, knowing now that the former are related to free energy changes, we should suspect that ΔG^0 values are also related.

The equation is a simple one

$$\Delta G^0 = -n\mathscr{F}\mathscr{E}^0$$

where ΔG^0 is the standard free energy change for the reaction, n is the total number of electrons transferred between oxidizing and reducing agent, $\mathscr{F}$ is the Faraday constant (96,490 coulomb mole^{-1}) and $\mathscr{E}^0$ is the standard cell potential for the reaction.

Note that we use overall cell potentials in this equation, not half-cell potentials. It is difficult to give a meaning to ΔG^0 for a half-cell reaction, because of the way in which half-cell potentials are assigned.

Example 16.12   Given that $\mathcal{E}^0 = +1.10$ V, calculate ΔG^0 for the reaction

$$Zn + Cu^{2+} \longrightarrow Zn^{2+} + Cu$$

SOLUTION   This is a two-electron transfer, so

$$\Delta G^0 = -2 \times 96{,}490 \times 1.10 = -2.12 \times 10^5 \text{ J}$$
$$= -212 \text{ kJ}$$

This is a large, negative value. As we have already seen (Example 15.10), the corresponding value of K is very large:

$$\log K = \frac{-\Delta G^0}{2.303 RT} = \frac{+2.12 \times 10^5}{2.303 \times 8.314 \times 298} = 37.3 \qquad K = 2 \times 10^{37}$$

Not surprisingly, since we're using the same approach, this is the same value that we obtained previously from the Nernst equation.

Example 16.13   If $\mathcal{E}^0 = 2.00$ V, what is ΔG^0 for the reaction

$$2MnO_4^- + 5C_2O_4^{2-} + 16H^+ \longrightarrow 2Mn^{2+} + 10CO_2 + 8H_2O$$

SOLUTION   In this case, $n = 10$ (total electrons transferred according to the equation) so we have

$$\Delta G^0 = -10 \times 96{,}490 \times 2.00 = -1.93 \times 10^6 \text{ J}$$
$$= -1930 \text{ kJ}$$

Obviously, from the form of the equation, a positive value will always give a negative ΔG^0 value, so we have consistency between our two criteria for forecasting the direction in which a chemical reaction will spontaneously proceed.

The relationship between $\mathcal{E}^0$ values and ΔG^0 for a reaction enables us to solve a question that we left unanswered in Chapter 15; how can we combine two half-cell reactions to find $\mathcal{E}^0$ for a reaction in which electrons gained and lost do *not* cancel out? We know now that ΔG values are always additive when two or more reactions are combined, so, if we calculate ΔG values for the two half-cell reactions, we can combine them to calculate ΔG for the combined reaction and then find $\mathcal{E}^0$ for the overall reaction from this.

Consider the general case of the reduction of some species A to give first B and then C in a two-stage process:

(i) $A + xe^- \longrightarrow B$ $\mathcal{E}_1^0$ volt

(ii) $B + ye^- \longrightarrow C$ $\mathcal{E}_2^0$ volt

Combining these gives the overall reduction

(iii) $A + (x + y)e^- \longrightarrow C$ $\mathcal{E}_3^0$ volt

The values of ΔG for each step are

(i) $$\Delta G_1^0 = -x\mathcal{F}\mathcal{E}_1^0$$

(ii) $$\Delta G_2^0 = -y\mathcal{F}\mathcal{E}_2^0$$

(iii) $$\Delta G_3^0 = -(x + y)\mathcal{F}\mathcal{E}_3^0$$

Knowing that $\Delta G_1^0 + \Delta G_2^0 = \Delta G_3^0$ we can now relate $\mathcal{E}_1^0$, $\mathcal{E}_2^0$, and $\mathcal{E}_3^0$:

$$-x\mathcal{F}\mathcal{E}_1^0 + (-y\mathcal{F}\mathcal{E}_2^0) = -(x + y)\mathcal{F}\mathcal{E}_3^0 \qquad \mathcal{E}_3^0 = \frac{x\mathcal{E}_1^0 + y\mathcal{E}_2^0}{x + y}$$

We see from the form of this equation that $\mathcal{E}_3^0$ is *not* equal to the sum of $\mathcal{E}_1^0$ and $\mathcal{E}_2^0$ in a process of this type.

Example 16.14   Calculate $\mathcal{E}^0$ for the process

$$Fe^{3+} + 3e^- \longrightarrow Fe(s)$$

given the following reduction potentials

$$Fe^{3+} + e^- \longrightarrow Fe^{2+} \qquad \mathcal{E}_1^0 = +0.77V$$
$$Fe^{2+} + 2e^- \longrightarrow Fe(s) \qquad \mathcal{E}_2^0 = -0.44V$$

SOLUTION   We can use the formula derived above, where $x = 1$, $\mathcal{E}_1^0 = +0.77$ V; $y = 2$, $\mathcal{E}_2^0 = -0.44$ V.
For the overall reduction of Fe^{3+} to $Fe(s)$,

$$\mathcal{E}^0 = \frac{0.77 + 2(-0.44)}{1 + 2} = \frac{0.77 - 0.88}{3} = \frac{-0.11}{3} = -0.037 \text{ V}$$

In this chapter, we have considered the relationship between free energy change, equilibrium constant, and standard cell potential. Rather than think of these as separate quantities, you should regard them as different numerical ways of measuring the same quantity—namely, the overall equilibrium situation in a chemical system.

KEY WORDS

free energy

free energy change, ΔG

entropy

entropy change, ΔS

STUDY QUESTIONS

1. What makes a chemical reaction proceed spontaneously?

2. Is it possible for an endothermic process to proceed spontaneously?

3. What characteristic of a system does entropy measure?

4. How does entropy change as a substance goes from solid to liquid to gas?

5. How is the free energy change for a process related to the changes in enthalpy and entropy?

6. Of the above three quantities, which are roughly constant as temperature changes and which varies?

7. What can you say about a reaction for which ΔG is (a) positive, (b) zero, and (c) negative?

8. How is the equilibrium constant for a reaction related to ΔG^0?

9. How does the equilibrium constant change as we increase temperature for a reaction in which ΔH^0 is (a) positive and (b) negative?

10. What is the minimum number of values for equilibrium constants for a reaction at different temperatures that we need in order to find ΔH^0?

11. How is the ΔG^0 for a reaction related to its overall standard cell potential, $\mathcal{E}^0$? How do we relate our criteria of "spontaneous change" using these two figures?

12. How do we use ΔG values to add together $\mathcal{E}^0$ values for half-reactions?

PROBLEMS

1. For the reaction

$$Fe(s) + Cl_2(g) \longrightarrow FeCl_2(s)$$

calculate ΔG^0 at 25°C, given that ΔH_f^0 for $FeCl_2$ is -341 kJ mole^{-1} and S^0 values are 27.2, 223.0 and 119.7 J K^{-1} mole^{-1} for Fe, Cl_2, and $FeCl_2$ respectively.

2. For the reaction

$$3Fe_2O_3 \longrightarrow 2Fe_3O_4 + \frac{1}{2}O_2$$

(a) calculate ΔH^0, ΔS^0, and ΔG^0 at 25°C.
(b) At what temperature does $\Delta G^0 = 0$?

	ΔH_f^0 (kJ mole^{-1})	S^0 (J K^{-1} mole^{-1})
Fe_2O_3:	-822	90
Fe_3O_4:	-1117	146
O_2:	—	205

3. For the reaction

$$2H_2S(g) + SO_2(g) \longrightarrow 3S(s) + 2H_2O(g)$$

$\Delta H^0 = -146.3$ kJ and $\Delta G^0 = -90.7$ kJ at 298 K. Calculate ΔS^0 for this process.

4. Nitrogen dioxide decomposes reversibly on heating:

$$NO_2(g) \rightleftharpoons NO(g) + \frac{1}{2}O_2(g)$$

(a) Calculate ΔG^0 and K_p at 400°C, 500°C, and 600°C.
(b) At what temperature does $\Delta G^0 = 0$?

	ΔH_f^0 (kJ mole^{-1})	S^0 (J K^{-1} mole^{-1})
NO_2:	$+33.9$	240.5
NO:	$+90.4$	210.6
O_2:	—	205

5. Copper sulfate crystallizes from water at 25°C as the familiar blue pentahydrate, $CuSO_4 \cdot 5H_2O$. On heating to 110°C, this is converted into the white monohydrate, $CuSO_4 \cdot H_2O$. Use the following data to explain these results:

	ΔH_f^0 (kJ mole^{-1})	S^0 (J K^{-1} mole^{-1})
$CuSO_4 \cdot 5H_2O$:	-2278	305
$CuSO_4 \cdot H_2O$:	-1084	150
$H_2O(g)$:	-242	189

6. Group II metal carbonates decompose on heating according to

$$MCO_3(s) \rightleftharpoons MO(s) + CO_2(g)$$

where M represents Mg, Ca, etc. From the following data, calculate which metal carbonate is more easily decomposed by heating. (*Hint*: a good criterion would be the temperature at which the equilibrium pressure of CO_2 reaches one atmosphere.)

	ΔH_f^0 (kJ mole^{-1})	S^0 (J K^{-1} mole^{-1})
$CaCO_3(s)$:	-1207	92.9
$CaO(s)$:	-636	39.8
$MgCO_3(s)$:	-1113	65.7
$MgO(s)$:	-602	26.8
$CO_2(g)$:	-394	213.6

7. For the reaction

$$Ca(OH)_2(s) \rightleftharpoons CaO(s) + H_2O(g)$$

the equilibrium constant, K_p, changes from 0.302 at 400°C to 3.77 at 500°C. Calculate ΔH^0 for this reaction.

8. A reaction has $\Delta H^0 = -48$ kJ. What will be the effect on the equilibrium constant of changing the temperature from 0°C to 120°C?

9. For the equilibrium

$$2SO_2 + O_2 \rightleftharpoons 2SO_3$$

K_p is 1.05×10^8 at 300°C and 3.84×10^6 at 350°C. What is ΔH^0 for this reaction?

10. For the gaseous equilibrium

$$H_2(g) + CO_2(g) \rightleftharpoons H_2O(g) + CO(g)$$

ΔH^0 is $+41$ kJ. If the value of K_p at 300°C is 3.11×10^{-2}, what will it be at 400°C?

11. Calculate ΔG^0 at 298 K for the following cell reactions
    (a) $2Fe^{3+} + Sn^{2+} \longrightarrow 2Fe^{2+} + Sn^{4+}$ $\mathcal{E}^0 = +0.62$V
    (b) $Cr_2O_7^{2-} + 3Zn + 14H^+ \longrightarrow 2Cr^{3+} + 3Zn^{2+} + 7H_2O$ $\mathcal{E}^0 = +2.09$V
    (c) $Ag^+ + Fe^{2+} \longrightarrow Ag(s) + Fe^{3+}$ $\mathcal{E}^0 = +0.03$V

    (d) $Al(s) + 3H^+ \longrightarrow Al^{3+} + \frac{3}{2}H_2$ $\mathcal{E}^0 = +1.66$V

12. For the decomposition of water

$$H_2O(l) \longrightarrow H_2(g) + \frac{1}{2}O_2(g)$$

$\Delta G^0 = +237$ kJ at 25°C. Use this to evaluate $\mathcal{E}^0$ for the half-reaction

$$H_2O(l) \longrightarrow 2H^+ + \frac{1}{2}O_2 + 2e^-$$

and compare your answer with the value in Table 15.1. (*Hint*: Combine the last equation with another well-known half-reaction to give an equation for the electrolysis of water.)

13. In the following cases, add together the half-reactions given and calculate the ε^0 value of the process represented by the resulting equation.

(a) $Co^{3+} + e^- \longrightarrow Co^{2+} \qquad \varepsilon_1^0 = +1.82V$
$Co^{2+} + 2e^- \longrightarrow Co(s) \qquad \varepsilon_2^0 = -0.277V$

(b) $MnO_4^- + 4H^+ + 3e^- \longrightarrow MnO_2(s) + 2H_2O \qquad \varepsilon_1^0 = +1.70V$
$MnO_2(s) + 4H^+ + 2e^- \longrightarrow Mn^{2+} + 2H_2O \qquad \varepsilon_2^0 = +1.23V$

(c) $H_2O_2 + 2H^+ + 2e^- \longrightarrow 2H_2O \qquad \varepsilon_1^0 = +1.77V$
$O_2 + 2H^+ + 2e^- \longrightarrow H_2O_2 \qquad \varepsilon_2^0 = +0.682V$

SOLUTIONS TO PROBLEMS

1. $\Delta S^0 = -130.5 \text{ J K}^{-1}$
$\Delta G^0 = -341 + (298 \times 0.1305) = -302 \text{ kJ}.$

2. (a) $\Delta H^0 = 2(-1117) - 3(-822) = +232 \text{ kJ}$

$\Delta S^0 = \dfrac{205}{2} + (2 \times 146) - (3 \times 90) = 124.5 \text{ J K}^{-1}$

at 298 K: $\Delta G^0 = 232 - (298 \times 0.1245) = +195 \text{ kJ}$

(b) $\Delta G^0 = 0$ when $T = \dfrac{232}{0.1245} = 1863 \text{ K} \qquad (1590°C).$

3. $T\Delta S^0 = \Delta H^0 - \Delta G^0 = -146.3 - (-90.7) = -55.6 \text{ kJ}$

$\Delta S^0 = \dfrac{-55.6}{298} = -0.186 \text{ kJ K}^{-1} \qquad (-186 \text{ J K}^{-1})$

4. (a) $\Delta H^0 = +56.5 \text{ kJ} \qquad \Delta S^0 = 72.6 \text{ J K}^{-1}$
Using $\Delta G^0 = \Delta H^0 - T\Delta S^0 = 56.5 - 0.0726T$ and $\log K_p = -\Delta G^0/2.303RT$
at 400°C (673 K): $\Delta G^0 = +7.64 \text{ kJ}, K_p = 0.255 \text{ atm}^{1/2}$
at 500°C (773 K): $\Delta G^0 = +0.38 \text{ kJ}, K_p = 0.943 \text{ atm}^{1/2}$
at 600°C (873 K): $\Delta G^0 = -6.88 \text{ kJ}, K_p = 2.58 \text{ atm}^{1/2}$
(b) $\Delta G^0 = 0$ when $T = 778 \text{ K} (505°C).$

5. The reaction is

$$CuSO_4 \cdot 5H_2O \rightleftharpoons CuSO_4 \cdot H_2O + 4H_2O(g)$$

for which $\Delta H^0 = +226 \text{ kJ}, \qquad \Delta S^0 = 601 \text{ J K}^{-1}$
at 298 K (25°C): $\Delta G^0 = +47 \text{ kJ} \qquad K_p = 6.0 \times 10^{-9} \text{ atm}^4$
at 383 K (110°C): $\Delta G^0 = -4.2 \text{ kJ} \qquad K_p = 3.7 \text{ atm}^4$
At 25°C, the equilibrium favors the pentahydrate, whereas at 110°C it favors the monohydrate.

6. For the reaction

$$CaCO_3 \rightleftharpoons CaO + CO_2$$

$\Delta H^0 = +177 \text{ kJ}$ and $\Delta S^0 = 160.5 \text{ J K}^{-1}$

$\Delta G^0 = 0$ when $T = \dfrac{177}{0.1605} = 1103 \text{ K} (830°C)$

For the reaction

$$MgCO_3 \rightleftharpoons MgO + CO_2$$

$$\Delta H^0 = +117 \text{ kJ} \quad \text{and} \quad \Delta S^0 = 174.7 \text{ J K}^{-1}$$

$$\Delta G^0 = 0 \quad \text{when} \quad T = \frac{117}{0.1747} = 670 \text{ K } (397°C)$$

For these reactions, $K_p = [CO_2]$, so the CO_2 pressure reaches one atmosphere when $K_p = 1$. This occurs when $\Delta G^0 = 0$, so $MgCO_3$ is clearly decomposed at a lower temperature than $CaCO_3$.

7. $\log\left(\dfrac{0.302}{3.77}\right) = \dfrac{\Delta H^0}{2.303R}\left(\dfrac{1}{773} - \dfrac{1}{673}\right)$, whence $\Delta H^0 = +109$ kJ.

8. If the equilibrium constant is K_1 at 0°C and K_2 at 120°C, then

$$\log\left(\frac{K_1}{K_2}\right) = \frac{-48 \times 10^3}{2.303R}\left[\frac{1}{393} - \frac{1}{273}\right] = +2.80 \qquad \frac{K_1}{K_2} = 637$$

The equilibrium constant *decreases* by a factor 637 on going from 0°C to 120°C.

9. $\log\left(\dfrac{1.05 \times 10^8}{3.84 \times 10^6}\right) = \dfrac{\Delta H^0}{2.303R}\left(\dfrac{1}{623} - \dfrac{1}{573}\right)$, whence $\Delta H^0 = -196$ kJ

10. Call the constant K_2 at 400°C, then

$$\log\left(\frac{3.11 \times 10^{-2}}{K_2}\right) = \frac{41 \times 10^3}{2.303R}\left(\frac{1}{673} - \frac{1}{573}\right) = -0.556$$

$$\frac{3.11 \times 10^{-2}}{K_2} = 0.278 \qquad K_2 = 0.112$$

11. (a) a two-electron reaction: $\Delta G^0 = -2 \times 96,490 \times 0.62 = -120$ kJ
    (b) a six-electron reaction: $\Delta G^0 = -6 \times 96,490 \times 2.09 = -1210$ kJ
    (c) a one-electron reaction: $\Delta G^0 = -96,490 \times 0.03 = -3$ kJ
    (d) a three-electron reaction: $\Delta G^0 = -3 \times 96,490 \times 1.66 = -481$ kJ

12. If the unknown half-reaction potential is $\mathcal{E}_x^0$, then we may combine that reaction with the standard hydrogen electrode $(\mathcal{E}^0 = 0.0 \text{ V})$

$$H_2O(l) \longrightarrow 2H^+ + \frac{1}{2}O_2 + 2e^- \qquad \mathcal{E} = \mathcal{E}_x^0$$

$$\underline{2H^+ + 2e^- \longrightarrow H_2 \qquad\qquad\qquad\qquad \mathcal{E} = 0.0}$$

overall: $\quad H_2O(l) \longrightarrow H_2 + \frac{1}{2}O_2 \qquad\qquad \mathcal{E} = \mathcal{E}_x^0$

This is the reaction for which we are given $\Delta G^0 = +237$ kJ, and it is a two-electron change

$$237 = \frac{-2 \times 96,490 \times \mathcal{E}_x^0}{1000}$$

$$\mathcal{E}_x^0 = -1.23 \text{ V}$$

As we hoped, this is the accepted value as given in Table 15.1.

13. Overall reaction: $Co^{3+} + 3e^- \longrightarrow Co(s)$

$$\varepsilon^0 = \frac{\varepsilon_1^0 + 2\varepsilon_2^0}{1 + 2} = \frac{1.82 + 2(-0.277)}{3} = \frac{1.82 - 0.554}{3} = \frac{1.266}{3} = 0.422 \text{ V}$$

(b) Overall reaction: $MnO_4^- + 8H^+ + 5e^- \longrightarrow Mn^{2+} + 2H_2O$

$$\varepsilon^0 = \frac{3\varepsilon_1^0 + 2\varepsilon_2^0}{3 + 2} = \frac{(3 \times 1.70) + (2 \times 1.23)}{5} = \frac{5.10 + 2.46}{5} = \frac{7.56}{5} = 1.51 \text{ V}$$

(c) Overall reaction: $O_2 + 4H^+ + 4e^- \longrightarrow 2H_2O$

$$\varepsilon^0 = \frac{2\varepsilon_1^0 + 2\varepsilon_2^0}{2 + 2} = \frac{(2 \times 1.77) + (2 \times 0.682)}{4} = \frac{3.54 + 1.36}{4} = \frac{4.90}{4} = 1.23 \text{ V}$$

Appendix A
Common Mathematical Operations

It is just these very simple things which are extremely likely to be overlooked.
THE SIGN OF FOUR

The mathematical operations required to solve problems in general chemistry are very straightforward. We need a little algebra to put the unknown and known quantities together in an equation, then enough competence in simple arithmetic to solve the equation and obtain a numerical answer. Experience shows, however, that many students have trouble with the purely numerical aspects of problems, and these notes are offered in an attempt to help them.

Five topics are included:

1. Significant figures
2. Exponential notation
3. Logarithms
4. Ratio and proportion
5. Linear and quadratic equations

A.1   Significant Figures

Numbers used in chemistry may be divided into two categories, *exact* and *inexact*. Exact numbers are those whose value is precisely fixed by definition (that is, by conventional agreement among chemists). They usually have integral values.

An obvious use of exact numbers would be to relate two quantities within the same system of units, for example,

$$1\ m = 100\ cm = 1000\ mm = 1 \times 10^{10}\ Å$$

$$1\ h = 60\ min = 3600\ s$$

There is no question of error or uncertainty in these numbers.

Nonintegral exact numbers may arise when a defined value of a quantity replaces a previous experimental value. For example, the defined calorie is exactly 4.184 joule, replacing the previous definition of the calorie in terms of heating up water. Similarly, the standard atmosphere in SI units is exactly 1.01325×10^5 pascal, replacing (but equivalent to) the old definition of 760 mm of mercury.

Another use of exact numbers is in counting discrete objects (10 beakers, 2 students, etc.). In chemistry, this extends to moles, and we would write 2HCl, $3NH_3$, etc. to mean exactly 2 moles, 3 moles, and so on.

Inexact numbers arise from experimental measurements, which always introduce some degree of error or uncertainty. Some quantities are known with very great accuracy, such as the velocity of light—$(2.99792458 \pm 0.00000001) \times 10^8$ m s^{-1}—where the uncertainty is less than one part per million, but there is always the possibility that further experiments will force us to change an accepted value.

In very accurate work, we can indicate the limits of error on a measured quantity (that is, the upper and lower limits within which we are fairly confident that the "true" value actually lies) by putting ± 0.01, ± 0.0002, etc. after the number. In ordinary work, we do this by the use of *significant figures*, that is, by the number of digits that we use in writing out a number. Our choice of how many significant figures to use will of course be based on our estimate of the maximum probable error in the original data on which the result is based.

We generally take the probable error implied by our use of significant figures to be "5" in the decimal place following the last figure given, for example,

$$\text{mass} = 3.6521 \text{ g} \quad \text{implies} \quad 3.6521 \pm 0.00005 \text{ g}$$

We believe the true value to be between 3.65205 and 3.65215 g, and the value 3.6521 g expresses this to five significant figures. Obviously, if we had determined the mass on a less accurate balance, we would record it to fewer figures, such as 3.65 g (probable error ± 0.005 g, three significant figures).

Note that the position of the decimal point is irrelevant in determining the number of significant figures. 365, 36.5, 3.65 and 0.365 all have *three* figures. Students are often confused with the use of zero in a number. Can a zero ever be a significant figure? The answer is a qualified "yes"; a zero is taken as a significant figure *unless* it is being used merely to position the decimal point. Putting this another way, leading zeros are never significant; we must start our count of significant figures with the first digit (reading from left to right) which is *not* zero.

$$\left. \begin{array}{l} 3.652 \\ 0.3652 \\ 0.03652 \\ 0.003652 \end{array} \right\} four \text{ significant figures in each}$$

$$3.6520 \quad five \text{ significant figures}$$

The last number implies 3.6520 ± 0.00005, and the last zero is significant. Similarly, 3.65200 would have six significant figures.

Ambiguity may arise when zeros occur at the end of a number. Are they intended to be significant, or are they being used merely to position the decimal point? Thus, the population of a city is given as 237,000. How do we interpret this? Three significant figures, or four, five or six?

Common sense tells us that the above figure is intended to be accurate to the nearest thousand, and this is readily indicated by writing it in exponential form

$$237,000 = 2.37 \times 10^5 \quad \text{(three s.f.)}$$

If we knew that this was accurate to the nearest hundred, we would add a zero which would be significant

$$2.370 \times 10^5 \quad \text{(four s.f.)}$$

Example A.1 How many significant figures are present in the following numbers:

(a) 0.0231 (b) 4.010×10^8 (c) 9.0981×10^{-15}
(d) 0.00005 (e) 410 (f) Six million

SOLUTION

(a) three (zeros not significant)
(b) four (both zeros significant)
(c) five
(d) one (zeros not significant)
(e) three (This could be ambiguous, but, in the absence of other information, the assumption would be an implied uncertainty of ± 0.5.)
(f) one (Giving a quantity in this way implies "to the nearest million" unless we imply greater accuracy by saying "exactly six million," etc.)

Although there is a slight ambiguity (as in Example A.1(e)) in numbers ending in zero, it would be a little clumsy to write numbers under one thousand in exponential form simply in order to indicate the number of significant figures. We usually assume, therefore, that a quantity such as 80, 100, 650, etc. has been expressed with an implied accuracy of ± 0.5 (two or three significant figures), unless stated otherwise, and this convention has been followed in this book.

Numbers over one thousand should always be written in exponential form, using the appropriate number of significant figures, to avoid ambiguity, that is,

$$1,580 \quad \text{should be} \quad 1.58 \times 10^3 \quad \text{or} \quad 1.580 \times 10^3$$

Manipulating inexact numbers gives rise to many problems, particularly with the advent of electronic calculators. Those little numbers coming up in colored lights look very impressive, but do they actually mean anything? The basic rule is very simple; *no calculation can give a result that is more accurate than the data on which it was based.* Thus, it is easy to show on a calculator that

$$1.731 \times 1.14 = 1.97334$$

and as a piece of arithmetic, this is correct; but as a practical scientific calculation it is *wrong* because the answer implies a greater degree of accuracy than is justified by the data given. It should be written 1.97, using three *significant figures* only.

Let's look at the use of significant figures in common arithmetical operations:

Addition and Subtraction

The absolute precision of the answer must be related to the absolute precision of *each* of the numbers used. Its error can be deduced from the absolute error of the least precise of these. The following examples in addition will make this clear.

15.51	(4 s.f.)	0.6631	(4 s.f.)
0.065	(2 s.f.)	0.04113	(4 s.f.)
0.001	(1 s.f.)	0.0223	(3 s.f.)
15.58	(4 s.f.)	0.7265	(4 s.f.)

In the first example, the least precise number, on an absolute basis, is the top one (15.51), where the accuracy ends at the second decimal place. Although the other two numbers have fewer significant figures, the absolute accuracy (to the third decimal place) is greater. The second example is similar.

Note the manner of "rounding off" which has been used. The additions give 15.576 and 0.72653 respectively. When these are rounded off to the last significant digit, 15.576 is "rounded up" to 15.58 and 0.72653 is "rounded down" to 0.7265. If the digit following the last significant figure is 5 or above, we round up; if it is below 5, we round down. Examples:

$$1.2813 \quad \text{to} \quad 3 \text{ s.f.} \quad \text{becomes} \quad 1.28$$
$$5.2671 \quad \text{to} \quad 3 \text{ s.f.} \quad \text{becomes} \quad 5.27$$
$$1.375 \times 10^8 \quad \text{to} \quad 3 \text{ s.f.} \quad \text{becomes} \quad 1.38 \times 10^8$$
$$4.1914 \times 10^{-7} \quad \text{to} \quad 4 \text{ s.f.} \quad \text{becomes} \quad 4.191 \times 10^{-7}$$

In subtraction, we follow the same principles:

141.8	(4 s.f.)	169.673	(6 s.f.)
56.12	(4 s.f.)	48.1	(3 s.f.)
85.7	(3 s.f.)	121.6	(4 s.f.)

It frequently happens that the answer contains fewer significant figures than the original data as we take the difference between two numbers that are close together:

30.9953	(5 s.f.)	103.185	(6 s.f.)
30.9815	(5 s.f.)	103.103	(6 s.f.)
0.0138	(3 s.f.)	0.082	(2 s.f.)

Multiplication and Division

Generally speaking, the answer will carry the same number of significant figures as the *smallest* number of significant figures present in the original data. Thus, in the example given at the beginning of this section, we should write

$$1.731 \times 1.14 = 1.97$$
$$(4 \text{ s.f.}) \quad (3 \text{ s.f.}) \quad (3 \text{ s.f.})$$

If the numbers had both been given to four figures, then the answer would also have had four figures.

$$1.731 \times 1.146 = 1.983726 = 1.984 \qquad \text{(4 s.f., rounded up)}$$

The same applies to division:

$$\frac{1.731}{1.14} = 1.518 = 1.52 \qquad \text{(3 s.f., rounded up)}$$

$$\frac{1.731}{1.146} = 1.510 \qquad \text{(4 s.f.)}$$

$$\frac{8.7285 \times (1.133 \times 10^{-6}) \times 5.81}{47.36 \times (8.99 \times 10^{-4})} = 1.35 \times 10^{-3}$$

(*Note*: Only 3 s.f. are justified, because two numbers in the calculation are only given to 3 significant figures accuracy.)

Common sense dictates a slight departure from the above rules in certain cases. Consider the following:

$$99.3 \times 1.079 = 107.1$$

$$\frac{103.8}{1.108} = 93.7$$

In the first example, although strictly speaking 99.3 has only three significant figures, its implied accuracy is better than 0.1 in 99, or about 0.1%. We are therefore justified in giving the answer to the nearest 0.1 (still about 0.1%), although this now needs 4 significant figures. The second example shows the reverse situation, where we have to drop from 4 significant figures in our data to 3 in our answer to keep the same degree of accuracy (0.1%). It would be wrong to give the answer as 93.68, implying an accuracy of 0.01%.

Example A.2   Carry out the following calculations, giving your answer with the appropriate number of significant figures:

(a) $511.61 + 10.3289 + 161.5$ (b) $1.042 - 0.012$ (c) $6.875 - 0.04314$

(d) 4.932×2.06 (e) $13.73 \times 64.915 \times 0.0120$ (f) $\dfrac{0.485}{103.68}$

(g) $\dfrac{0.8900 \times 4.116}{2.9311 \times 0.06875}$

SOLUTION

(a) 683.4389 rounded to 683.4 (4 s.f.). The accuracy cannot be taken beyond the first decimal place, because that was the limit of accuracy in one piece of data (161.5).

(b) 1.030 (4 s.f.). Both figures were given to three decimal places. Note that the final zero in the answer is significant, and should be retained.

(c) 6.83186 rounds to 6.832 (4 s.f.)

(d) 10.15992 rounds to 10.2 (3 s.f.)

(e) 10.6954 rounds to 10.7 (3 s.f.) In this case, 3 significant figures are justified because 0.0120 has 3 significant figures (including the final zero).

(f) 0.00468 (3 s.f.)

(g) 18.18 (4 s.f.) All data are given to at least 4 significant figures, including 0.8900 (two final significant zeros).

When a calculation involving several steps is carried out, small errors can sometimes be introduced by rounding off in the intermediate stages. For example, suppose we want to calculate the molarity of a solution of 6.62 g of NaOH in 0.200 liter. Knowing the molar mass of NaOH to be 40.0 g, we write:

$$\text{molarity} = \frac{6.62 \text{ g}}{(40.0 \text{ g mole}^{-1})(0.200 \text{ liter})} = 0.8275 M = 0.828 M \text{ (3 s.f.)}$$

If we do the same calculation in two stages, we could write

$$\frac{6.62 \text{ g}}{40.0 \text{ g mole}^{-1}} = 0.1655 \text{ mole} = 0.166 \text{ mole (3 s.f.)}$$

$$\frac{0.166 \text{ mole}}{0.200 \text{ liter}} = 0.830 M \text{ (3 s.f.)}$$

Obviously, the first answer $(0.828 M)$ is correct.

Although the difference is small, a significant error can arise if it happens that several "rounding off" corrections in a calculation are all in the same direction. This can be avoided either by doing all the numerical calculation in a problem in one operation, or by carrying an extra digit on the intermediate figures in a calculation, and only rounding off in the final answer.

In this book, for the sake of clarity in showing the course of a calculation, the numbers in intermediate steps have been shown rounded off to the appropriate number of significant figures, but the final answer has usually been calculated from the original data to avoid introducing progressive errors.

We finally note, in connection with significant figures, that, in a calculation involving both exact numbers and inexact numbers, the exact numbers have no effect on the number of significant figures in the answer. This is commonly seen in stoichiometry problems, e.g., in Example 3.24 concerning a permanganate–oxalate titration, the answer is obtained from the calculation

$$\frac{10^3 \times 0.188 \times 2}{134 \times 5 \times 17.5} = 3.20 \times 10^{-2}$$

in which the figures 10^3, 2, and 5 are exact. The correct number of significant figures (3) is determined by the remaining data, which are given to 3 significant figures. (Use of significant figures in logarithms is mentioned at the end of the section on logarithms.)

A.2 Exponential Notation (Scientific Notation)

Exponential notation is widely used in scientific work when dealing with very large or very small numbers. (A further use, which we noted in the previous section, was in the presentation of the correct number of significant figures in a quantity without ambiguity.)

By "exponential," we mean that we have expressed a number by writing it as the product of two terms, one a quantity of reasonable size (usually between 1 and 10) and the other an exact power of 10. The second term has the effect of making the product very large (if it is a positive power of 10) or very small (if it is a negative power).

In this way, we avoid the necessity of writing down a large number of zeroes in a number. A few examples will make this clear:

$$653 = 6.53 \times 10^2 \qquad\qquad 653{,}000 = 6.53 \times 10^5$$

$$0.00653 = 6.53 \times 10^{-3} \qquad\qquad 653 \text{ million} = 653 \times 10^6 = 6.53 \times 10^8$$

The exponent of 10 in these expressions gives us the number of places by which the decimal point has to be moved to get the required number. Thus, in going from 6.53×10^5 to the extended form 653,000 we move the decimal point five places to the *right* as we make a big number. In going from 6.53×10^{-3} to 0.00653, we move it three places to the *left* as we make a small number.

Example A.3 Express the following in exponential notation in the form $a \times 10^n$, where a is a quantity between 1 and 10 and n is an integer:

(a) 0.00004870   (b) 237.11   (c) 60,000,000   (d) 29.8   (e) 0.3010

SOLUTION

(a) 4.870×10^{-5} (decimal point is moved 5 places)
(Note the preservation of the final significant zero.)

(b) 2.3711×10^2 (decimal moves two places)
(c) 6×10^7 (In the absence of any specific indication, we assume that the zeroes were being used to position the decimal point and are not significant.)
(d) 2.98×10 (n in this case is of course 1, but we do not bother to write 10^1.)
(e) 3.010×10^{-1}

We must know how to handle numbers written in exponential form when carrying out the common arithmetical operations.

In *multiplication*, the numbers are multiplied together in the ordinary way, but the exponents are *added*. In other words, we split the calculation into two parts:

$$(2 \times 10^4) \times (3 \times 10^2) = (2 \times 3) \times 10^{(4+2)} = 6 \times 10^6$$

You can easily verify the above result by multiplying 20,000 by 300 by longhand. The same rule applies when one or both of the exponents is negative:

$$(2 \times 10^4) \times (3 \times 10^{-2}) = (2 \times 3) \times 10^{(4-2)} = 6 \times 10^2$$

$$(2 \times 10^{-4}) \times (3 \times 10^2) = (2 \times 3) \times 10^{(-4+2)} = 6 \times 10^{-2}$$

$$(2 \times 10^{-4}) \times (3 \times 10^{-2}) = (2 \times 3) \times 10^{(-4-2)} = 6 \times 10^{-6}$$

This rule enables us to raise exponential numbers to powers (squares, cubes, etc.). If we have to raise $(a \times 10^n)$ to the power x we raise the number a to the power x, but we *multiply* the exponent n by x.

$$(a \times 10^n)^x = a^x \times 10^{nx}$$

$$(3 \times 10^4)^2 = 3^2 \times 10^{(4 \times 2)} = 9 \times 10^8$$

$$(3 \times 10^{-2})^3 = 3^3 \times 10^{(-2 \times 3)} = 27 \times 10^{-6}$$

In division, the principle is very similar. We divide the two numbers in the usual way, but we *subtract* one exponent from the other:

$$\frac{5 \times 10^6}{2 \times 10^3} = \frac{5}{2} \times 10^{(6-3)} = 2.5 \times 10^3 \qquad \frac{8 \times 10^{-8}}{2 \times 10^4} = \frac{8}{2} \times 10^{(-8-4)} = 4 \times 10^{-12}$$

Be careful when subtracting negative exponents, remembering that minus (minus) equals plus:

$$\frac{5 \times 10^6}{2 \times 10^{-3}} = \frac{5}{2} \times 10^{6-(-3)} = 2.5 \times 10^9 \qquad \frac{8 \times 10^{-8}}{2 \times 10^{-4}} = \frac{8}{2} \times 10^{-8-(-4)} = 4 \times 10^{-4}$$

Obviously we can extend this to the taking of reciprocals:

$$\frac{1}{10^n} = 10^{-n} \quad \text{as in} \quad \frac{1}{10^7} = 10^{-7} \quad \text{or} \quad \frac{1}{10^{-12}} = 10^{12} \quad \text{etc.}$$

$$\frac{1}{5 \times 10^{-5}} = \frac{1}{5} \times 10^5 = 0.2 \times 10^5 \qquad \frac{1}{4 \times 10^6} = \frac{1}{4} \times 10^{-6} = 0.25 \times 10^{-6}$$

Remember that $10^0 = 1$, so if the sum or difference of the exponents comes to zero, we can drop the power of ten:

$$(2 \times 10^{-3}) \times (6 \times 10^3) = (2 \times 6) \times 10^{-3+3} = 12 \qquad \frac{6 \times 10^6}{4 \times 10^6} = \frac{6}{4} \times 10^{(6-6)} = 1.5$$

It frequently happens that we want to change the exponent in a quantity, i.e., to change the way of writing it, without changing the actual value of the quantity. For example, we obtained above answers of 27×10^{-6} and 0.2×10^5, which are not in our usual format where the number part is between 1 and 10.

When the way of writing a quantity is changed, we must *increase* one part by a factor of 10 and, at the same time, *decrease* the other part. In that way, the value of their product will stay the same. In other words, we are moving a factor of 10 across from one half of the product to the other. Returning to the above examples

$$27 \times 10^{-6} = (2.7 \times 10) \times 10^{-6} = 2.7 \times 10^{-5} \qquad 0.2 \times 10^5 = (2 \times 10^{-1}) \times 10^5 = 2 \times 10^4$$

It's often necessary to go the other way round when taking square roots or cube roots, because we should arrange our number so that the exponent of 10 is exactly divisible by 2 or 3 respectively.

To take the square root, the square root of the number part is taken in the usual way (often by using logarithms, see next section), while the exponent is divided by two:

$$\sqrt{4 \times 10^8} = \sqrt{4} \times 10^{8/2} = 2 \times 10^4$$

$$\sqrt{9 \times 10^{-16}} = \sqrt{9} \times 10^{-16/2} = 3 \times 10^{-8}$$

$$\sqrt{2.5 \times 10^7} = \sqrt{25 \times 10^6} = \sqrt{25} \times 10^{6/2} = 5.0 \times 10^3$$

$$\sqrt{4.9 \times 10^{-9}} = \sqrt{49 \times 10^{-10}} = \sqrt{49} \times 10^{-10/2} = 7.0 \times 10^{-5}$$

Note carefully the way in which the number was rearranged in the last two examples so that we had an even exponent of ten.

For cube roots, take the cube root of the number and divide the power of ten by three. Again, some rearrangement may be necessary:

$$\sqrt[3]{8 \times 10^{-6}} = \sqrt[3]{8} \times 10^{-6/3} = 2 \times 10^{-2}$$

$$\sqrt[3]{2.7 \times 10^4} = \sqrt[3]{27 \times 10^3} = \sqrt[3]{27} \times 10^{3/3} = 3.0 \times 10$$

$$\sqrt[3]{1.25 \times 10^{-7}} = \sqrt[3]{125 \times 10^{-9}} = \sqrt[3]{125} \times 10^{-9/3} = 5.0 \times 10^{-3}$$

Addition and subtraction of quantities written in exponential form is straightforward. If the exponent on both quantities is the same, we simply add or subtract the numbers:

$$(7.2 \times 10^{-3}) + (5.9 \times 10^{-3}) = (7.2 + 5.9) \times 10^{-3} = 13.1 \times 10^{-3} = 1.31 \times 10^{-2}$$

$$(9.62 \times 10^6) - (9.51 \times 10^6) = (9.62 - 9.51) \times 10^6 = 0.11 \times 10^6 = 1.1 \times 10^5$$

If the exponents are different, we have to rewrite the numbers, as described previously, so that they have the same exponent, then add or subtract them:

$$(7.23 \times 10^{-3}) + (5.91 \times 10^{-4}) = (7.23 \times 10^{-3}) + (0.591 \times 10^{-3})$$

$$= (7.23 + 0.591) \times 10^{-3} = 7.82 \times 10^{-3}$$

$$(5.21 \times 10^7) - (7.93 \times 10^5) = (5.21 \times 10^7) - (0.0793 \times 10^7)$$

$$= (5.21 - 0.0793) \times 10^7 = 5.13 \times 10^7$$

(*Note*: Remember to watch significant figures in these calculations!)

A.3 Logarithms

Many students are introduced to logarithms as aids to doing multiplication and division. Having acquired an electronic calculator or a slide-rule, they tend to think of logarithms as something of no practical use. This is a mistaken and unfortunate attitude, and we will find several occasions in this book (and in other science courses) where logarithms are vitally important. Every student should therefore take the short time required to master this concept.

What is a logarithm? It represents a functional relationship between two quantities, that is, it tells us the manner in which one is dependent on the other. Consider the equation

$$y = A^x$$

where y and x are variables and A is a constant. This represents an exponential relationship, because x is the exponent of A (the power to which A is to be raised), and y varies exponentially with x.

Suppose the constant A is equal to 10

$$y = 10^x$$

We can easily evaluate y for integral values of x. If $x = 1$, $y = 10$; $x = 2$, $y = 100$; $x = 3$, $y = 1000$, etc. We see that, as x increases, y increases very rapidly, which is a characteristic of exponential growth.

Another method of expressing a relationship of this type is to write it in logarithmic form. In the equation

$$y = A^x$$

x is, by definition, the *logarithm* of y to the base A, that is to say, the power to which A must be raised to give y. For any value of y, the value of x will obviously depend on the value of our constant A, and 10 is a convenient value to choose. We refer to A as the "base" of our logarithmic system, so if $A = 10$, we say we are dealing with logarithms to the base 10, or common logarithms. So we can write two equations

$$y = 10^x \qquad \log_{10} y = x$$

Both of these equations denote exactly the same relationship, and we refer to them as "the exponential form" and "the logarithmic form" respectively.

The subscript 10 following the log symbol indicates the base we are using, but it is not usually included. In the absence of any other indication, the abbreviation "log" means "logarithm to the base 10," so we write $\log y = x$.

Returning to the numbers we worked out for the exponential equation, we can see that $\log 10 = 1$, $\log 100 = 2$, $\log 1000 = 3$, etc., or in general $\log(10^x) = x$. Remembering that $10^0 = 1$, it follows that $\log 1 = \log(10^0) = 0$.

Suppose we have a number that is not an exact power of 10; how should we find its logarithm? We have just shown that $\log 1 = 0$ and $\log 10 = 1$, so it follows that *the logs of all numbers between 1 and 10 must lie between 0 and 1*. To find them, we use a set of log tables, which generally give an accuracy of four significant figures.

The figures in a set of four-figure log tables are the logs of quantities between 1 and 10, and all are to be read as four digits following the decimal point. Thus, opposite "65" we see "8129," which means that the common log of 6.5 (*not* 65) is 0.8129. To find the log of a number outside the range 1–10, we write it in exponential notation (i.e., as a product) and use the fact that the log of a product is the *sum* of the logs of the terms in the product. Thus, to take the log of 65, we split this into 6.5×10, then take the log of 6.5 (0.8129) and the log of 10(1.0000) and add them together, obtaining $\log 65 = 1.8129$. Similarly, $\log 650 = 2.8129$, $\log 650{,}000 = 5.8129$, etc.

Strictly speaking, these logs greater than 1 are in two parts; the figures *before* the decimal (the "characteristic") and the figures *after* the decimal (the "mantissa"). When reversing the process to find the "antilog," it is essential to split the log at the decimal point, then treat each part separately. That part *before* the decimal becomes simply a power of 10, that part *after* the decimal is converted by using the log table "backwards" (or using "anti-log" tables, but these are not generally available) and becomes a number *between 1 and 10*.

Example A.4   Evaluate the common logs of the following:

(a) 2.38   (b) 238   (c) 4.72×10^6

SOLUTION

(a) From the table, $\log 2.38 = 0.3766$
(b) $\log 238 = \log(2.38 \times 10^2) = \log 2.38 + \log(10^2) = 0.3766 + 2 = 2.3766$
(c) $\log(4.72 \times 10^6) = \log 4.72 + \log(10^6) = 0.6739 + 6 = 6.6739$

Example A.5   Of what numbers are these the logarithms?

(a) 0.4871   (b) 2.8921   (c) 18

SOLUTION

(a) Directly from the table, 0.4871 is the log of 3.070.
(b) Splitting this into two parts,
     $2.8921 = 0.8921 + 2.0000$
     antilogs: 7.80×10^2

(c) Don't be confused when you have to take the antilog of a quantity with nothing after the decimal point. 18 is the log of 10^{18} (you could write 1×10^{18}).

The same rules apply to the logs of small quantities. As we have seen, logs decrease as the numbers become smaller, until the log of 1 has reached zero. As we take the logs of smaller quantities, they become *negative*.

The log of a quantity between zero and one is always negative.

It is not possible to take the log of a quantity less than zero (i.e., a negative number). Applying this to some examples, what is the log of 0.5? First write in exponential form

$$0.5 = 5 \times 10^{-1}$$

This is a product, so we take the logs of the two parts and add them

$$\log 5 = 0.6990$$
$$\log(10^{-1}) = -1$$
$$\log 0.5 = 0.6990 - 1 = -0.3010$$

Although this is quite correct, it is often more convenient to write it in a form which leaves the part following the decimal (the mantissa) in a positive form, putting a bar over the characteristic to show that it is negative

$$\log 0.5 = \bar{1}.6990 \text{ (bar-one-point-six-nine-nine-zero)}$$

Be very careful when using this system that you remember the significance of the bar sign. It make the characteristic negative, but leaves the mantissa positive.

Obviously in taking the antilog of a number expressed in this way we should "split it" at the decimal point and treat the characteristic and the mantissa separately.

Example A.6   Evaluate the common logs of the following:

(a) 0.25   (b) 6.88×10^{-3} (c) 0.00441

SOLUTION

(a) $0.25 = 2.5 \times 10^{-1}$
    take logs: $\log 2.5 = 0.3979$ $\log 10^{-1} = -1$
    $\log 0.25 = 0.3979 - 1 = \bar{1}.3979$
    (This could also be written as -0.6021.)
(b) $\log 6.88 = 0.8376$ $\log 10^{-3} = -3$
    $\log(6.88 \times 10^{-3}) = 0.8376 - 3 = \bar{3}.8376$
    (This could be written as -2.1624.)
(c) $0.00441 = 4.41 \times 10^{-3}$
    $\log(4.41 \times 10^{-3}) = 0.6444 - 3 = \bar{3}.6444$ (or -2.3556)

Example A.7   Of what numbers are these the logs:

(a) $\bar{2}.73$ (b) $\bar{5}.9011$ (c) -4.811

SOLUTION

(a) antilog $\bar{2} = 10^{-2}$ antilog $0.73 = 5.37$ antilog $\bar{2}.73 = 5.37 \times 10^{-2}$

(b) antilog $\bar{5} = 10^{-5}$ antilog $0.9011 = 7.963$ antilog $\bar{5}.9011 = 7.963 \times 10^{-5}$

(c) In this example, the whole number is negative. To convert it to the "bar" notation, we must add an integer large enough to make the number positive, then subtract the same integer so that the total value of the quantity is unchanged. Obviously in this case the integer is 5, so we simultaneously add and subtract 5:

$$-4.811 = -5 + (+5 - 4.811) = -5 + 0.189 \quad \text{or} \quad \bar{5}.189$$

$$\text{antilog}(-5) = 10^{-5} \qquad \text{antilog } 0.189 = 1.545$$

$$\text{antilog}(-4.811) = \text{antilog}(\bar{5}.189) = 1.545 \times 10^{-5}$$

As you should realize, the reason for keeping the mantissa positive is that our four-figure log tables only give positive log values. We cannot directly look up the antilog of a negative quantity.

When we use logarithms to do the common operations of multiplication and division, we have to be extra careful when using the bar notation. Always bear in mind the exact significance of the bar; it makes the characteristic negative while leaving the mantissa positive.

Example A.8   Use logarithms to evaluate the following:

(a) 84.19×76.38

(b) $(4.81 \times 10^{-3}) \times (6.18 \times 10^{-2})$

(c) $304.1 \times (4.97 \times 10^{-8})$

(d) $\dfrac{76.3}{851}$

(e) $\dfrac{1.66 \times 10^{-2}}{3.75 \times 10^{-4}}$

SOLUTION

(a) log 84.19 is   1.9253
    log 76.38 is   1.8830
    _____

    add together 3.8083
    antilog 3.8083 is 6.431×10^3

(b) log 4.81×10^{-3} is $\bar{3}.6821$
    log 6.18×10^{-2} is $\bar{2}.7910$
    _____

    add together $\bar{4}.4731$

Note carefully how the addition is done. The sum of the positive parts following the decimal point is $+1.4731$, so we carry $+1$ over to the sum of the two negative characteristics, giving $-3 - 2 + 1 = -4$ for the final characteristic

$$\text{antilog } \bar{4}.4731 \text{ is } 2.97 \times 10^{-4}$$

There is an alternative way of doing such calculations, in which we do the multiplication of the numbers by using logs (or an electronic calculator) and treat the powers of ten separately by mental arithmetic, that is:

$$(4.81 \times 10^{-3}) \times (6.18 \times 10^{-2})$$

$$= (4.81 \times 6.18) \times (10^{-3} \times 10^{-2})$$

$$= 29.7 \times 10^{-5} = 2.97 \times 10^{-4}$$

Try this method for yourself on the remaining parts of this example

(c) log 304.1 is   2.4830
    log 4.97×10^{-8} is $\bar{8}.6964$

    add together $\bar{5}.1794$ (Remember: $+2 - 8 + 1 = -5$.)
    antilog $\bar{5}.1794$ is 1.51×10^{-5}

(d) log 76.3 is 1.8825
    log 851 is 2.9299

    subtract $\bar{2}.9526$
    On subtracting, the difference between the two mantissa parts gives
    $0.8825 - 0.9299 = -1 + 0.9526$
    The -1 is included in the characteristics, giving $+1 - 2 - 1 = -2$
    antilog of $\bar{2}.9526$ is 8.97×10^{-2}

(e) log 1.66×10^{-2} is $\bar{2}.2201$
    log 3.75×10^{-4} is $\bar{4}.5740$

    subtract   1.6461
    (the characteristic is $-2 - (-4) - 1 = +1$)
    antilog 1.6461 is 4.43×10

Example A.9   Use logarithms to take the reciprocals of the following

(a) 2.50   (b) 6.881×10^{7} (c) 1.493×10^{-5}

SOLUTION   Taking a reciprocal is a division in which we divide the number into 1.0000. Remembering that log 1 = 0, we see that, using logs, we shall be subtracting the log from zero, which will give us *minus* the value of the log. In other words

$$\log\left(\frac{1}{x}\right) = -\log x$$

(a) log 2.50 = 0.3979

$$\log\left(\frac{1}{2.50}\right) = -0.3979 = -1 + 0.6021 \qquad (\bar{1}.6021)$$

antilog $\bar{1}.6021 = 4.00 \times 10^{-1} = 0.400$

(b) $\log(6.881 \times 10^{7}) = 7.8377$
    change sign $-7.8377 = -8 + 0.1623$ $(\bar{8}.1623)$
    antilog $\bar{8}.1623 = 1.453 \times 10^{-8}$

(c) $\log(1.493 \times 10^{-5}) = \bar{5}.1741$
    in changing sign, remember the significance of the bar. We have $-5 + 0.1741$, so when we change the sign of *both* parts we get $+5 - 0.1741 = 4.8259$.
    antilog $4.8259 = 6.698 \times 10^{4}$

Logarithms are very useful for calculating powers and roots. The general relationship is

$$\log(x^n) = n \log x$$

The value of n may be an integer (2, 3, etc.) in which case we are squaring, cubing, etc. or it may be nonintegral, so we can use this relationship to raise a number to the power 2.7 or 4.18, etc.

To raise a number to the power n, look up the log of the number, *multiply* the log by n, look up the antilog.

In evaluating roots (square roots, cube roots, etc.) we use the relationship

$$\sqrt[n]{x} = x^{(1/n)} \qquad \log x^{(1/n)} = \frac{1}{n} \log x$$

To take the nth root of a number ($n = 2$, 3 etc.), look up the log of the number, *divide* the log by n, look up the antilog.

Example A.10   Use logarithms to evaluate the following:

(a) $2.00^{1.5}$ (b) $(4.8 \times 10^{-3})^3$ (c) $(7.5 \times 10^{-6})^{2.91}$ (d) $\left(\frac{1}{2}\right)^{7.69}$

SOLUTION
(a) Look up log 2.0 (0.3010), multiply by 1.5 (0.4515), look up the antilog (2.83). When you multiply by the exponent (1.5 in this case), you may carry out the multiplication by using logs if the numbers are awkward. This means of course, that you would take the log of a log, which is not difficult if you remember what you are doing:

$$\log 0.3010 = \bar{1}.4786$$
$$\log 1.5 \quad = 0.1761$$
$$\overline{}$$
$$\text{sum} = \bar{1}.6547$$
$$\text{antilog } \bar{1}.6547 = 0.4515$$
$$\text{antilog } 0.4515 = 2.83$$

In this particular calculation, we could have taken a short cut by noting that

$$(2.0)^{1.5} = \sqrt{2.0^3} = \sqrt{8.0} = 2.83$$

(b) With small quantities, where the logs are negative, the multiplication has to be done carefully, bearing in mind the significance of the "bar" notation

$$\log 4.8 \times 10^{-3} = \bar{3}.6812$$

Multiply by 3, remembering that this quantity is $-3 + 0.6812$

$$(-3 + 0.6812) \times 3 = -9 + 2.0436 = -7 + 0.0436$$

So our product is $\bar{7}.0436$, whose antilog is 1.11×10^{-7}.

An alternative way of doing calculations of this type is to treat the power of ten separately, i.e., to write:

$$(4.8 \times 10^{-3})^3 = (4.8)^3 \times (10^{-3})^3 = (4.8)^3 \times 10^{-9}$$

Look up log 4.8 (0.6812), multiply by 3 (2.0436), look up antilog (111),

$$(4.8)^3 \times 10^{-9} = 111 \times 10^{-9} = 1.11 \times 10^{-7}$$

However, this will only work when raising to an integral power. In the following example this separating technique is inapplicable.

(c) This is similar to the previous example, but we have to take logs twice because the numbers are awkward to multiply.

$$\log(7.5 \times 10^{-6}) = \bar{6}.8751 = -6 + 0.8751$$

Multiply the two parts by 2.91 separately, using logs (you can't look up the log of -6, so remove the minus sign and look up log 6. Remember to replace the minus sign *after* doing the multiplication!)

$$
\begin{array}{ll}
\log 6 \ \ = \ 0.7782 & \log 0.8751 = \bar{1}.9421 \\
\log 2.91 = \ 0.4639 & \log 2.91 \ \ \ = 0.4639 \\
\hline
\text{sum} = \ 1.2421 & \text{sum} = 0.4060 \\
\text{antilog} = 17.462 & \text{antilog} = 2.547
\end{array}
$$

$$(-6 + 0.8751) \times 2.91 = -17.462 + 2.547$$
$$= -14.915 = \overline{15}.085$$
$$\text{antilog } \overline{15}.085 = 1.22 \times 10^{-15}$$

so we find

$$(7.5 \times 10^{-6})^{2.91} = 1.22 \times 10^{-15}$$

Obviously this calculation could be made easier by doing the intermediate multiplications on an electronic calculator or a slide-rule.

(d) This is a "half-life" problem (Example 10.10). To evaluate $\left(\dfrac{1}{2}\right)^{7.69} = (0.5)^{7.69}$, look up log 0.5 ($\bar{1}.699$) and multiply its two parts separately by 7.69. Logs may, of course, be used to do these multiplications

$$(\bar{1}.699) \times 7.69 = (-1 + 0.699) \times 7.69$$
$$= -7.69 + 5.375 = -2.315 = \bar{3}.685$$
$$\text{antilog } \bar{3}.685 = 4.84 \times 10^{-3}$$

Example A.11   Use logarithms to evaluate the following:

(a) $\sqrt{8.6 \times 10^3}$ (b) $\sqrt[3]{2.1 \times 10^4}$ (c) $\sqrt{6.2 \times 10^{-5}}$ (d) $\sqrt[3]{1.8 \times 10^{-7}}$

SOLUTION

(a) To obtain an even power of ten, write 8.6×10^3 as 86×10^2 (we *increase* the number by a factor of 10, we *decrease* the exponent by one, their *product* is unchanged).

$$\sqrt{86 \times 10^2} = \sqrt{86} \times \sqrt{10^2} = \sqrt{86} \times 10$$

To find $\sqrt{86}$, take the log (1.9345), divide by 2 (0.9673), look up the antilog (9.27).

$$\sqrt{8.6 \times 10^3} = 9.27 \times 10 = 92.7$$

(b) Here we have to adjust the power of 10 to be divisible by 3. It's better to do this by *increasing* the number and *decreasing* the exponent (to avoid the number being less than 1) so:

$$\sqrt[3]{2.1 \times 10^4} = \sqrt[3]{21 \times 10^3} = \sqrt[3]{21} \times 10$$

Look up log 21 (1.3222), divide by 3 (0.4406), look up antilog (2.76)

$$\sqrt[3]{2.1 \times 10^4} = 2.76 \times 10 = 27.6$$

(c) Exactly the same principles are followed when the exponent is negative

$$\sqrt{6.2 \times 10^{-5}} = \sqrt{62 \times 10^{-6}} = \sqrt{62} \times 10^{-3}$$

look up log 62 (1.7924), divide by 2 (0.8962), look up antilog (7.87)

$$\sqrt{6.2 \times 10^{-5}} = 7.87 \times 10^{-3}$$

(d) $\sqrt[3]{1.8 \times 10^{-7}} = \sqrt[3]{180 \times 10^{-9}} = \sqrt[3]{180} \times 10^{-3}$

look up log 180 (2.2553), divide by 3 (0.7518), look up antilog (5.65)

$$\sqrt[3]{1.8 \times 10^{-7}} = 5.65 \times 10^{-3}$$

All of the above calculations have been carried out using common (base 10) logarithms. There is, however, another very important and widely used base, the indeterminate number e (2.71828 ...). Logarithms based on e are called *natural* (sometimes Naperian) logarithms, and abbreviated to ln to distinguish them from common logs. The significance of the number e will not be discussed here, but you will meet it again if you ever take a course in calculus.

So we have again two equations showing the same relationship

$$\text{exponential: } y = e^x \qquad \text{logarithmic: } \ln y = x$$

In chemistry, we meet a relationship of this type in the Arrhenius equation (Chapter 10).

$$k = e^{(-E_a/RT)} \quad \text{or} \quad \ln k = \frac{-E_a}{RT}$$

Both of these represent exactly the same relationship.

Although it is possible to look up the values of natural logarithms in tables, these are not generally available, so it is convenient to know how to convert from natural logarithms to common logarithms.

Suppose $\log_{10} y = p$, what is $\ln y$? Converting to exponential form, we have $y = 10^p$. Now take the *natural* log of both sides

$$\ln y = \ln(10^p) = p \ln 10$$

but $p = \log y$, therefore

$$\ln y = (\log y) \times (\ln 10)$$

The natural log of a number is always related to its common log by a constant factor $\ln 10$, or 2.303. Hence we can always evaluate a natural log by first looking up the common log, then multiplying it by 2.303.

$$\log 2 = 0.3010 \qquad \ln 2 = 0.3010 \times 2.303 = 0.6931$$
$$\log 514 = 2.711 \qquad \ln 514 = 2.711 \times 2.303 = 6.242$$

Conversely, to convert from natural to common logs, we *divide* by 2.303. Returning to the Arrhenius equation, using natural logs

$$\ln k = \frac{-E_a}{RT}$$

Divide both sides by 2.303 to give common logs

$$\log k = \frac{-E_a}{2.303RT}$$

If we rewrite the above equations, we may calculate values of e^x (again, tables of this function are published, but may not always be at hand).

If $y = e^x$, taking natural logs gives $\ln y = x$. Dividing by 2.303 converts to common logs

$$\log y = \frac{x}{2.303}$$

So to evaluate e^x, we divide the value of x by 2.303 and look up the antilog of the quotient. Thus, to evaluate $e^{2.5}$

$$\frac{2.5}{2.303} = 1.086 \qquad \text{antilog } 1.086 = 12.18$$

The value of x may be negative, in which case e^x will be less than 1. Thus, to evaluate $e^{-6.83}$

$$\frac{-6.83}{2.303} = -2.966 = \bar{3}.034 \qquad \text{antilog } \bar{3}.034 = 1.081 \times 10^{-3}$$

To end this discussion of logarithms, we should briefly mention the use of significant figures in log calculations. It is difficult to decide in all cases how many significant figures are justified in the logarithm of a number, and the following rules are suggested:

1. In using four-figure tables to look up logs, always record all four figures in the mantissa (i.e., to four decimal places), regardless of how many figures were in the original data.
2. Do *not* "round off" logs below four decimal places as the calculation proceeds.
3. When the final answer is obtained by taking an antilog, round it off to the appropriate number of significant figures, according to the accuracy of the original data, in the usual way.
4. If the log is *itself* the final answer to the problem (as in pH calculations, Chapter 13) it should be rounded off in the manner usual for that particular type of answer. Values of pH, for example, are usually given to two decimal places.
5. If it is necessary to work with data that are given to five or more significant figures, then four-figure tables cannot be used. This does not often happen in simple chemical calculations, but an example will be found in the section on mass defect in this book (Chapter 4, Example 4.13). These calculations must be done by hand (unless a calculator is available) because six-figure accuracy is needed.

A.4   Ratio and Proportions

This section may seem very elementary to you, but experience shows that some instruction on these points is often needed.

One apple costs 10 cents. How much do 3 apples cost?
O.K. so far? Now try this:
One mole of substance weighs 120 g. How much do 3 moles weigh?
Still with me? Now we'll get tricky:
One apple costs 10 cents. How much does 0.5 apple cost?
All agree on 5 cents? Now back to chemistry:
One mole of substance weighs 120 g. How much does 0.5 mole weigh?
All those who replied "240 g" had better stick to apples.

By now you should realize that, in principle, each of the above calculations is *exactly* the same. We are dealing with two quantities which are directly related to each other. When one of them (number of apples; number of moles) changes in a certain ratio, the other (cost; weight) must change in the *same ratio*.

We multiply the number of moles by two; we multiply the weight by two. We multiply the number of moles by 0.5, we multiply the weight by 0.5. For some reason, many people have a mental blackout when numbers below one appear. Think about it carefully. To go from 1.00 to 0.50 you must *multiply* by 0.50 (If you *divide* 1.00 by 0.50 the result will be 2.00.)

Seen in the abstract with simple numbers, it all seems quite straightforward, doesn't it? But panic sets in when the numbers get tricky:

A solution contains 0.0831 g liter^{-1} of compound Q. How much Q is contained in a volume of (a) 12.00 liter (b) 0.375 liter?

(a) 1 liter contains 0.0831 g
   12.00 liter contains $12.00 \times 0.0831 = 0.997$ g
   (We multiply both volume and mass by 12.00.)
(b) 1 liter contains 0.0831 g
   0.375 liter contains $0.0831 \times 0.375 = 0.0312$ g
   (We multiply both volume and mass by 0.375.)
   One is directly proportional to the other.

If you like to make a mathematical equation out of this type of calculation, look at it this way:

The concentration of the solution is constant.
Concentration is a ratio of mass to volume.
Let the unknown mass be x g.
Then the ratio of x g to 0.375 liter is the same as the ratio of 0.0831 g to 1 liter

$$\frac{x}{0.375} = \frac{0.0831}{1}$$

Multiply both sides by 0.375

$$x = \frac{0.0831 \times 0.375}{1} = 0.0312 \text{ g}$$

Let's try this calculation the other way round. Suppose the concentration is 0.0831 g liter^{-1}. What volume contains 0.0500 g? By proportionality:

0.0831 g is present in 1 liter
1 g is present in 1/0.0831 liter
(divide *both* by 0.0831)
0.0500 g is present in $(0.0500 \times 1)/0.0831 = 0.602$ liter
(multiply *both* by 0.0500)

In making up an equation, the concentration is still the constant factor. If the required volume is x liter, we can say the ratio of 0.0500 g to x liter is the same as the ratio of 0.0831 g to 1 liter, or in an equation:

$$\frac{0.0500}{x} = \frac{0.0831}{1}$$

Multiply both sides by x and divide both sides by 0.0831

$$\frac{0.0500}{0.0831} = x \qquad x = 0.602 \text{ liter}$$

The manipulations required above are the most elementary type of algebra, and such operations should become second nature to you. Remember that chemical problems require the same type of simple logic that is needed in any other calculation. Let's look at a final example, involving a sequential calculation, to emphasize this point.

Daniel bought 5 bags of apples for $2.00. Maria ate 3 apples. "That's 24 cents worth of apples you've eaten, Maria" said Daniel. How many apples were in each bag?

24 cents buys 3 apples

1 cent buys $\dfrac{3}{24}$ apples

$2.00 buys $\dfrac{200 \times 3}{24} = 25$ apples

5 bags contain 25 apples

1 bag contains $\dfrac{25}{5} = 5$ apples

Simple proportion throughout. Can you see any similarity in the following calculation?

A solution is known to have a concentration of 6.75 g liter^{-1}. Experiment shows that 25 ml of solution contains 8.5×10^{-4} moles. What is the molecular weight?

25 ml contain 8.5×10^{-4} mole

1 ml contains $\dfrac{8.5 \times 10^{-4}}{25}$ mole

1000 ml contains $\dfrac{8.5 \times 10^{-4}}{25} \times 1000 = 3.4 \times 10^{-2}$ mole

3.4×10^{-2} mole is equal to 6.75 g

1 mole is equal to $\dfrac{6.75}{3.4 \times 10^{-2}} = 1.98 \times 10^2$ g

The molecular weight is 198. Two divisions and one multiplication. Easy.

A.5   Linear and Quadratic Equations

Many physical and chemical properties are interrelated by a linear dependence, that is, one is directly proportional to the other. Other relationships may be written in a linear manner by a mathematical adjustment in the writing of the equation. There are many examples in this book, and you will often encounter such relationships in laboratory work. For example:

(a) freezing-point depression:

$T = mK_f$ (T is directly proportional to K_f.)

(b) Boyle's law:

$P = \dfrac{nRT}{V}$ (Pressure is directly proportional to $1/V$.)

(c) Arrhenius' equation:

$k = Ae^{-E_a/RT}$ (an exponential equation)

$\log k = \log A - \dfrac{E_a}{2.303RT}$ (a linear equation)

When we talk of a "linear" or "straight-line" relationship, we are not implying that an actual graph is to be drawn showing the variation of one quantity as a function of the other. All we are suggesting is that the equation is to be handled in a manner appropriate to linear relationship. However, it is convenient to consider what a graphical presentation would look like, and what we could deduce from it.

The general equation to a straight line is $y = ax + b$. The coefficients a and b may have any value, positive or negative, and these will determine the appearance of the line. Figure A.1 shows several possible straight lines. In lines (i), (ii), and (iii), the slope is positive, that is, the line runs uphill from left to right. This happens when coefficient a is positive, and the value of a is called the slope of the line. It may be measured anywhere along the line (see Figure A.2). The three lines in the illustration are parallel, so the slope must be

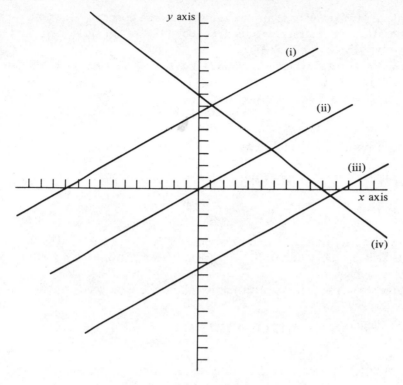

Figure A.1

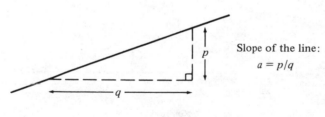

Slope of the line:

$$a = p/q$$

Figure A.2

the same for each of them. Line (iv) slopes downward, so a must be negative in the equation to this line.

Line (ii) goes through the origin (the point $x = 0$, $y = 0$). This tells us that the co-efficient b in its equation must be zero, that is, its equation is simply $y = ax$. Many relationships are of this type.

Line (i) crosses the y axis at a positive value of y. We say its *intercept* on the y axis is positive. Since $x = 0$ along the y axis, the intercept tells us the value of b in the equation

$$y = ax + b \quad \text{when} \quad x = 0, y = b$$

If the intercept is positive, b must be positive. Similarly with line (iii), the intercept with the y axis is negative, so b must be negative.

The ideal gas laws provide a simple example of these concepts. Suppose we plot volume against temperature in degrees celsius for one mole of an ideal gas, at 1 atmosphere pressure, what will the graph look like?

We know $PV = nRT$ where $R = 0.0821$ when T is in Kelvin. If we call the celsius temperature T_c, then $T = T_c + 273$, so

$$PV = nR(T_c + 273)$$

Putting $P = 1$ atm and $n = 1$ mole gives

$$V = R(T_c + 273) = 0.0821(T_c + 273)$$

$$= 0.0821T_c + 22.4$$

This is the equation to a straight line of slope 0.0821. (Figure A.3.) The intercept on the y (vertical) axis will be the value of V at $T_c = 0$. Clearly, from the equation, when $T_c = 0$, $V = 22.4$ liter.

The intercept on the x (horizontal) axis will be the value of T_c when $V = 0$.

$$\text{If } V = 0, \quad 0.0821T_c + 22.4 = 0 \qquad T_c = \frac{-22.4}{0.0821} = -273$$

As expected, V falls to zero at the absolute zero of temperature, $-273°C$.

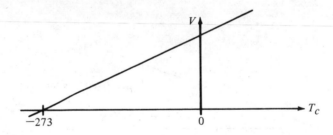

Figure A.3

To take another example, returning to Arrhenius' equation mentioned above

$$\log k = \log A - \frac{E_a}{2.303RT}$$

We could experimentally determine k at various different values of T. Plotting $\log k$ against $1/T$ would give a straight line of slope $-E_a/2.303R$ making an intercept equal to $\log A$ on the axis $1/T = 0$, so we could determine both E_a and A directly from such a graph.

Many chemical problems end up in a quadratic equation in the unknown quantity x, that is, an equation including terms in x^2. Quite often these can be simplified by noting that the equation is a perfect square, so we can reduce it from a quadratic to a linear equation by taking the square root of both sides. The commonest case of this is the equation $x^2 = A$ (where A is some number), but we sometimes find equations of the type

$$\frac{x^2}{(x + B)^2} = A$$

(see Example 11.2). When a quadratic equation is of this type, there are always two solutions, since A has two square roots, $+\sqrt{A}$ and $-\sqrt{A}$. Invariably, only one of these will be chemically realistic, that is to say, only one value of x will give physically possible values of the quantities we are finding. Thus, in Example 12.2, we have $[Ag^+] = \sqrt{7.7 \times 10^{-13}}$. Obviously we want the positive root $(+8.8 \times 10^{-7})$, since $[Ag^+]$ cannot be negative.

A quadratic equation that is not a perfect square may sometimes be solved by making a slight approximation which makes it into this form by removing some of the terms. This is not mathematically exact, of course, but it is quite permissible, provided that any errors introduced are less than the precision to which the data in the problem are given. In chemical systems, this situation is frequently encountered in connection with weak acids and bases, and is discussed in Chapter 13 (see Example 13.1).

When no approximations are possible, the quadratic equation must be solved rigorously using the well-known formula. After reducing the equation to the form

$$ax^2 + bx + c = 0$$

we can write

$$x = \frac{-b \pm \sqrt{b^2 - 4ac}}{2a}$$

This will always give two values for x, unless $b^2 - 4ac = 0$, and these values may be either positive, negative, or one of each, depending on the values of a, b, and c. If $b^2 < 4ac$, the roots of the equation will be complex (that is, they can only be expressed in terms of the square root of a negative quantity). Obviously this cannot happen when x corresponds to some real physical quantity, so a complex result indicates that you have made a mistake in the problem, either in the arithmetic, or by trying to impose some physically impossible conditions on a system.

Let's illustrate this method of solution by considering Example 11.4. The data given in the problem produce the relationship

$$\frac{(1.544 + 2x)^2}{(0.428 - x)(0.228 - x)} = 45.9$$

This is not a perfect square and no approximation is possible, so we have to multiply it out

$$(1.544 + 2x)^2 = 45.9(0.428 - x)(0.228 - x)$$

$$2.384 + 6.17x + 4x^2 = 45.9(0.0976 - 0.656x + x^2)$$

$$2.384 + 6.17x + 4x^2 = 4.480 - 30.11x + 45.9x^2$$

Collecting terms gives $41.9x^2 - 36.29x + 2.095 = 0$. This is a quadratic equation with $a = 41.9$, $b = -36.29$, and $c = 2.095$. From the formula

$$x = \frac{-b \pm \sqrt{b^2 - 4ac}}{2a}$$

$$= \frac{+36.29 \pm \sqrt{(-36.29)^2 - (4 \times 41.9 \times 2.095)}}{2 \times 41.9}$$

$$= \frac{36.29 \pm \sqrt{1317 - 351}}{83.8}$$

$$= \frac{36.29 \pm \sqrt{966}}{83.8}$$

$$= \frac{36.29 + 31.1}{83.8} \quad \text{or} \quad \frac{36.29 - 31.1}{83.8}$$

$$= \frac{67.39}{83.8} \quad \text{or} \quad \frac{5.19}{83.8}$$

$$= 0.804 \quad \text{or} \quad 0.0619$$

Either of these numbers is mathematically acceptable as a solution to the quadratic equation. To decide which is applicable in our problem, we have to look back at the data, and we see in this case that the concentration of one component, H_2, was given as $(0.428 - x)$. This means that the solution $x = 0.804$ must be rejected, since it would lead to a negative value of $[H_2]$, which is physically impossible. The other solution, $x = 0.062$, is the one needed here.

It will never happen that two different solutions of a quadratic equation lead to two different answers to a problem both of which are physically possible.

PROBLEMS

Note: It is strongly recommended that you go through these problems, including those on logarithms, *even if you are using an electronic calculator.*

1. Carry out the following calculations and express your answer with the correct number of significant figures:
   (a) $(1.73 \times 10^{-4}) \times 0.162$ (b) 691.8×317.14
   (c) $519.6 + 21.63 + 104.8$ (d) $1.098 \times 10^7/1.213 \times 10^4$
   (e) $86.118 - 0.0695$ (f) $19.181 \times (2.70 \times 10^{-5})$

2. Evaluate the common logs of the following:
   (a) 7.839   (b) 0.1221   (c) 37.42   (d) 6.83×10^5 (e) 4.986×10^{-10}

3. Evaluate the natural logs of the following:
   (a) 4.11   (b) 0.0161   (c) 3.8×10^4 (d) 6.22×10^{-8}

4. Of what numbers are these the common logs?
   (a) 23   (b) 0.6914   (c) -2 (d) $\bar{3}.81$ (e) 15.8172   (f) -4.333

5. Use logs to evaluate the following:
   (a) $\sqrt{8.62 \times 10^7}$ (b) $\sqrt[3]{9.14 \times 10^{-11}}$ (c) $\sqrt{1.25 \times 10^{-9}}$

   (d) $\sqrt[3]{4.83 \times 10^{-7}}$ (e) $(5.66)^{1.8}$ (f) $(7.23)^{4.5}$

   (g) $\left(\frac{1}{2}\right)^{14.5}$ (h) $(1.8 \times 10^{-6})^{3.75}$ (i) e^2

   (j) e^{-3} (k) $e^{4.55}$ (l) $e^{-3.73}$

6. One liter of a gas weighs 2.38 g. What do 27.0 ml weigh?

7. 48.3 ml of gas weigh 0.561 g. What do 22.4 liter weigh?

8. A solution contains 6.40 g liter^{-1} of dissolved compound. How much is contained in 0.143 ml?

9. What volume of the above solution will contain (a) 7.50 g and (b) 0.112 g?

10. 1.81×10^{-3} mole weigh 5.14 g. What is the molecular weight?

11. A compound has *M.W.* 671. How many mole in (a) 12.5 g and (b) 1.8×10^{-2} g?

12. A solution contains 34.2 g liter^{-1}. We find 25 ml to contain 2.14×10^{-2} mole. Calculate:
    (a) the concentration in units of mole liter^{-1}
    (b) the *M.W.*

Additional problems involving the concept of ratio and proportions will be found following Chapter 3, and throughout this book.

SOLUTIONS TO PROBLEMS

1. (a) 2.80×10^{-5} (b) 2.194×10^5 (c) 646.0
   (d) 905   (e) 86.048   (f) 5.18×10^{-4}

2. (a) 0.8943   (b) $\bar{1}.0867$ [or -0.9133]   (c) 1.5731
   (d) 5.8344   (e) $\overline{10}.6977$ [or -9.3023]

3. (a) 1.413   (b) -4.129 (or $\bar{5}.871$)   (c) 10.545   (d) -16.593 (or $\overline{17}.407$)

4. (a) 1.0×10^{23} (b) 4.914   (c) 1.0×10^{-2}
   (d) 6.456×10^{-3} (e) 6.564×10^{15} (f) 4.645×10^{-5} (remember, $-4.333 = \bar{5}.667$)

5. (a) 9.28×10^3 (b) 4.50×10^{-4} (c) 3.54×10^{-5}
    (d) 7.85×10^{-3} (e) 22.7   (f) 7.35×10^3
    (g) 4.3×10^{-5} (h) 2.9×10^{-22} (i) 7.39
    (j) 0.0498   (k) 94.6   (l) 0.0240

6. 6.43×10^{-2} g   7. 260 g   8. 9.15×10^{-4} g

9. (a) 1.17 liter   (b) 17.5 ml

10. 2.84×10^3

11. (a) 1.86×10^{-2} mole   (b) 2.69×10^{-5} mole

12. (a) 0.856 mole liter^{-1} (b) 40.0

Appendix B
Useful Data

Data! data! data! I can't make bricks without clay!   THE ADVENTURE OF THE
COPPER BEECHES

B.1 Units, Conversion Factors, and Constants

SI Prefixes

10^{-1}	deci	d	10	deca	da
10^{-2}	centi	c	10^{2}	hecto	h
10^{-3}	milli	m	10^{3}	kilo	k
10^{-6}	micro	μ	10^{6}	mega	M
10^{-9}	nano	n	10^{9}	giga	G
10^{-12}	pico	p	10^{12}	tera	T

SI Base Units

Quantity	Name of unit	Abbreviation
length	metre*	m
mass	kilogram	kg
time	second	s
electric current	ampere	A
thermodynamic temperature	kelvin	K
amount of substance	mole	mol

* The approved spelling of the SI unit is "metre," but the commonly accepted
North American spelling "meter" has been used in this book.

SI Derived Units

Quantity	Name of unit	Abbreviation	SI base units
frequency	hertz	Hz	s^{-1}
energy	joule	J	$kg\ m^2\ s^{-2}$
force	newton	N	$kg\ m\ s^{-2} = J\ m^{-1}$
power	watt	W	$kg\ m^2\ s^{-3} = J\ s^{-1}$
pressure	pascal	Pa	$kg\ m^{-1}\ s^{-2} = N\ m^{-2}$
electric charge	coulomb	C	$A\ s$
electric potential difference	volt	V	$kg\ m^2\ s^{-3}\ A^{-1} = J\ C^{-1}$

Useful Conversion Factors

Quantity	Name of unit	Abbreviation	
length	angstrom unit	Å	10^{-10} m = 10 nm $= 10^{-8}$ cm
volume	cubic meter	m^3	
	cubic decimeter	dm^3	$10^{-3}\ m^3$
	liter	l	$10^{-3}\ m^3$*
energy	erg		10^{-7} J
	calorie	cal	4.184 J
	electron volt	eV†	1.602×10^{-19} J
	liter atmosphere	l atm	101.3 J = 24.22 cal
force	dyne		10^{-5} N
pressure	atmosphere	atm	1.013×10^5 N m^{-2} $= 1.013 \times 10^5$ Pa $= 1.013 \times 10^6$ dyne cm^{-2}

* "Liter" should be thought of as an alternative name for 1 dm^3.
† 1 eV per particle = 96.49 kJ $mole^{-1}$ = 23.06 kcal $mole^{-1}$

B.2   Atomic Weights

Atomic weights are given to four significant figures, based on $^{12}C = 12.000$. Elements with no stable isotope are not included. Z is the atomic number.

Element	Symbol	Z	At.wt	Element	Symbol	Z	At.wt
Aluminum	Al	13	26.98	Molybdenum	Mo	42	95.94
Antimony	Sb	51	121.8	Neodymium	Nd	60	144.2
Argon	Ar	18	39.95	Neon	Ne	10	20.18
Arsenic	As	33	74.92	Nickel	Ni	28	58.70
Barium	Ba	56	137.3	Niobium	Nb	41	92.91
Beryllium	Be	4	9.012	Nitrogen	N	7	14.01
Bismuth	Bi	83	209.0	Osmium	Os	76	190.2
Boron	B	5	10.81	Oxygen	O	8	16.00
Bromine	Br	35	79.90	Palladium	Pd	46	106.4
Cadmium	Cd	48	112.4	Phosphorus	P	15	30.97
Calcium	Ca	20	40.08	Platinum	Pt	78	195.1
Carbon	C	6	12.011	Potassium	K	19	39.10
Cerium	Ce	58	140.1	Praseodymium	Pr	59	140.9
Cesium	Cs	55	132.9	Rhenium	Re	75	186.2
Chlorine	Cl	17	35.45	Rhodium	Rh	45	102.9
Chromium	Cr	24	52.00	Rubidium	Rb	37	85.47
Cobalt	Co	27	58.93	Ruthenium	Ru	44	101.1
Copper	Cu	29	63.55	Samarium	Sm	62	150.4
Dysprosium	Dy	66	162.5	Scandium	Sc	21	44.96
Erbium	Er	68	167.3	Selenium	Se	34	78.96
Europium	Eu	63	152.0	Silicon	Si	14	28.09
Fluorine	F	9	19.00	Silver	Ag	47	107.9
Gadolinium	Gd	64	157.3	Sodium	Na	11	22.99
Gallium	Ga	31	69.72	Strontium	Sr	38	87.62
Germanium	Ge	32	72.59	Sulfur	S	16	32.06
Gold	Au	79	197.0	Tantalum	Ta	73	180.9
Hafnium	Hf	72	178.5	Tellurium	Te	52	127.6
Helium	He	2	4.003	Terbium	Tb	65	158.9
Holmium	Ho	67	164.9	Thallium	Tl	81	204.4
Hydrogen	H	1	1.0079	Thorium	Th	90	232.0
Indium	In	49	114.8	Thulium	Tm	69	168.9
Iodine	I	53	126.9	Tin	Sn	50	118.7
Iridium	Ir	77	192.2	Titanium	Ti	22	47.90
Iron	Fe	26	55.85	Tungsten	W	74	183.9
Krypton	Kr	36	83.80	Uranium	U	92	238.0
Lanthanum	La	57	138.9	Vanadium	V	23	50.94
Lead	Pb	82	207.2	Xenon	Xe	54	131.3
Lithium	Li	3	6.941	Ytterbium	Yb	70	173.0
Lutetium	Lu	71	175.0	Yttrium	Y	39	88.91
Magnesium	Mg	12	24.31	Zinc	Zn	30	65.38
Manganese	Mn	25	54.94	Zirconium	Zr	40	91.22
Mercury	Hg	80	200.6				

COMMON LOGARITHMS OF NUMBERS

N	0	1	2	3	4	5	6	7	8	9	Proportional parts								
											1	2	3	4	5	6	7	8	9
10	0000	0043	0086	0128	0170	0212	0253	0294	0334	0374	4	8	12	17	21	25	29	33	37
11	0414	0453	0492	0531	0569	0607	0645	0682	0719	0755	4	8	11	15	19	23	26	30	34
12	0792	0828	0864	0899	0934	0969	1004	1038	1072	1006	3	7	10	14	17	21	24	28	31
13	1139	1173	1206	1239	1271	1303	1335	1367	1399	1430	3	6	10	13	16	19	23	26	29
14	1461	1492	1523	1553	1584	1614	1644	1673	1703	1732	3	6	9	12	15	18	21	24	27
15	1761	1790	1818	1847	1875	1903	1931	1959	1987	2014	3	6	8	11	14	17	20	22	25
16	2041	2068	2095	2122	2148	2175	2201	2227	2253	2279	3	5	8	11	13	16	18	21	24
17	2304	2330	2355	2380	2405	2430	2455	2480	2504	2529	2	5	7	10	12	15	17	20	22
18	2553	2577	2601	2625	2648	2672	2695	2718	2742	2765	2	5	7	9	12	14	16	19	21
19	2788	2810	2833	2856	2878	2900	2923	2945	2967	2989	2	4	7	9	11	13	16	18	20
20	3010	3032	3054	3075	3096	3118	3139	3160	3181	3201	2	4	6	8	11	13	15	17	19
21	3222	3243	3263	3284	3304	3324	3345	3365	3385	3404	2	4	6	8	10	12	14	16	18
22	3424	3444	3464	3483	3502	3522	3541	3560	3579	3598	2	4	6	8	10	12	14	15	17
23	3617	3636	3655	3674	3692	3711	3729	3747	3766	3784	2	4	6	7	9	11	13	15	17
24	3802	3820	3838	3856	3874	3892	3909	3927	3945	3962	2	4	5	7	9	11	12	14	16
25	3979	3997	4014	4031	4048	4065	4082	4099	4116	4133	2	4	5	7	9	10	12	14	15
26	4150	4166	4183	4200	4216	4232	4249	4265	4281	4298	2	3	5	7	8	10	11	13	15
27	4314	4330	4346	4362	4378	4393	4409	4425	4440	4456	2	3	5	6	8	9	11	13	14
28	4472	4487	4502	4518	4533	4548	4564	4579	4594	4609	2	3	5	6	8	9	11	12	14
29	4624	4639	4654	4669	4683	4698	4713	4728	4742	4757	1	3	4	6	7	9	10	12	13
30	4771	4786	4800	4814	4829	4843	4857	4871	4886	4900	1	3	4	6	7	9	10	11	13
31	4914	4928	4942	4955	4969	4983	4997	5011	5024	5038	1	3	4	5	7	8	10	11	12
32	5051	5065	5079	5092	5105	5119	5132	5145	5159	5172	1	3	4	5	7	8	9	11	12
33	5185	5198	5211	5224	5237	5250	5263	5276	5289	5302	1	3	4	5	7	8	9	10	12
34	5315	5328	5340	5353	5366	5378	5391	5403	5416	5428	1	2	4	5	6	8	9	10	11
35	5441	5453	5465	5478	5490	5502	5514	5527	5539	5551	1	2	4	5	6	7	9	10	11
36	5563	5575	5587	5599	5611	5623	5635	5647	5658	5670	1	2	4	5	6	7	8	10	11
37	5682	5694	5705	5717	5729	5740	5752	5763	5775	5786	1	2	4	5	6	7	8	9	10
38	5798	5809	5821	5832	5843	5855	5866	5877	5888	5899	1	2	3	5	6	7	8	9	10
39	5911	5922	5933	5944	5955	5966	5977	5988	5999	6010	1	2	3	4	5	7	8	9	10
40	6021	6031	6042	6053	6064	6075	6085	6096	6107	6117	1	2	3	4	5	6	8	9	10
41	6128	6138	6149	6160	6170	6180	6191	6201	6212	6222	1	2	3	4	5	6	7	8	9
42	6232	6243	6253	6263	6274	6284	6294	6304	6314	6325	1	2	3	4	5	6	7	8	9
43	6335	6345	6355	6365	6375	6385	6395	6405	6415	6425	1	2	3	4	5	6	7	8	9
44	6435	6444	6454	6464	6474	6484	6493	6503	6513	6522	1	2	3	4	5	6	7	8	9
45	6532	6542	6551	6561	6571	6580	6590	6599	6609	6618	1	2	3	4	5	6	7	8	9
46	6628	6637	6646	6656	6665	6675	6684	6693	6702	6712	1	2	3	4	5	6	7	7	8
47	6721	6730	6739	6749	6758	6767	6776	6785	6794	6803	1	2	3	4	5	5	6	7	8
48	6812	6821	6830	6839	6848	6857	6866	6875	6884	6893	1	2	3	4	5	5	6	7	8
49	6902	6911	6920	6928	6937	6946	6955	6964	6972	6981	1	2	3	4	4	5	6	7	8
50	6990	6998	7007	7016	7024	7033	7042	7050	7059	7067	1	2	3	3	4	5	6	7	8
51	7076	7084	7093	7101	7110	7118	7126	7135	7143	7152	1	2	3	3	4	5	6	7	8
52	7160	7168	7177	7185	7193	7202	7210	7218	7226	7235	1	2	3	3	4	5	6	7	7
53	7243	7251	7259	7267	7275	7284	7292	7300	7308	7316	1	2	2	3	4	5	6	6	7
54	7324	7332	7340	7348	7356	7364	7372	7380	7388	7396	1	2	2	3	4	5	6	7	7
N	0	1	2	3	4	5	6	7	8	9	1	2	3	4	5	6	7	8	9

COMMON LOGARITHMS OF NUMBERS

N	0	1	2	3	4	5	6	7	8	9	1	2	3	4	5	6	7	8	9
											Proportional parts								
55	7404	7412	7419	7427	7435	7443	7451	7459	7466	7474	1	2	2	3	4	5	5	6	7
56	7482	7490	7497	7505	7513	7520	7528	7536	7543	7551	1	2	2	3	4	5	5	6	7
57	7559	7566	7574	7582	7589	7597	7604	7612	7619	7627	1	1	2	3	4	5	5	6	7
58	7634	7642	7649	7657	7664	7672	7679	7686	7694	7701	1	1	2	3	4	4	5	6	7
59	7709	7716	7723	7731	7738	7745	7752	7760	7767	7774	1	1	2	3	4	4	5	6	7
60	7782	7789	7796	7803	7810	7818	7825	7832	7839	7846	1	1	2	3	4	4	5	6	6
61	7853	7860	7868	7875	7882	7889	7896	7903	7910	7917	1	1	2	3	3	4	5	6	6
62	7924	7931	7938	7945	7952	7959	7966	7973	7980	7987	1	1	2	3	3	4	5	6	6
63	7993	8000	8007	8014	8021	8028	8035	8041	8048	8055	1	1	2	3	3	4	5	5	6
64	8062	8069	8075	8082	8089	8096	8102	8109	8116	8122	1	1	2	3	3	4	5	5	6
65	8129	8136	8142	8149	8156	8162	8169	8176	8182	8189	1	1	2	3	3	4	5	5	6
66	8195	8202	8209	8215	8222	8228	8235	8241	8248	8254	1	1	2	3	3	4	5	5	6
67	8261	8267	8274	8280	8287	8293	8299	8306	8312	8319	1	1	2	3	3	4	5	5	6
68	8325	8331	8338	8344	8351	8357	8363	8370	8376	8382	1	1	2	3	3	4	4	5	6
69	8388	8395	8401	8407	8414	8420	8426	8432	8439	8445	1	1	2	3	3	4	4	5	6
70	8451	8457	8463	8470	8476	8482	8488	8494	8500	8506	1	1	2	3	3	4	4	5	6
71	8513	8519	8525	8531	8537	8543	8549	8555	8561	8567	1	1	2	3	3	4	4	5	5
72	8573	8579	8585	8591	8597	8603	8609	8615	8621	8627	1	1	2	3	3	4	4	5	5
73	8633	8639	8645	8651	8657	8663	8669	8675	8681	8686	1	1	2	2	3	4	4	5	5
74	8692	8698	8704	8710	8716	8722	8727	8733	8739	8745	1	1	2	2	3	4	4	5	5
75	8751	8756	8762	8768	8774	8779	8785	8791	8797	8802	1	1	2	2	3	3	4	5	5
76	8808	8814	8820	8825	8831	8837	8842	8848	8854	8859	1	1	2	2	3	3	4	5	5
77	8865	8871	8876	8882	8887	8893	8899	8904	8910	8915	1	1	2	2	3	3	4	4	5
78	8921	8927	8932	8938	8943	8949	8954	8960	8965	8971	1	1	2	2	3	3	4	4	5
79	8976	8982	8987	8993	8998	9004	9009	9015	9020	9025	1	1	2	2	3	3	4	4	5
80	9031	9036	9042	9047	9053	9058	9063	9069	9074	9079	1	1	2	2	3	3	4	4	5
81	9085	9090	9096	9101	9106	9112	9117	9122	9128	9133	1	1	2	2	3	3	4	4	5
82	9138	9143	9149	9154	9159	9165	9170	9175	9180	9186	1	1	2	2	3	3	4	4	5
83	9191	9196	9201	9206	9212	9217	9222	9227	9232	9238	1	1	2	2	3	3	4	4	5
84	9243	9248	9253	9258	9263	9269	9274	9279	9284	9289	1	1	2	2	3	3	4	4	5
85	9294	9299	9304	9309	9315	9320	9325	9330	9335	9340	1	1	2	2	3	3	4	4	5
86	9345	9350	9355	9360	9365	9370	9375	9380	9385	9390	1	1	2	2	3	3	4	4	5
87	9395	9400	9405	9410	9415	9420	9425	9430	9435	9440	1	1	2	2	3	3	3	4	4
88	9445	9450	9455	9460	9465	9469	9474	9479	9484	9489	0	1	1	2	2	3	3	4	4
89	9494	9499	9504	9409	9513	9518	9523	9528	9533	9538	0	1	1	2	2	3	3	4	4
90	9542	9547	9552	9557	9562	9566	9571	9576	9581	9586	0	1	1	2	2	3	3	4	4
91	9590	9595	9600	9605	9609	9614	9619	9624	9628	9633	0	1	1	2	2	3	3	4	4
92	9638	9643	9647	9652	9657	9661	9666	9671	9675	9680	0	1	1	2	2	3	3	4	4
93	9685	9689	9694	9699	9703	9708	9713	9717	9722	9727	0	1	1	2	2	3	3	4	4
94	9731	9736	9741	9745	9750	9754	9759	9763	9768	9773	0	1	1	2	2	3	3	4	4
95	9777	9782	9786	9791	9795	9800	9805	9809	9814	9818	0	1	1	2	2	3	3	4	4
96	9823	9827	9832	9836	9841	9845	9850	9854	9859	9863	0	1	1	2	2	3	3	4	4
97	9868	9872	9877	9881	9886	9890	9894	9899	9903	9908	0	1	1	2	2	3	3	4	4
98	9912	9917	9921	9926	9930	9934	9939	9943	9948	9952	0	1	1	2	2	3	3	4	4
99	9956	9961	9965	9969	9974	9978	9983	9987	9991	9996	0	1	1	2	2	3	3	3	4
N	0	1	2	3	4	5	6	7	8	9	1	2	3	4	5	6	7	8	9

PERIODIC TABLE OF THE ELEMENTS

Group I	Group II				Transition Elements								Group III	Group IV	Group V	Group VI	Group VII	Group VIII
1 **H** 1.00797																		2 **He** 4.0026
3 **Li** 6.939	4 **Be** 9.0122												5 **B** 10.811	6 **C** 12.01115	7 **N** 14.0067	8 **O** 15.9994	9 **F** 18.9984	10 **Ne** 20.183
11 **Na** 22.9898	12 **Mg** 24.312												13 **Al** 26.9815	14 **Si** 28.086	15 **P** 30.9738	16 **S** 32.064	17 **Cl** 35.453	18 **Ar** 39.948
19 **K** 39.102	20 **Ca** 40.08	21 **Sc** 44.956	22 **Ti** 47.90	23 **V** 50.942	24 **Cr** 51.996	25 **Mn** 54.9380	26 **Fe** 55.847	27 **Co** 58.9332	28 **Ni** 58.71	29 **Cu** 63.54	30 **Zn** 65.37		31 **Ga** 69.72	32 **Ge** 72.59	33 **As** 74.9216	34 **Se** 78.96	35 **Br** 79.909	36 **Kr** 83.80
37 **Rb** 85.47	38 **Sr** 87.62	39 **Y** 88.905	40 **Zr** 91.22	41 **Nb** 92.906	42 **Mo** 95.94	43 **Tc** (99)	44 **Ru** 101.07	45 **Rh** 102.905	46 **Pd** 106.4	47 **Ag** 107.870	48 **Cd** 112.40		49 **In** 114.82	50 **Sn** 118.69	51 **Sb** 121.75	52 **Te** 127.60	53 **I** 126.9044	54 **Xe** 131.30
55 **Cs** 132.905	56 **Ba** 137.34	57 **La*** 138.91	72 **Hf** 178.49	73 **Ta** 180.948	74 **W** 183.85	75 **Re** 186.2	76 **Os** 190.2	77 **Ir** 192.2	78 **Pt** 195.09	79 **Au** 196.967	80 **Hg** 200.59		81 **Tl** 204.37	82 **Pb** 207.19	83 **Bi** 208.980	84 **Po** (210)	85 **At** (210)	86 **Rn** (222)
87 **Fr** (223)	88 **Ra** (226)	89 **Ac#** (227)																

*Lanthanide Series

58 **Ce** 140.12	59 **Pr** 140.907	60 **Nd** 144.24	61 **Pm** (147)	62 **Sm** 150.35	63 **Eu** 151.96	64 **Gd** 157.25	65 **Tb** 158.924	66 **Dy** 162.50	67 **Ho** 164.930	68 **Er** 167.26	69 **Tm** 168.934	70 **Yb** 173.04	71 **Lu** 174.97

#Actinide Series

90 **Th** 232.038	91 **Pa** (231)	92 **U** 238.03	93 **Np** (237)	94 **Pu** (242)	95 **Am** (243)	96 **Cm** (247)	97 **Bk** (247)	98 **Cf** (249)	99 **Es** (254)	100 **Fm** (253)	101 **Md** (256)	102 **No** (253)	103 **Lw** (257)

A number in parentheses represents the mass number of the most stable isotope of a radioactive element.

BCDEFGHIJ-A-7987